Frontiers and Best Practices in Bio, Circular, and Green Growth and Eco-Innovation

Frontiers and Best Practices in Bio, Circular, and Green Growth and Eco-Innovation

Editors

Elkhan Richard Sadik-Zada
Andrea Gatto

MDPI • Basel • Beijing • Wuhan • Barcelona • Belgrade • Manchester • Tokyo • Cluj • Tianjin

Editors
Elkhan Richard Sadik-Zada
Ruhr-University Bochum
Bochum
Germany

Andrea Gatto
Wenzhou-Kean University
Wenzhou
China

Editorial Office
MDPI
St. Alban-Anlage 66
4052 Basel, Switzerland

This is a reprint of articles from the Special Issue published online in the open access journal *Sustainability* (ISSN 2071-1050) (available at: https://www.mdpi.com/journal/sustainability/special_issues/green_ecoinnovation).

For citation purposes, cite each article independently as indicated on the article page online and as indicated below:

LastName, A.A.; LastName, B.B.; LastName, C.C. Article Title. *Journal Name* **Year**, *Volume Number*, Page Range.

ISBN 978-3-0365-7564-3 (Hbk)
ISBN 978-3-0365-7565-0 (PDF)

Cover image courtesy of Getty Images Suite

Contents

About the Editors

Elkhan Richard Sadik-Zada

Dr. Elkhan Richard Sadik-Zada is a senior research and teaching fellow at the Institute of Development Research and Development Policy at Ruhr University of Bochum in Germany. At the same time, he is an honorary member of the Center of Economic Development and Social Change in Naples, Italy and a visiting lecturer at the Institute of Social Development at the Institute of the Western Cape (UWC) in South Africa. Dr. Sadik-Zada has a publication record in international peer-reviewed journals—including Resources Policy, Renewable Energy, Mineral Economics, Post-Communist Economies, Review of Development Economics, The European Journal of Development Research, Socio-Economic Planning Sciences, and The Atlantic Economic Journal. He is an author of three monographs.

Andrea Gatto

Prof. Andrea Gatto is Senior Assistant Professor at Wenzhou Kean University (US-China). He is a visiting research fellow and lecturer at the Natural Resources Institute (NRI) at the University of Greenwich. Dr. Gatto has a publication record in international peer-reviewed journals—including The Annual Review of Environment and Resources, The Journal of Cleaner Production, Resources, Conservation and Recycling, Energy Policy, and Business Strategy and the Environment. An economic and business consultant, Andrea is President of the CED—Center for Economic Development and Social Change and Research Fellow at the Centre for Studies on the European Economy at UNEC. Andrea sits on several Q1 international peer-reviewed journals' editorial boards. He has conducted fieldwork and worked for longer appointments in twelve countries worldwide, including positions for the European Commission, CalPoly, CREATES at Aarhus University, and the NCH at Northeastern University. He is fluent in Italian, English, French, and Portuguese, and speaks some Spanish.

Preface to "Frontiers and Best Practices in Bio, Circular, and Green Growth and Eco-Innovation"

This book project accommodates a substantial number of contributions on the novel business models, technology advances and public regulations that are favorable for the development of a circular economy and green economic growth in the developing and transition economies, and in the structurally challenged areas within the industrialized world. The positive examples make clear that the frontiers of circular and bio-economy development, and green growth are by no means confined to economically advanced countries. This means that developing economies do not have to repeat the inverted U-shaped income–environment relationship and enter the carbon-saving sustainable growth and development track commencing from the initial phases of economic modernization. The positive examples in this book project show that the pathways to sustainable development in the Global South are real. There is, however, no guarantee that developing countries would join them. To assure this, advanced countries must rethink their international development cooperation and enhanced development policies in the realm of international climate action.

Elkhan Richard Sadik-Zada and Andrea Gatto
Editors

sustainability

Editorial

Grow First, Clean Up Later? Dropping Old Paradigms and Opening Up New Horizons of Sustainable Development

Elkhan Richard Sadik-Zada [1,3,*] and Andrea Gatto [2,3,4,*]

1 Institute of Development Research and Development Policy, Ruhr-University Bochum, 44801 Bochum, Germany
2 Faculty of Economics, Wenzhou-Kean University, Wenzhou 325015, China
3 Center for Studies on Europe, Azerbaijan State University of Economics, Baku AZ1001, Azerbaijan
4 Center for Economic Development and Social Change (CED), 80128 Naples, Italy
* Correspondence: elkhan.r.sadik-zada@ruhr-uni-bochum.de (E.R.S.-Z.); agatto@kean.edu (A.G.)

Abstract: After almost two decades of continuous development in bio, circular, and green economy, it is time to assess the major achievements and challenges that private and public enterprises face today for further enhancing global sustainability concepts. To this end, the present thematic issue accommodates twenty articles on different topics related to circular economy development and green growth, proposing a contribution to the field of environmental economics and policy. The central feature of this Special Issue is the focus on the best practices and challenges in terms of green growth and eco-innovation in developing and transitioning structurally challenged areas. Hence, the study elaborates on the pathways of bio, circular, and green growth and eco-innovation in the context of countries with relatively low per capita income. By doing this, the collection shows that the empirically established environmental Kuznets curve—i.e., the inverted U-shaped income-environment nexus—can and must be critically questioned, at least in the contexts mentioned within the framework of our Special Issue. Hence, the geographic frontiers of environmental upgrading, carbon-saving bioeconomic development, and green growth are not limited to the economically advanced areas.

Keywords: circular economy; bioeconomy; green growth; eco-innovation; developing countries; environmental Kuznets curve; EKC; environmental economics and policy; ecological economics; resource management; energy policy; sustainable development

Citation: Sadik-Zada, E.R.; Gatto, A. *Grow First, Clean Up Later*? Dropping Old Paradigms and Opening Up New Horizons of Sustainable Development. *Sustainability* **2023**, *15*, 3595. https://doi.org/10.3390/su15043595

Received: 2 February 2023
Accepted: 6 February 2023
Published: 15 February 2023

1. Introduction: Grow & Perish?

Five decades have passed since the publications of the *Limits to Growth* of the Club of Rome and the "Only the Earth" in Stockholm—the first UN Conference on the limitations in the absorptive capacity of the global environment [1]. Both events marked the start of the discussion of global environmentalism and putting the environment on the political agenda of the dialogue between industrialized and developing countries [2].

The 1972 UN Conference on the Environment in Stockholm was the first truly global conference on the environment that, for the first time in history, elevated environmental protection to the top of the global political agenda. Twenty years later, the Earth Summit in Rio de Janeiro in 1992 was an important caesura in understanding the importance of developing countries in protecting the global environment. The Paris Agreement of 2015 was probably the greatest milestone and the first truly global treaty on climate action. Despite generous nationally-determined pledges within the agreement, there have been plenty of drawbacks [3]. The undertaken climate action is nowhere near enough to achieve the set climate targets [4].

Despite increasing environmental awareness in developing, transitioning, and emerging economies and impressive nationally determined pledges within the Paris Agreement, economic growth, job creation, and poverty alleviation have been the central focus of the public policies in these countries [5]. The Global North has not been taking sufficient responsibility for the development of carbon-saving technology solutions in the Global South. This explains why fifty years after the publication of *The Limits to Growth* we do not only have an empirical validation of its pessimism regarding the doubling of global resource use, but also the evidence that the business-as-usual rushes human civilization to its own destruction. Fifty years later, mankind is on the brink of the socioeconomic abyss and climate catastrophe [4]. Hence, mankind can no longer afford the "grow first, clean up later" model, which continues to dominate the growth and development patterns in the developing, transitioning, and structurally challenged areas worldwide [6].

The inverted U-shaped relationship between per capita income (PCI) and per capita greenhouse gas (GHG) emissions, i.e., the environmental Kuznets curve (EKC) hypothesis is probably empirically the most established regularity within the framework of environmental macroeconomics. It has been validated for advanced, developing, and transitional economies [7,8]. The EKC is nothing but a regularity of growing first and cleaning up later and it is at odds with the ongoing climate crisis and the related climate targets within the framework of the Paris Agreement [9].

The augmentation of the income-environment perspective by the pollution haven hypothesis, which has been also empirically validated in a number of recent studies, shows that even the progress in terms of environmental upgrading in OECD countries are attributed to the relocation of the dirty industries to the countries of the Global South [9]. Hence, following the empirically validated patterns of environmental development is not congruent with the energy and climate futures we want [9–12].

Does this mean that there are no alternatives to the environmental Kuznets curve and the pollution haven hypotheses? Is humanity predestined to lose the climate emergency race? How are grave disruptions such as COVID-19, geopolitical turmoil, and major shocks affecting our societies, economies, and ecosystems? Shall we accept a grow-and-perish paradigm, opting for a trade-off between environmental, social, and economic targets [13–37]?

2. Content of the Thematic Issue

The thematic issue accommodates three reviews and sixteen research papers. Breaugh et al. [37], provide a systematic literature review on innovation scaling and the role of innovation in overcoming global issues such as climate change, sustainable management of natural resources, and economic inequality. They outline the ontological differences between the "diffusion" and "scaling" of innovations and establish a research agenda for the assessment of innovation scaling and by doing so, frame the present thematic issue conceptually [37,38]. The rest of the works, including the two review papers, Solaymani [39] and Abbas et al. [32], address the individual country cases of developing or structurally challenged areas of Southern Europe [39].

Abbas et al. [32] analyze the effect of COVID-19 on the Pakistani economy and find that the pandemic severely hit different sectors of the economy and large swaths of the population. However, the poorest in society, especially the daily-wage earners and the runners of small businesses, were disproportionately hit in Pakistan. Wang and Chen [14] address an interesting question of the efficiency and effectiveness of the consumption coupons in China that were introduced in the face of the COVID-19 pandemic. They propose an improved minimum-cost maximum flow approach that could optimize consumption coupon policies both in the Chinese and non-Chinese contexts.

Fernández Luiña et al. [15] elaborate on the issue of the community commitment to sustainability using the case study of forest protection policies in Guatemala. Here, the authors illustrate how Guatemala implements forest preservation policies through community concessions. The research paper contributes to the theoretical literature by extending the scope of the analysis to modern municipals' forest preservation practices. Based on the empirical findings, the authors recommend the government of Guatemala the employment of decentralized forest policies at the national level.

Bin-Nashwan et al. [16] conduct a cross-country analysis of the donor responses to the fundraising appeals for combating the epidemic. The study is based on a quantitative analysis of 801 donors from Kuwait and Bahrain. The authors find that both in Kuwait and Bahrain, humanitarian projects, internet technology, social networking sites, and religiosity are significant determinants of donor attitudes toward online donations.

Mammadli [17] and Zeynalova and Namazova [18] analyze the environmentally relevant behavior of enterprises and end-consumers in the context of transition economies. Both research articles focus on the case studies of Azerbaijan. In the former publication, the environmentally responsible approaches of Azerbaijani companies correspond with the size of the private firms in terms of sales and the number of their employees [17]. In the latter of these studies, the willingness to pay for green technology-based products is predicated on the statistical analysis of 536 structured interviews in the greater area of Baku [18]. In contrast to [17] which analyzes enterprises and [18] which analyzes consumer behavior, Montakhabi et al. [19] explore the behavior of the prosumers and their barriers to the enhancement of sustainable business models in electricity markets. The work contributes to the theory of the firm by applying the resource-based approach in an entirely novel context of energy economics [20–22].

Leal Filho et al. [23] analyze the role of renewable energy for fostering energy security of small island developing states. Their study is based on a comprehensive quantitative literature review and policy analysis that has been conducted by means of VOSviewer, a prominent software for bibliometric analyses. The authors also find that renewable energy contributes in a decisive way to the food security of the small island developing states.

Kennedyd et al. [24] address the question of economic sustainability in China. The inquiry explores the underlying factors that influence Chinese online consumers' acceptance of online shopping platforms. The authors of the study assess 691 interviews of Chinese online shoppers. Embedding the data into the theoretical model yields interesting results with regard to the choice patterns of the e-commerce platforms and consumer attitudes.

Marjerison et al. [25] analyze the creation of sustainable organizations through knowledge sharing and organizational agility in China. The study analyzes 720 online questionnaires and finds that working groups that are more agile have a greater capacity of harnessing the knowledge-sharing culture. The validity of these findings spans various classes of businesses. Li et al. [26] complement the investigation of Marjerison et al. [25] in a meaningful way. They propose a novel multi-factor three-step feature selection and deep learning framework for regional GDP prediction in the Chinese provinces [26].

Chen [27] compares two panel data sets of developing and advanced economies for a time frame spanning between 1990 and 2021 and shows that economic competitiveness and increased commodity prices lead to the increasing adoption of green technologies and decarbonization of the entire economy. The study of Xu et al. [28] address a similar issue by investigating the changes in firm innovation in the aftermath of the Sino-U.S. trade grievance. Their analysis is based on the unique microdata of the Chinese technology firms. The authors find that Sino-U.S. trade friction has led to a significant increase of the innovations of Chinese exporters.

Zhao and Chen [29] also address the case of China and examine the effect of family lifestyle and neighborhood on the willingness of households to sort their waste. The authors inspect China Labor Force Dynamics Survey for 2016 and show that lifestyle and neighborhood have a significant impact on households' waste management patterns. Abdelsalam et al. [30] also address the problem of waste management, but in a totally different setting, focusing on the Libyan public hospitals. This publication finds that organizational culture and structure play a decisive role in the waste management practices of the Libyan hospitals.

Based on structured interviews of 359 consumers, Civero et al. [40] scrutinize consumer behavior in the city of Naples in Southern Italy. The authors show that also in the structurally challenged areas, the consumers' purchasing choices of food staples are not based only on purely economic but also on ethical and environmental factors. The authors employ a factor analysis that enables the classification of the consumers in five homogenous clusters. Selected pathways of using the emerging trend of new social consumption are predicated for the formulation of better public policies to foster ecological sustainability.

Sadik-Zada [37] analyze the perspectives of green hydrogen rollout from the lens of both international and development economics. The author shows that the development of the green hydrogen value chain and the substitution of fossil fuels by green hydrogen could lead to clean and circular energy systems in a number of countries of the Global North. In addition, the review of the national and European hydrogen strategies indicates that fostering of international development cooperation, especially international energy cooperation of the EU with the solar energy-abundant countries of Africa could accelerate the green hydrogen take-off in Europe. Furthermore, the study also shows that investing in the development of green hydrogen production in Africa could also substantially contribute to socio-economic development and energy transition in Sub-Saharan Africa.

Solaymani [38] analyzes the environmental and energy sustainability problems of fossil fuels that rich countries are using through the case study of Iran. The research paper is of particular interest because of the growing significance of petroleum-exporting countries in the global carbon footprint [7]. There is also empirical evidence that oil-rich developing and transition economies do not follow the inverted U-shaped, but rather a monotonically increasing average income-environment relationship [8]. The author employs advanced co-integration techniques to the data from Iran and detects a unidirectional causality between non-renewable energy production and the growth of the renewables in the energy mix. This implies that Iranian growth, which almost entirely relies on petroleum exports, has substantially contributed to decarbonization of Iranian electricity mix.

3. Concluding Remarks: Striving to Be More Sustainable

Together with the editorial, the present Special Issue accommodated twenty contributions, whereby three of them are literature reviews and sixteen of them are research papers. Both review and research papers focus on the analysis of developing, transition, emerging economies, and structurally challenged areas of Southern Europe. "Frontiers and Best Practices in Bio, Circular, and Green Growth" make clear that geographic frontiers of environmental upgrading, carbon-saving bioeconomy development, and green growth are not confined to economically advanced areas and shall be tackled as global ecological and sustainable development priorities.

The presented positive examples from this group of countries in this thematic issue do not contest the empirical validity of the environmental Kuznets curve and pollution haven hypotheses. By showing positive examples, it rather shows that based on novel management approaches, the mentioned empirical regularities could be successfully obviated. Economic growth of developing, transition, and emerging economies can be aligned with environmental sustainability. This confirms that, even in developing, transitioning, and emerging economies, positive experiences denote a global strive towards sustainability. These shreds of evidence also suggest that the economic, environmental, and social components of development shall not be liable to trade-offs and a holistic and pluralistic sustainable development purview needs to be enforced globally.

Institutional Review Board Statement: Not applicable.

Informed Consent Statement: Not applicable.

Conflicts of Interest: The authors declare no conflict of interest.

References

1. Meadows, D.H.; Meadows, D.L.; Randers, J.; Behrens, W.W. *The Limits to Growth: A Report for the Club of Rome's Project on the Predicament of Mankind*; Universe Books: New York, NY, USA, 1972.
2. UNFCCC. Fifty Years of Environmental Action. Available online: https://unfccc.int/blog/fifty-years-of-environmental-action (accessed on 3 June 2022).

3. Unmüßig, B. Global Environmental Policy between Successes and Failures–Fifty Years after Stockholm 1972. Heinrich-Böll Foundation. 2022. Available online: https://www.boell.de/en/2022/05/28/global-environmental-policy-between-successes-and-failures-fifty-years-after-stockholm (accessed on 1 June 2022).
4. Herrigton, G. The Limits to Growth Model: Still Prescient 50 Years Later. EARTH4ALL: DEEP-DIVE PAPER 02. Available online: https://www.clubofrome.org/publication/earth4all-ltg-model (accessed on 1 June 2022).
5. Loewenstein, W.; Bender, D. Labour Market Failure, Capital Accumulation, Growth and Poverty Dynamics in Partially Formalised Economies: Why Developing Countriess Growth Patterns are Different. 2017. Available online: https://doi.org/10.2139/ssrn.3022146 (accessed on 1 June 2022).
6. Azadi, H.; Verheijke, G.; Witlox, F. Pollute first, clean up later? *Glob. Planet. Chang.* **2011**, *78*, 77–82. [CrossRef]
7. Grossman, G.; Krueger, A. *Environmental Impacts of a North American Free Trade Agreement*; NBER Working Paper No. 3914; National Bureau of Economic Research, Inc.: Cambridge, MA, USA, 1991. Available online: https://econpapers.repec.org/paper/nbrnberwo/3914.htm (accessed on 1 June 2022).
8. Panayotou, T. *Empirical Tests and Policy Analysis of Environmental Degradation at Different Stages of Economic Development*; [ILO Working Paper]; International Labour Organization: Geneva, Switzerland, 1993. Available online: https://econpapers.repec.org/paper/iloilowps/992927783402676.htm (accessed on 1 June 2022).
9. Dasgupta, P.; Mäler, K.-G. Net national product, wealth, and social well-being. *Environ. Dev. Econ.* **2000**, *5*, 69–93. [CrossRef]
10. Nordhaus, W.D. The ghosts of climates past and the specters of climate change future. *Energy Policy* **1995**, *23*, 269–282. [CrossRef]
11. Aldieri, L.; Gatto, A.; Vinci, C.P. Is there any room for renewable energy innovation in developing and transition economies? Data envelopment analysis of energy behaviour and resilience data. *Resour. Conserv. Recycl.* **2022**, *186*, 106587. [CrossRef]
12. Koundouri, P.; Reppas, D.; Souliotis, I. A retrospective on The Allocation of Energy Resource by William D. Nordhaus. *J. Nat. Resour. Policy Res.* **2015**, *7*, 173–176. [CrossRef]
13. Nordhaus, W.D. The allocation of energy resources. *Brook. Pap. Econ. Act.* **1973**, *4*, 529–576. [CrossRef]
14. Wang, S.; Chen, Y. Consumption Coupons, Consumption Probability and Inventory Optimization: An Improved Minimum-Cost Maximum-Flow Approach. *Sustainability* **2022**, *14*, 7759. [CrossRef]
15. Luiña, E.F.; Ordóñez, S.F.; Wang, W.H. The Community Commitment to Sustainability: Forest Protection in Guatemala. *Sustainability* **2022**, *14*, 6953. [CrossRef]
16. Bin-Nashwan, S.A.; Sarea, A.; Al-Daihani, M.; Ado, A.B.; Begum, H.; Alosaimi, M.H.; Abdul-Jabbar, H.; Abdelsalam, M.K. Fundraising Appeals for the COVID-19 Epidemic Fight: A Cross-Country Study of Donor Responses. *Sustainability* **2022**, *14*, 6486. [CrossRef]
17. Mammadli, M. Environmentally Responsible Business Approaches in Azerbaijan. *Sustainability* **2022**, *14*, 6227. [CrossRef]
18. Zeynalova, Z.; Namazova, N. Revealing Consumer Behavior toward Green Consumption. *Sustainability* **2022**, *14*, 5806. [CrossRef]
19. Montakhabi, M.; Van Zeeland, I.; Ballon, P. Barriers for Prosumers' Open Business Models: A Resource-Based View on Assets and Data-Sharing in Electricity Markets. *Sustainability* **2022**, *14*, 5705. [CrossRef]
20. Nordhaus, W.D. An Optimal Transition Path for Controlling Greenhouse Gases. *Science* **1992**, *258*, 1315–1319. [CrossRef] [PubMed]
21. Dasgupta, P.; Heal, G. The Optimal Depletion of Exhaustible Resources. The Review of Economic Studies. In Proceedings of the Symposium on the Economics of Exhaustible Resources, Edinbugh, Scotland, 1974; Volume 41, pp. 3–28.
22. Kenneth, A.; Dasgupta, P.; Goulder, L.; Daily, G.; Ehrlich, P.; Heal, G.; Levin, S.; Mäler, K.-G.; Schneider, S.; Starrett, D.; et al. Are We Consuming Too Much? *J. Econ. Perspect.* **2004**, *18*, 147–172.
23. Filho, W.L.; Balogun, A.-L.; Surroop, D.; Salvia, A.L.; Narula, K.; Li, C.; Hunt, J.D.; Gatto, A.; Sharifi, A.; Feng, H.; et al. Realising the Potential of Renewable Energy as a Tool for Energy Security in Small Island Developing States. *Sustainability* **2022**, *14*, 4965. [CrossRef]
24. Kennedyd, S.I.; Marjerison, R.K.; Yu, Y.; Zi, Q.; Tang, X.; Yang, Z. E-Commerce Engagement: A Prerequisite for Economic Sustainability—An Empirical Examination of Influencing Factors. *Sustainability* **2022**, *14*, 4554. [CrossRef]
25. Marjerison, R.K.; Andrews, M.; Kuan, G. Creating Sustainable Organizations through Knowledge Sharing and Organizational Agility: Empirical Evidence from China. *Sustainability* **2022**, *14*, 4531. [CrossRef]
26. Li, Q.; Yan, G.; Yu, C. A Novel Multi-Factor Three-Step Feature Selection and Deep Learning Framework for Regional GDP Prediction: Evidence from China. *Sustainability* **2022**, *14*, 4408. [CrossRef]
27. Chen, W.-J. Toward Sustainability: Dynamics of Total Carbon Dioxide Emissions, Aggregate Income, Non-Renewable Energy, and Renewable Power. *Sustainability* **2022**, *14*, 2712. [CrossRef]
28. Xu, Z.; Zhong, X.; Zhang, Z. Does the Sino–US Trade Friction Promote Firm Innovation? The Role of the Export Grab Effect. *Sustainability* **2022**, *14*, 2709. [CrossRef]
29. Zhao, L.; Chen, H. Exploring the Effect of Family Life and Neighbourhood on the Willingness of Household Waste Sorting. *Sustainability* **2021**, *13*, 13653. [CrossRef]
30. Abdelsalam, M.K.; Egdair, I.M.M.; Begum, H.; Jadi, D.; Al Issa, H.-E.; Abrika, O.S.S.; Alam, A.S.A.F. The Key Organizational Factors in Healthcare Waste Management Practices of Libyan Public Hospitals. *Sustainability* **2021**, *13*, 12785. [CrossRef]
31. Sadik-Zada, E.R. Political Economy of Green Hydrogen Rollout: A Global Perspective. *Sustainability* **2021**, *13*, 13464. [CrossRef]

32. Abbass, K.; Begum, H.; Alam, A.S.A.F.; Awang, A.H.; Abdelsalam, M.K.; Egdair, I.M.M.; Wahid, R. 2022. Fresh Insight through a Keynesian Theory Approach to Investigate the Economic Impact of the COVID-19 Pandemic in Paki-stan. *Sustainability* **2020**, *14*, 1054. [CrossRef]
33. Mišík, M. The EU needs to improve its external energy security. *Energy Pol.* **2022**, *165*, 112930. [CrossRef]
34. Cattivelli, V.; Rusciano, V. Social Innovation and Food Provisioning during COVID-19: The Case of Urban–Rural Initiatives in the Province of Naples. *Sustainability* **2020**, *12*, 4444. [CrossRef]
35. Rusciano, V.; Gatto, A. Effects of the COVID-19 Outbreak on the Use and Perceptions of Metropolitan Agricultural Parks—Evidence from Milan and Naples of Urban and Environmental Resilience. *Sustainability* **2022**, *14*, 7509. [CrossRef]
36. Ghufran, M.; Ali, S.; Ariyesti, F.R.; Nawaz, M.A.; Aldieri, L.; Xiaobao, P. Impact of COVID-19 to customers switching intention in the food segments: The push, pull and mooring effects in consumer migration towards organic food. *Food Qual. Preference* **2022**, *99*, 104561. [CrossRef]
37. Panarello, D. Economic insecurity, conservatism, and the crisis of environmentalism: 30 years of evidence. *Socio-Economic Plan. Sci.* **2020**, *73*, 100925. [CrossRef]
38. Breaugh, J.; McBride, K.; Kleinaltenkamp, M.; Hammerschmid, G. Beyond Diffusion: A Systematic Literature Review of Innovation Scaling. *Sustainability* **2021**, *13*, 13528. [CrossRef]
39. Solaymani, S. A Review on Energy and Renewable Energy Policies in Iran. *Sustainability* **2021**, *13*, 7328. [CrossRef]
40. Civero, G.; Rusciano, V.; Scarpato, D.; Simeone, M. Food: Not Only Safety, but Also Sustainability. The Emerging Trend of New Social Consumers. *Sustainability* **2021**, *13*, 12967. [CrossRef]

sustainability

MDPI

Article

Consumption Coupons, Consumption Probability and Inventory Optimization: An Improved Minimum-Cost Maximum-Flow Approach

Shunlin Wang [1] **and Yifang Chen** [2,*]

[1] School of Business and Administration, Ningbo Polytechnic, Ningbo 315800, China; shunlin_wangk@163.com
[2] The Dean's Office, Ningbo Polytechnic, Ningbo 315800, China
* Correspondence: yifang_chenw@163.com; Tel.: +86-574-8689-1274

Abstract: The issuance of consumption coupons during the epidemic period to stimulate the economy must take full account of the level of probabilistic consumption and inventory optimization. In this paper, an improved minimum-cost maximum-flow model is constructed to dynamically adjust the inventory capacity of node enterprises with the change of probabilistic consumption level, and three scenarios are simulated by numerical assumptions. The results show that: (1) The model can better solve the problem of consumption coupons, probabilistic consumption and inventory optimization; (2) Consumer welfare remains unchanged, the largest number of government consumption coupons is issued, and the number of enterprise inventories reaches the lowest; (3) Enterprise inventories are minimized with different decisions on consumer probability consumption, and the government's issuance of consumption coupons and the satisfaction of consumer demand have reached a dynamic balance. Corresponding suggestions are put forward, hoping to better help the government to implement the consumption coupons policy to stimulate the economy.

Keywords: consumption coupons; purchase probability; inventory optimization; minimum-cost maximum-flow

Citation: Wang, S.; Chen, Y. Consumption Coupons, Consumption Probability and Inventory Optimization: An Improved Minimum-Cost Maximum-Flow Approach. *Sustainability* **2022**, *14*, 7759. https://doi.org/10.3390/su14137759

Academic Editors: Elkhan Richard Sadik-Zada and Andrea Gatto

Received: 16 May 2022
Accepted: 21 June 2022
Published: 25 June 2022

1. Introduction

Since March 2022, a new round of COVID-19 has broken out in China [1]. Consumption continues to be sluggish, and downward pressure on the economy is increasing. Provinces and cities in China have launched their own economic incentive policies in the hopes of accelerating the expansion of consumption and expanding domestic demand. Consumption coupons as an economic stimulus have been repeatedly reported in the press. For example, in April 2022, Shenzhen city issued 500 million yuan of consumption coupons to its consumers, and Ningbo issued 300 million yuan to consumers [2]. Despite different views [3], the effect in practice [4,5] has gradually gained widespread attention.

Consumers will be affected by various factors when consuming. The issuance of consumption coupons by the government to consumers does not mean that they will meet 100 percent of their consumption needs with those coupons [6]. Rational consumers will selectively use consumption coupons based on their personal preferences, resulting in a consumers' probability consumption scenario. The probabilistic selling [7] strategies used by merchants also strengthen the trend of probabilistic consumer spending. This also urges the government to fully consider the probability consumption of consumers when using coupons to incentivize consumer demand.

In the era of commodity economy, the level of inventory management not only affects the cost of enterprises, but also plays a key role in stabilizing and coordinating consumer demand [8]. The dynamic changes in consumer demand preferences determine that enterprises must always control their inventory levels, neither out of stock, nor too much inventory, resulting in backlogs. Consumer behavior using coupons will inevitably challenge the traditional level of inventory management.

Therefore, it is of great significance to analyze the probabilistic consumption of consumers and the inventory optimization level of merchants in the process of using consumption coupons, which can provide some reference for the subsequent scientific development of incentive policies of consumption coupons. In this study, an improved minimum-cost and maximum-flow model was constructed (see Section 3.1), and the characteristic of the model was to dynamically adjust the inventory capacity of node enterprises with the level of probabilistic consumption (see Section 3.2). Through assumptions, the paper simulated three scenarios and performed calculus (see Section 4.2). The results showed that this study could better solve the problems of consumption coupons, probabilistic consumption and inventory optimization (see Section 5).

2. Literature Review

In the face of the global recession and the impact of the COVID-19 pandemic on the economy, the economies of the world are facing enormous challenges. The economic contraction [9] has been triggered by the pandemic, and social and economic losses continue to increase in many countries [10]. The effect of investment pulling is not obvious in the short term. The pandemic has reduced investments and the fiscal revenues of governments, thus increasing uncertainty. A demand-side shock was triggered by global and national restrictions to limit the spread of COVID-19, such as lockdown measures and travel bans [11]. The revitalization of consumption has become an important tool to stabilize the economic development. For consumers, coupons mean the increase of their temporary incomes. Such fiscal policy would increase the demand for consumption and simulate aggregate demand [12,13]. Stimulating the economy by issuing consumption coupons and activating market demand can better alleviate market failures, promote economic development in the short term, and help economic recovery [14].

Consumption coupons are a temporary incentive for the government to stimulate consumption [15], and are also a special coupon, which can be used exclusively by dividing the consumption categories of designed products [16]. The government uses consumption coupons to boost consumer confidence [17], which, in turn, will increase consumers' willingness to buy more goods and thus stimulate market dynamics, thus stimulating market vitality. Experts and scholars have various views on the role of consumption coupons, such as that they can effectively improve consumption, promote economic growth and increase employment [18]. Coupons can enhance the experience and outcomes for participants, benefiting more low-income households [19]. Consumption coupons are effective in stimulating short-term consumption levels. It will reduce the welfare level of society as a whole if the savings plan is not optimal [20]. The substitution effect of consumption coupons makes them difficult to pull effective demand for a long time [21]. The multiplier brought by consumption coupons in China during the 2020 epidemic is around 5–25, which has a significant effect on the GDP in the short term [22]. Consumption coupons are conducive to activating the frozen domestic demand [23] and market dynamics under the epidemic and make up for the short-term gap in overseas markets due to the global spread of the epidemic.

The difference in consumer purchasing behavior mainly comes from the difference in shopping decisions, and there is uncertainty [24] due to the influence of various factors in the process of purchasing behaviors [25], which is generally expressed by probability. Uncertainties faced by consumers in their purchase decisions include preference uncertainty and product performance uncertainty [26]. People may be non-indifferent towards the timing of the resolution of consumption uncertainty [27]. Probability, as an expression of purchase intention [28], can produce more accurate consumption predictions and may elicit different responses than standard purchase intention scales. Businesses must pay attention to consumer dynamics [29] and make appropriate marketing decisions based on an accurate understanding of consumers. Consumer dynamics are defined as temporal changes in consumer attitudes and behaviors. Today, probabilistic selling [30,31] is increasingly becoming a sales strategy for businesses and growing in popularity. Probabilistic selling could generally increase a firm's expected profit, and the firm will increase the price and the order quantity for the component product [32]. From a business perspective, adequate and controllable inventory can help companies better gain profits while reducing total costs [33]. Stock-outs will lead directly to lost sales, reduced profits, and the potential loss of customers [34]. Maintenance of inventories is a significant concern for any business organization due to decay or deterioration over time [35].

As one of the core problems of network optimization, the minimum-cost maximum-flow problem has a wide range of applications in reality, and the common special cases include the shortest path problem, the maximum-flow problem, and the assignment problem. A prognostic decision-making strategy is proposed to solve in real time the electric vehicle dynamic stochastic shortest path problem, aiming at the simultaneous utilization of historical and real-time traffic data [36]. A network is called uncertain if the arc capacities of the network are uncertain variables. Uncertainty theory is an efficient tool to deal with non-deterministic information, especially expert data and subjective estimation [37]. Scholars have conducted multidimensional studies based on different application scenarios [38–40], combined with the minimum-cost maximum-flow network model, and have achieved fruitful results.

Compared with existing studies, the main contributions of this study are as follows: (1) A minimum-cost and maximum-flow model is constructed, in which the capacity value on each arc will vary with the change of consumption probability value on its own arc. This helps to more effectively regulate the relationship among coupons, probabilistic consumption and inventory optimization. (2) On the basis of satisfying the government to issue sufficient consumption coupons and to stimulate consumer demand, it can help companies minimize ineffective inventory. (3) With the dynamic change of consumption probability, the inventory of enterprises will also vary, and the needs of consumers have been fully satisfied. (4) Based on the model derivation, corresponding countermeasures are proposed.

3. Materials and Methods

3.1. The Improved Idea of a Minimum-Cost and Maximum-Flow Model, in Which the Capacity Value on Each Arc Would Vary with Its Own Probability Value

In a directed network $D = (V, A, C)$ with only one single source node (V_s) and one single sink node (V_t), each arc (v_i, v_j) belonged to the capacity set (C), and the cost per unit of flow on each arc $b(v_i, v_j)$ was greater than or equal to 0. Under the premise of finding a maximum flow f in this network D and minimizing $b(f) = \sum_{(v_i,v_j)\epsilon A} b_{ij} f_{ij}$, the objective was to keep the sum of C over each arc in this directed network to be minimum.

First, a table of capacity and probability values on each arc was constructed, as shown in Figure 1.

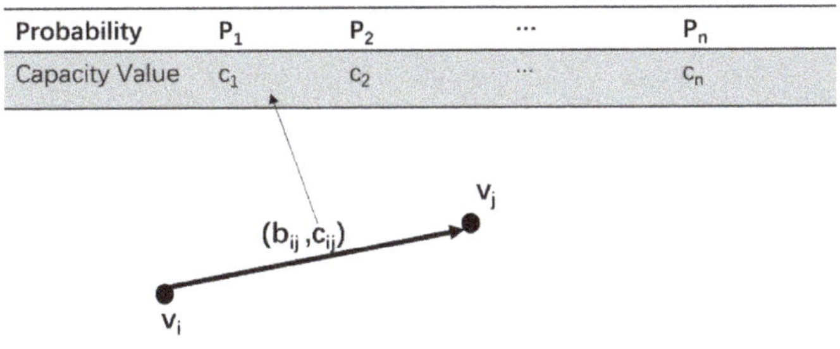

Figure 1. A figure of the capacity and probability values on the arc.

P_1 was the initial probability corresponding to the capacity c_1 on an arc. As the probability value increased from P_1 to $P_2, \ldots, P_n$, capacity value also synchronously increased from c_1 to $c_2, \ldots, c_n$. P_n was the maximum probability, and c_n represented the maximum capacity value for the maximum probability. The capacity value under the initial probability P_1 was chosen to generate the directed network under this initial probability. Based on the cost unit of flow (b_{ij}), a shortest path from V_s to V_t was calculated using Dijkstra's method to find the corresponding augmented chain. Taking the initial flow f on this augmented chain as 0, the minimum value of the difference between the capacity value and the flow value of each arc on this augmented chain was used as the adjustment value to adjust the flow f. Next, the weighted digraph was constructed. We recalculated the shortest circuit, found out the corresponding augmented chain, and then adjusted the capacity value according to the probability value on each arc. We repeated the above steps until you could not find a shortest path from V_s to V_t, then ended the loop. At this point, the sum of capacity c on all arcs in the directed network was the minimum, and the maximum flow and minimum cost were satisfied.

3.2. Calculation Steps to the Model

As shown in Figure 2, the calculation steps could be expressed as:

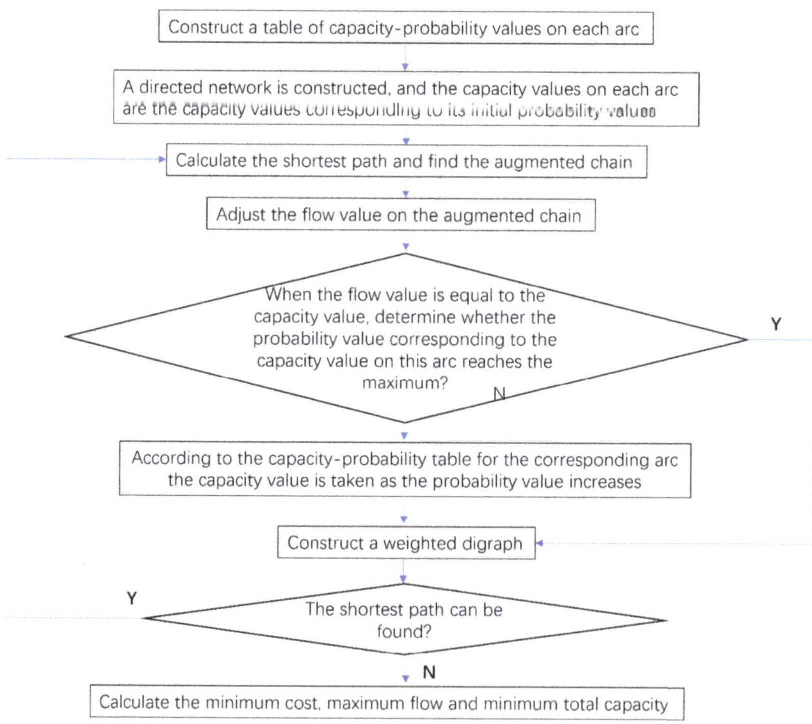

Figure 2. Calculation steps to the model.

Step 1: A table of capacity-probability values on each arc was constructed. Specifically, the maximum capacity to be controlled on the arc was used as the capacity value corresponding to the maximum probability value. Different probability values were set, and different probability values correspond to different capacities to form a table of capacity-probability values. This model assumed that the probability values in the capacity-probability table showed a single-term increasing trend, and the capacity value also grew as the probability value increased.

Step 2: We constructed a directed network where the capacity value on each arc was the capacity value corresponding to its initial probability value. Depending on the purpose of the study, we determined the initial probability values corresponding to different arcs. The corresponding control capacity c_{ij} were selected by the probability values, and the directed network was constructed by combining the cost per unit of flow on each arc $b(v_i,v_j)$.

Step 3: We calculated the shortest path and found the augmented chain. Using Dijkstra's method, a shortest path from V_s to V_t was calculated based on the cost per unit of flow, $b(v_i,v_j)$, on each arc, and the corresponding augmented chain was determined.

Step 4: We adjusted the flow value on the augmented chain. If the flow value on the augmented chain was calculated for the first time, the initial flow value f_{ij} on that augmented chain could be assumed to be 0. The minimum of all capacity values on that chain was used as the adjustment value θ for flow f_{ij}, and then the flow value on each arc plus θ was used as the new flow value f_{ij}. If there was already a flow value f_{ij} on an arc on the augmented chain, the minimum value of the difference between the capacity value and the flow value on each arc in the incremental chain was used as the adjustment value θ to adjust the flow f_{ij} when calculating the adjustment value, and the flow value on each arc plus θ was used as the new flow value f_{ij}.

Step 5: When the flow value was equal to the capacity value of an arc on the augmented chain, we determined whether the probability value corresponding to the existing capacity value on that arc had reached the maximum probability value. If not, we proceeded to step 6. Otherwise, we proceeded to step 7.

Step 6: We looked up the table of capacity-probability values for the corresponding arc, selected the corresponding capacity values along the probability growth direction, and adjusted the capacity values for the corresponding arcs on the augmented chain. Because the probability was one-way increasing, the adjacent probability value could be found, and we then determined the corresponding capacity control value c_{ij} by that adjacent probability value, replacing the capacity value on the existing arc.

Step 7: We constructed a weighted digraph. On the augmented chain, a weighted digraph identical to the current cost and flow network graph was constructed, and each arc in the original network was decomposed into two arcs (v_i, v_j) and (v_j, v_i) that were inverse to each other, and the weights were recorded as ω_{ij} and ω_{ji}, respectively. The rules were as follows:

$$\omega_{ij} = \begin{cases} b_{ij} \; when \; f_{ij} < c_{ij} \\ +\infty \; when \; f_{ij} = c_{ij} \end{cases}$$

$$\omega_{ji} = \begin{cases} -b_{ij} \; when \; f_{ij} > 0 \\ +\infty \; when \; f_{ij} = 0 \end{cases}$$

Step 8: We determined whether the shortest circuit could be found or not. If the shortest path could be found, it meant that further adjustment was needed, then we went back to step 3. Otherwise, it meant that the augmented chain at this time was already the shortest path, there was no need to adjust, optimization was over, and step 9 was entered.

Step 9: Based on the results of step 8, the minimum cost $\sum_{(v_i,v_j)\epsilon A} b_{ij}f_{ij}$ was calculated, while the maximum network flow was determined as the sum of the outgoing flow values at node v_s. The sum of the capacity values on all arcs was then calculated as the minimum total capacity on that network.

3.3. Feasibility Analysis of an Improved Model for the Application of Consumption Coupons, Probabilistic Consumption and Inventory Optimization

With the recurrence of COVID-19, it has become imperative for governments to take different approaches to stimulating their economies. As mentioned earlier, governments have already started to practice in many countries by issuing consumption coupons to stimulate the economy. In this scenario, three subjects would be involved, which were government, enterprises and consumers. In the minimum-cost maximum-flow graph shown in Figure 3, node V_s could be viewed as the government, node V_c as the consumer, and the intermediate nodes as the firms that satisfy the different needs of the consumer. It was assumed that the government-issued consumer coupons did not restrict consumers from purchasing any goods. It was only stipulated that only one consumption coupon could be used on one commodity.

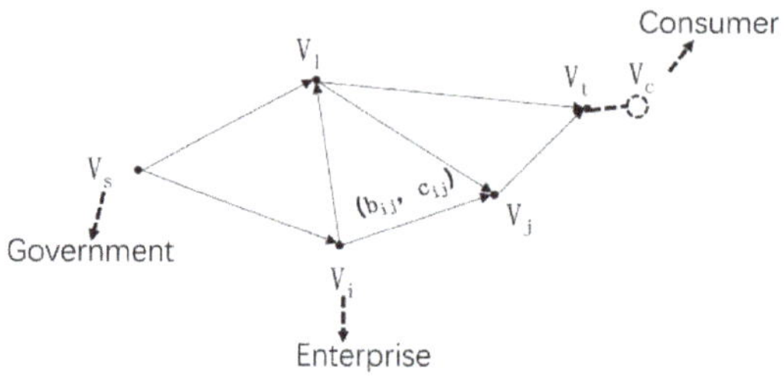

Figure 3. Feasibility analysis of an improved model.

It was assumed that the government-issued consumer coupons did not restrict consumers from purchasing any goods. It was only stipulated that only one consumption coupon could be used on one commodity. The flow in the network could be regarded as the consumption coupons issued by the government to purchase goods from node enterprises to meet the needs of consumers. The use of coupons would not change the price of products while stimulating consumer demand. Therefore, the cost unit of flow on each arc could be regarded as the commodity price, and the usage of coupons could be analogous to the quantity of goods purchased. Even if consumers held enough consumer coupons, there was a probabilistic purchase problem when faced with a specific demand. It did not exactly follow the individual's real demand to purchase the maximum goods, so the capacity on the arc also needed to change with the probability of consumer purchase. For node enterprises in the network, the capacity on the arc could be compared to the stock quantities of the enterprise. Only adequate inventory could ensure that the enterprise met the needs of consumers. Excessive inventory would inevitably increase the cost of enterprises and was not conducive to competition, which would inevitably drive enterprises to seek minimum inventory. For the government, what was needed was that the coupons paid out met the consumer demand to the greatest extent. For consumers, what was needed was to meet the maximum consumption on the basis of the minimum cost, that is, to be able to find the shortest path from node v_s to node v_t. In the specific calculation, the connection between node v_t and node v_c could be deleted, and the improved minimum-cost maximum-flow model could be directly used for calculation. According to the above analysis, it could be seen that the improved minimum-cost maximum-flow model could better complete the research and had a strong practical guidance.

4. Results and Discussion

4.1. Simulation Problem and Data Descriptions

It was supposed that the government agency (node v_s) in a region was ready to issue a number of consumption coupons to stimulate consumer demand, and consumers needed to satisfy four demands during that incentive period, which were borne by four different enterprises (four nodes, v_1, v_2, v_3 and v_t, as shown in Figure 4). When purchasing goods, consumers could only use one consumption coupon per product, and the consumption coupon would not change the original price of the product. Enterprises needed to prepare the maximum inventory according to the maximum number of coupons issued to cope with the consumer's purchase. There were various strategies for consumers to use consumer coupons to purchase goods;; for example, the arc (v_2, v_1) indicated that the consumer used the consumer coupon to purchase goods from the node enterprise v_1 after passing through node enterprise v_2. The weight group value (2, 50) on the arc, where 2 indicated that the unit price of commodities purchased by consumers from node v_1 with coupons is 2; 50 represented the maximum amount of inventory prepared by node enterprise v_1 to cope with the consumption demand coming from node v_2 is 50.

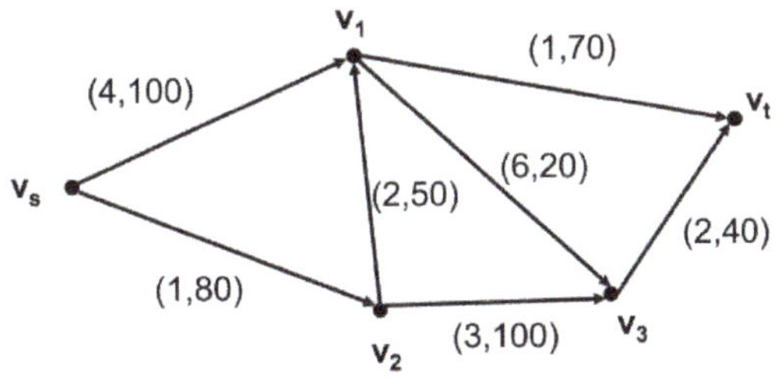

Figure 4. Directed graphs for simulation problem.

The problem now was to find out what was the maximum number of consumption coupons that the government could issue, so that consumers could obtain the maximum consumption satisfaction with the minimum cost, and the inventory of enterprises could reach the minimum.

4.2. Simulation Scenario Calculations

4.2.1. Scenario 1: Calculation Using the Traditional Minimum-Cost Maximum-Flow Model

In the first step, the Dijkstra method was used to solve the shortest path from v_s to v_t, and the augmented chain $u = (v_s, v_2, v_1, v_t)$ was obtained. In the second step, the flow value on the augmented chain u was adjusted, and the adjusted flow figure I shown in Figure 5 was obtained.

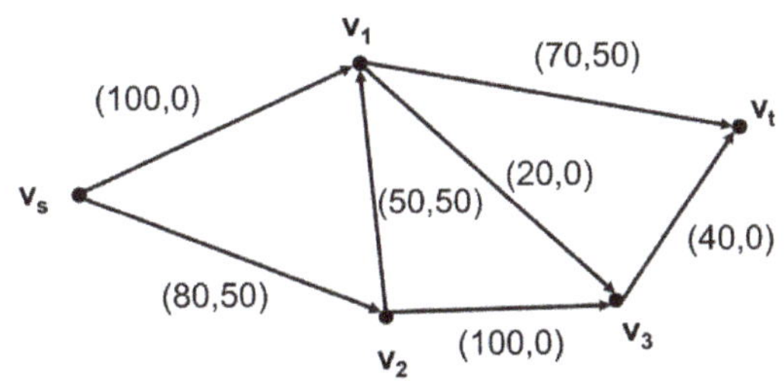

Figure 5. The adjusted flow figure in scenario 1.

In the third step, a weighted directed graph was constructed to obtain the weighted digraph I shown in Figure 6.

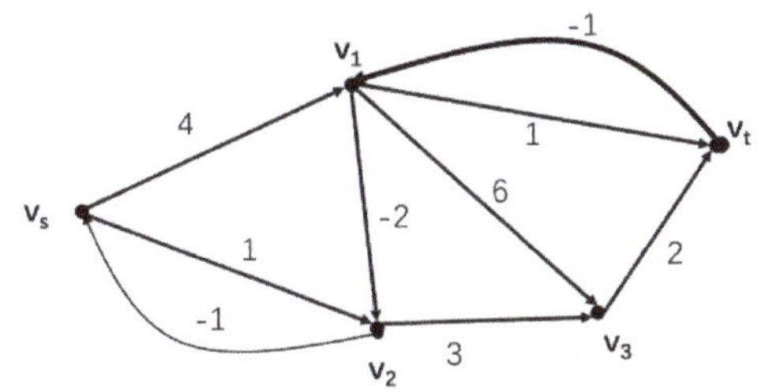

Figure 6. The weighted digraph in the third step.

Repeating the above steps, the final shortest path obtained was shown in Figure 7.

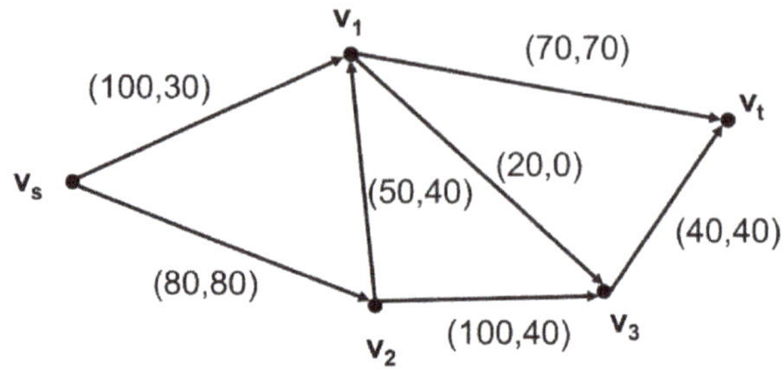

Figure 7. The shortest path of scenario 1.

The maximum number of coupons issued from node *Vs* was: 30 + 80 = 110.

The total cost for consumers to satisfy their own consumption was: 4 × 30 + 1 × 80 + 2 × 40 + 3 × 40 + 6 × 0 + 1 × 70 + 2 × 40 = 550.

The total inventory of enterprise stock was: 100 + 80 + 70 + 50 + 100 + 20 + 40 = 460.

4.2.2. Scenario 2: Improved Model Calculation with Probability Change from 50% to 100%

Step 1: A table of capacity-probability values on each arc was constructed. The capacity on each arc in Scenario 1 was taken as the capacity value corresponding to the maximum probability value of 100% for each arc, while the capacity value corresponding to each probability value was assumed to be the maximum capacity multiplied by the corresponding probability value. We took the probability values of 50% and 80%, respectively, to obtain the capacity-probability value table shown in Table 1.

Table 1. The capacity-probability value I.

arc	Probability Value		
	50%	**80%**	**100%**
(v_s, v_1)	50	80	100
(v_s, v_2)	40	64	80
(v_2, v_1)	25	40	50
(v_2, v_3)	50	80	100
(v_1, v_t)	35	56	70
(v_1, v_3)	10	16	20
(v_3, v_t)	20	32	40

Step 2: The capacity value corresponding to 50% of the initial probability on each arc was chosen to construct the directed network graph, as shown in Figure 8.

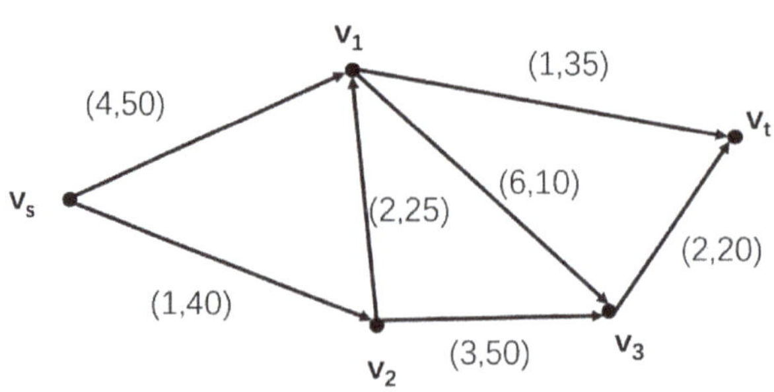

Figure 8. The directed graph corresponding to an initial probability of 50%.

Step 3: Using Dijkstra's method, the shortest path from v_s to v_t was solved to obtain the augmented chain $u = (v_s, v_2, v_1, v_t)$.

Step 4: We adjusted the flow value on the augmented chain to obtain the adjusted flow figure II shown in Figure 9.

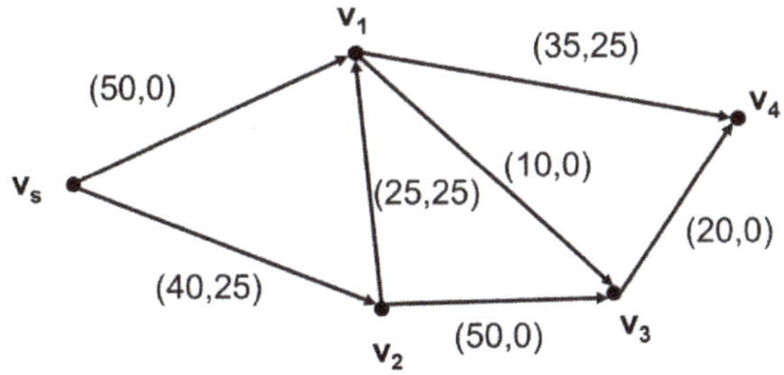

Figure 9. The adjusted flow figure in step 4.

Step 5: The flow and capacity values on arc (v_2, v_1) were both equal to 25, the current probability value was 50% and the maximum 100% was not reached. We then went to step 6; otherwise, to step 7.

Step 6: The flow value and capacity value on the arc (v_2, v_1) were equal, so from Table 1, it could be found that the next probability value was 80%, and the corresponding capacity value was 40. We replaced the existing capacity value with 40.

Step 7: We constructed the weighted directed graph, and according to the weighted rules, obtained the weighted digraph shown in Figure 10.

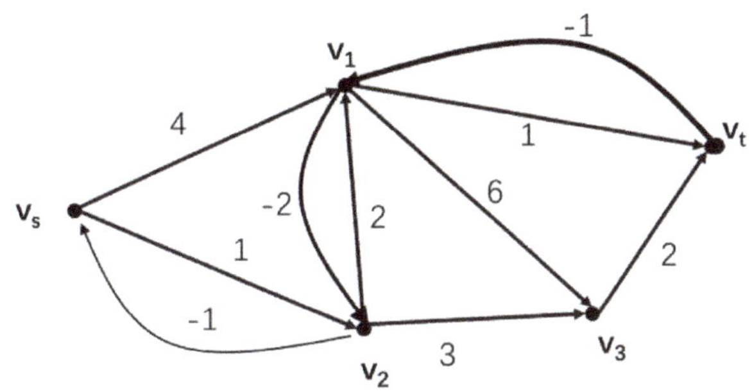

Figure 10. The weighted digraph in step 7.

Step 8: From Figure 10, a shortest path could be found with the Dijkstra method, so we went back to step 3 and repeated the above steps, until the directed graph could not find a shortest path; at this point it could indicate that the optimization is complete and we could go to step 9.

Step 9: By looping multiple times, the final shortest path was shown in Figure 11.

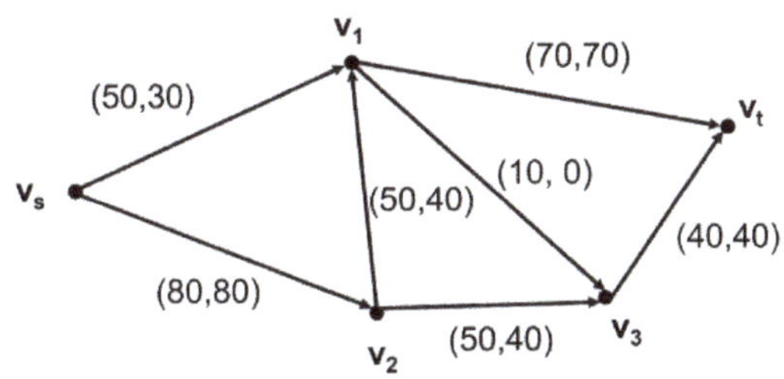

Figure 11. The shortest path of scenario 2.

The maximum number of coupons issued from node *Vs* was: 30 + 80 = 110.

The total cost for consumers to satisfy their own consumption was: 4 × 30 + 1 × 80 + 2 × 40 + 3 × 40 + 6 × 0 + 1 × 70 + 2 × 40 = 550.

The total inventory of enterprise stock was: 50 + 80 + 70 + 50 + 50 + 10 + 40 = 350.

4.2.3. Scenario 3: Improved Model Calculation with Probability Value on Part of the Arc Not Reaching 100%

The capacity on each arc in Scenario 1 was taken as the capacity value corresponding to the maximum probability value of 100% for each arc, while the capacity value corresponding to each probability value was assumed to be the maximum capacity multiplied by the corresponding probability value. Take the probability values of 50% and 80%, respectively, to obtain the capacity-probability value table shown in Table 2, where the maximum probability value s of arc (v_2, v_1) and arc (v_1, v_t) were taken as 80%.

Table 2. The capacity-probability value II.

arc	Probability Value		
	50%	80%	100%
(v_s, v_1)	50	80	100
(v_s, v_2)	40	64	80
(v_2, v_1)	25	40	-
(v_2, v_3)	50	80	100
(v_1, v_t)	35	56	-
(v_1, v_3)	10	16	20
(v_3, v_t)	20	32	40

Note: "-" meant that the probability value in this cell did not need to be set according to the study, that was, the probability value on the corresponding arc could not reach 100%.

By following the steps shown in Figure 2, through multiple loops, the final shortest path is shown in Figure 12.

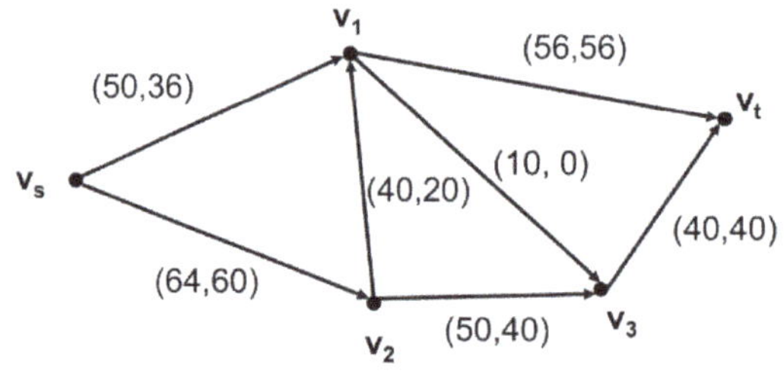

Figure 12. The shortest path of scenario 3.

The maximum number of coupons issued from node *Vs* was: 36 + 60 = 96.

The total cost for consumers to satisfy their own consumption was: 4 × 36 + 1 × 60 + 2 × 20 + 3 × 40 + 6 × 0 + 1 × 56 + 2 × 40 = 500.

The total inventory of enterprise stock was: 50 + 64 + 40 + 56 + 50 + 10 + 40 = 310.

4.3. Improved Simulation Results

From the contents of Section 4, it could be seen that the algorithm of the improved minimum-cost maximum-flow model was effective and feasible in solving the above problem, as demonstrated by the derivation steps and results in Sections 4.2.2 and 4.2.3.

Consumer welfare remained unchanged, the largest number of government consumption coupons was issued, and the number of enterprise inventories reached the lowest. In Section 4.2.1, the minimum-cost maximum-flow model was used to obtain that the maximum number of consumption coupons issued by the government was 110 and the minimum total cost for consumers to satisfy their own consumption was 550. In Section 4.2.2, the maximum number of government-issued consumption coupons using the improved minimum-cost maximum-flow model was 110, and the minimum total cost of consumers to meet their own consumption is 550. Comparing these results, it was found that the number of consumption coupons issued by the government and the total cost spent by consumers remained unchanged. In Section 4.2.1, the total inventory was 460; in Section 4.2.2, the total inventory was 350, which was lower than the inventory in Section 4.2.1. It could be concluded that the use of the improved minimum-cost maximum-flow model could effectively help enterprises reduce the amount of ineffective inventory stocked to respond to consumer demand and brought it to a minimum, while keeping the number of government-issued consumption coupons and the total cost spent by consumers constant.

Enterprise inventories were minimized with different decisions on consumer probability consumption, and the government's issuance of consumption coupons and the satisfaction of consumer demand have reached a dynamic balance. The improved minimum-cost maximum-flow model worked equally well when different node enterprises used different values of termination probability. In Section 4.2.3, there were two arcs with a termination probability value of 80% each, the maximum number of consumption coupons issued by the government was 96, the minimum total cost of consumers to meet their own consumption was 500, and the total inventory of enterprises was 310. Compared with the equal termination probability value of enterprises at different nodes in Section 4.2.2, the number of consumption coupons issued by the government, the minimum cost of consumers and the inventory of enterprises to prepare goods have shown a downward trend. This suggested that, as consumption probability decreased, the minimum total cost spent by the consumer decreased, and the number of government-issued consumption coupons and enterprise inventories would also drop. In this process, according to the consumer's consumption intention, the needs of consumers at all nodes have been met. From the model point of view, that was to say, the shortest path has not changed, but the flow value and inventory on the arc connecting the various node enterprises has changed dynamically with the probability value. From the perspective of practical practice, it was more realistic for enterprises at different nodes to take different consumption probabilities for inventory control.

5. Conclusions and Policy Recommendations

In this paper, an improved minimum-cost maximum-flow model was constructed to study the optimization problems of consumption coupons, probabilistic consumption and inventory. With the objectives of maximizing the number of coupons issued and minimizing the enterprises' inventory, different results of consumers' probabilistic consumption were studied. Simulation study showed that enterprise inventories were minimized with different decisions on consumer probability consumption, and the government's issuance of consumption coupons and the satisfaction of consumer demand have reached a dynamic balance. From the perspective of consumer demand, consumer demand is satisfied at each node by using different probabilistic consumption. From the perspective of enterprise inventory control, the optimal amount of inventory can be achieved by dynamically changing the enterprise inventory with the value of probabilistic consumption. From the perspective of government issuance of coupons to stimulate the economy, consumer welfare remains unchanged, the largest number of government consumption coupons can be issued, and the number of enterprise inventories reaches the lowest.

In order to better promote consumers' probabilistic consumption and enterprises' inventory optimization level in the process of using consumption coupons, the following policy recommendations are made at three levels: government-issued consumption coupons, enterprise inventory optimization and probabilistic consumption.

First, keeping a close eye on consumer demand, the government should strictly control the issuance of consumption coupons. When adopting consumer coupons to stimulate the economy, the government should not simply aim to issue the maximum number of coupons. What should be more of a concern is whether these consumption coupons can really stimulate consumers' potential demand desire and promote consumption. Increasingly rational consumers, in the face of the uncertainty of the future brought about by the epidemic, will inevitably pursue consumption decisions at the lowest cost of consumption when consuming. Therefore, the government needs to prudently design consumption coupons so that consumers are clear about the strength of the cost discount brought by the coupons. Second, the government should fully cooperate with third-party market research companies to investigate the potential consumption needs of the target group, and determine the optimal number of consumption coupons with the help of the minimum-cost maximum-flow model, rather than issuing consumption coupons arbitrarily according to personal wishes.

Next, to help enterprises minimize inventory to the maximum extent. Supply and demand are the two ends of the economy, and only when both are smooth can we ensure sustainable economic development. In most cases, enterprises will prepare more inventory to cope with consumption and avoid losses caused by out-of-stocks. When there is a stimulus policy acting on the consumption side, this extra inventory preparation by supply-side enterprises is artificially amplified. When the stimulus ends, the growth in inventory to cope with consumption increases the cost of the business. Therefore, helping enterprises minimize inventory is a factor that must be considered when the government adopts incentive policies. The minimum inventory can be determined with the help of the minimum-cost maximum-flow models in the paper, which is not only effective but also feasible.

Last, to track consumer demand in real time and dynamically adjust consumption probability, consumers, as a fickle population, will involve a variety of factors when satisfying their own consumption. In the face of the consumption of a certain product, consumers will inevitably have a probability consumption tendency. This determines that, when launching consumption coupons to stimulate the economy, it is necessary to use modern scientific and technological means and consumption big data to track consumers' purchase demand in real time and dynamically push the consumption probability to enterprises in real time, so as to ensure that enterprises adjust their inventory according to the purchase probability in time, thus ensuring that the inventory reaches the lowest level while meeting the consumption demand.

In the future, this paper will conduct further research on the following scenarios. First is the application of different types of consumption coupons to different goods. The second is the impact on inventories of cooperation between competing companies in the supply of goods. In the follow-up study, further cooperation will be carried out with relevant institutions such as the government, and the model will be empirically explored using real-life examples.

Author Contributions: S.W. mainly put forward the idea of this paper, designed a research model and analyzed the data, and he is one of the directors of the funding project that supported this paper. Y.C. is the corresponding author of this article; she contributed to the literature review of the study, investigation and wrote the manuscript with input from all authors, and she is also one of the directors of the funding project that supported this paper. All authors have read and agreed to the published version of the manuscript.

Funding: This work was supported by the Soft Science Research Program of Zhejiang Province in 2022 (Grant Number: 2022C25071) and the General Project of Humanities and Social Sciences, Ministry of Education, China (Grant Number: 16YJA630054).

Institutional Review Board Statement: Not applicable.

Informed Consent Statement: Not applicable.

Data Availability Statement: The datasets used and/or analyzed under the current study are available from the corresponding author on reasonable request.

Conflicts of Interest: The authors declare no conflict of interest.

References

1. China Battles Multiple COVID-19 Outbreaks, Driven by Stealth Omicron. Available online: https://www.npr.org/2022/03/14/1086394167/china-battles-multiple-outbreaks-driven-by-stealth-omicron (accessed on 14 June 2022).
2. 2.43 Billion Yuan of Consumer Coupons are Issued by 13 Places to Benefit Wine. Available online: https://3g.163.com/jiu/article/H75E5H1I008297TK.html (accessed on 12 May 2022).
3. Lin, F.L.; Chen, W.Y. Did the consumption voucher scheme stimulate the economy? Evidence from smooth time-varying cointegration analysis. *Sustainability* **2020**, *12*, 4895. [CrossRef]
4. Guan, X.; Atlas, S.A.; Vadiveloo, M. Targeted retail coupons influence category-level food purchases over 2-years. *Int. J. Behav. Nutr. Phys. Act.* **2018**, *15*, 111. [CrossRef] [PubMed]
5. Wu, Y.L.; Zhang, Y. Effect evaluation of consumer vouchers based on psm-did model. *World Surv. Res.* **2021**, *1*, 14–24.
6. Hess, J.D.; Gerstner, E. Double couponing: Pricing and consumer perspectives. *Mark. Lett.* **1993**, *4*, 153–163. [CrossRef]
7. Wu, Y.; Jin, S. Joint pricing and inventory decision under a probabilistic selling strategy. *Oper. Res. Int. J.* **2022**, *22*, 1209–1233. [CrossRef]
8. Jauhari, W.A.; Wangsa, I.D. A Manufacturer-Retailer Inventory Model with Remanufacturing, Stochastic Demand, and Green Investments. *Process Integr. Optim. Sustain.* **2022**, *6*, 253–273. [CrossRef]
9. Salvatore, D.E.; Ksenia, K.; Marcos, P.R. Medium-term fiscal multipliers during protracted economic contractions. *J. Macroecon.* **2018**, *56*, 35–52.
10. Wu, F.; Liu, G.; Guo, N.; Li, Z.; Deng, X. The impact of COVID-19 on China's regional economies and industries. *J. Geogr. Sci.* **2021**, *31*, 565–583. [CrossRef]
11. Manal, S. Modeling long-term impacts of the COVID-19 pandemic and oil price declines on Gulf oil economies. *Econ. Model.* **2022**, *112*, 105849.
12. Kan, K.; Peng, S.-K.; Wang, P. Understanding consumption behavior: Evidence from consumers' reaction to shopping vouchers. *Am. Econ. J. Econ. Policy* **2017**, *9*, 137–153. [CrossRef]
13. Dawid, H.; Harting, P.; Neugart, M. Fiscal transfers and regional economic growth. *Rev. Int. Econ.* **2018**, *26*, 651–671. [CrossRef]
14. Madson, J.; Arendt, S. Women, infants, and children (wic) participants' intention to use wic farmers' market coupons in illinois. *J. Am. Acad. Nutr. Diet.* **2019**, *119*, A70. [CrossRef]
15. Reichhart, P.; Pescher, C.; Spann, M. A comparison of the effectiveness of e-mail coupons and mobile text message coupons for digital products. *Electron Markets* **2013**, *23*, 217–225. [CrossRef]
16. Muk, A. Perceptions of mobile coupons: A cross-national study. *J. Direct Data Digit. Mark. Pract.* **2012**, *13*, 311–324. [CrossRef]
17. Hampson, D.P.; Ma, S.S.; Wang, Y.; Han, M.S. Consumer confidence and conspicuous consumption: A conservation of resources perspective. *Int. J. Consum. Stud.* **2021**, *45*, 1392–1409. [CrossRef]
18. Ruo, K.; Wang, D.H. Effect Assessment on the Policy of China's Consumption Coupon Issuance under the Financial Crisis: Based on Mundell-Fleming Model. *Zhejiang Soc. Sci.* **2010**, *2*, 46–51+126.
19. Caron-Roy, S.; Lee, Y.Y.; Sayed, S.A.; Lashewicz, B.; Milaney, K.; Dunn, S.; O'Hara, H.; Leblanc, P.; Prowse, R.J.; Fournier, B.; et al. Experiences and Perceived Outcomes of Low-Income Adults During and After Participating in the British Columbia Farmers' Market Nutrition Coupon Program: A Longitudinal Qualitative Study. *J. Acad. Nutr. Diet.* 2022, *in press*. [CrossRef]
20. Blake, D. Nudges and Networks: How to Use Behavioural Economics to Improve the Life Cycle Savings-Consumption Balance. *J. Risk Financ. Manag.* **2022**, *15*, 217. [CrossRef]
21. Karpyn, A.; McCallops, K.; Wolgast, H.; Glanz, K. Improving Consumption and Purchases of Healthier Foods in Retail Environments: A Systematic Review. *Int. J. Environ. Res. Public Health* **2020**, *17*, 7524. [CrossRef]
22. Gao, X.Y. A Retrospective Study of the Economic Effects of Consumption Coupons during the Epidemic. *J. Qujing Norm. Univ.* **2022**, *41*, 81–89.
23. Wang, W.; Liu, L.; Yang, Y. Spatial Matching Analysis and Development Strategies of County Night-Time Economy: A Case of Anning County, Yunnan Province. *Sustainability* **2022**, *14*, 4891. [CrossRef]
24. Nao, A.; Man, B.; Mha, C.; Ssa, D. The panic buying behavior of consumers during the COVID-19 pandemic: Examining the influences of uncertainty, perceptions of severity, perceptions of scarcity, and anxiety. *J. Retail. Consum. Serv.* **2021**, *62*, 102600.
25. Wunderlich, S.; Gatto, K.; Smoller, M. Consumer knowledge about food production systems and their purchasing behavior. *Environ. Dev. Sustain.* **2018**, *20*, 2871–2881. [CrossRef]
26. Roberts, J.H.; Urban, G.L. Modeling Multiattribute Utility, Risk, and Belief Dynamics for New Consumer Durable Brand Choice. *Manag. Sci.* **1988**, *34*, 167–185. [CrossRef]
27. Meissner, T.; Pfeiffer, P. Measuring Preferences Over the Temporal Resolution of Consumption Uncertainty. *J. Econ. Theory* **2015**, *200*, 105379. [CrossRef]
28. Wright, M.; MacRae, M. Bias and variability in purchase intention scales. *J. Acad. Mark. Sci.* **2007**, *35*, 617–624. [CrossRef]
29. Zhang, J.Z.; Chang, C.W. Correction to: Consumer dynamics: Theories, methods, and emerging directions. *J. Acad. Mark. Sci.* **2021**, *49*, 197. [CrossRef]
30. Spiliotis, E.; Makridakis, S.; Kaltsounis, A.; Assimakopoulos, V. Product sales probabilistic forecasting: An empirical evaluation using the M5 competition data. *Int. J. Prod. Econ.* **2021**, *240*, 108237. [CrossRef]
31. Fay, S. Selling an opaque product through an intermediary: The case of disguising one's product. *J. Retail.* **2008**, *84*, 59–75. [CrossRef]

32. Palanivel, M.; Uthayakumar, R. A production-inventory model with promotional effort, variable production cost and probabilistic deterioration. *Int. J. Syst. Assur. Eng. Manag.* **2017**, 290–300. [CrossRef]
33. Maji, A.; Bhunia, A.K.; Mondal, S.K. A production-reliability-inventory model for a series-parallel system with mixed strategy considering shortage, warranty period, credit period in crisp and stochastic sense. *OPSEARCH* **2022**, 1–46. [CrossRef]
34. Avlijas, G.; Simicevic, A.; Avlijas, R.; Prodanovic, M. Measuring the impact of stock-keeping unit attributes on retail stock-out performance. *Oper. Manag. Res.* **2015**, *8*, 131–141. [CrossRef]
35. Adak, S.; Mahapatra, G.S. Effect of reliability on multi-item inventory system with shortages and partial backlog incorporating time dependent demand and deterioration. *Ann. Oper. Res.* **2020**, 1–21. [CrossRef]
36. Rozas, H.; Muñoz-Carpintero, D.; Doris Saéz, D.; Marcos, E. Orchard. Solving in real-time the dynamic and stochastic shortest path problem for electric vehicles by a prognostic decision making strategy. *Expert Syst. Appl.* **2021**, *184*, 115489. [CrossRef]
37. Han, S.W.; Peng, Z.X.; Wang, S.Q. The maximum flow problem of uncertain network. *Inf. Sci.* **2014**, *265*, 167–175. [CrossRef]
38. Büsing, C.; Koster, A.M.C.A.; Schmitz, S. Robust minimum cost flow problem under consistent flow constraints. *Ann. Oper. Res.* 2021. [CrossRef]
39. Goldberg, A.V.; Hed, S.; Kaplan, H.; Tarjan, R.E. Minimum-Cost Flows in Unit-Capacity Networks. *Theory Comput. Syst.* **2017**, *61*, 987–1010. [CrossRef]
40. Hirai, H.; Ikeda, M. A cost-scaling algorithm for minimum-cost node-capacitated multiflow problem. *Math. Program.* **2021**, 1–33. [CrossRef]

sustainability

MDPI

Article

The Community Commitment to Sustainability: Forest Protection in Guatemala

Eduardo Fernández Luiña [1,*], Santiago Fernández Ordóñez [2] and William Hongsong Wang [1,*]

1 Instituto de Estudios Políticos y Relaciones Internacionales, Universidad Francisco Marroquín (Madrid Campus), Calle de Arturo Soria 245, 28033 Madrid, Spain
2 Instituto de Estudios Políticos y Relaciones Internacionales, Universidad Francisco Marroquín, Calle Manuel F. Ayau (6 Calle Final), Zona 10, Guatemala 01010, Guatemala; sfo@ufm.edu
* Correspondence: efernandez@ufm.edu (E.F.L.); hongsongwang@ufm.edu (W.H.W.)

Abstract: This article covers the current research vacuum on how Guatemala partially conducts forest preservation through community concessions. Our paper starts its analysis by synthesizing the private property-rights approach environmentalist theory and the community concession theory. It is argued that the shared common private property as a community arrangement can turn conflicts into potential opportunities for the involved parties to solve the existing environmental problems by win-win games. Based on the above theoretical views, our study extends the scope to the modern and democratic municipals' forest preservation in Guatemala, as previous research mainly focused on how the Guatemalan traditional indigenous communities have conducted forest preservation. Our empirical results show that the in-force forest concessions in the Maya Biosphere Reserve have achieved the Guatemalan government's forest conservation target in recent years. However, as the Guatemalan forest concession arrangements are just usufructs and the state still owns forest titles, the current Guatemalan forest concession could reverse the result of the limited, decentralized forest reform. In this regard, we suggest that Guatemala state should privatize all these forests to the concessions' communities and firms. If the results are positive, we propose the Guatemalan government further apply the decentralization forest policy to the whole country.

Keywords: Guatemala; forest; property rights; community concession; entrepreneurship; forest sustainability; economic development; free-market environmentalism; economic development; Latin America

Citation: Fernández Luiña, E.; Fernández Ordóñez, S.; Wang, W.H. The Community Commitment to Sustainability: Forest Protection in Guatemala. *Sustainability* **2022**, *14*, 6953. https://doi.org/10.3390/su14126953

Academic Editors: Elkhan Richard Sadik-Zada and Andrea Gatto

Received: 16 March 2022
Accepted: 30 May 2022
Published: 7 June 2022

Publisher's Note: MDPI stays neutral with regard to jurisdictional claims in published maps and institutional affiliations.

1. Introduction

As a third-wave democracy, Guatemala has suffered an internal armed conflict of considerable magnitude like other republics in Latin America [1]. Since its democratization, the country has been unable to find a clear direction toward the long-awaited economic development [2,3]. Despite the above problems, Guatemala still has highlights in environmental protection affairs. Although it lacks economic growth [4], the community-based rural, peripheral, and indigenous world conducts successful forest conservation and provides highly qualified agricultural and forest exportations [5]. The community identity plays a vital role in Guatemala's environmental protection issues [6,7].

This article aims to cover the current research vacuum on how Guatemala partially conducts forest preservation through community concessions, as previous empirical research mainly focused on how the Guatemalan traditional indigenous communities have conducted forest preservation. We consider that the in-force forest concessions in the Maya Biosphere Reserve (RBM, for its acronym "Reserva de la Biósfera Maya" in Spanish) have achieved the Guatemalan government's forest conservation target. It shows how forest concessions as community arrangements help local communities and local firms not only protect the forests, but also make their economy competitive.

Our paper starts its analysis by combining the private property-rights approach environmentalist (PAE) theory and the community concession (CC) theory. Wang et al. argued that the traditional top-down environmental policy assumes that central planning possesses or can acquire the relevant knowledge and prices to design a prosperous economy [8]. They considered that this assumption was initially challenged one hundred years ago by Ludwig von Mises [9] and later by the 1974 Nobel Prize in Economic Sciences winner Fredrich A. Hayek [10]. He contended that a central body such as the government could not concentrate the relevant knowledge for economic success and the use of resources. In fact, knowledge is scattered among agents in society due to its very local, tacit, subjective, exclusive, and personal nature [11]. Additionally, prices and production are coordinated in the decentralized market economy [9]. This reasoning breaks with any top-down and centralized policy program, advocating a bottom-up and decentralized view of policymaking, respecting the latter initiative's role in providing environmental protection and sustainability. Moreover, Wang et al. argued that many authors had approached the study of environmental issues based on this Hayekian approach [12–17]. Both theory and empirical evidence have shown how decentralized alternatives achieve environmental transformations and, conversely, how centralized policies fail to attain the expected and planned policy objectives.

The PAE theory defines environmentalism as the science that studies human beings' relations with each other and their environment [15]. The PAE theory considers that the existing decentralized and spontaneous market process propelled by the creative entrepreneurship coordinates better with and adjusts better for the rest of the species and elements of the natural environment than the centralized planned economy [15].

The CC theory was developed during the second half of the 20th century by Vincent Ostrom [18] and the 2009 Nobel Prize in Economic Sciences winner Elinor Ostrom [19]. The CC theory sees community concession as a polycentric reality and the third way between nationalization and privatization [18]. Elinor Ostrom's work has been interpreted as a reaction to Garret Hardin's *The Tragedy of the Commons* [20]. The latter's work recommended privatization as an easy, viable, and quick solution to the famous tragedy of the commons. Vincent and Elinor Ostrom react against this claim, arguing that, in some situations, community property management may be more effective—and more beneficial and legitimate—than simply privatization itself. Based on the contribution of the community concession approach, this paper discusses concessions as usufructs for community use. Although the Ostroms did not consent that the community arrangement could be a potential private property, we consider that the CC theory can cooperate with the PAE criteria if the concession is treated as a transition towards a shared private property. Moreover, we reckon that as the state still owns forest titles, the current Guatemalan forest concession could reverse the result of the limited, decentralized forest reform.

The article is organized in the following structure. Section 2 provides our research methodology and background. Section 3 studies the modern forest concession institutions in the Maya Biosphere Reserve (RBM). Section 4 shows our research results. Section 5 is our discussion of future research proposals. Section 6 is the conclusion.

2. Research Methodology and Background

This section introduces our research methodology and background. Section 2.1 provides the fundamental theoretical tools we use in this research: the private property-rights approach environmentalist theory and the community concession theory. Section 2.2 provides the fundamental concepts involving forest conservation. Section 2.3 reviews previous research on the Guatemalan communities' forest management. Section 2.4 illustrates Guatemala's centralized political institution and its failed rationalization of land policy. Section 2.5 reviews Guatemala's traditional indigenous communities' forest arrangements.

2.1. Property-Rights Approach Environmentalist Theory and the Community Concession Theory

2.1.1. Private Property-Rights Approach Environmentalist Theory

The PAE theory, also named free-market environmentalism, has been continuously developed by the Austrian school economists represented by Ludwig von Mises, Fredrich von Hayek, and Jesús Huerta de Soto since the early 20th century. It defines environmentalism as the science that studies human beings' relations with each other and their environment [15]. The PAE theory considers that the entrepreneurship-based market process coordinates better with and adjusts better for the rest of the species and elements of the natural environment than the centralized planned economy [15].

The PAE theory considers three fundamental problems with any centrally planned environmental policy. The first is the impossibility of economic calculation through central planning. The exact identification of private property rights gives the property owner the incentive to protect the environment in which he lives and sue anyone who violates his property's environment. In contrast, the lack of private property generates the tragedy of the commons, and the environment is polluted without the incentive to protect it [13,21–27]. When property rights are violated, human beings cannot act as they want because the necessary information and price signals are disturbed. Therefore, even the most radical environmentalists cannot ensure that their centrally planned proposals would not cause even more environmental damage [15].

Secondly, nationalizing natural resources as public property prevents economic calculation and undermines entrepreneurship [15]. As the market economy's driving force [28,29], based on price signals, the entrepreneurs make better decisions and allocate resources more efficiently to protect the environment than the central planning of governments. However, when natural resources are nationalized, it becomes impossible for entrepreneurs to make economic calculations. Therefore, the related environmental-friendly products might not be produced due to the missing role of entrepreneurship.

Thirdly, zero-sum games are created through public policies and legislative decisions, while the market might solve these problems. Governmental orders substitute voluntary contracts and actions [15,30–32]. Conflicts might be solved by voluntary negotiations. However, state legislation might cause the unexpected "one party wins, and the other loses" consequence. Furthermore, incomprehensible legislation could cause the inefficiency of resource allocation through interventionism and regulation. Then, there is no way for the consumers and producers to internalize the costs and benefits of environmental protection-related production, and a zero-sum game is created by state legislation.

2.1.2. The Community Concession Theory

The CC theory sees community concession as a polycentric reality and a third way between complete nationalization of the resource and privatization [18]. It was developed during the second half of the 20th century by the 2009 Nobel Prize in Economics winner Elinor Ostrom [19] and her husband Vincent Ostrom [18]. Their works have been interpreted as a reaction to Garret Hardin's *The Tragedy of the Commons* [20]. The latter's work recommended privatization as an easy, viable, and quick solution to the famous tragedy of the commons. Vincent and Elinor Ostrom challenged this claim by pointing out that, in some situations, common-pool resources (CPR) management may be more effective, beneficial, and legitimate than privatization itself.

The CC theory considers that the rules of local governance and institutions could turn potential conflicts into opportunities by applying the 1974 Nobel Prize in Economic Sciences winner Fredrich A. Hayek's insight on the function of local knowledge [33,34]. E. Ostrom argued that any scientific knowledge has its limitation as any constricted model cannot fully explain the diversity of CPR problems [19] (p. 24). Therefore, it is essential to study how different local and self-organizations solve conflicts without top-down planning [19] (pp. 24–25).

Furthermore, the CC theory emphasizes the importance of empirical studies on how the community arrangements solve environmental-related problems [19,34]. E. Ostrom herself stressed the importance of community arrangements in developing countries [35,36]. Peter Boettke contended that E. Ostrom "has demonstrated in a variety of historical circumstances and within a diversity of institutional environments how individuals can craft rules so that they can live better together in their communities and realize the gains from social cooperation under the division of labor" [33]. Some excellent and widely cited cases of E. Ostrom's studies include the mountain grazing in Switzerland and irrigation systems in Spain. She argued that it was local internal rules and monitoring arrangements that disciplined temptations to violate community rules, ensuring robust conformity to those rules of governing the CPR [19] (pp. 58–102, cited by Boettke [33]).

Therefore, it is the community arrangements that turn conflicts into potential opportunities for both parties to solve the existing environmental problems by win-win games. The Hardinian "tragedy of the commons" then could be solved. Interestingly, in his later work, Hardin changed his mind, recognizing that the tragedy of the commons could be managed and avoided [37]. Although he did not mention the role of private property or the concept of local knowledge directly, he recognized that the common pool could be managed well by its group owners if there is an advance in the scientific capacity of common good management.

2.1.3. The Common Grounds of the Two Norms

Although the CC theory treats the common good criteria as a third way to solve the dilemma between the government solution and the pure private-property approach, it is still considered that the CC theory cooperates with the PAE criteria. In the first place, private ownership is the first common ground between the above two norms. The private-property approach economist Ludwig von Mises defined private ownership as "full control of the services that can be derived from a good" [28] (p. 678). Hence, once a person can fully restrain on a particular property, he is the de facto owner of this property. Walter Block reckoned that private property right does not mean that *only a single individual* has control over a definite resource [38]. He argued that if a common good is *fully controlled* by a group of the pool users, it is considered a *jointly owned* private property by these users. Boettke further emphasized that E. Ostrom's common-pool resources theory cooperates with the private property approach [33]. He argued that what Ostrom has demonstrated [39] is: *the rules in use* determine practice rather than *the rules in form*. Therefore, the function of private property could be served by different forms of rules. Furthermore, although some types of common-pool resources are owned by the state as the so-called "public goods," if in practice, the group of users of this property has *full control* of the services derived from it, this common-pool good is the de facto private property. In this regard, we consider that both the PAE criteria and the CC theory cooperate with each once the common-pool joint users have full control of this common property.

Secondly, in environmental protection teams, both the PAE and CC recognize the role of decentralized political institutions and oppose the nationalization of natural resources and the conflicts caused by legislative decision-making. The PAE criteria shown in Section 2.1.1 above implies that central planning institutions cannot acquire the relevant knowledge and prices to design a prosperous economy. If the central planning bodies cannot coordinate knowledge and prices, it might create discoordination or even zero-sum games among the involved parties as the conflicts could have been solved by them without the central government's intervention. The above argumentation discontents with any top-down and centralized policy program, advocating a bottom-up and decentralized view of policymaking. Therefore, both theories opposed the economic inefficiency caused by nationalizing natural resources and precluding the zero-sum games created by legislative bodies. Hence, the above insights cooperate with the CC's proposal on decentralization and community arrangements.

2.1.4. The Guatemalan Community Concession: A Step towards a Jointly Owned Private Property

Our above analysis has shown that the CC theory and the PAE criteria can cooperate with each other if the community has full control of this jointly owned property. However, as we have illustrated above, E. Ostrom's proposal does not exclude another scenario of CC: the community can only make some limited decisions for the common good, if the state still regards the common property as its "public good". The Guatemalan CC case that we analyze in this paper matches the traditional public-good scenario CC. Due to the Guatemalan Forest Law [40], the forest concession is defined as a usufruct through which the state authorizes the use of its partial territories to an individual, a company, or a community. This usufruct arrangement does not refer to the privatization of property to an individual or an enterprise, but a kind of rent for the local community to manage the natural resource in a sustainable way. In this regard, we consider that the current Guatemalan forest concession is a transitional step towards a jointly owned private property.

2.2. Key Concepts Involving Forest Conservation

This section provides some key concepts related to forest conservation. Forest conservation (forest preservation) is defined as "the practice of planting and maintaining forested areas for the benefit and sustainability of future generations" [41]. The concept also aims at a quick shift in the composition of tree species and age distribution. In forestry, forest protection is defined as a concept "concerned with the prevention and control of damage to forests arising from the action of people or livestock, of pests and abiotic agents" [42]. Therefore, forests can be preserved sustainably for future generations by preventing and controlling forest damage. In our case, although the Guatemalan state's National Council of Protected Areas (CONAP) did not provide a clear definition of what is reforestation or the improvement of forest condition, it can be assumed from the same document that the concept refers to the natural regeneration of forests to the places that did not have it before the regrowth process [43]. Forest degradation (forest loss or deforestation) is defined as a condition "when forest ecosystems lose their capacity to provide important goods and services to people and nature" [44].

2.3. Previous Research on the Guatemalan Communities Forest Management

Previous research has studied Guatemala's communities' forest management, providing a substantial base for further analysis. Pacheco et al. pointed out that spontaneous indigenous societies had started to protect forests in the later 19th century in Latin American countries [45]. Veblen argued that Guatemala's forest degradation accelerated in the 1930s and the 1940s due to planned development projects and colonization [46]. On the other hand, he argued that the indigenous communities conducted forest preservation better as they were far from cities, and the forest was in inaccessible areas. Hess studied Guatemala's traditional indigenous forest conservation, arguing the venerability and forest degradation of these institutional arrangements [47]. He discussed that the overuse of carpentry livelihoods, the state regulations, and the lack of management capacity prevent the aboriginal communities from efficiently managing their forest conservation. He proposed the necessity of educating the indigenous communities on how to allocate natural resources in an eco-friendly way. Although the author cited two of E. Ostrom's works on ecological protection [35,36], he did not apply Ostrom's CC theory to the case of Guatemalan forest preservation.

By applying Ostrom's CC theory [19], Elias and Wittman argued that local institutions (including NGOs) played an important role in the management and administration of communal forest resources inside the indigenous communities [48]. However, they discovered that the lack of funding and the unclearness of property titles had impeded forest conservation. On the other hand, they criticized Guatemala's centralized political institutions, since the 1996 Forest Laws [40] provide very limited space for decentralization. They reckoned that the Law only assigns responsibilities to municipalities, but is not linked to other decentralization initiatives, such as the more general decentralization laws (The Decentralization Law [49] and The Municipal Code [50]) passed in 2002. These laws require the implementation of urban and rural development councils. In another paper, Elias reckoned that the indigenous peoples have the right to control the collective natural resource management, proposing to create new governance policies to respect the collective groups [51]. Wittman and Geisler further warned that the current Guatemalan political decentralization actually centralized political power at the local level, generating corruption and weakening successful village-level forest governance structures and local livelihoods [52]. By using the quantitative method, Priebe et al. discovered [53] that in some higher population and road densities areas, deforestation and reforestation rates increased with a net forest cover growth after the forest management decentralization since 1996. They also argued that the number of employees dedicated to forestry activities is the most significant social variable in reforestation efforts during the post-decentralization era. Reddy claimed that the local indigenous forest management is the de facto common private property [54]. While the above literature provides us with many insightful observations on the local indigenous arrangements of forest and natural resource protection in Guatemala, there is still a lack of relevant research on modern communities' arrangements for forest preservation. This affair then becomes the focus of our paper.

2.4. Guatemala's Centralized Political Institution with a Failed Forest Nationalization

Guatemala is considered a highly centralized presidential republic. Its constitution has defined its centralist and unitary political features since its democratization in 1985 [55]. For fiscal policy, the Guatemalan central government has an important decision-making role. Article 257 of the Constitution requests its central government's executive branch to distribute only 10% of its annual general budget to its 340 municipalities [55]. In other words, 90% of the state income is consumed by the central government [56].

Guatemala's forest policy also reflects the country's centralist political institutions. It has a high degree of forest nationalization and concentration [56]. Its 2002-03 National Forest Inventory estimated that the possession of Guatemala's forests was: 34% of national property, 8% of municipal property, 38% of private property, 15% of communal property, and 5% not determined [57]. Furthermore, 77% of smallholders (with less than 7 hectares per person) work on only 15% of productive land [57]. Therefore, the smallholders only own around 30% of the country's whole land.

The above forest policy features are deeply embedded in its long-standing history of the interventionist natural-resource policy. The model can even be traced back to the 19th century after its independence in 1821. The Guatemalan ecologist Prado-Córdova considered that the country's land policy was being shaped by the decision-making of the country's ruling vested interest groups at the beginning of the Guatemalan independence [56]. In the name of privatization during the last quarter of the 19th century, these groups acquired the aboriginals' lands by forces and political mandate, confiscating them as their private properties [41]. The Catholic Church accumulated and expropriated lands through legislative justification, utilizing the lands by forced peasant labors. Among them, most laborers were the indigenous people [56]. Therefore, Guatemala's land policy results from state-interventionist policy and power games, not the normative privatization. The latter standard should be executed based on voluntary actions instead of state coercion [11].

However, Guatemala continued its state-interventionist land reform, no matter whether in the name of privatization or nationalization. In 1952, President Jacobo Árbenz and his government carried out the famous Guatemalan Agrarian Reform (Decree 900). The reform planned to favor a more equitable land distribution to local farmers (mostly the indigenous people) through a state coercive land redistribution policy. Due to Decree 900, any uncultivated land larger than 673 acres (2.72 km^2) was taken by the government. If the estate was between 672 acres (2.72 km^2) and 224 acres (0.91 km^2) in size, only those with less than two-thirds of the cultivated area would be confiscated [58] (pp. 149–164). Landlords would receive government bonds (equal to the values of the confiscated land calculated through the landowner's tax return in 1952) as compensation (pp. 149–164). The local governments would set up a government committee to decide the redistribution to laborers (pp. 149–164). Of the nearly 350,000 pieces of private land, only 1,710 were confiscated (pp. 149–164). However, not only did Decree 900 spark domestic discontent in Guatemala, but it also eventually led to the CIA's involvement as the U.S. government worried about the consequence of the communism-based land reform (pp. 222–225). The Árbenz government was overthrown in 1954 by a U.S.-led coup, while Decree 900 and the land reform were canceled.

It is argued that the 1952 Agrarian Reform caused conflicts between Guatemalan social classes and has caused long-run consequences for the country till today, influencing its environmental policies. Guatemalan sociologist Torres-Rivas argued that land reform was one of the triggers for the 30-year Civil War (1960–1996) [59]. The World Bank considered that the 19th land reform, Decree 900, and the upcoming Civil War created barriers for a large party of the Guatemalan citizens to receive the benefits of globalization: they could not use natural capital and resources that existed in their communities as both the state and a small group of elites have controlled the majority of land resources [60]. Therefore, most Guatemalans did not participate in the decision-making regarding the use of natural capital and did not receive the benefits of the ecological protection goal proposed by the Guatemalan government [60]. To get more opinions, we interviewed two Guatemalan specialists who thoroughly studied the country's land policy. Historian Rodrigo Fernández reckoned that the above policies have led to unclear land ownership, increasing social conflict, and political instability. As property titles are not settled, different individuals and interest groups have been contesting to obtain the titles, causing a zero-sum game due to public legislative decision-making. Ecologist Óscar Rojas believed the conflicting land policies have ecological consequences. Deforestation in Guatemala has been ramped up because of the tragedy of the commons or the transformation of the rural sector (from agriculture to extensive livestock farming), resulting in uncontrolled resource use.

2.5. The Traditional Indigenous Communities' Forest Arrangements

Apart from the above unsuccessful land nationalization, as we have mentioned in Section 2.3, several traditional indigenous communities conduced forest conservation even before the independence of Guatemala. To better compare the conventional indigenous forest preservation model and the modern communities' arrangements, this section introduces the well-studied case of Totonicapán shortly, an executive department in the west of the country [46–48,51–54]. Totonicapán is one of the Guatemalan departments with the lowest territorial extension in the country (1061 km^2), the highest population density (256 inhabitants/km^2), and the highest rate of smallholding [61]. Furthermore, it is a department with the highest percentage of forest cover (60%), respecting its territorial extension [61].

Totonicapán had become a bottom-up forest preservation case even when it was still a part of the Spanish Crown. In the era of the Spanish Conquest, it was one of the most densely settled areas in Middle America [46] The local cantons bought the exploitation and conservation rights from the Spanish Kingdom in 1811, ten years before the Guatemalan independence. Since that time, instead of the central authority, it is the cantons (through a Board of Directors) that control sustainability and regulate the exploitation of the resource [62]. The whole forest consisted of more than 220 km^2 [63], being considered a well-protected forest area compared with many regions in Middle America [46]. Since 1985, Guatemala's democratic constitution (see its Articles 66, 67, 68, and 69 [55]) also recognizes the indigenous people's autonomy inside their settled executive departments. This recognition protects the indigenous heritage from a legal perspective.

Four factors are considered the most important reasons why Totonicapán has been conducting forest conservation more efficiently [46]. First, interpersonal relationships are face-to-face there. As the indigenous people generally recognize one another or each family, the absence of anonymity has incentivized the locals' respect for the established communal forest boundaries. Second, as a powerful special interest group, the carpenters acknowledged that if all of the forests in Totonicapán were transformed into pasture or agricultural land, there would be no other local natural resource as an alternative. Therefore, they encourage the indigenous communities to maintain vigilance over the communal forest use and punish those who cut wood illegally. Third, in general, the Totonicapán indigenous population is willing to retain its independent identity and control the local affairs as much as possible. The motivation for self-sufficiency encourages the maintenance of traditional forest resource exploitation. Fourth, the decision-making in the Totonicapán department is decentralized. This structure helps the local indigenous communities negotiate their common interest more efficiently.

The forest management structure of the Baquiax faction of Totonicapán is an excellent example. The Guatemalan State has recognized the Baquiax faction as a form of autonomous community organization since 1953. Unlike other democratic administrative divisions in Guatemala, family lineages (with paternal kinship ties) make the decision-making of forest conservation in Baquiax. Their governing roles include: (1) the use and defense of forests; (2) strengthening forestry administration; and (3) monitoring forestry activities [61]. Furthermore, the Baquiax faction has only a few members who make executive decisions on the Board of Natural Resources Directors (Junta de Condueños in Spanish). The Board aims to protect forest resources and construct sustainable forest exploitation that benefits the 48 cantons in Totonicapán. As of 2007, there were 44 members of the Board, including 14 women and 30 men [61]. The limited members in the Board have shortened the policymaking process with fewer arguments and more efficiency [61].

Women play important roles in the community decision-making process. In the case above, the proportion of female Board members was almost half of the male ones. Moreover, in the Baquiax faction, 14 of the 44 community owners of the forest are women [61]. Some of them have become the top leaders of the indigenous environmental protection organizations in the last 20 years. In 2012, Andrea Ixíu, an indigenous activist and communicator, was elected as the Board's first female president. The above facts have shown clearly that the Baquiax faction, as a traditional indigenous institution, respects the female leadership even though it does not implement the modern concept of gender equality.

As we have discussed above, the social-political organizations of Totonicapán are the result of an indigenous institutional evolution through three centuries [48,64]. They have conducted forest preservation in an efficient way inside the indigenous communities. However, their success is not reproducible for the rest of Guatemala's Hispanic population who live in the modern democratic institutions. These institutions are hardly exportable as the indigenous world enjoys a particular evolution and institutions that have favored their community's natural resources management. More applicable institutional arrangements should be discussed to solve the forest conservation matter in Guatemala's modern democratic institutions.

3. The Modern Forest Concession Institutions: The Case of Maya Biosphere Reserve

Compared with the traditional indigenous forest preservation arrangements discussed above, it is necessary to pay attention to the modern forest concession institutions in Guatemala. As they are based on constitutional and democratic mechanisms, their model could be applied to other Guatemalan executive departments and even other Latin American countries who are facing similar forest conservation challenges. This section discusses RBM as Guatemala's modern forest concession institution. Section 3.1 introduces Guatemala's state regulations on forest concessions. Section 3.2 shows the active and inactive concessions within the RBM. Sections 3.3 and 3.4 separately deal with RBM's forest degradation and conservation. This section provides some essential and previously undisclosed official data related to the RBM forest that the official CONAP provided to us in October 2019.

3.1. Guatemala's State Regulations on Forest Concessions

This subsection introduces Guatemala's state regulations on its forest preservation purposes, its regulated forest types, and its state regulatory body. As we have discussed in Section 2.3, as a result of a decentralization movement, two laws were passed in 2002 to authorize the decentralized use of land: (1) The Decentralization Law [49] and (2) The Municipal Code [50]. Even though the decentration of power is limited, as we have shown above, these regulations still allow Guatemala's local communities to conduct forest preservation with their autonomy.

Our research also considers that Forest Law (1996) [40] addresses more legal clauses related to forest concessions. As the Law is strongly connected with the current communities' roles in forest preservation, it is crucial to provide a detailed introduction. Article 3 of the Law defines forest concession as "a political power that the State grants to the Guatemalans, individual or juridical, so that—at their own risk—they carry out forest exploitation in state-owned forests, with the rights and obligations agreed upon in their granting" [40]. Therefore, this concession arrangement is a *usufruct* that the Guatemalan state authorizes the use of state lands to a specific individual, company, or community.

The Forest Law also defines the purpose of forest concession as the following [40]:

1. Reduce forest degradation and advance the agricultural frontier by increasing land use due to its vocation without omitting the characteristics of soil, topography, and climate.
2. Promote reforestation and provide the forest products that the Guatemalan state requires.
3. Increase the productivity of existing forests, subject them to rational and sustained management according to their biological and economic potential; promote the use of industrial systems and equipment that help reach the most significant added value to forest products.
4. Support, promote, and encourage public and private investment in forestry activities to increase forest resources' production, marketing, diversification, industrialization, and conservation.
5. Conserve the country's forest ecosystems by developing programs and strategies that promote compliance with the respective legislation.
6. Improve the communities' living standards by increasing the provision of goods and services from the forest to meet the needs of firewood, housing, rural infrastructure, and food.

Moreover, the Forest Law *regulates two types of communities' forest exploitation* in Article 1 [40]. The first type is *the commercial use* of the forest. The Law allows this wood category to obtain monetary benefits by selling the communities' forest products. The second type is *non-commercial use*. The Law classifies the non-commercial use of the forest as (a) for scientific research and technological development purposes and (b) for families' energy consumption (i.e., fuel, fence posts, and constructions). The regulation will determine the maximum permissible volumes. The maximum volumes are five cubic meters of standing wood, while the volumes can be increased if the regulations are modified.

Article 5 and Article 30 of the Law indicate that the National Forest Institute (INAB) is *a governmental regulatory body* that grants concessions, monitors their operation, and suppresses the forest concessions if the communities' contracts are not fulfilled. Due to the Articles, the Guatemalans can get the concessions through the following three ways: (1) the community that they live in, (2) their enterprises, and (3) as an individual who wishes to obtain it [40]. The above institutions must request concessions from the National Forest Institute (INAB). Once the INAB approves the concession request, the above identities can obtain the usufruct of the forest.

The concession monitoring period carried out by INAB will take place at least once a year. Furthermore, due to Article 30, the concessions will last between 25 and 50 years [40]. The same article also indicates that the Guatemalan state should send inspectors to the communities to review whether the concessions' forests are well protected each year. If the communities do not match the state regulations, their usufruct will be canceled.

As can be seen, due to the bottom-up and decentralized nature of the legal architecture in force, the concessions provide the local communities initiatives to manage their forest as the communities' members could receive the economic and non-economic benefits from their own working from the woods. However, it should be noticed that, as the communities' members do not have full control of forests due to the usufruct nature of the concessions, we consider that the current Guatemalan forest concessions are a transitional step towards a pure common private property approach for forest preservation. This affair will be discussed more in Section 4.

3.2. The Active and Inactive Concessions within RBM

Table 1 shows the active and inactive concessions within the RBM. As of 2022, there have been 14 concessions in Guatemala. Each management unit (MU) belongs to its specific and unique concessionary organization. According to the Forest Law (see Section 3.2) and our field interview in the RBM, two groups of actors play essential roles in forest conservation, management, and timber harvest. The first group is the communities committed to the concessions. They are responsible for exploiting the forest and, at the same time, ensuring its sustainability. Among them, both local autonomous communities and private firms have the right to use the forest concessions. The second group is the Guatemalan government. It monitors the above communities every two years due to the concession regulations that we mentioned above.

Table 1. Active and inactive concessions within the Mayan Biosphere Reserve.

No.	Management Unit	Concessionary Organization	Total Area (ha)	Number of Partners	Year of Contract	Type of Concession	Situation
1	Carmelita	Cooperativa Carmelita	53.797	160	1997	Community	In force
2	Chosquitán	Sociedad Civil Laborantes del Bosque	19.39	73	2000	Community	In force
3	La Unión	Sociedad Civil Custodios de la Selva	21.176	85	2002	Community	In force
4	Las Ventanas	Sociedad Civil Árbol Verde	64.973	340	2001	Community	In force
5	Río Chanchich	Sociedad Civil Impulsores Suchitecos	12.217	21	1998	Community	In force
6	San Andrés	Asociación Forestal Integral San Andrés	51.94	169	2000	Community	In force
7	Uaxactún	Sociedad Civil Organización, Manejo y Conservación	83.558	210	2000	Community	In force
8	Yaloch	Sociedad Civil El Esfuerzo	25.386	30	2002	Community	In force
9	Cruce a La Colorada	Asociación Forestal Integral Cruce a la Colorada	20.469	87	2001	Community	In force
10	La Pasadita	Asociación de Productores la Pasadita	18.817	137	1997	Community	Management plan suspended
11	San Miguel	Asociación Agroforestal de San Miguel	7.039	-	1994	Community	Canceled
12	La Colorada	Asociación Forestal Integral La Colorada	22.067	-	2001	Community	Canceled
13	La Gloria	BAREN Comercial, S.A.	66.548	-	1999	Industrial	In force
14	Paxbán	GIBOR, S.A.	65.755	-	1999	Industrial	In force

Source: Own translation from CONAP [43] (p. 24).

The current general condition of each MU diversifies. Uaxactún has the largest total area of 83,558 hectares, while the MU San Miguel has the lowest total area of 7.039 hectares. La Ventanas has the biggest number of partners (the number of individuals who have participated in the concession unit), 340, and the management unit Río Chanchich has the least partners, which has a total of 21. The partner numbers of the MUs San Miguel, La Colorada, La Gloria, and Paxbán are not available.

Due to various reasons, such as the willingness to cooperate and economic factors, three MUs are not running currently. The management plan of La Pasadita was suspended, and the projects of San Miguel and La Colorada were canceled due to a breach of contract. The remaining 11 MUs are still running as of 2022. This result shows the success and effectiveness of the concession policy in general.

Among all of the MUs, the first contract started in 1997, and the last was initiated in 2002. As of 2022, all of the 14 MUs have been under the concession for at least 20 years. There are, in total, two types of concessions. In total, 12 of the 14 MUs belong to the community type, while only two belong to the industrial type. They are La Gloria of the BAREN Comercialand Paxbán of GIBOR. It is worth mentioning that both BAREN Comercial and Paxbán are private enterprises. In other words, both local communities, and private firms have the right to use forest lands. This dynamic shows the diversity of Guatemalan concessionary organizations. We argue that this is positive for the further development of the concessionary organizations.

3.3. The Control of Deforestation in the RBM

Table 2 shows the control of deforestation in the RBM from 1989 to 2017. The data show that the concessions granted have served to control deforestation in the RMB. It is calculated that during the period 2000–2010, deforestation in Guatemala took place at a rate of 1.2% of the forest mass, while in the RBM, it was 1.4% [65]. The deforestation rate in the RBM concession areas is less than the rest of the country. The average deforestation rate in the concession area forests has been only 0.024% [65]. Due to the Forest Law [40] and the CONAP report [66], this deforestation rate shows that these concessions, compared with the general forest condition in Guatemala and other areas of the RBM, have reached the Guatemalan state's requirements of reducing forest degradation and conserving the country's forest ecosystems. However, due to the lack of data, there have been no other quantitative method to show the specific economic benefits that the Forest Law requires. We consider this vacancy should be the scope analysis of further research.

Table 2. Deforestation suffered in the RBM, 1989–2017.

Year	Forest Coverage (H)	Difference %
1989	477,864	-
2001	477,466	−0.08%
2002	477,370	−0.02%
2003	477,182	−0.03%
2004	477,139	−0.009%
2010	477,165	0.005%
2013	476,905	−0.05%
2014	476,819	−0.01%
2015	476,582	−0.04%
2016	476,555	−0.005%
2017	476,537	−0.003%
Average	−0.024%	
Total deforestation	1327	0.27%

Source: Own elaboration through CONAP [66].

As we have argued above, although the Guatemalan state monitors the forest concessions, the local communities play an essential role in forest conservation, management, and timber harvest. They are responsible for exploiting the forest and, at the same time, ensuring its sustainability. Among them, both local autonomous communities and private firms have the right to use the forest concessions. As Section 3.1 has shown, due to the Forest Law [40], the communities can use the wood to obtain monetary benefits by selling the communities' forest products. Furthermore, they are also allowed to use the forest for scientific research, technological development, and families' energy consumption (i.e., fuel, fence posts, and constructions). As autonomous communities and private enterprises have the above economic and non-economic incentives, they pay special attention to forest preservation.

3.4. Forest Conservation and its Economic Impacts in the RBM

Figures 1–3 show the forest conservation flux in the RBM from 2000 to 2017. The three maps include both the areas controlled by the concession units and the destruction of the other regions where concession policy is not applied. The concessions include the abovementioned MUs: Carmelita, Choquistán, La Unión, Las Ventanas, Chanchich River, San Andrés, Uaxactún, Yaloch, and Cruce de la Colorada. As the three maps in the three Figures show together, the official CONAP marked the above concessions as medium-high broadleaved forest (in green color). Due to the Forest Law [40] and the CONAP report [66], these concessions have reached the forest conservation target by the CONAP. These result show that these concessions have effectively conserved the existing forest mass in the area; therefore generalizing the future increasing economic opportunities for the citizens that inhabit there.

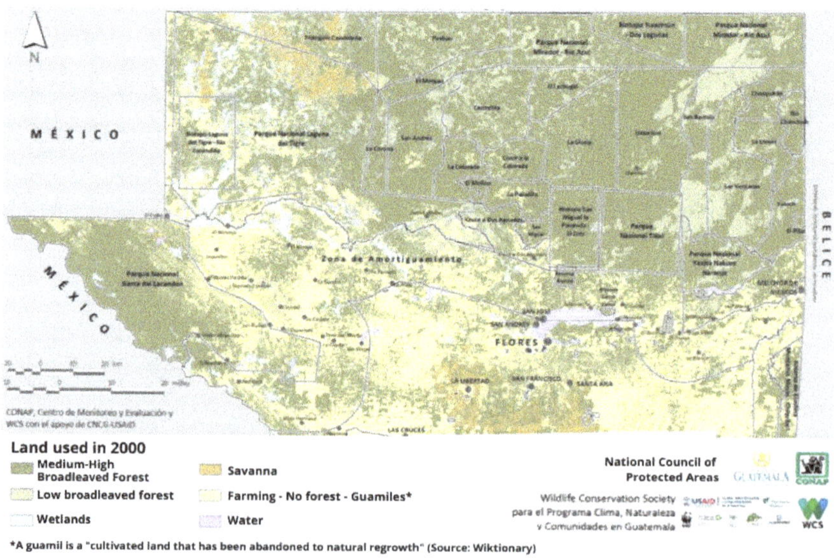

Figure 1. Map 1: Forest cover in the Maya Biosphere Reserve in 2000. Source: Forest cover in the Maya Biosphere Reserve in 2016–2017. Source: Own translation from CONAP [43] (p. 44).

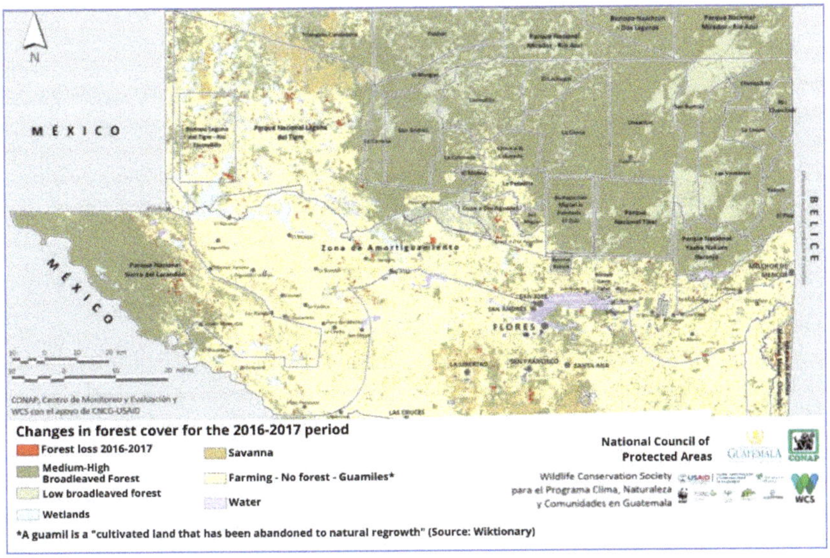

Figure 2. Map 2: Forest cover in the Maya Biosphere Reserve in 2016–2017. Source: Own translation from CONAP [43] (p. 46).

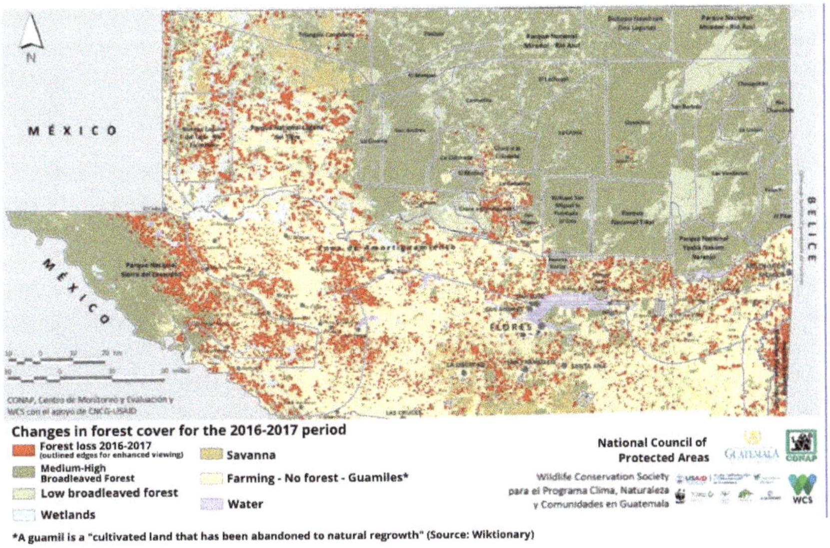

Figure 3. Map 3: Forest cover in the period 2016–2017. Source: Forest cover in the Maya Biosphere Reserve in 2016–2017. Source: Own translation from CONAP [43] (p. 46).

As indicated, due to the Forest Law [40] and the CONAP report [66], these concessions have reached the Guatemalan state's requirements of reducing forest degradation and conserving the country's forest ecosystems, compared to the deforestation rate in other areas of the country. Table 1 and Figures 1 and 2, when considered together, show that deforestation in the concession areas is almost invisible. The image shows a very different reality in the adjacent regions, which suffer the problems related to the land use, being victims, in some cases, of the tragedy of the commons. Together, the three maps show that other areas where concession policy is not applied have a growing tendency of forest destruction. In Figures 2 and 3, the CONAP marked them as forest degradation areas (red color). The forest degradation areas also include the abovementioned two concessions, San Miguel and La Colorada, as concession policy has been canceled for breaching the usufruct contract.

Apart from the two canceled concessions that did not fulfill forest conservation targets, the rest have been conducting forest conservation efficiently and effectively. As the above data shows, community co-responsibility in resource management and the bottom-up dynamics promoted by the Guatemalan government have worked to fulfill its forest environmental protection objectives.

Furthermore, as we have indicated, the success of the concessions is not limited to forest preservation itself, but benefits the inhabitants there in general. As a result of the concession arrangements, inhabitants have created community forestry companies (EFC). These EFCs have gained importance by exploiting the forests responsibly. In addition to the wood exploitation to produce goods of high added value (for the manufacture of soils, etc.), the EFCs have diversified their activities by participating in active tourism and adventure.

Furthermore, the concessions also bring about incomes for the inhabitants living there. As of 2010, the income resulting from the forest community management represented between 11 and 63% of the revenue generated by the families living in the concessions. The average earnings that came directly from the forest represent 38% of the family income in the 292 owners in the area [65]. In this regard, the concession arrangements have positively helped the local inhabitants take their own initiative and responsibility to protect the forest area and make their own resource allocation.

As a result of the concession policy, the business and sustainable use of the forest—has also increased. It has generated some 58 million USD through 111 small and medium companies employing 8800 individuals [65]. This result shows that the concession policy not only helps to conserve forests but also is profitable. We consider that this outcome indicates that a more market-based institutional arrangement brings a better result than the traditional state planning that Guatemala has adopted since the 19th century. In this regard, we encourage the Guatemalan state to take action to expand the concession policy to the rest of the territories of the country.

In short, compared to other realities in Guatemalan territory, the current concession model in RBM has been conducted effectively. In addition, unlike the traditional, consuetudinary, and ancestral system that Totonicapán adopts (see Section 2.5), the RBM model allows both community and private firms to rule the forests. As mentioned in Section 2.4, the RBM is a system resulting from the Forest Law and the other relevant state regulations. As the institutional concession arrangements have been written as state statutes, they can be easier implemented throughout the country than any system that lacks the rule of law. The above institutional arrangements aim to promote community participation in public policy, enhance economic activities, and protect and even generate an asset in the rich natural territory in Guatemala.

4. Results

This article aims to cover the current research vacuum on how Guatemala partially conducts forest preservation through community concessions. Our paper starts its analysis with the synthesis of the private property-rights approach environmentalist (PAE) theory and the community concession (CC) theory. The PAE argues that as private ownership is the full control of a property, it provides the good's owner with incentives to protect the environment. Therefore, the shared common property could also be a type of private property.

On the other hand, the CC theory emphasizes the importance of empirical studies on how the community arrangements solve environmental-related problems [19,34]. Due to CC theory, community arrangements can turn conflicts into potential opportunities for both parties to solve the existing environmental issues by win-win games. The Hardinian "tragedy of the commons" could be solved.

The Guatemalan CC case that we have analyzed in this paper matches the traditional public-good scenario CC. Due to the Guatemalan state Forest Law [40], the forest concession is defined as an usufruct through which the state authorizes the use of its partial territories to an individual, a company, or a community, part of its territory. Therefore, in our paper, forest concessions are considered usufructs for community use. It is not the traditional privatization to an individual or an enterprise, but a kind of rent for the local community, aiming to include the community in the sustainable management of the resource. In this regard, we consider that the current Guatemalan forest concession is a transitional step towards a jointly owned private property. Table 3 below summarizes the political institutions, forest policies, and communities' wood arrangements in Guatemala. It shows that both the traditional indigenous settled department Totonicapán and the modern community concessions in RBM have efficiently conducted forest preservation.

Table 3. Political institutions, forest policies, and communities' forest arrangements in Guatemala.

Political Institutions	Forest Policies	Community Arrangements
A highly centralized presidential republic with unitary political features since its democratization in 1985 [55]. In the example of its fiscal policy, the central government's executive branch distributes only 10% of its annual general budget to the country's 340 municipalities [55,56]. Historically, the country's ruling vested interest groups confiscated the indigenous people's private lands in 1821 by political mandate, causing ethnic conflicts till now [41,56]. Guatemala's land policy results from state-interventionist policy and power games, not the normative privatization, as the latter standard should be executed based on voluntary actions instead of state coercion [11].	Due to centralized policies, both the state and the traditional private property owners process at least one-third of their forest lands. In contrast, the smallholders (less than 7 hectares per person, work on only 15% of productive land) only own the rest, 30% altogether [57]. The decentralized use of land started in 2002 due to the Decentralization Law [49] and the Municipal Code [50]. Limited forest decentralization still allows communities to conduct forest preservation with their own autonomy. The Forest Law (1996) [40] addressed more legal clauses related to forest concessions. The Guatemalan state is still the owner of the forest concessions while the communities have the usufruct of them.	Totonicapán, as a traditional indigenous settled department, has the highest percentage of forest cover (60%), respecting its territorial extension [61]. The state INAB grants concessions, monitors their operation, and suppresses the forest concessions after the local communities signed the contracts. The Guatemalans can get the concessions through (1) the community that they live in, (2) their enterprises, and (3) as an individual who wishes to obtain it [40]. As of 2010, the deforestation in Guatemala was at a rate of 1.2% of the forest mass, while in the RBM's 11 in-force concessions, it was only 0.024%, the lowest in the whole country [65]. As a result of the concession policy, the business and sustainable use of the forest has also increased. It has generated some 58 million USD through 111 small and medium companies employing 8800 individuals [65].

5. Discussion and Proposals

The above analysis has shown the successful application of the community concession in protecting the Guatemalan forest and its sustainability. However, we consider that several aspects should be discussed and could be applied for further theoretical and empirical studies.

First, it is necessary to synthesize further the PAE criteria and the CC theory to make their application clear. Some of the traditional private property rights theories criticize the community concession approach as it seems the latter one is not a pure private property solution [12]. It is admitted that the CC criteria do not conduct traditional privatization when environmental issues are considered. However, given that entrepreneurship is the driving force of the market economy, both the communities and firms could take more initiatives when making decisions for the benefit or profit of their own organization under the community concession solution. The theoretical proposal also cooperates with the empirical results that our paper deal with, as two firms have participated in the concessions successfully.

We consider that the CC theory can help create a better community or firm-level decision-making process by respecting private property rights. In contrast, a further release of property titles to the community and firms in the mid-run would improve forest and environmental protection. This argumentation is based on the three principles of PAE theory that we have shown in Section 2.1, as Wang et al. argued [8]. The first is the impossibility of economic calculation through centrally planned government policy. The exact identification of private property rights gives the property owner the incentive to protect the environment in which he lives and sue anyone who violates his property's environment. In contrast, the lack of private property generates the tragedy of the commons, and the environment is polluted without the incentive to protect it [13,21–27]. Therefore, as the state still holds the property titles of the forest or the land, they can legitimately intervene in the decision-making of the local communities and firms, making economic calculation impossible and revere the current result of forest and land preservation. Furthermore, as the state owns property titles, they could nationalize the woods or the land easier than conducting the policy under a pure private property right criterion. This result could prevent economic calculation and undermines entrepreneurship [15]. State ownership could still create a zero-sum scenario when the state wants to intervene in the local arrangements [15,30–32]. Conflicts might be solved by voluntary negotiations. However, state legislation might cause the unexpected "one party wins, and the other loses" consequence.

Secondly, based on the PAE principles, community arrangements could also be privately owned under a pure private property scenario. Furthermore, full protection and respect for private property rights will allow communities to compete with each other fairly and avoid cronyism, providing better forest protection and output mechanisms through market mechanisms. In this regard, further privatization of the existing state-title-based concession does not change the right of use but strengthens the private ownership. Therefore, the theoretical research needs to be further deepened, as the conflict between existing theories (the PAE and the CC) is not enough to provide better theoretical support for the announcement policy.

Thirdly, relevant empirical research should also be strengthened. We believe that further empirical research is reflected in the following aspects. (1) Our paper shows that the community concession policy has helped Guatemala conduct better forest conservation since the middle 1990s when the policy was adopted. However, the policy has not been applied in the rest of the country's territories. The research on related policies should be deepened for the current community concessions. These policies should include a detailed analysis of state-interventionist policies and the community concessions. Our study only proposes a preliminary direction for policy research related to Guatemala, and the specific content must be deepened in future research. (2) The relevant empirical research should be extended to a broader range of countries, especially developing regions. Given the different economic conditions faced by developing countries and the characteristics of developing countries, it is necessary to examine the specific economic conditions of various developing countries and their gains and losses in forest and land protection policies separately.

Finally, it is both theoretically and empirically important to discuss how to understand the Environmental Kuznets Curve (EKC) from the PAE-CC criteria, as it is an essential theoretical foundation for much previous mainstream environmental research [67–69]. The EKC argues that economic development initially detorts the environment while the society later perceives the importance of ecological protection, stating to reduce environmental degradation [70]. Although it is unnecessary to start protecting the environment after it has been polluted by human beings, whether the PAE-CC criteria could contribute to and modify the EKC to benefit further environmental economics studies remains a question. In the cases of Guatemala and other developing countries, it is necessary to study how to avoid environmental and forest degradation even when these countries start their initial steps of economic development. In this regard, the PAE-CC criteria can benefit the related research.

6. Conclusions

Based on the private property-rights approach environmentalist theory and the community concession theory, this paper discusses how community concessions (both local executive communities and private firms) as usufructs in RBM of Guatemala partially effectively conducted forest preservation. Previously, research studied how the traditional indigenous communities in Totonicapán have successfully conducted forest preservation. Our study extends the scope to Guatemala's modern and democratic municipals' forest preservation. Our empirical results show that the 11 in-force concessions in RBM had only 0.024% forest deforestation in recent years, which was the lowest rate in the whole country (while deforestation in Guatemala was at a rate of 1.2%) [65]. Due to the Forest Law [40] and the CONAP report [66], this deforestation rate shows that these concessions, compared with the general forest condition in Guatemala and other areas of the RBM, have reached the Guatemalan state's requirements of reducing forest degradation and conserving the country's forest ecosystems. However, due to the lack of data, there have been no other quantitative result to show the specific economic benefits that each concession can bring about. This vacancy should be the scope analysis of further research.

The concession arrangements also have generated business, and the sustainable use of the forest has also increased. It has generated some 58 million USD through 111 small and medium companies employing 8800 individuals [65]. Yet, as the accessible data are limited currently, we suggest that it is necessary to conduct more detailed quantitative studies on the Guatemalan forest concession arrangements. In contrast, the Guatemalan government should provide more transparency on forest data issues.

On the theoretical side, our research reckons that community property as a shared good could also be treated as a common private property if the owners could have full control of it. However, the Guatemalan forest concession arrangements are just usufructs, which means the state still owns the property. Therefore, the current Guatemalan forest concession arrangement is just a third way: neither traditional private ownership nor the state's full planning of forest preservation. We argue that state ownership could reverse the result of the current and limited decentralized forest reform. As the state still holds the property titles of the forest or the land, they can legitimately intervene in the decision-making of the local communities and firms, making economic calculation impossible and revere the current result of forest and land preservation. Furthermore, as the state owns property titles, they could nationalize the woods or the land easier than conducting the policy under a pure private property right criterion. In this regard, we suggest that Guatemala state should privatize all of these forests to the current concessions' communities and firms. If the results are positive, we propose the Guatemalan government further apply the decentralization forest policy to the whole country.

Author Contributions: Conceptualization, E.F.L. and S.F.O.; methodology, E.F.L., S.F.O. and W.H.W.; software, E.F.L., S.F.O. and W.H.W.; validation E.F.L. and W.H.W.; formal analysis, E.F.L., S.F.O. and W.H.W.; investigation, E.F.L., S.F.O. and W.H.W.; resources, E.F.L., S.F.O. and W.H.W.; data curation, E.F.L., S.F.O. and W.H.W.; writing—original draft preparation, E.F.L. and S.F.O.; writing—review and editing, E.F.L., S.F.O. and W.H.W.; visualization, W.H.W.; supervision, E.F.L. and W.H.W.; project administration, W.H.W.; funding acquisition, W.H.W. All authors have read and agreed to the published version of the manuscript.

Funding: Instituto de Estudios Políticos y Relaciones Internacionales at the Francisco Marroquin University funded this project.

Institutional Review Board Statement: Not applicable.

Informed Consent Statement: Not applicable.

Data Availability Statement: Not applicable.

Acknowledgments: This article is based on the study by Wang et al. [8] on free-market environmentalism. We gratefully acknowledge the helpful comments of Guillermo Rodriquez del Valls and the three anonymous referees. Any remaining errors are the responsibility of the authors.

Conflicts of Interest: The authors declare no conflict of interest.

References

1. Huntington, S.P. Democracy's third wave. *J. Democr.* **1991**, *2*, 12–34. [CrossRef]
2. Erbsen de Maldonado, K. Guatemala: Danzando con las crisis económica y política. *Rev. Cienc. Política* **2010**, *30*, 335–360. [CrossRef]
3. Zarzalejos, J.; Fernández Luiña, E. *América Latina: Una Agenda de Libertad 2018*; FAES—Fundación para el Análisis y los Estudios Sociales: Madrid, Spain, 2018; ISBN 8492561424.
4. PNUD. Informe de Desarrollo Humano Guatemala. Available online: https://desarrollohumano.org.gt/desarrollo-humano/calculo-de-idh/ (accessed on 18 April 2022).
5. Marroquín Gramajo, A.; Noel Alfaro, L. Protestant Ethic and Prosperity: Vegetable Production in Almolonga, Guatemala. In *Political Economy, Neoliberalism, and the Prehistoric Economies of Latin America*; Matejowsky, T., Wood, D.C., Eds.; Research in Economic Anthropology; Emerald Group Publishing Limited: Bradford, UK, 2012; Volume 32, pp. 85–107. ISBN 978-1-78190-059-8.
6. Latinobaromero. Available online: http://www.latinobarometro.org/latOnline.jsp (accessed on 18 April 2022).
7. Azpuru, D.; Mariana, R.; Zechmeister, E. *Cultura Política de la Democracia en Guatemala y en las Américas 2016–2017. Un Estudio Comparado sobre Democracia y Gobernabilidad*; Instituto de Estudios Peruanos: Jesus Maria, Peru, 2017.
8. Wang, W.H.; Moreno-Casas, V.; de Soto, J. A free-market environmentalist transition toward renewable energy: The cases of Germany, Denmark, and the United Kingdom. *Energies* **2021**, *14*, 4659. [CrossRef]
9. Von Mises, L. *Socialism: An Economic and Sociological Analysis*; Liberty Fund: Carmel, IN, USA, 2012.
10. Hayek, F.A. The use of knowledge in society. *Am. Econ. Rev.* **1945**, *35*, 519–530.
11. Huerta de Soto, J. *Socialism, Economic Calculation and Entrepreneurship*; Edward Elgar Publishing: Cheltenham, UK, 2010.
12. Block, W. Environmentalism and economic freedom: The case for private property rights. *J. Bus. Ethics* **1998**, *17*, 1887–1899. [CrossRef]
13. Anderson, T.L.; Baden, J.; Block, W.; Borcherding, T.; Chant, J.; Dolan, E.; Mc Fetridge, D.; Rothbard, M.N.; Smith, D.; Shaw, J.; et al. *Economics and the Environment: A Reconciliation*; Block, W., Ed.; The Fraser Institute: Vancouver, BC, Canada, 1990.
14. Anderson, T.L.; Leal, D.R. *Free Market Environmentalism*; Revised Edition; Palgrave Macmillan: New York, NY, USA, 2001.
15. Huerta de Soto, J. Entrepreneurship and the Theory of Free Market Environmentalism. In *The Theory of Dynamic Efficiency*; Routledge: New York, NY, USA, 2009; pp. 94–99.
16. Cordato, R. Toward an Austrian theory of environmental economics. *Q. J. Austrian Econ.* **2004**, *7*, 3–16. [CrossRef]
17. Bradley, R.L. Resourceship: An Austrian theory of mineral resources. *Rev. Austrian Econ.* **2007**, *20*, 63–90. [CrossRef]
18. Ostrom, V. *Polycentricity*; Indiana University Press: Bloomington, IN, USA, 1972.
19. Ostrom, E. *Governing the Commons: The Evolution of Institutions for Collective Action*; Political Economy of Institutions and Decisions; Cambridge University Press: Cambridge, UK, 1990.
20. Hardin, G. The tragedy of the commons. *Science* **1968**, *162*, 1243–1248. [CrossRef]
21. Rothbard, M.N. Law, property rights, and air pollution. *Cato J.* **1982**, *2*, 55–99.
22. Reimers, P. Monarchy, democracy and private property order how human rights have been violated and how to protect them a response to Hans H. Hoppe, F. A. Hayek, and Elinor Ostrom. *Rev. Procesos Merc.* **2019**, *16*, 177–212. [CrossRef]
23. Fuller, E.W. The non-aggression principle: A short history. *Rev. Procesos Merc.* **2019**, *16*, 31–88. [CrossRef]
24. Reimers, P. An Austrian school view on Eucken's ordoliberalism. Analyzing the roots and concept of German ordoliberalism from the perspective of Austrian school economics. *Rev. Procesos Merc.* **2020**, *17*, 13–53. [CrossRef]
25. Espinosa, V.I.; Wang, W.H.; Zhu, H. Israel kirzner on dynamic efficiency and economic development. *Procesos Merc. Rev. Eur. Econ. Política* **2020**, *17*, 283–310. [CrossRef]
26. Kämpe, J. Individual secession and extraterritoriality. *Rev. Procesos Merc.* **2013**, *10*, 195–239. [CrossRef]
27. Hoppe, H.-H. A realistic libertarianism. *Rev. Procesos Merc.* **2015**, *12*, 203–229. [CrossRef]
28. Von Mises, L. *Human Action: A Treatise on Economics*; Ludwig von Mises Institute: Auburn, AL, USA, 1998.
29. Foss, N.J.; Klein, P.G. *Organizing Entrepreneurial Judgment: A New Approach to the Firm*; Cambridge University Press: Cambridge, UK, 2012; ISBN 9781139021173.
30. Merino, J.B. Teoría del intercambio. Propuesta de una nueva teoría de los cambios interpersonales basada en tres elementos más simples. *Rev. Procesos Merc.* **2015**, *12*, 143–176. [CrossRef]
31. Boettke, P.J.; Coyne, C.J.; Leeson, P.T. Institutional stickiness and the new development economics. *Am. J. Econ. Sociol.* **2008**, *67*, 331–358. [CrossRef]
32. Boettke, P. Economics and public administration. *South. Econ. J.* **2018**, *84*, 938–959. [CrossRef]
33. Boettke, P. Is the only form of 'reasonable regulation' self regulation? Lessons from Lin Ostrom on regulating the commons and cultivating citizens. *Public Choice* **2010**, *143*, 283–291. [CrossRef]
34. Tabarrok, A. Elinor Ostrom and the Well-Governed Commons. Available online: https://marginalrevolution.com/marginalrevolution/2009/10/elinor-ostrom-and-the-wellgoverned-commons.html (accessed on 18 April 2022).
35. Elinor, O. A general framework for analyzing sustainability of social-ecological systems. *Science* **2009**, *325*, 419–422. [CrossRef]
36. Ostrom, E. Background on the institutional analysis and development framework. *Policy Stud. J.* **2011**, *39*, 7–27. [CrossRef]
37. Garrett, H. Extensions of "The Tragedy of the Commons". *Science* **1998**, *280*, 682–683. [CrossRef]

38. Block, W.; Jankovic, I. Tragedy of the partnership: A critique of Elinor Ostrom. *Am. J. Econ. Sociol.* **2016**, *75*, 289–318. [CrossRef]
39. Ostrom, E. *Understanding Institutional Diversity*; Princeton University Press: Princeton, NJ, USA, 2009; ISBN 9781400831739.
40. El Congreso de la República de Guatemala. Ley Forestal. 1996. Available online: http://www.sice.oas.org/investment/natleg/gtm/forestal_s.pdf (accessed on 18 April 2022).
41. Pawar, K.V.; Rothkar, R.V. Forest conservation & environmental awareness. *Procedia Earth Planet. Sci.* **2015**, *11*, 212–215. [CrossRef]
42. General Multilingual Environment Thesaurus Forest Protection. Available online: https://www.eionet.europa.eu/gemet/en/concept/10706 (accessed on 23 April 2022).
43. Consejo Nacional de Áreas Protegida. *Monitoreo de la Gobernabilidad en la Reserva de la Biosfera Maya*; Gobierno de la Republica de Guatemala: Guatemala, Guatemala, 2018.
44. International Union for Conservation of Nature Deforestation and Forest Degradation. Available online: https://www.iucn.org/resources/issues-briefs/deforestation-and-forest-degradation (accessed on 23 April 2022).
45. Pacheco, P.; Ibarra, E.; Cronkleton, P.; Amaral, P. Políticas Públicas que Afectan el Manejo Forestal Comunitario. In *Manejo Forestal Comunitario en América Latina: Experiencia, Lecciones Aprendidas y Retos Para el Futuro*; Centro para la Investigación Forestal (CIFOR): Bogor Barat, Indonesia, 2008; pp. 201–230.
46. Veblen, T.T. Forest preservation in the western highlands of Guatemala. *Geogr. Rev.* **1978**, *68*, 417–434. [CrossRef]
47. Hess, K. Contextual vulnerability of the communal forests and population of Totonicapán, Guatemala. *Espac. Desarro.* **2018**, *31*, 117–146. [CrossRef]
48. Elías, S.; Wittman, H. State, Forest and Community: Decentralization of forest administration in Guatemala. In *The Politics of Decentralization: Forests, Power, and People*; Colfer, C.J.P., Capistrano, D., Eds.; Routledge: London, UK, 2005; pp. 282–295. ISBN 9781138995109.
49. Congreso de la República. *Ley General de Descentralización*; El Congreso de la República de Guatemala: Guatemala, Guatemala, 2002.
50. Congreso de la República. *El Código Municipal*; El Congreso de la República de Guatemala: Guatemala, Guatemala, 2002.
51. Elías, S. From communal forests to protected areas. *Conserv. Soc.* **2012**, *10*, 151–160. [CrossRef]
52. Wittman, H.; Geisler, C. Negotiating locality: Decentralization and communal forest management in the Guatemalan highlands. *Hum. Organ.* **2005**, *64*, 62–74. [CrossRef]
53. Paulson, P.M.E.; Evans, T.; Andersson, K.; Castellanos, E. Decentralization, forest management, and forest conditions in Guatemala. *J. Land Use Sci.* **2015**, *10*, 425–441. [CrossRef]
54. Reddy, S.P. Communal forests, political spaces: Territorial competition between common property institutions and the state in Guatemala. *Sp. Polity* **2002**, *6*, 271–287. [CrossRef]
55. Constitución Política de la República. *Constitución Política de la República*; El Tribunal Constitucional de Guatemala: Guatemala, Guatemala, 1985.
56. Garoz, B.; Gauster, S.; Sigüenza, P.; Dür, J. Algunas Implicaciones e la Concentración de la Tierra para la Gestión de los Recursos Naturales y el Territorio de Guatemala. In *Territorios*; Instituto de Estudios Agrarios y Rurales: Guatemala, Guatemala, 2006; pp. 1–14.
57. Transparencia Forestal. 2012. Available online: http://www.transparenciaforestal.info/background/forest-transparency/33/la-medici-n-de-la-transparencia-del-sector-forestal (accessed on 18 April 2022).
58. Gleijeses, P. *Shattered Hope: The Guatemalan Revolution and the United States, 1944–1954*; Princeton University Press: Princeton, NJ, USA, 1991; ISBN 0691025568.
59. Torres-Rivas, E. *Revoluciones sin Cambios Revolucionarios*; F&G editores: Guatemala, Guatemala, 2011.
60. Bank, W. *Republic of Guatemala Country Environmental Analysis: Addressing the Environmental Aspects of Trade and Infrastructure Expansion*; The World bank: Washington, DC, USA, 2006.
61. Huertas, I.A.G. Gestión Colectiva y su Incidencia en la Conservación y Utilización del Bosque Comunal, Parcialidad Baxquiax, Cantón Juchanep, Municipio de Totonicapán, Totonicapán. Ph.D. Thesis, Universidad de San Carlos de Guatemala, Guatemala, Guatemala, 2007.
62. Ixíu, A. Totonicapán, un Bosque. Plaza Pública. 2013. Available online: https://www.plazapublica.com.gt/content/un-bosque (accessed on 18 April 2022).
63. Gamazo, A.C. Totonicapán. El Poder Político de un Bosque. 7K. 2016. Available online: https://www.naiz.eus/en/hemeroteca/7k/editions/7k_ (accessed on 18 April 2022).
64. López, R.T.; Hierro, P.G. *Los Bosques Comunales de Totonicapán: Historia, Situación Jurídica y Derechos Indígenas*; Facultad Latinoamericana de Ciencias Sociales (FLACSO-Guatemala): Guatemala, Guatemala, 2002; Volume 4, ISBN 9992266600.
65. Rojas, Ó.E. *Concesiones Forestales Comunitarias–Experiencia Petén*; ACOFOP: Guatemala, Guatemala, 2020.
66. Consejo Nacional de Áreas Protegidas Consejo Nacional de Áreas Protegidas. Available online: https://www.conap.gob.gt/ (accessed on 15 March 2022).
67. Sadik-Zada, E.R.; Ferrari, M. Environmental policy stringency, technical progress and pollution haven hypothesis. *Sustainability* **2020**, *12*, 3880. [CrossRef]
68. Gatto, A.; Busato, F. Energy vulnerability around the world: The global energy vulnerability index (GEVI). *J. Clean. Prod.* **2020**, *253*, 118691. [CrossRef]

69. Sadik-Zada, E.R.; Gatto, A. The puzzle of greenhouse gas footprints of oil abundance. *Socioecon. Plann. Sci.* **2021**, *75*, 100936. [CrossRef]
70. Kuznets, S. Economic growth and income inequality. *Am. Econ. Rev.* **1955**, *45*, 1–28.

sustainability

MDPI

Article

Fundraising Appeals for the COVID-19 Epidemic Fight: A Cross-Country Study of Donor Responses

Saeed Awadh Bin-Nashwan [1,2], Adel Sarea [3], Meshari Al-Daihani [4], Abdullahi Bala Ado [5], Halima Begum [6,7,*], Mushari Hamdan Alosaimi [8], Hijattulah Abdul-Jabbar [9] and Mohammed Khalifa Abdelsalam [6]

1 Institute of Shariah Governance and Islamic Finance, Islamic Business School, Universiti Utara Malaysia, Sintok 06010, Kedah, Malaysia; s.nashwan233@gmail.com
2 College of Administrative Sciences, Seiyun University, Seiyun 70981, Yemen
3 College of Business and Finance, Ahlia University, Manama 10878, Bahrain; adelsarea@yahoo.com
4 Academy of Islamic Studies, University of Malaya, Kuala Lumpur 50603, Malaysia; meshari89.r@gmail.com
5 Department of Accounting, AL-Qalam University, Katsina 820241, Nigeria; abdullahibalaado@auk.edu.ng
6 School of Economics, Finance and Banking, Universiti Utara Malaysia, Sintok 06010, Kedah, Malaysia; m.khalifa.abdelsalam@uum.edu.my
7 Centre for Studies on Europe, Azerbaijan State University of Economics (UNEC), Baku 1007, Azerbaijan
8 Department of Accounting, Umm Al-Qura University, Makkah 24382, Saudi Arabia; mhosaimi@uqu.edu.sa
9 Tunku Puteri Intan Safinaz School of Accountancy, Universiti Utara Malaysia, Sintok 06010, Kedah, Malaysia; hijat@uum.edu.my
* Correspondence: dr.halima.begum@uum.edu.my; Tel.: +6-04-928-6566

Abstract: This research explores the intrinsic and extrinsic motivations driving donors to engage in fundraising appeals launched through social networking sites (SNSs) to mitigate COVID-19's impact on vulnerable communities from a cross-national perspective. The research adopted a quantitative approach through a web-based survey; a total of 801 donors were obtained from Kuwait and Bahrain and were useable for the analysis. Smart PLS structural equation modelling was used to validate the research model and derive significant insights. In the Kuwaiti sample, we found that humanitarian projects, internet technology, SNSs and religiosity significantly drive donor attitudes towards online donation. All these relationships are indirectly related to the intention to give via SNS through the mediating role of attitudes. As for the Bahraini sample, humanitarian projects, non-profit organizations (NPOs), SNSs, and religiosity significantly influence donors' attitudes. Attitudes, on the other hand, have a visible mediating role in the relationships between these predictors and behavioral intentions. The findings could be useful for the development of appropriate policies that boost online monetary donations to support emergency aid for communities crushed by the pandemic. This research differs from the existing literature in that its multi-national study scrutinizes the incorporation of both internal and external predictors of fundraising activities into a distinctive related context such as SNSs, particularly in a time of epidemiological crises such as COVID-19.

Keywords: fundraising campaigns; COVID-19; social networking sites; NPO; humanitarian response; Kuwait; Bahrain

Citation: Bin-Nashwan, S.A.; Sarea, A.; Al-Daihani, M.; Ado, A.B.; Begum, H.; Alosaimi, M.H.; Abdul-Jabbar, H.; Abdelsalam, M.K. Fundraising Appeals for the COVID-19 Epidemic Fight: A Cross-Country Study of Donor Responses. *Sustainability* **2022**, *14*, 6486. https://doi.org/10.3390/su14116486

Academic Editors: Elkhan Richard Sadik-Zada and Andrea Gatto

Received: 6 April 2022
Accepted: 19 May 2022
Published: 25 May 2022

1. Introduction

Although the COVID-19 crisis has led to substantial and unprecedented impacts on all walks of life, the labour market and socio-economic contexts have garnered a high degree of interest, as they have been dramatically disrupted. This trend is projected to continue in the near future, as indicated by the International Labour Organization (ILO) [1]. Within the labour market context, considering the extraordinary measures taken by governments worldwide to save humankind from the pandemic, almost all of the workers across the planet—some 94%—are living in countries that have some kind of strict workplace closure procedures [1]. In the latest analysis of the ILO on economies and labour impacted by the pandemic, corporations and employees are facing immense shocks in both industrialised and emerging economies. Massive losses in working hours and wages have been reported; it has been estimated that about 1.25 billion members of the global workforce are threatened in their livelihoods [2].

Shocked by the pandemic and oil price plunge, the wealthy Gulf Cooperation Council (GCC) governments are experiencing the biggest economic challenge in history [3]. Within this, most, if not all, business activities from different sectors—e.g., hospitality, plumbing, blacksmithing, health, trade, transportation, construction, and other activities—are facing massive shift shocks in demand, supply chains, mobility, transportation, and worker protection. Those worst hit are low-wage expatriate workers, where the GCC hosts some 35 million migrant workers, and a significant number of them have lost their jobs, been stranded, and been rendered unemployed as a result of the economic lockdown triggered by the epidemic [1,4]. Kuwait and Bahrain are among the GCC countries which are most reliant on migrant workers. In 2018, migrant workers made up approximately 70% and 55% of the total population in both countries, respectively [5,6]. It could be surmised that although the GCC authorities possess financial reserves to maintain the flow of money to their citizens, the ongoing COVID-19 catastrophe is showing severe and unfortunate consequences, not only for national economies through sharp contraction and fiscal deficit in GDP, but also for the welfare of migrant workers and their families [7,8].

In addition to governments' efforts to fight the pandemic, the humanitarian community has a huge role to play in responding to the extraordinary needs that this pandemic is generating. However, the humanitarian sector worldwide is itself facing a critical threat from this global crisis as a result of funding constraints, forcing non-profit organisations (NPOs) to downsize and lay-off their workforce at the precise moment when their work is becoming vitally important [9]. In normal times, there could be uncertainty about how to bridge the growing funding gap year-in and year-out (Figure 1). In the wake of COVID-19's arrival in the world, securing the necessary funds has become a greater challenge than ever. The UN has asked for $10.3 billion to fight the virus, which is its biggest-ever fundraising call, because it is causing a lot of people to go hungry [10].

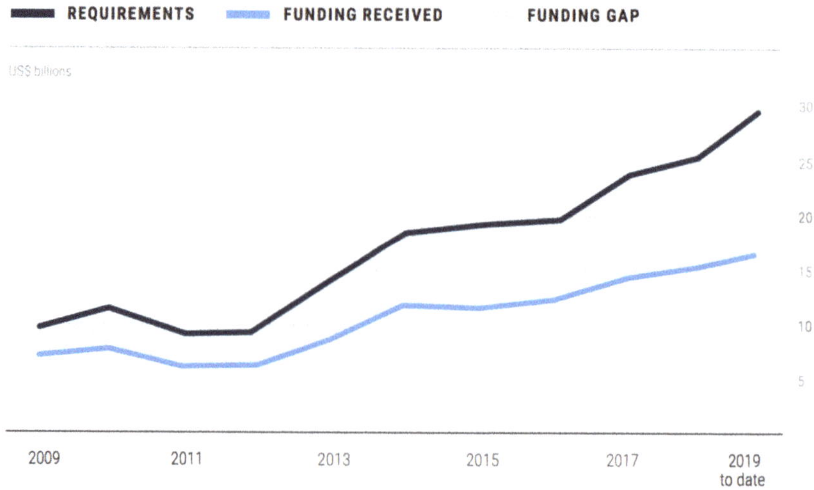

Figure 1. Humanitarian funding gap over the 2009–2019 period. Source [11].

In a rapid and strategic response to mitigate the impact of the pandemic, some authorities worldwide—including GCC like Kuwait and Bahrain—have launched fundraising campaigns to provide timely and effective assistance to the developmental and humanitarian sector [5,11]. Within the framework of containment efforts of the pandemic, as in a collective community, a number of non-profits are promoted into the fundraising appeal through online platforms (i.e., social networking sites). In Kuwait, for example, donations are being solicited, and the total amount raised has surpassed USD30 million, contributing in response to the crisis of COVID-19 to emergency relief operations [12]. The funds raised are used to provide immediate needs for those who are deeply affected by the pandemic, such as migrant workers, health centres and logistical support, frontline healthcare personnel and quarantine facilities [5,11]. However, in this unprecedented crisis, given the inevitable exacerbation of existing inequalities and uncertainty about the future, governments' efforts to confront the outbreak may be undermined by insufficient funds. It is important to determine what needs to be done to encourage donors to provide sufficient contributions. Therefore, by conducting a multi-national study of donors' responses amid the pandemic, this research strives to discover and provide better insights into the behavioural intentions of donors with regard to online monetary donations in the time of COVID-19-related social distancing, through social networking sites (SNSs), in order to help the most vulnerable populations affected by the crisis. Following the recent breakout of COVID-19 and the tactical shutdown of cities throughout the world, platforms for social networking were adopted more intensively and broadly than anticipated. In response to the pandemic, SNSs have been used to make pressing aid appeals to possible contributors.

Some prior research [13–16] has explored behavioural intentions to give money by applying the theory of planned behaviour (TPB) as the most predictive and viable theory for understanding donors' intentions and behaviours. Within the context of monetary donations made online, prior research has relied on extrinsic determinants influencing attitudes towards online donation (e.g., charities or NPOs, charitable activities, technologies, and SNSs), and specifically on how these factors predict general attitudes of donors towards online donations, as this will boost their behavioural intention to make a monetary donation online [15–17]. Nevertheless, intrinsic determinants, such as religious faith, that could drive attitudes towards the online donation model have been neglected. Given the unique nature of SNSs (e.g., Instagram, Snapchat, Twitter, and Facebook), the current study has practical importance, in that it investigates both extrinsic and intrinsic drives that influence donor attitudes towards donating online, fuelling the propensity to donate money, especially via SNSs.

This study attempts to discover cross-national differences and similarities by undertaking a multi-national study on fundraising campaigns via SNSs during the COVID-19 outbreak. It is expected that the findings of this research will not only shed light on donors' motivations for these campaigns, but also investigate the consistency of the findings and motivations in such countries as Kuwait and Bahrain, which are deemed to be among the most generous countries in the World Giving Index (see Table 1) by the Charities Aid Foundation [18]. Although previous research has investigated how SNSs can aid in the promotion of internet-based donations and how NPOs control the use of SNSs to boost giving behaviours and humanitarian activities [19,20], the extrinsic and intrinsic drivers that shape SNS-driven donations to encounter the unavoidable consequences of epidemic diseases have yet to be fully investigated.

Table 1. Culture characteristics: Kuwait and Bahrain.

	Criteria	Kuwait	Bahrain
1.	Area (capital)	17,818 sq. km (Kuwait City).	760 sq. km (Manama).
2.	Population (migrants %)	4.2 million (70%).	1.7 million (55%).
3.	Religion (language)	Islam (Arabic).	Islam (Arabic).
4.	Economic status	- A petroleum-based economy. - GDP (PPP) $289.7 billion, World rank 58th. - GDP (nominal) $120. 7 billion, World rank 57th. - GPD growth 1.2%.	- The fastest-growing economy in the Arab world. - GDP (PPP) $71.17 billion, World rank 100th. - GDP (nominal) $$35.33 billion, World rank 91st. - GPD growth 1.8%.
5.	Internet penetration	99.1%, approximately 4.1 million of the population.	94.9%, approximately 1.6 million of the population.
6.	SNSs penetration	Social networking penetration 99% Facebook 79%, Twitter 61%, Instagram 55%, Snapchat 54%.	Social networking penetration 84% Snapchat 63%, Facebook 60%, Twitter 58%, Instagram 51%.
7.	Monetary donation and giving indices	- World Giving Index 2018 ranking 33rd (43% of the population donate). - Helping strangers ranking 3rd (80%). - Volunteering time ranking 120th (11%).	- World Giving Index 2018 ranking 10th (53% of the population donate). - Helping strangers ranking 6th (74%). - Volunteering time ranking 24th (33%).

Source: Criteria 1–4 [5,6,21]; 5–6 [22,23]; 7 [18].

2. Theoretical Basis

2.1. Cross-National Context: Kuwait and Bahrain

Based on the scenario (people donate money using SNSs in the time of COVID-19-related social distancing), this study intends to carry out an online survey in order to explore the differences in donors' responses to fundraising appeals for the COVID-19 fight via SNSs between Kuwaitis and Bahrainis. In both countries, Islam and the Arabic language are dominant and widely used. Kuwait and Bahrain have shown steady economic growth in the Middle East over the past few years. In terms of internet penetration, they have high penetration rates of over 90%. Similarly, the rate of SNS penetration across the countries is high (over 80%). Worldwide, there are some indices which are used to measure and capture giving behaviour and donating practices. The World Giving Index is arguably the most prevalent measure of global giving behaviour. As exhibited in Table 1, Bahrain and Kuwait were ranked, in 2018, among the most generous countries in the world. Of the 144 countries surveyed, Bahrain and Kuwait were ranked 10th and 33rd in the World Giving Index 2018. Admirably, Kuwait and Bahrain have occupied the third and sixth positions among the top 10 countries through their participation in helping strangers. However, according to the proportion of people who volunteer, Bahrain and Kuwait have low scores, with the 24th and 120th positions, respectively [18].

It could be argued that previous comparative research on online fundraising has been undertaken between Western countries, such as the comparison of the United States and Netherlands [24], and East Asian countries, such as the comparison between South Korea and Malaysia [16]. What has not yet been explored, however, is comparison research on online fundraising and donating activities between Middle Eastern countries, e.g., Kuwait and Bahrain. The researchers' motivation to embark on this study is borne out of the need for potential meaningful insights into donation intentions via SNSs in countries located in similar Middle Eastern and Arabian Gulf culture zones, as well as with expected differences in responses to fundraising appeals to defeat the pandemic.

2.2. Monetary Donation Intentions via SNSs for the COVID-19 Epidemic Fight

Social responsibility, ethics and social finance innovation play a crucial role in attaining the wellbeing of all during crises such as COVID-19 [25,26]. In the context of donation, the burgeoning literature on giving behaviour has identified various forms of donations, such as blood [27], organs [28], money [17] or alms [8,29]. In order to survive and thrive as non-profit organisations and their development agendas, financial solutions and funding from donors (monetary donations) have a crucial role to play in breathing fresh life into the association between the surplus funds and the funding needs [30]. Donor financial contributions can be made not only in physical form but also electronically (for example, through online donation). However, one of the drivers of charitable donation within the context of online donations is basically the act of collecting monetary donations [16,31]. SNSs, or social networking sites, are regarded as the most effective tool for achieving this. Because they are viewed as a social venue, through the concept of bridging and bonding, users can grow networks and retain interactions and social relationships [31]. According to Kietzmann et al. [32], SNSs are critical interactive tools built on network and mobile technology that enable societies and individuals to create, modify, share, and analyse user-generated information.

Previous studies have found that a significant number of funders give priority to the usage of a range of social media platforms to communicate with humanitarian organisations [16,31,33,34]. For this reason, humanitarian organisations have successfully exploited social networking sites (SNSs) to extend their outreach to current and future donors. However, Lovejoy and Saxton [35] and Kim and Lee [36] indicated that the NPOs are aware that certain users of their websites may not be compelled to donatee. They regarded SNSs as a promotional and marketing tool to be used to encourage and increase donation rates.

Previous studies have shown considerable interest in the ways in which social networking sites aid and facilitate online fundraising activities, and particularly how humanitarian organisations can leverage and manipulate SNSs to promote their charitable efforts [19,20,35–37]. However, a small number of studies have explored the behavioural intentions of people to contribute to charity by using SNSs, particularly for the reduction of the socioeconomic effects of coronavirus epidemics. The literature puts emphasis on the extrinsic factors influencing the behavioural intentions to contribute through social media channels, not taking into consideration the impact of internal predictors (e.g., religious faith) that might be a compelling reason to donate through the SNSs. Consequently, the current research seeks to address the intrinsic (religiosity) and the extrinsic (humanitarian activities, SNS features, NPOs, internet technologies, IT) determinants influencing the attitude of donors, as well as their intention to give via SNSs in response to the COVID-19 crisis.

2.3. Intrinsic and Extrinsic Motivations

In donation research, the TPB is a predictive and unique theory for the assessment of individuals' intentions to donate online, relying primarily on extrinsic determinants influencing donors' attitudes toward online donation: the internet, the type of project, and the characteristics of charities. In accordance with Treiblmaier and Pollach's [17] recommendations, Sura et al. [16] empirically discovered the influence of extrinsic factors on attitude and intention to donate online, incorporating social network site features into the model as an essential aspect of online donations, Users' profiles are included in social networking sites, forums for discussion, real-life broadcasting, photos, and videos, closely associating with the stakeholders (beneficiaries and donors) and the general public. This can play a major role in the evaluation of the value of stakeholder engagements and online communications at a time when SNSs continue to increase. Even though researchers believe that the intrinsic features of donation behaviour are the primary motives to boost donations and help the disadvantaged [38,39], they have received little attention in prior research, specifically in the context of internet-based donations using SNSs. Consequently, this current research aims to uncover new facts about the influence of both internal and external factors on attitudes and intentions to make online donations using SNSs, mobilising financial resources for vulnerable populations who are immediately hit by a reduction in job income and opportunities.

2.3.1. Humanitarian Projects

The type of humanitarian or charitable activity itself has a crucial role in social work and the process of donations and fundraising practices [40]. A humanitarian project can be explained as the charitable cause and event that NPOs support and fund [17,41]. In this study, humanitarian projects are about the social and charitable activities, events, and causes undertaken by individuals or organisations in order to improve human welfare and support those in need. These projects vary among those necessitated by poverty, natural disasters (e.g., epidemics and earthquakes) and man-made disasters (e.g., crimes, civil disorders, and wars). The COVID-19 pandemic poses a dire threat to human lives and livelihoods. In addition to the rapid outbreak of the virus and the alarming number of infected and dead people, unprecedented restrictive measures adopted by governments worldwide to contain the outbreak have inevitably triggered social and financial disruption, creating immense human suffering. In the Gulf region, for instance, there are some 35 million migrant workers, who form over 10% of all migrants globally [2]. In this climate, millions of low-wage workers have been affected, and have endured a great deal of pain; they have been stranded, laid off, and are in need of emergency aid and relief. The GCC authorities and NPOs have, in response, launched urgent humanitarian projects and urged donors to contribute to funding these projects, which are geared towards mitigating COVID-19's consequences.

Prior research has argued that donors' feelings towards humanitarian projects may build their attitudes towards donation, reinforcing the willingness to donate online [16,17,33]. To be clear, the literature has indicated that prior to funding, donors are likely to be inclined to evaluate the features of humanitarian or charitable projects, such as the project's location, type, and connection to and attachment to the humanitarian project itself. Therefore, it can be assumed that donors would be willing to donate online in order to help people and families crushed by the pandemic if they feel connected and attached to the humanitarian projects involved.

Hypothesis 1 (H1). *Donors' perceptions of humanitarian projects positively influence their attitudes towards online monetary donations for COVID-19-affected people.*

2.3.2. NPOs

Since the late 1970s, NPOs have played an increasingly vital role in the social, financial, and development sectors. NPOs are formal entities and foundations that arise when a group of people organise themselves into a social unit that is established with the express purpose of achieving certain goals [42]. The goal of NPOs is almost always linked to development issues, that is, to issues surrounding the social, cultural, and economic order of a region. For NPOs to survive, they usually seek out funding from existing and potential donors, with the aim of properly managing the resources and processes involved. However, NPOs need to maintain a high level of stakeholder confidence, as the voluntary sector's unique status depends on the legitimacy granted to it by public interest.

The changed situation of the COVID-19 crisis and coping with travel restrictions and extraordinary lockdown and health concerns, has triggered an acute funding crisis for charities and non-profit organisations across the humanitarian community [9]. To help address such a challenge, this study attempts to provide emerging ideas and insights by empirically examining donors' attitudes and perceptions of NPOs, providing a sound understanding of the level to which donors are inclined to grant financial support to front the surge in capacity needed to fight this crisis. The literature has claimed that the way NPOs approach donors can shape their attitudes and propensity to donate online [19,31]. This would lead to good communication and coherent relationships between donors and non-profit organisations. In order to rapidly respond to humanitarian needs and relief,

NPOs have mostly harnessed various social networking sites to improve and promote online fundraising campaigns and funding solutions [16,31]. Previous studies have pointed out some important characteristics that can positively trigger donors' attitudes and intentions to give money, such as financial management [17], the affinity and attachment to donors [30,34], image and reputation [30,43], and the management of donation-related information [16,44].

Hypothesis 2 (H2). *Donors' perceptions of NPOs positively influence their attitudes towards online monetary donations for COVID-19-affected people.*

2.3.3. IT Features

In recent years, the conditions of the third non-profit sector have changed dramatically as a result of prevailing social trends. Although NPOs have been considered among the early adopters of IT, it is, at the same time, relatively unattended by scientific research [16]. Digital technology is being employed to streamline and improve charity activities. A study by Quinton and Fennemore [20] claimed that donations and fundraising appeals can be democratized using technological solutions, e.g., SNSs, by gaining mass audience engagement, while simultaneously connecting audiences on a personal scale. IT can be extremely beneficial in terms of increasing donations, productivity, and efficiency, as well as being secure, simple, and time saving [17,31,33,34].

Researchers have indicated that digital technology features (reliability, security, trustworthiness, effectiveness, and privacy) have much to offer in boosting donor attitudes to donating money online [16,34]. According to Sura et al. [16], attitudes towards online fundraising were positively shaped by IT. Nevertheless, Bandyopadhyay [45] argued that a donor is more concerned about digital technology-related security and privacy matters than actual fundraising. As such, this can forge either a positive or negative attitude towards online fundraising. Thus, overall satisfaction with IT features could strengthen the online donation processes to be both effective and well-received. Success in grasping donors' beliefs regarding digital technology is likely to significantly motivate their attitude towards online fundraising to support the COVID-19 emergency response.

Hypothesis 3 (H3). *Donors' perceptions of IT positively influence their attitudes towards online monetary donations for COVID-19-affected people.*

2.3.4. SNS Features

Over the last few years, the application of SNSs (e.g., Instagram, Snapchat, Twitter and Facebook) has grown in popularity, as many people use these platforms on a daily basis. Social networking sites (SNSs) are often regarded as the most effective online platforms for communication and other engagement. They provide one-to-one communication, and offer content personalization, own interest, customization and sharing. Likewise, they can accurately describe target information and facilitate communities with the same interests [46,47]. As reported by Wong and Jusoff [33], the strength of SNSs may be leveraged to sustain and attract existing and prospective donors. SNSs have been identified as a positive motivator for strengthening attitudes towards online donations and, as a result, increasing humanitarian fundraising [47]. Social networking sites can provide a perceived neutral environment in which humanitarian organisations and donors may engage, particularly with benefactors and other major stakeholders [20].

Prior research in this area shows that donors are more likely to prefer social media channels to interact with humanitarian organisations, and are more inclined to donate in this way to fundraising efforts [20,36]. The audience's attention cannot be sustained by communication alone; as such, some elements of social network platforms (sharing videos and photos in related charity events) are required in order to encourage and attract prospective donors [33]. Following the recent outbreak of COVID-19 and the tactical shutdown of cities throughout the world, platforms for social networking were adopted more intensively and broadly than anticipated. As claimed by Xiang et al. [48], the SNSs—such as WeChat, which is used China—were utilised extensively for supervision, administration, communication and the sharing of information for the purpose of providing additional advantages to the ecosystem in order to deal with the epidemic. Along the same line, using donations to mitigate COVID-19's severe socio-economic impact and enormous public health consequences, in order to support relief operations and vulnerable communities, SNSs have been used to make pressing aid appeals to possible contributors. Therefore, the usage of SNS components is predicted to have an important impact on donor attitudes regarding monetary donations in these scenarios.

Hypothesis 4 (H4). *Donors' perceptions of SNS features positively influences their attitudes towards online monetary donation for COVID-19 affected people.*

2.3.5. Religiosity

The propensity to donate is seen to be a complicated pattern, and religious beliefs should not be underestimated by scholars when this labyrinthine matter is scrutinized [49]. Charitable practices and religious beliefs can often go hand in hand; in the known main religions—Islam, Judaism, Christianity, Hinduism and Buddhism—donations are determined to be a common practice of social behaviours [50,51]. Religious commands have always stimulated philanthropists to lend a helping hand to the underprivileged through charitable activities. In Islam, for example, annual charitable donations that stem from religious teachings [52], such as voluntary charity (*Sadgah*) and mandatory alms (*Zakat*), are imposed on Muslims from all walks of life. Likewise, other religions like Hinduism and Christianity emphasize the significance of giving and generosity. Therefore, religiosity has been deemed to be a decisive motive in boosting donors' inclination to donate, as religion itself is simply the basis on which the idea of donation is conceived and nurtured [49,50,52–54].

Some credible evidence has claimed that religiosity and donation behaviour are positively related. For instance, Ranganathan and Henley's [51] research of 214 students in the United States revealed that religiosity exerts a significant effect on donors' attitudes towards helping others and behavioural intentions as well. They showed that an individual with a high degree of religiosity possesses altruistic behaviour, and consequently would have a positive attitude toward charitable activities and positive intentions to donate. Similarly, Abreu et al. [50] conducted a study of 612 charitable donors in Portugal and found that the religiosity levels of donors positively influence their donation behaviour. However, despite the abundance of empirical studies on the association between donation practices and religiosity, research on the effect of religious beliefs on donor attitudes and intentions to donate online, especially via SNSs, is relatively scarce. In a collective Muslim community (e.g., Bahrain or Kuwait), we could envisage that religious donors would demonstrate a robust attitude and intention towards donating money online, especially in light of the ongoing crisis of COVID-19. That is, the greater the level of religiosity the donors have, the stronger their attitudes and intentions towards online fundraising.

Hypothesis 5 (H5). *Donors' religiosity positively influences their attitudes towards online monetary donations for COVID-19-affected people.*

Hypothesis 6 (H6). *Donors' religiosity positively influences their monetary donation intentions via SNS for the COVID-19 epidemic fight.*

2.3.6. Attitudes towards Online Monetary Donation

Attitude, in this study, can be labelled as the degree of a donor's positive feelings towards helping to support and fund the fundraising campaigns to fight the pandemic using social media-based donation platforms. Prior studies dedicated to behavioural intentions in general [14,55,56], and the most feasible behavioural model of TPB in particular [57–59], have deduced that attitude is an important and vital factor in shaping behavioural intentions. More so, the growing body of studies on social work has highlighted the significance of attitudes in the identification of positive outcomes of social interaction. It is critical to foster positive attitudes toward charitable initiatives, which will strengthen donation behaviour intention [52]. Other findings showed that the views of individuals are directly related to their intentions to contribute to charity [15,55,56].

In connection with an online donation, a study by Treiblmaier and Pollach [17] claimed that the degree to which donors are knowledgeable of fundraising websites has a significant impact on the contribution process. It showed that when there is a favourable attitude towards online donations, donors' inclination to make donations online would be stronger, by administering an online survey to 258 potential contributors from Malaysia and South Korea. However, Sura et al. [16] argued that attitudes have a crucial role in influencing people's behavioural intentions about donating practises through social networking sites. This conclusion was further confirmed by Mittelman and Rojas-Méndez [41]. As COVID-19 rapidly spreads across the globe, the main battle is around ensuring that social protection reaches people with low incomes, and around defending their welfare and livelihoods. Fundraising campaigns are one of the economic support packages and financial interventions implemented by many governments around the world. However, the underlying issue is not only related to these fundraising appeals; understanding donors' responses and attitudes towards these appeals might be the lingering obstacle. We think that donors who are more set in their minds about giving money online will be more likely to give money through social networking sites.

Hypothesis 7 (H7). *Donors' attitudes positively affect monetary donation intentions via SNSs for the COVID-19 epidemic fight.*

In addition to the direct effect of extrinsic and intrinsic factors on donors' inclination to give money via SNSs, the study investigates the role of attitudes in mediating relationships between extrinsic factors (i.e., internet technologies, NPOs, humanitarian projects, and SNSs) and intrinsic motivation (religiosity) and the behavioural intention to donate online in order to alleviate acute financial strains and provide urgent relief for those affected by the virus. Attitudes as a mediator may help boost or increase the likelihood that donors will contribute using SNSs. Scholars [16,51,57] proposed that altruism (the desire to help others) has a direct impact on behavioural intention, but it also influences attitudes toward philanthropic endeavours, technology characteristics, and NPOs, where donors are more enthusiastic and are actively encouraged to make online donations. Therefore, for the mediation analysis in this research model, the following hypotheses were developed:

Hypothesis 8 (H8). *The relationships between humanitarian projects and monetary donation intentions via SNSs is mediated by online donation attitudes.*

Hypothesis 9 (H9). *The relationships between NPOs and monetary donation intentions via SNSs is mediated by online donation attitudes.*

Hypothesis 10 (H10). *The relationships between internet technologies and monetary donation intentions via SNSs is mediated by online donation attitudes.*

Hypothesis 11 (H11). *The relationships between SNSs and monetary donation intentions via SNSs is mediated by online donation attitudes.*

Hypothesis 12 (H12) *The relationships between levels of religiosity and monetary donation intentions via SNSs are mediated by online donation attitudes.*

3. Materials and Methods

3.1. Sampling and Data Collection

As there is a pressing need to understand the factors driving donors' attitudes and behavioural intentions to donate through SNSs for funding relief efforts aimed at containing the COVID-19 crisis, the study followed a multi-national perspective among donors in two of the most generous societies in the world (Kuwait and Bahrain), as indicated by the World Giving Index of 2018 [18]. Drawing on the back-translation procedure (English vs. Arabic), the study relied on an online survey [60], and was distributed based on snowball sampling by sharing and posting its URL via online platforms (SNSs such as Instagram, Snapchat, Twitter, and Facebook) among potential donors in both countries. The survey was conducted between 1 May and 31 July 2020. A final set of 801 questionnaires was retrieved from respondents in Kuwait (565 responses) and Bahrain (236 responses), and was used for the subsequent analysis, after some incomplete responses were discarded.

The characteristics of the sample indicated that the gender split was fairly even in both samples, demonstrating that two-thirds (63%) of the respondents from Kuwait were female. In contrast, 56% of the Bahraini respondents were male. Coincidentally, the age groups in both samples were similar: 84% of the participants in both samples were aged 20–40 years, while the rest (16%) were aged 41 years and over (see Table 2). The study found that a larger proportion of respondents reached a higher level of education (bachelor's degree and above) in the Bahrain sample than in the Kuwait sample (82% and 73%, respectively). Of a total of 801 potential donors, 90% and 70% of Kuwaitis and Bahrainis, respectively, identified their financial situation as equal to or higher than the average. According to the World Bank's [61] statistics, USD2000 and USD1480 are the average monthly incomes in Kuwait and Bahrain, respectively. In general, we can deduce that both samples from these two countries appear to be well-diversified with regard to gender, age, educational background and income level. This could be an indication that the respondents of this study possess at least a certain level of educational background, and the basic knowledge to respond appropriately to the research survey.

Table 2. Sample profile.

Variable	Categories	Kuwaiti Sample		Bahrain Sample	
		n = 565	%	*n* = 236	%
Gender	Male	209	37	133	56
	Female	356	63	103	44
Age	20–30 years	330	59	136	57
	31–40 years	142	25	63	27
	41–50 years	51	9	21	9
	51- above	42	7	16	7
Education level	Diploma certificate	152	27	43	18
	Bachelor's degree	343	61	124	53
	Master's degree	52	9	38	16
	PhD degree	18	3	31	13
Income level	Less than average	59	10	68	29
	Average *	266	47	81	34
	Higher than average	240	43	87	37

Note(s): * The average monthly incomes per capita in Kuwait and Bahrain were KD620 and BD560 (i.e., USD2000 and USD1480, respectively) in July 2020.

3.2. Measurement and Analysis

As elaborated in Section 3.1, the survey instrument began with questions to gather demographic information from the participants. Parts 2–7 of the survey were designed for the observable variables (humanitarian projects, NPOs, IT features, SNS features, religiosity, attitude, and monetary donation intention). In order to measure these extrinsic and intrinsic constructs, a 5-point Likert scale for 35-item questions was developed (see all measurement items in Supplementary Materials). Table 3 summarizes the conceptualization and operationalization sources of all of the constructs in this study. In order to ensure the appropriateness of the instrument, we surveyed a pilot study of 30 sampled respondents. Accordingly, we decided to conduct the main distributions of the survey after the syntax of sentences was checked and amended, and after we ensured their readability, reliability, and validity.

Using the SmartPLS variance-based structural modelling equation (PL-SEM), the research model and hypotheses of this study were built and tested. Unlike covariance-based SEM, SmartPLS has the advantage that it is a more suitable technique and a more effective tool for the prediction-oriented strategy that characterises this study. In contrast to covariance-based-SEM, PLS-SEM performs better and possesses more statistical power for non-normal data, as reported by Hair et al. [62]. In order to ensure that the most appropriate data analysis tool was employed, we performed Mardia's analysis for potential multivariate normality using online software called Web Power [62,63]. We found that the data in both samples seem to be non-normally distributed. For the Kuwaiti data, the multivariate skewness and kurtosis measures were 9.755 and 79.948, respectively, at significance level of 0.000, with similar results for Bahrain (skewness $\beta = 11.505$; kurtosis $\beta = 80.351$, $p = 0.000$). Consequently, as a non-parametric analytical instrument, we continued to use PLS-SEM.

Table 3. Conceptualization and source of the variables.

Variable	Conceptualization	No. of Item	Source
Humanitarian projects	Humanitarian endeavors and logistical assistance that provide unique and urgent responses and activities intended to mitigate the impact of COVID-19 on the most vulnerable populations.	5	[17]
NPOss	NPOs or Non-profit organizations set-up by citizens with the aim of support social protection services, human welfare, and seeking out financial generosity from existing and potential donors as well as well-managed fundraising.	5	[16,41]
IT-features	Features of digital technology and information, e.g., effectiveness, privacy, trustworthiness, reliability, security, that assist philanthropic activities.	5	[16]
SNS-features	Features of social network sites and platforms, such as information presented, familiar SNS, user-friendly, communication and interactive features, that are harnessed for charity and donation.	5	[36]
Religiosity	The level to which donors adhere to the religious teachings, including helping vulnerable groups and those affected by the pandemic.	6	[8]
Attitude	Feelings or evaluations of donors, whether favorable or unfavorable toward online monetary donation to fight the pandemic.	4	[64]
Monetary donations intentions via SNSs	Donors' willingness to donate via SNSs to financially support relief efforts and respond positively to the current fundraising appeals.	5	[64]

4. Results

4.1. Model Evaluation

Reliability and discriminant and convergent validity are the important criteria for the evaluation of the model's reflective measurements. The composite reliability (CR) was tested in order to assess the internal consistency of all of the latent constructs. The findings of both samples show that the CR values in the online donation model well exceeded the recommended value of 0.50, affirming the reliable constructs. Most of the factor-standardized loadings of both samples were above the benchmark of 0.60, and some loaded in the range of 0.40–0.60, while a few items loaded poorly with values less than 0.40, and were discarded in accordance with Hair et al. [62]. In the data from Kuwait, the average variance extracted (AVE) values varied from 0.505 to 0.833, whereas in the data from Bahrain, the values ranged from 0.628 to 0.897. This result establishes the convergent validity, as all of the values were above 0.50 (see Table 4).

Table 4. Measurement model assessment.

Constructs	Items	Kuwaiti Sample n = 565			Items	Bahraini Sample n = 236		
		Loadings	AVE	CR		Loadings	AVE	CR
Humanitarian projects (HUP)	HUP1	0.820	0.556	0.832	HUP3	0.701	0.833	0.628
	HUP2	0.700			HUP4	0.900		
	HUP3	0.816			HUP5	0.763		
	HUP4	0.630						
Non-profit organisations (NPO)	NPO1	0.782	0.658	0.885	NPO1	0.575	0.505	0.797
	NPO2	0.816			NPO2	0.537		
	NPO3	0.842			NPO4	0.868		
	NPO4	0.804			NPO5	0.805		
IT-features (ITF)	ITF1	0.814	0.767	0.943	ITF1	0.502	0.554	0.857
	ITF2	0.908			ITF2	0.663		
	ITF3	0.919			ITF3	0.852		
	ITF4	0.879			ITF4	0.820		
	ITF5	0.854			ITF5	0.824		
SNS-features (SNS)	SNS1	0.728	0.533	0.849	SNS1	0.637	0.527	0.846
	SNS2	0.852			SNS2	0.688		
	SNS3	0.796			SNS3	0.803		
	SNS4	0.647			SNS4	0.660		
	SNS5	0.596			SNS5	0.822		
Religiosity (REL)	REL1	0.584	0.549	0.827	REL1	0.623	0.533	0.850
	REL4	0.783			REL3	0.693		
	REL5	0.852			REL4	0.818		
	REL6	0.719			REL5	0.768		
					REL6	0.734		
Attitude toward online monetary donation (ATT)	ATT1	0.731	0.713	0.908	ATT2	0.654	0.576	0.801
	ATT2	0.824			ATT3	0.860		
	ATT3	0.916			ATT4	0.749		
	ATT4	0.894						
Monetary donation intentions via SNSs for COVID-19 epidemic fight (INT)	INT1	0.889	0.764	0.942	INT1	0.809	0.687	0.897
	INT2	0.920			INT2	0.903		
	INT3	0.885			INT3	0.876		
	INT4	0.892			INT4	0.716		
	INT5	0.776						

As shown in Table 5, the study also scrutinized the heterotrait–monotrait (HTMT) ratio to check the discriminant validity. In both samples, we concluded that the HTMT ratios were considerably less than the indicated 0.85 threshold for all latent constructs [65]. This shows that the discriminant validity was affirmed. Generally, the results of the measurement model evaluations for the Kuwaiti and Bahraini data confirm that the measurement scales of the online donation model are satisfactory, valid, and reliable. Thus, we can move on to the testing of the structural model.

Table 5. Discriminant validity—HTMT criterion.

	1	2	3	4	5	6	7
			Kuwaiti Sample				
1. HUP							
2. NPO	0.438						
3. ITF	0.310	0.460					
4. SNS	0.269	0.391	0.133				
5. REL	0.280	0.288	0.209	0.109			
6. ATT	0.526	0.358	0.470	0.223	0.296		
7. INT	0.405	0.208	0.268	0.345	0.214	0.538	–
			Bahraini sample				
1. HUP							
2. NPO	0.411						
3. ITF	0.332	0.377					
4. SNS	0.317	0.368	0.420				
5. REL	0.257	0.142	0.288	0.492			
6. ATT	0.539	0.562	0.357	0.507	0.590		
7. INT	0.378	0.282	0.230	0.246	0.542	0.425	–

Note(s): HUP = humanitarian projects; NPO = non-profit organizations; ITF = IT features; SNS = social networking sites; REL = religiosity; ATT = attitude; INT = intention.

4.2. Hypothesis and Significance Tests

To validate the study model, we estimated the coefficient of determination (R^2) of all of the latent variables in order to assess the predictive capability of the model. The online donation model for both the Kuwaiti and Bahraini data reveals that the endogenous construct of attitude toward donating online has an adequate R^2 value of 0.319 and 0.428, respectively, implying that the five exogenous variables (humanitarian projects, NPOs, IT features, SNS features and religiosity) can account for around 32% and 43% of the difference in the Kuwaiti and Bahraini donor attitude towards online donations for the most vulnerable individuals affected by the COVID-19 pandemic. Moreover, the second endogenous construct's R^2 values (monetary donation intention) for both samples were 0.240 and 0.235, showing that religiosity and attitude have the potential to play a critical role in forecasting both Kuwaiti and Bahraini donors' intentions to donate via SNSs. Furthermore, we evaluated the online donation model's predictive relevance (Q^2) using the blindfolding procedure in PLS-SEM. The findings from the endogenous variable Q^2 analysis (attitude and monetary donation intention) in both countries were above zero. This indicates that the online donation model developed for this study has the ability to predict the observed values of the items used to measure either endogenous or exogenous constructs. Hair et al. [62] suggested that a Q^2 value above zero implies that the model is able to predict the original data accurately.

In both samples, bootstrapping—a technique in PLS-SEM that allows the determination of the significance of direct and indirect proposed hypotheses, was employed with 5000 bootstrap samples. For direct relationships in the Kuwaiti sample (Table 6), the results show that all of the relationships between independent constructs and dependents are accepted, except for two paths (NPOATT and RELINT). Therefore, the analysis supports H1, H3, H4, H5, and H7 as $p < 0.05$, while it does not support H2 and H6. For Bahrain, all of the direct relationships are supported ($p < 0.05$), except for internet technology features ($p > 0.05$).

Table 6. Structural model results.

Hypothesized Relationship		Kuwaiti Sample				Bahraini Sample			
		β	t	p	Sig?	β	t	p	Sig?
Direct effects									
H1	HUP → ATT	0.279	6.388	0.000	Yes	0.168	2.626	0.009	Yes
H2	NPO → ATT	0.019	0.421	0.674	No	0.356	5.704	0.000	Yes
H3	ITF → ATT	0.307	6.925	0.000	Yes	0.014	0.251	0.802	No
H4	SNS → ATT	0.128	3.756	0.000	Yes	0.150	2.081	0.037	Yes
H5	REL → ATT	0.128	3.071	0.002	Yes	0.305	4.666	0.000	Yes
H6	REL → INT	0.073	1.685	0.092	No	0.362	5.881	0.000	Yes
H7	ATT → INT	0.467	12.001	0.000	Yes	0.203	4.022	0.000	Yes
Indirect effects									
H8	HUP → ATT → INT	0.130	5.558	0.000	Yes	0.034	2.057	0.040	Yes
H9	NPO → ATT → INT	0.009	0.417	0.676	No	0.072	2.810	0.005	Yes
H10	ITF → ATT → INT	0.143	5.773	0.000	Yes	0.003	0.237	0.812	No
H11	SNS → ATT → INT	0.060	3.430	0.001	Yes	0.030	2.093	0.036	Yes
H12	REL → ATT → INT	0.060	2.901	0.004	Yes	0.062	2.968	0.003	Yes

Note(s): HUP = humanitarian projects; NPO = non- profit organizations; ITF = IT features; SNS = social networking sites; REL = religiosity; ATT = attitude; INT = intention.

In particular, the attitudes of donors toward donating online are influenced by humanitarian initiatives. In Bahrain but not in Kuwait, NPOs had a positive effect on attitudes. In Kuwait, both the SNS and IT features exhibited a strong relationship with the online attitudes towards donation. Similarly, Bahraini donors' attitudes were significantly influenced by SNSs, while technology failed to reveal statistical evidence. The religiosity level and attitude towards online donation in both countries revealed a positive relationship. Religiosity had a positive relationship with the intentions of Bahraini donors to give via SNSs to support the humanitarian efforts to fight the pandemic, but not in Kuwait. In both countries, the willingness to donate through SNSs was found to be linked to how people felt about online donations.

Given the recommendations made by Rungtusanatham et al. [66] regarding the indirect (mediation) effects, as shown in Table 6, the current study investigated the mediating effects of attitudes towards monetary donations on the relationship between the intrinsic and extrinsic factors, and the intentions of donors to give through SNSs. In Kuwait, the findings from this study show that the relationships of humanitarian activities → intentions, internet technology → intentions, SNS features → intentions, and religiosity → intentions through attitudes were seen to be statistically significant. As a result, H8, H10, H11, and H12 are supported. On the other hand, attitudes did not mediate the relationship of NPOs → intentions, rejecting H9. As for Bahrain, all of the indirect relationships (for the paths HUP→ATT→ INT, NPO→ATT→ INT, SNS →ATT→ INT, and REL→ATT→ INT) were accepted, with one exception (for the path ITF→ATT→ INT).

5. Discussion

The pandemic has wreaked havoc on the world's health, social, and economic systems [67–71]. Within the framework of the contradictory debates about whether extrinsic or intrinsic determinants influence donor attitudes and behavioural intent to donate online, a more extensive online fundraising model was empirically tested and adopted in this study, covering both the extrinsic (the humanitarian initiatives, NPOs, IT features, and SNS features) and intrinsic (religiosity) factors that drive donor attitudes towards online monetary donations, and which thus eventually shape the inclination to donate via SNS.

This work is pioneering research on the potential donors of two different markets, i.e., Kuwaiti and Bahraini. The findings support the conclusions reached by Sura et al. [16] and Chen et al. [56], demonstrating the influence of attitude in shaping online donation behaviour, which is among the three main components of TPB theory, and should not be disregarded. The consistency of the outcomes in this study shows that the model established might accurately reflect the donors' tendencies for donations and fundraising activities through SNSs. This can eventually be translated into effective financial solutions to deal with threats like COVID-19.

The outcomes from this research have provided interesting insights into donors' behavioural makeups, especially amid a crisis of uncertainty such as the COVID-19 epidemic. The nature of humanitarian projects (i.e., project location, type, connection, and attachment to project) to mitigate the social impact of COVID-19 had a positive and significant effect on donor attitudes to donation online, according to both Kuwaiti and Bahraini donors. This outcome is in agreement with the findings of research conducted by Ahn et al. [72], suggesting that prior to funding humanitarian projects, donors are very concerned about the features of such projects and obtaining complete and explicit related information. Bahraini respondents believe that NPOs play an important role in predicting their attitude towards online donation; however, in Kuwait, donors did not consider that NPOs can shape their attitudes towards giving money. We could attribute this discovery to the nature of a collectivist society with a low individualism rate like Kuwait society [73]. Moreover, as Kuwaitis are wholly Muslims, apart from fundraisers or NPOs, we could argue that donors are likely to obey and adhere to the Islamic religious guidelines, which to make donations and help the needy, through legitimate entities such as the government. As for the Kuwaiti technology features such as privacy, trustworthiness, reliability and security, it was found that the attitude toward giving money online was positively and significantly related to these IT features. This finding is in line with those of other researchers [17,34], but Bahrainis had different views of technology insignificantly related to attitudes. This could be linked to people's intense enthusiasm for donations and helping the vulnerable who are greatly affected by the outbreak; when people are highly willing to participate in fundraising and donations, this could make them less concerned about the internet features.

Internet-based platforms, such as social networking sites, provide the framework for NPOs to function and perform properly. Remarkably, the study discovered that the SNS characteristics—such as the information provided and fully used, interactive, user-friendly, and familiar features—have a beneficial and critical role across both countries in determining attitudes towards online donations for less-privileged segments affected by the pandemic. The effect of these extrinsic factors was also observed in conjunction with that of the intrinsic perspective of religiosity in positively triggering donors' attitudes toward contributing to the COVID-19 fundraising campaigns. As was found in research by Ranganathan and Henley [51], religious beliefs are an important motive in encouraging donation behaviour. However, the religiosity level of donors from Kuwait did not show a strong relationship with the intention to donate using SNS, indicating the robustness of religious faith in influencing attitudes, as opposed to behavioural intention. Bahraini respondents showed that the greater the levels of religiosity, the stronger the attitudes as well as the intent to donate via the SNS.

However, the online donation attitude in the model had the greatest effect on people's intentions to donate using SNSs. This corresponds to the findings of the literature [16,56]. From a theoretical point of view, the result is also in line with the TPB model developed by Ajzen [57], which argues that attitude exerts a strong positive relationship with human behavioural intention. Moreover, according to the analysis of the Kuwaiti data, attitudes towards internet-based donations significantly mediate the pathways that connect humanitarian initiatives, the technology features, SNSs, and religious beliefs to the intention to donate using SNSs. Even though religious belief had no direct effect on behavioural intent, we found that when religiosity is coupled with donor attitude, the intention to donate via social media platforms can be strengthened. As for Bahrain, attitude mediated all of the relationships between the predictors examined and the intentions to donate via SNS but failed with the association between IT and intentions. These results can offer important implications and insights into policy and theory.

5.1. Theoretical Implications

To the best of the authors' knowledge, to date, no study has examined online fundraising and donation behaviours that are driven by internal and external determinants in a cross-country context, especially among Middle Eastern or Arabian Gulf countries. Thus, the analysis of similarities and differences between Kuwait and Bahrain in this study significantly enriches the extant literature on donation behaviour. The results contribute by exploring donors' behaviours and motivations for donating and responding to the rapidly escalating humanitarian needs caused by the global unprecedented crisis of COVID-19.

The empirical investigations have mainly focused on the extrinsic motivations of donors to donate online [16,17]. However, it is somewhat surprising that few studies have been devoted to internal perspectives (e.g., religious beliefs) that encourage online donations. Therefore, this study adds to the existing body of knowledge by expanding our understanding of donors' psychological behavioural makeups by providing a broader perspective of both intrinsic and extrinsic motives, particularly in a unique and novel related setting like SNSs. Meanwhile, in contrast to the empirical results provided by Sura et al. [16] in the Malaysian and South Korean contexts, which failed to discover a statistically significant relationship between attitude and external variables, some significant and interesting results emerged in the present work. We conclusively found that both intrinsic and extrinsic predictors have much to offer in fostering monetary donations via SNSs among both Kuwaiti and Bahraini donors. The study adds to and expands on previous research by looking at the mediating effects of online monetary donation attitudes on pathways from external and internal factors to the donor inclination to donate through SNSs in order to help poverty-stricken segments and those affected by the pandemic.

5.2. Policy Implications

The pandemic of COVID-19 is seen as an unprecedented event in terms of need and scope. Almost all nations, sectors, organisations, communities, and households are reeling from the impacts of this crisis on the economy, society, and health. Although the global humanitarian community is at the forefront of responding to the extraordinary inequalities and needs that this catastrophe is generating, this sector is facing acute funding constraints and threats from the pandemic at the moment when it is needed most. With this in mind, this study sought to generate better insights into the internal and external makeups of donor responses to the recent fundraising appeals to alleviate the pandemic impacts on the welfare of vulnerable groups. Based on the results of this study, some appropriate proactive strategies could be formulated for the humanitarian community (NPOs and donors) and governments.

Now more than ever, it is important to bring people together and to help innovate an effective humanitarian financial model. We need to set up humanitarian projects with inclusive information and statistics in terms of location and type, and connected to human welfare, which are geared towards protecting vulnerable populations. Donors could also be encouraged to donate money via SNSs by their attitudes and perceptions of NPOs, as in the case of Bahraini respondents. The general views of NPOs may indicate that they are prone to wasting financial resources [74,75]. It is essential for all local NPOs to carefully manage their brand image and perception of efficiency. Creating transparency in all related processes and maintaining consistent communication with current and potential donors could be vitally important. The results also suggest another fundamental aspect to consider is digital technology in the charity sector. NPOs have to put in place transformative platforms so that donors are able to give money easily, and to find out where and how to help those on the receiving end. SNSs could help to boost donor attitudes towards contributing, which may help guide policymakers to come up with promotional exposure campaigns using the SNSs to raise the general public's attitude toward making online donations. As a crucial internal driver, the positive religious beliefs of donors are significant in tailoring the policies that match donors' attitudes and religious doctrines. Highly religious donors may feel that helping people affected by the pandemic is their duty, and they would be rewarded with a sense of satisfaction. To this end, boosting attitudes towards monetary donations using digital initiatives, such as SNSs, could serve the humanitarian community as a means of developing feasible financial solutions at scale to address immediate needs.

6. Conclusions

In the wake of COVID-19, the third non-profit sector is facing a complicated hoard of unprecedented financial challenges. The humanitarian and NPO communities have seen a catastrophic drop in liquidity and pressure to rapidly downsize operational capacity. With the hope of finding and developing all of the available solutions, this study aimed to develop a clearer understanding of the way in which donors respond to the current fundraising appeals in order to replenish the financial resources that philanthropic efforts crave. Based on a cross-national perspective, we found some interesting differences between donors' attitudes and perceptions of fundraising campaigns to fight the pandemic across Kuwait and Bahrain. Overall, there is agreement across the samples that donor's attitudes and intentions towards monetary donation via SNSs can be fostered by both intrinsic and extrinsic motivations. Humanitarianism, well-defined projects, transparent NPOs, advanced IT-based infrastructure, user-friendly and more interactive SNSs, and favourable religious beliefs can all have an important role in future fundraising prospects amid COVID-19 and beyond.

Despite the ambitious aspirations of this research, certain limitations are worth noting. The research time horizon is cross-sectional; as such, changes may be missed in opinions over time, particularly during the uncertain and rapidly changing times resulting from the pandemic. Therefore, it is strongly suggested that future studies should use longitudinal data. Finally, future studies based on this model are proposed to take into account additional external and internal elements, such as significant social media influencers and knowledge roles in online donation behaviour.

Supplementary Materials: The following supporting information can be downloaded at: https://www.mdpi.com/article/10.3390/su14116486/s1, File is including measurement items.

Author Contributions: Conceptualization, S.A.B.-N.; Funding acquisition, S.A.B.-N., A.S., M.A.-D., M.H.A. and A.B.A.; Investigation, A.S. and M.A.-D.; Methodology, S.A.B.-N.; Project administration, S.A.B.-N.; Writing—original draft, S.A.B.-N.; Writing—review and editing, M.H.A., A.B.A., H.A.-J., H.B. and M.K.A. All authors have read and agreed to the published version of the manuscript.

Funding: This research received no external funding.

Institutional Review Board Statement: Not applicable.

Informed Consent Statement: Not applicable.

Data Availability Statement: Not applicable.

Conflicts of Interest: The authors declare no conflict of interest.

References

1. International Labour Organization (ILO). *COVID-19 Impact on Labour Market Statistics*; International Labour Organization: Geneva, Switzerland, 2020; pp. 1–28.
2. International Labour Organization (ILO). *ILO: COVID-19 Causes Devastating Losses in Working Hours and Employment*; International Labour Organization: Geneva, Switzerland, 2020.
3. Ajami, R. Globalization, the challenge of COVID-19 and Oil price uncertainty. *J. Asia Pacific Bus.* **2020**, *21*, 1–3. [CrossRef]
4. Young, K. How COVID-19 Will Change GCC Labor Markets. 2020. Available online: www.alonitor.com/pulse/originals/2020/04/gulf-covid19-coronavirus-change-gcc-labor-markets-crisis.html (accessed on 2 April 2020).
5. Information & eGovernment Authority (iGA). *RHF and Iga Launch Registration Service for Those Entitled for Support from the 'Feena Khair' Campaign*; Information & eGovernment Authority (iGA): Manama, Bahrain, 2020.
6. Kuwait Central Statistical Bureau. *Annual Statistical Abstract in Kuwait*; Kuwait Central Statistical Bureau: Kuwait City, Kuwait, 2018.
7. Seznec, J.-F. *The Economic Outlook for MENA Amid COVID-19 and the Oil Price Drop*; Middle East Institute: Washington, DC, USA, 2020.

8. Bin-Nashwan, S.A.; Abdul-Jabbar, H.; Aziz, S.A.; Haladu, A. Zakah compliance behavior among entrepreneurs: Economic factors approach. *Int. J. Ethic- Syst.* **2020**, *36*, 285–302. [CrossRef]
9. O'Connell, S. *How to Reform NGO Funding So We Can Deal with Threats like COVID-19*; World Economic Forum: Geneva, Switzerland, 2020.
10. United Nations. *UN Response to COVID-19*; United Nations: New York, NY, USA, 2020.
11. The United Nations Office for the Coordination of Humanitarian Affairs (OCHA). *Global Humanitarian Overview 2020*; United Nations: New York, NY, USA, 2020.
12. Ministry of Social Affairs and Labor. *Kuwaiti Donation Campaign*; Ministry of Social Affairs and Labor: Kuwait City, Kuwait, 2020.
13. Van Der Linden, S. Charitable Intent: A Moral or Social Construct? A Revised Theory of Planned Behavior Model. *Curr. Psychol.* **2011**, *30*, 355–374. [CrossRef]
14. Smith, J.R.; McSweeney, A. Charitable giving: The effectiveness of a revised theory of planned behaviour model in predicting donating intentions and behaviour. *J. Commun. Appl. Soc. Psychol.* **2007**, *17*, 363–386. [CrossRef]
15. Knowles, S.R.; Hyde, M.K.; White, K. Predictors of Young People's Charitable Intentions to Donate Money: An Extended Theory of Planned Behavior Perspective. *J. Appl. Soc. Psychol.* **2012**, *42*, 2096–2110. [CrossRef]
16. Sura, S.; Ahn, J.; Lee, O. Factors influencing intention to donate via social network site (SNS): From Asian's perspective. *Telemat. Inform.* **2017**, *34*, 164–176. [CrossRef]
17. Treiblmaier, H.; Pollach, I. A framework for measuring people's intention to donate online. In Proceedings of the Tenth Pacific Asia Conference on Information Systems, Kuala Lumpur, Malaysia, 6–9 July 2006; pp. 808–817.
18. Charities Aid Foundation (CAF). *CAF World Giving Index 2018: A Global View of Giving Trends*; Charities Aid Foundation: London, UK, 2018.
19. Zhong, Z.J.; Lin, S. The antecedents and consequences of charitable donation heterogeneity on social media. *Int. J. Nonprofit Volunt. Sect. Mark.* **2018**, *23*, e1585. [CrossRef]
20. Quinton, S.; Fennemore, P. Missing a strategic marketing trick? The use of online social networks by UK charities. *Int. J. Nonprofit Volunt. Sect. Mark.* **2013**, *18*, 36–51. [CrossRef]
21. *The World Factbook*; The World Factbook: Kuwait City, Kuwait; Manama, Bahrain, 2020.
22. Kemp, S. We Are Social/Hoot Suite. *Glob. Digit. Statshot*. 2020, 2020. Available online: https://datareportal.com/ (accessed on 22 June 2021).
23. Internet World Stats. *Internet World Stats: Usage and Population Statistics*; Internet World Stats: Bogota, Colombia, 2020.
24. Leushuis, R. Continuous Donation to and Trust in a Charitable Organization. Master's Thesis, University of Twente, Twente, The Netherlands, 2012.
25. Gatto, A.; Polselli, N.; Bloom, G. Empowering gender equality through rural development Rural markets and micro-finance in Kyrgyzstan. In *L'Europa e la Comunità Internazionale Difronte alle Sfide dello Sviluppo*; Giannini: Naples, Italy, 2016; pp. 65–89.
26. Gatto, A.; Sadik-Zada, E.R. Revisiting the East Asian Financial Crises: Lessons from Ethics and Development Patterns. In *Economic Growth and Financial Development*; Springer: Cham, Switzerland, 2021; pp. 23–31.
27. Williams, L.A.; Sun, J.; Masser, B. Integrating self-determination theory and the theory of planned behaviour to predict intention to donate blood. *Transfus. Med.* **2017**, *29*, 59–64. [CrossRef]
28. Ju, M.; Sim, M.; Son, S. A Study on Nursing Students' Knowledge, Attitude, and Educational Needs for Brain-Death Organ Transplantation and Donation and Intent to Donate Organs. *Transplant. Proc.* **2018**, *50*, 1187–1191. [CrossRef]
29. Bin-Nashwan, S.A.; Abdul-Jabbar, H.; Dziegielewski, S.F.; Aziz, S.A. Moderating Effect of Perceived Behavioral Control on Islamic Tax (*Zakah*) Compliance Behavior among Businessmen in Yemen. *J. Soc. Serv. Res.* **2021**, *47*, 292–302. [CrossRef]
30. Beldad, A.; Snip, B.; van Hoof, J. Generosity the second time around: Determinants of individuals' repeat donation intention. *Nonprofit Volunt. Sect. Q.* **2014**, *43*, 144–163. [CrossRef]
31. Li, Y.; Yu, L. Factors Influencing Social Media Users' Continued Intent to Donate. *Sustainability* **2020**, *12*, 879. [CrossRef]
32. Kietzmann, J.H.; Hermkens, K.; McCarthy, I.P.; Silvestre, B.S. Social media? Get serious! Understanding the functional building blocks of social media. *Bus. Horiz.* **2011**, *54*, 241–251. [CrossRef]
33. Wong, D.M.; Jusoff, K. Social networking in charity advocacy. *World Appl. Sci. J.* **2011**, *12*, 65–72.
34. Shier, M.L.; Handy, F. Understanding online donor behavior: The role of donor characteristics, perceptions of the internet, website and program, and influence from social networks. *Int. J. Nonprofit Volunt. Sect. Mark.* **2012**, *17*, 219–230. [CrossRef]
35. Lovejoy, K.; Saxton, G.D. Information, Community, and Action: How Nonprofit Organizations Use Social Media. *J. Comput. Commun.* **2012**, *17*, 337–353. [CrossRef]
36. Kim, Y.; Lee, W.-N. Networking for Philanthropy: Increasing Volunteer Behavior via Social Networking Sites. *Cyberpsychology Behav. Soc. Netw.* **2014**, *17*, 160–165. [CrossRef]
37. Curtis, L.; Edwards, C.; Fraser, K.L.; Gudelsky, S.; Holmquist, J.; Thornton, K.; Sweetser, K.D. Adoption of social media for public relations by nonprofit organizations. *Public Relat. Rev.* **2010**, *36*, 90–92. [CrossRef]
38. Sargeant, A.; Ford, J.B.; West, D.C. Perceptual determinants of nonprofit giving behavior. *J. Bus. Res.* **2006**, *59*, 155–165. [CrossRef]
39. Lee, Y.K.; Chang, C.-T. Intrinsic or Extrinsic? Determinants Affecting Donation Behaviors. *Int. J. Educ. Adv.* **2008**, *8*, 13–24. [CrossRef]
40. Bagheri, A.; Chitsazan, H.; Ebrahimi, A. Crowdfunding motivations: A focus on donors' perspectives. *Technol. Forecast. Soc. Chang.* **2019**, *146*, 218–232. [CrossRef]

41. Mittelman, R.; Rojas-Méndez, J. Why Canadians give to charity: An extended theory of planned behaviour model. *Int. Rev. Public Nonprofit Mark.* **2018**, *15*, 189–204. [CrossRef]
42. Banks, N.; Hulme, D.; Edwards, M. NGOs, States, and Donors Revisited: Still Too Close for Comfort? *World Dev.* **2015**, *66*, 707–718. [CrossRef]
43. Kinsbergen, S.; Tolsma, J. Explaining monetary donations to international development organisations: A factorial survey approach. *Soc. Sci. Res.* **2013**, *42*, 1571–1586. [CrossRef]
44. Van Iwaarden, J.; Van Der Wiele, T.; Williams, R.; Moxham, C. Charities: How important is performance to donors? *Int. J. Qual. Reliab. Manag.* **2009**, *26*, 5–22. [CrossRef]
45. Bandyopadhyay, S. Consumers' online privacy concerns: Causes and effects. *Innov. Mark.* **2012**, *8*, 32–39.
46. Hallikainen, P. Why people use social media platforms: Exploring the motivations and consequences of use. In *From Information to Smart Society*; Springer: Berlin/Heidelberg, Germany, 2015; pp. 9–17.
47. Castillo, M.; Petrie, R.; Wardell, C. Fundraising through online social networks: A field experiment on peer-to-peer solicitation. *J. Public Econ.* **2014**, *114*, 29–35. [CrossRef]
48. Xiang, Y.T.; Li, W.; Zhang, Q.; Jin, Y.; Rao, W.W.; Zeng, L.N.; Lok, G.K.; Chow, I.H.; Cheung, T.; Hall, B.J. Timely research papers about COVID-19 in China. *Lancet* **2020**, *395*, 684–685. [CrossRef]
49. Stephenson, M.T.; Morgan, S.; Roberts-Perez, S.D.; Harrison, T.; Afifi, W.; Long, S.D. The Role of Religiosity, Religious Norms, Subjective Norms, and Bodily Integrity in Signing an Organ Donor Card. *Health Commun.* **2008**, *23*, 436–447. [CrossRef]
50. De Abreu, M.E.; Laureano, R.M.S.; da Silva, R.V.; Dionísio, P. Volunteerism, compassion and religiosity as drivers of donations practices. *Int. J. Nonprofit Volunt. Sect. Mark.* **2015**, *20*, 256–276. [CrossRef]
51. Ranganathan, S.K.; Henley, W.H. Determinants of charitable donation intentions: A structural equation model. *Int. J. Nonprofit Volunt. Sect. Mark.* **2008**, *13*, 1–11. [CrossRef]
52. Bin-Nashwan, S.A.; Abdul-Jabbar, H.; Aziz, S.A.; Viswanathan, K.K. A socio-economic model of *Zakah* compliance. *Int. J. Sociol. Soc. Policy* **2020**, *40*, 304–320. [CrossRef]
53. Bin-Nashwan, S.A.; Abdul-Jabbar, H.; Aziz, S.A. Does trust in zakat institution enhance entrepreneurs' zakat compliance? *J. Islam. Account. Bus. Res.* **2021**, *12*, 768–790. [CrossRef]
54. Reitsma, J.; Scheepers, P.; Grotenhuis, M.T. Dimensions of individual religiosity and charity: Cross-national effect differences in European countries? *Rev. Relig. Res.* **2006**, *47*, 347–362.
55. McIntyre, E.; Saliba, A.J.; Wiener, K.K.K.; Bishop, F.L. Predicting the intention to use herbal medicines for anxiety symptoms: A model of health behaviour. *J. Ment. Health* **2019**, *28*, 589–596. [CrossRef]
56. Chen, Y.; Dai, R.; Yao, J.; Li, Y. Donate Time or Money? The Determinants of Donation Intention in Online Crowdfunding. *Sustainability* **2019**, *11*, 4269. [CrossRef]
57. Ajzen, I. The Theory of Planned Behavior. *Organ. Behav. Hum. Decis. Process.* **1991**, *50*, 179–211. [CrossRef]
58. Oosterhof, L.; Heuvelman, A.; Peters, O. Donation to disaster relief campaigns: Underlying social cognitive factors exposed. *Evaluation Program Plan.* **2009**, *32*, 148–157. [CrossRef]
59. Ng, J.C.; Lai, K.K.; Chen, Y. Individual's Charity Intention in Hong Kong: An Empirical Study on Charitable Behavior. In Proceedings of the Fourth International Joint Conference on Computational Sciences and Optimization, Kunming and Lijiang City, China, 15–19 April 2011; pp. 436–440.
60. Evans, J.R.; Mathur, A. The value of online surveys. *Internet Res.* **2005**, *15*, 195–219. [CrossRef]
61. The World Bank. *World Bank Open Data*; The World Bank: Washington, DC, USA, 2018.
62. Hair, J.F., Jr.; Hult GT, M.; Ringle, C.M.; Sarstedt, M. *A Primer on Partial Least Squares Structural Equation Modeling (PLS-SEM)*, 2nd ed.; Sage Publications: London, UK, 2017.
63. Cain, M.K.; Zhang, Z.; Yuan, K.-H. Univariate and multivariate skewness and kurtosis for measuring nonnormality: Prevalence, influence and estimation. *Behav. Res. Methods* **2017**, *49*, 1716–1735. [CrossRef]
64. Ajzen, I. Constructing a TPB Questionnaire: Conceptual and Methodological Considerations. 2002. Available online: https://people.umass.edu/aizen/pdf/tpb.measurement.pdf (accessed on 24 October 2016).
65. Henseler, J.; Ringle, C.M.; Sarstedt, M. A new criterion for assessing discriminant validity in variance-based structural equation modeling. *J. Acad. Mark. Sci.* **2015**, *43*, 115–135. [CrossRef]
66. Rungtusanatham, M.; Miller, J.; Boyer, K.K. Theorizing, testing, and concluding for mediation in SCM research: Tutorial and procedural recommendations. *J. Oper. Manag.* **2014**, *32*, 99–113. [CrossRef]
67. Gatto, A.; Drago, C.; Ruggeri, M. On the Frontline—A bibliometric Study on Sustainability, Development, Coronaviruses, and COVID-19. *Environ. Sci. Pollut. Res.* **2022**, 1–17. [CrossRef]
68. Bin-Nashwan, S.A. Toward diffusion of e-Zakat initiatives amid the COVID-19 crisis and beyond. *Foresight* **2022**, *24*, 141–158. [CrossRef]
69. Begum, H.; Abbas, K.; Alam, A.F.; Song, H.; Chowdhury, M.T.; Ghani, A.B.A. Impact of the COVID-19 pandemic on the environment and socioeconomic viability: A sustainable production chain alternative. *Foresight* **2022**, *24*, 456–475. [CrossRef]
70. Begum, H.; Alam, A.S.A.F.; Filho, W.L.; Awang, A.H.; Ghani, A.B.A. The COVID-19 Pandemic: Are There Any Impacts on Sustainability? *Sustainability* **2021**, *13*, 11956. [CrossRef]

71. Abbass, K.; Begum, H.; Alam, A.S.A.F.; Awang, A.H.; Abdelsalam, M.K.; Egdair, I.M.M.; Wahid, R. Fresh Insight through a Keynesian Theory Approach to Investigate the Economic Impact of the COVID-19 Pandemic in Pakistan. *Sustainability* **2022**, *14*, 1054. [CrossRef]
72. Ahn, J.C.; Sura, S.; An, J.-C. Intention to donate via social network sites (SNSs): A comparison study between Malaysian and South Korean users. *Inf. Technol. People* **2018**, *31*, 910–926. [CrossRef]
73. Hofstede Insights. *Country Comparison: Kuwait*; Hofstede Insights: Helsinki, Finland, 2017.
74. Teah, M.; Lwin, M.; Cheah, I. Moderating role of religious beliefs on attitudes towards charities and motivation to donate. *Asia Pac. J. Mark. Logist.* **2014**, *26*, 738–760. [CrossRef]
75. Abdul-Jabbar, H.; Bin-Nashwan, S.A. Does deterrence-based enforcement matter in alms tax (Zakat) compliance? *Int. J. Soc. Econ.* **2022**, *49*, 710–725. [CrossRef]

sustainability

MDPI

Article

Environmentally Responsible Business Approaches in Azerbaijan

Mubariz Mammadli

Department of Economics and Business Management, UNEC Business School, Azerbaijan State University of Economics, Baku AZ1001, Azerbaijan; mubariz.mammadli@unec.edu.az

Abstract: In the study; the eco-entrepreneurship and environmental sensitivity of companies operating in Azerbaijan were examined within the scope of environmental practices. For this purpose, companies operating in the country were invited to participate in a survey, and their sensitivity to the environment was examined from four main dimensions—consumption, production, financial, and technology. The mentioned dimensions were analyzed according to four aspects; namely, business field of activity, business size, operating year, and the level of implementation of environmentally friendly policies. According to the results of a one-way analysis of variance, carried out by field of activity, operating year, and the level of implementation of eco-friendly policy; it was found that there was no group that significantly differed from the overall group mean. A difference was found only in terms of the size of the enterprise. The study found that large businesses were more aware of environmental sensitivity than small- and medium-sized businesses.

Keywords: green business; consumption dimension; production dimension; financial dimension; technological dimension

Citation: Mammadli, M. Environmentally Responsible Business Approaches in Azerbaijan. *Sustainability* **2022**, *14*, 6227. https://doi.org/10.3390/su14106227

Academic Editors: Elkhan Richard Sadik-Zada and Andrea Gatto

Received: 7 March 2022
Accepted: 16 May 2022
Published: 20 May 2022

1. Introduction

Today, one of the most fundamental problems of the world we live in is environmental pollution. Namely, different environmental problems such as acid rain, contamination of agricultural foods, GMO products, pollution of drinking water, and global warming problems endanger the lives of both people and all living things. Since the environment is the interaction of all living organisms with air, water, and soil, problems in the environment can naturally create different risks in the lives of living organisms [1].

Generally, different factors can be the source of environmental problems. Today, one of these main factors is business activity. In fact, when business activities and their results are taken into account, two main market participants should be considered, namely the consumer and the producer, because one of these two actors' decisions influence and shapes the activity of the other. In other words, both consumers and producers have a greater responsibility to reduce or prevent activities that exacerbate environmental problems [2–5]. Today's consumers may be more sensitive to environmental issues and may require businesses to recognize their responsibility and implement practices that will better protect nature [6–8].

However, this article will only discuss producers and their environmental behavior. Companies' activities affect not only their owners, managers, and shareholders but also various communities and other entities with which they have direct and indirect contact [9]. Increasing concern about environmental degradation and its impact on the environment has encouraged many businesses to adopt sustainable business models and environmental standards.

Many companies today are trying to solve environmental problems. According to Antolin-Lopez et al. [10], this is about entrepreneurs' emotional commitment to environmental issues and their commitment to being involved in sustainable business practices.

Thus, in the academic literature, there are approaches in which environmental problems are often associated with entrepreneurial practices [11–13]. The main reason for this is that enterprises continue their production using various resources obtained from the environment. Enterprises that attach great importance to environmental responsible activities, that respect nature and that are sensitive to society have a good image in society [14]. Therefore, enterprises must strive to meet the expectations and needs of society, while profiting from the goods or services they put on the market.

Thus, companies should improve resource efficiency and reduce the environmental impact of waste by focusing on cleaner production. To achieve this goal, companies need to identify, evaluate, and manage the waste stream at the stage of process design and production planning before starting their activities [15]. In other words, companies must first prevent environmental pollution through activities that do not harm the environment. For such initiatives, companies should use appropriate technologies [16]. Entrepreneurship can contribute to the implementation of these guiding principles in a variety of ways. Eco-entrepreneurs can change the environment in a sustainable way by effectively using innovative skills [17]. Innovation, an important driver of entrepreneurship [18], can be used to find ways to increase recycling, conserve resources, and minimize waste [19]. Thus, the obligations of companies to protect the environment have a significant impact on the management of companies; the use of technology, financial, and consumer decisions; and other aspects of business [20,21].

In the article, the green management approach is studied using the example of Azerbaijani producers, taking into account consumption, production, financial, and technological aspects. The survey assessed the prospects of companies with different characteristics in relation to green business.

2. Green Business

Defining green entrepreneurship is a difficult task. The concept itself is relatively new and has attracted increasing attention since the 1990s [22,23]. Even today, this topic is still being studied and divided into different branches [24,25]. The reason for this division is related to the fact that the subject includes urban studies, political economy, sociology, business ethics, and other different subjects [26,27].

Essentially, entrepreneurs are trying to identify problems in the marketplace, find visible or invisible needs, and satisfy them through their own efforts. That is why they have been classified as prime movers of innovation [18,28]. In other words, entrepreneurs must be aware of gaps in areas that other market participants need and must engage in activities to fill them.

Entrepreneurs are the driving force behind the next industrial revolution [29]. Therefore, their impact on the environment must always be monitored. There are many scientific studies in the literature on the environmental aspect of business [30–41].

Green businesses take a more holistic approach to environmental aspects. In general, "green" business approaches differ from those of traditional entrepreneurs [42]. The main goals of environmental entrepreneurs are the planet and profit, and they focus on the future, while traditional entrepreneurs, on the contrary, are profit-oriented and look to the present [43]. It should be emphasized that, contrary to traditional opinion, reducing environmental problems does not decrease the economic benefits of the company. On the contrary, such behaviors increase enterprise profits [14,44].

Green business covers the entire process of enterprise. Various concepts, such as green marketing and green management, fall under the scope of green business and cover all business activities. Thus, green business, also called eco-business, considers the ecological environment as an important element in decision-making processes. This type of business is aimed at minimizing or completely eliminating environmental damage in its activities [45,46]. The activities carried out in this context change the various processes, from product design to packaging, and seek to instill the philosophy of environmental protection into the culture of enterprises.

The concept of eco-entrepreneurship, used in parallel with green business [47], was considered by Shaper [48] as a new type of entrepreneurship. The concept of eco-entrepreneu rship embraces eco-entrepreneurship and reflects the market success of businesses with eco-innovation focused on the talents and skills of entrepreneurs [49]. Simply put, eco-entrepreneurship includes activities that will provide a positive contribution to society while realizing the entrepreneurial goals of enterprises and minimizing the negative impact on people and the environment [48]. In other words, eco-entrepreneurship includes environment-oriented practices that include all stakeholders (customers, suppliers, business partners, employees, etc.) with whom businesses interact.

In his study "Ecopreneurship: Rationale, Current Issues and Future Challenges", Voleri [50] divided business into 2 groups. Firstly, these are "environmentally-conscious entrepreneurs." Eco-entrepreneurs included in this category are generally aware of environmental issues but do not operate in the environmental market. These entrepreneurs are taking advantage of business-oriented opportunities while taking into account the environmental aspect of their activities. In this context, they are trying to produce better goods and services and use fewer resources [51]. At the same time, they are trying to achieve eco-efficiency with less environmental impact. We see entrepreneurs included in this category in all sectors. The type of entrepreneurship known as "green entrepreneurship" falls into the second category. This group becomes aware of environmental issues. Their business ventures also take place in the environmental market. Green entrepreneurs are actively looking for eco-centric opportunities that offer good profit prospects.

In addition, green entrepreneurs refrain from activities that cause environmental problems and increase the social responsibility of the company through these activities. The main reason why we touched on the topic of social responsibility here is that the emphasis on social responsibility is a prerequisite for green behavior [52,53]. As is known, the products that enterprises put on the market, the resources and methods used in the production of these products [54,55], as well as the social responsibilities of the company, affect the behavior of consumers in the market [56–59]. Therefore, consumer behavior affects the determination of the market value of the company, its image, and its continuity. Society expects social benefits from business. As long as this benefit is produced, companies will continue their business in the long run and make a profit [60]. On the other hand, there is a growing understanding that environmental activities reduce corporate risks, such as waste management fees and fines for accidents [61]. Moreover, a good public reputation will help companies attract more customers and better employees. The business of companies that produce the values expected by society has continuity. That is why the number of studies in the field of sustainable entrepreneurship has increased today [62].

Companies striving for sustainability must also transform in accordance with the changing values and expectations of society, that is, keep pace with changing socio-cultural norms [63]. Innovation policies and various practices lead to an increase in shareholder profits [64,65]. Using their resources more efficiently, enterprises can take on a new role [66] and develop social and environmental policies due to their existing capabilities. At the same time, they can find innovative solutions to problems and increase their own profits by turning crises into opportunities. Finally, prevention is easier than cure. Therefore, it would be more beneficial for the company to address social and environmental issues before they arise or escalate.

Thus, there are different factors that push companies to implement environmentally friendly policies and turn to eco-entrepreneurship. We can generalize these factors as individual factors [8,67] (customers, employees, shareholders, etc.), institutional factors [67] (business environment, media, etc.), and legal factors [68] (laws, international agreements, decisions of local governments, etc.). In fact, these factors are not independent of each other. For example, social media seriously affect consumers, and they focus on the green activities of companies on social media. Studies in this area have found that the use of social media encourages pro-environmental action among consumers [69,70]. In this case, the media influences the environmental activities of the company through consumers.

Legal procedures are one of the main factors affecting the environmentally friendly activities of companies. Sometimes entrepreneurs use environmentally friendly methods because they are mandatory, i.e., required by the laws or regulatory systems in force in the country [68,71]. Environmental entrepreneurs want the majority of the population to support their business vision while enforcing the requirements of laws, regulatory systems, or financial pressure [72]. Today, the increase in information about the environment and continuing education has made people more conscious about this issue. People's lifestyles began to change positively towards environmentally friendly products [73,74]. As the demand for organic products increased, many new eco-entrepreneurs turned to this field [75].

Moreover, environmental awareness has affected people's consumption habits throughout the world, where everyone is a consumer. For example, environmentally conscious people avoid the misuse of electricity and oil and conserve resources by using products that consume less electricity [76]. That is, some consumers have an attitude towards the environment that we can accept as green purchasing behavior [77,78]. In other words, they strive for sustainable products and services that cause the least damage to the environment. All these have increased the importance of the environmental policy of companies in entrepreneurship and made it necessary for businesses to make radical changes in their own fields [79].

In addition, eco-entrepreneurship should use innovative technologies in its business models while taking into account consumer behavior [8]. Because another important factor that makes the emergence of green businesses important is related to the effective use of natural resources. In other words, the scarcity of natural resources makes efficiency important in the use of resources [80]. Efficiency makes the use of new technologies important. So, the eco-entrepreneur should realize fair development with the right management of natural resources in the business world [81] and create economic and social values by applying technological innovations from an environmental perspective [13]. Such behaviors will increase the performance of companies.

An empirical study by Zhang and Berhe shows the impact of green marketing and green investment on business performance. The researchers examined the impact of green marketing and green investment on the business performance of Ethiopian Chinese textile companies. Based on the responses of 237 participants, the researchers found that green marketing and green investment positively affect the firm's business performance [82]. For this reason, the environmental practices of companies provide them with many advantages. In other words, by providing access to specific markets, using environmentally friendly technologies such as pollution control, waste treatment, and reducing energy costs, companies either increase their revenues or reduce their costs [83] and improve their image.

An empirical study was conducted by Kushwaha and Sharma involving 306 young entrepreneur candidates. In this study, they found that green marketing factors, changing consumer behavior towards green products and favorable market conditions, have a significant and positive effect on ecological entrepreneurship [84]. One of the other important studies on this topic is the study by Silajic et al. for countries with economies in transition. According to this study, entrepreneurs in such countries are reluctant to invest in green activities. The lack of financial support is also one of the most important factors hindering green entrepreneurship [85]. Based on a survey of 12 nonprofit organizations, Gliedt and Parker [86] found that green entrepreneurship in an environmentally conscious country like Canada is driven by two factors. These factors are the loss of external government funding and the consequent market collapse. Moreover, three main factors contribute to the need for green entrepreneurship. These include external social capital flows, internal human capital stocks, and strategic partnerships. Another study by these authors [87] showed that green entrepreneurship accelerates in environmental organizations when companies face the risk of funding cuts.

In particular, SMEs face financial problems. According to the empirical results of a study conducted by Nizayeva and Coskun [88], the size and age of firms have a significant impact on the ability of SMEs to obtain financial resources. Other researchers also highlight the importance of funding green businesses. Noh [89] argues that governments should especially encourage private investment in green businesses.

In 2020, the European Commission and the OECD conducted a survey to examine current and planned green budget practices in member states of the two organizations. The results of the survey provided important information on environmental budgeting practices in member countries. That is, while about 14 out of 39 countries have a green budget practice, 9 of them plan to introduce some environmental practices in the future. The most frequently used tools in evaluating the questionnaire were ex-ante and ex-post environmental policy assessment, green budget labeling, environmental cost-benefit analysis, and carbon assessment. On the other hand, one of the important results of the study was that the main reason for the use of "green" budgets in countries was the fulfillment of international obligations. Many countries seek support from international organizations for various purposes [90].

3. Hypotheses

This section aims to develop a set of hypotheses to explore companies' approaches to eco-entrepreneurship in Azerbaijan from different angles. These parameters are consumption dimension, production dimension, financial dimension, and technological dimension, and they have been investigated depending on the size of the business, the field of activity, the duration of the activity, and whether the enterprise is carrying out environmentally friendly commercial activities. The study has four main hypotheses.

3.1. Responsible Business Varies by Business Size

The first hypothesis was aimed at determining whether a company's environmental performance varies depending on the size of the company. Some literature studies have shown that environmentally friendly activities vary by company size. According to a study by Wagner [91], there is a positive correlation with firm size and the likelihood of being classified as an eco-entrepreneur. So, the larger the firm, the more likely it is to become an eco-entrepreneur.

Thus, the first hypothesis can be derived, stating that:

Hypothesis 1 (H1). *The dimensions of an environmentally responsible business differ in terms of business size.*

Hypothesis 1a (H1a). *The consumption dimension of an environmentally responsible business differs in terms of business size.*

Hypothesis 1b (H1b). *The production dimension of an environmentally responsible business differs in terms of business size.*

Hypothesis 1c (H1c). *The financial dimension of an environmentally responsible business differs in terms of business size.*

Hypothesis 1d (H1d). *The technology dimension of an environmentally responsible business differs in terms of business size.*

3.2. Responsible Business Varies by Business Area

The strategies of successful companies must be compatible with the business environment in order to achieve high performance [92]. In other words, companies should provide eco-activities in accordance with the business environment of the areas they work in. According to Balatbat et al., this is not easy for construction companies because the construction sector is more uncertain and risky [93]. Meanwhile, the activities of the companies in the field they work in affect their incomes. For example, Goodwin and Francis analyzed the tourism industry in the United Kingdom and found that 59% of consumers are willing to pay more for some activities [94]. While 35% of consumers reported that they would support activities aimed at protecting the environment, 29% reported that employees would pay more to improve their working conditions. On the other hand, 21% of consumers reported that they want to support activities related to support. Such results can encourage companies to take more environmentally friendly steps in the field they work in.

Shin and Cho [95] have looked at the green activities of restaurants from different perspectives in their empirical research. Researchers have found that some of the restaurant's activities (information sharing and flexibility in arrangement) significantly improve their environmental performance. Furthermore, in the research conducted by Ceyhan and Ada [96], it was concluded that environmentally friendly business in Turkey's Kahramanmaraş province differed according to the company's field of activity. As a result, the study aimed to examine whether the dimensions of an environmentally friendly business differ according to the field of activity of the business. The hypotheses were developed as follows:

Hypothesis 2 (H2). *The dimensions of an environmentally responsible business differ in terms of the field of activity of the enterprise.*

Hypothesis 2a (H2a). *The consumption dimension of an environmentally responsible business differs in terms of the field of activity of the enterprise.*

Hypothesis 2b (H2b). *The production dimension of an environmentally responsible business differs in terms of the field of activity of the enterprise.*

Hypothesis 2c (H2c). *The financial dimension of an environmentally responsible business differs in terms of the field of activity of the enterprise.*

Hypothesis 2d (H2d). *The technology dimension of an environmentally responsible business differs in terms of the field of activity of the enterprise.*

3.3. Responsible Business Differ in Terms of the Operating Year of the Business

The year of operation of the enterprise is important in many respects. This indicator affects the financial strength of the company, the formation of its image, the level of acceptance of risks, and other factors. For example, according to the empirical results of the research conducted by Nizaeva and Coskun [88], firm age has a significant impact on a firm's capacity to obtain financial resources. On the other hand, brand trust and loyalty, which sometimes takes many years to develop, was described by Lerro et al. as a great influence on the decision of consumers [97].

Considering that such factors are important in its environmentally friendly practice, a third hypothesis can be derived, stating that:

Hypothesis 3 (H3). *The dimensions of an environmentally responsible business differ in terms of the operating year of the business.*

Hypothesis 3a (H3a). *The consumption dimension of an environmentally responsible business differs in terms of the operating year of the business.*

Hypothesis 3b (H3b). *The production dimension of an environmentally responsible business differs in terms of the operating year of the business.*

Hypothesis 3c (H3c). *The financial dimension of an environmentally responsible business differs in terms of the operating year of the business.*

Hypothesis 3d (H3d). *The technology dimension of an environmentally responsible business differs in terms of the operating year of the business.*

3.4. Responsible Business Differ in Terms of Whether the Business Is Operating Environmentally Sensitively

Today, the demand for corporate social responsibility has increased excessively. The world's leading companies state that they will not invest in companies that do not engage in ecological activities or companies that have high climate risks [98]. This means that investors are reviewing and evaluating companies' environmentally friendly activities. That is, if companies want to receive investment, they must have their own business model with environmentally friendly practices [40]. In relation to this, a second hypothesis can be formulated:

Hypothesis 4 (H4). *The dimensions of an environmentally responsible business differs depending on whether the business is operating environmentally sensitive.*

Hypothesis 4a (H4a). *The consumption dimension of an environmentally responsible business differs depending on whether the business is operating environmentally sensitive.*

Hypothesis 4b (H4b). *The production dimension of an environmentally responsible business differs depending on whether the business is operating environmentally sensitive.*

Hypothesis 4c (H4c). *The financial dimension of an environmentally responsible business differ depending on whether the business is operating environmentally sensitive.*

Hypothesis 4d (H4d). *The technology dimension of an environmentally responsible business differs depending on whether the business is operating environmentally sensitive.*

4. Material and Methods

In the study, primary data-collection techniques were used as a scientific research method. The research was sent to 75 companies operating in Azerbaijan. Owners, partners, or managers of businesses filled out the questionnaire prepared by us. Businesses operating in the textile, food, tourism, heavy industry, and other sectors participated in the surveys. The survey was conducted by the author between September and December 2021.

The IBM SPSS Statistics Version 26 (Baku, Azerbaijan) was used in the analysis and evaluation of the data. Before performing the ANOVA analysis, it was determined whether the data showed a normal distribution. One of the general assumptions regarding the normal distribution is that at least 30 ($n \geq 30$) samples are required, to ensure the normal distribution. This acceptance is based on the central limit theorem [99].

It is possible to evaluate whether a quantitative variable has a normal distribution or not according to different criteria. In this study, the normal distribution of the variables according to the skewness and kurtosis coefficients is discussed. Skewness indicates the degree of symmetry in the distribution of a variable. On the other hand, kurtosis is a measure that reflects whether the distribution is too high [100]. The fact that the skewness and kurtosis coefficients are in the range of −1 to +1 indicates that the variable has a normal distribution.

According to the data presented in Table 1, skewness and kurtosis values ranged from −1 to +1 for all variables. Since the data shows a normal distribution, we can do the ANOVA test.

Simple random sampling is used to meet the condition of random sampling. Simple random sampling is a type of sampling in which several different subjects are randomly selected so that each unit has an equal chance of being selected [101]. 57 out of 75 survey participants were randomly selected using a computer.

Table 1. Skewness and kurtosis coefficients of the variables.

Descriptives			
		Statistic	Std. Error
Consumption dimension	Skewness	0.509	0.346
	Kurtosis	−0.768	0.613
Production dimension	Skewness	−0.605	0.386
	Kurtosis	−0.834	0.629
Financial dimension	Skewness	0.703	0.416
	Kurtosis	−0.721	0.643
Technology dimension	Skewness	0.703	0.416
	Kurtosis	−0.721	0.643

4.1. Data Analysis and Findings

In the study, first of all, the reliability of the scales used in the analysis of the variables was measured with the Cronbach's alpha method. According to Cronbach's alpha method, for a scale to be considered reliable, its alpha value must be at least 0.70. Scales with a Cronbach value of less than 0.70 are classified as low-reliability or unreliable scales.

Table 2 shows that the reliability values of each variable are greater than the generally accepted value (0.70).

Table 2. Reliability statistics.

	Cronbach's Alpha	Cronbach's Alpha Based on Standardized Items	N of Items
Consumption_dimension	0.927	0.927	4
Production_dimension	0.706	0.719	4
Financial_dimension	0.754	0.754	4
Technology_dimension	0.748	0.759	4

4.2. Findings Regarding the Demographic Characteristics of the Survey Participants

In this section, the data obtained according to demographic variables such as gender, educational status, position in the enterprise, as well as business field, company scale, operating year, and environmentally friendly business practices of the companies, are presented.

According to the results presented in Table 3, 30.0% of the enterprises continue their activities for 1 to 5 years, 46.0% for 5 to 10 years, and 28% for 10 and more years.

Considering the spheres of activity of the surveyed companies, it can be seen that 25.0% operate in the food industry, 33.0% in the textile industry, 12.0% in heavy industry, 18.0% in the tourism industry, and 12.0% in other industries. Looking at the scale of enterprises, it can be seen that 19.0% of respondents are large companies, that 30.0% are mid-size enterprise, and that 51.0% are small companies.

When asked whether the surveyed enterprises are green entrepreneurs, 12% of the enterprises stated that they are green entrepreneurs, 42% that they are not green entrepreneurs, and 15% that they have partially worked in this field.

Table 3. Demographic characteristics.

	Number	Percent (%)
	Business field of activity	
Food industry	14	25
Textile industry	19	33
Heavy industry	7	12
Tourism	10	18
Other	7	12
	Business size	
Large enterprise	11	19
Mid-size enterprise	17	30
Small enterprise	29	51
	Operating year	
Up to 5 years	17	30
Between 5–10 years	24	42
More than 10 years	16	28
	I'm a green entrepreneur	
Yes	7	12
No	24	42
Partially	26	46
$N = 57$		

5. Results

The results of the ANOVA analysis of the environmentally friendly business dimensions according to different variables are given, and it is attempted to prove the hypotheses of the research.

5.1. Verification of Hypothesis 1

In order to prove the hypothesis, one by one, the approaches of the businesses to eco-entrepreneurship are handled one by one from different dimensions and examined in terms of business size.

Table 4 shows the ANOVA results related to the size of the firm and the consumption dimension.

When the ANOVA table is examined, it is seen that the "Sig." values are less than 0.05 for all components. Based on this result obtained by performing a one-way analysis of variance, the H_{1a} hypothesis is accepted. In other words, with 95% confidence, it has been found that the approaches to the consumption dimension according to the size of the enterprises are different from each other.

The one-way ANOVA test does not show us between which groups the difference is. In other words, it does not indicate how it changes depending on the size of enterprises. So, a Tukey post-hoc test is used.

According to the Tukey post-hoc test, there is a difference between the approaches of participants of large-scale companies and small-scale companies (see Appendix A Table A1). That is, although the participants of large-scale companies had a more positive view of the choices given in the consumption dimension, the participants of small-scale companies did not show the same attitude. This attitude of small businesses has also been observed for medium-sized companies. Namely, the choice of "The wishes of green consumers must be heeded" and "Promotions should be carried out to raise consumer awareness of eco-friendly products" was not considered favorably by medium-sized companies. Thus, H_{1a} was accepted depending on the result obtained.

As it can be seen in Table 5, since the values of "Sig." in the ANOVA analysis were less than 0.05 in all options, it was found that the production dimension also differed in terms of the size of the enterprise.

Table 4. ANOVA results regarding consumption dimension and business size.

ANOVA						
		Sum of Squares	**df**	**Mean Square**	**F**	**Sig.**
The wishes of green consumers must be heeded.	Between Groups	47.633	2	23.816	14.417	0.000
	Within Groups	89.209	54	1.652		
	Total	136.842	56			
Promotions should be carried out to raise consumer awareness of eco-friendly products.	Between Groups	27.466	2	13.733	8.577	0.001
	Within Groups	86.464	54	1.601		
	Total	113.930	56			
Green labeling method must be used.	Between Groups	37.329	2	18.664	10.630	0.000
	Within Groups	94.812	54	1.756		
	Total	132.140	56			
There should be laws regulating environmental awareness.	Between Groups	33.135	2	16.567	11.482	0.000
	Within Groups	77.918	54	1.443		
	Total	111.053	56			

Table 5. ANOVA results regarding production dimension and business size.

ANOVA						
		Sum of Squares	**df**	**Mean Square**	**F**	**Sig.**
Cooperation should be made with international and local organizations on eco-production.	Between Groups	28.594	2	14.297	11.919	0.000
	Within Groups	64.774	54	1.200		
	Total	93.368	56			
In the production process, environmental factors must be taken into account.	Between Groups	32.710	2	16.355	13.741	0.000
	Within Groups	64.273	54	1.190		
	Total	96.982	56			
Renewable energy sources must be used in production.	Between Groups	30.228	2	15.114	7.846	0.001
	Within Groups	104.017	54	1.926		
	Total	134.246	56			
Used products must be recycled.	Between Groups	23.628	2	11.814	6.173	0.004
	Within Groups	103.354	54	1.914		
	Total	126.982	56			

According to the results of the Tukey post-hoc test, participants of small-scale companies generally did not have a positive attitude towards environmentally friendly policies compared to participants of large-scale companies (see Appendix A Table A2). Participants of medium-sized companies, on the other hand, approached positively the "Renewable energy sources must be used in production" and "Used products must be recycled" options, while they did not show the same reaction to other options. Within the scope of these results, H_{1b} was accepted.

The next ANOVA analysis was performed in terms of financial dimension to prove Hypothesis $_{1c}$. The obtained results are given in Table 6.

Table 6. ANOVA results regarding financial dimension and business size.

ANOVA						
		Sum of Squares	df	Mean Square	F	Sig.
The production of eco-friendly products increases the company's revenues in the long run.	Between Groups	39.065	2	19.532	17.634	0.000
	Within Groups	59.813	54	1.108		
	Total	98.877	56			
The production of eco-friendly friendly products increases the chances of finding new investors.	Between Groups	32.601	2	16.301	14.925	0.000
	Within Groups	58.978	54	1.092		
	Total	91.579	56			
Financial support should be provided by the state for the green entrepreneurs.	Between Groups	40.874	2	20.437	21.065	0.000
	Within Groups	52.389	54	0.970		
	Total	93.263	56			
Green entrepreneurs should receive cheaper funding.	Between Groups	10.584	2	5.292	6.657	0.003
	Within Groups	42.925	54	0.795		
	Total	53.509	56			

As seen in Table 6, the "Sig." value took values less than 0.05 in all four cases. This reflects that the size of finance differs in terms of business size.

According to the results of the Tukey post-hoc test used to determine the direction of the difference, the main difference was observed between large enterprises and small enterprises (see Appendix A Table A3). The approach of small businesses in terms of environmental sensitivity has not been positive in general.

Medium-sized businesses, on the other hand, viewed approaches such as "the production of eco-friendly friendly products increases the chances of finding new investors", "the state should provide financial support for green entrepreneurs", and "green entrepreneurs should receive cheaper funding" favorably. H_{1c} was accepted within the scope of these results.

The relationship between technology dimension and business size is shown in Table 7.

Table 7. ANOVA results regarding technology dimension and business size.

ANOVA						
		Sum of Squares	Df	Mean Square	F	Sig.
Technologies that consume less energy should be used in production.	Between Groups	14.228	2	7.114	6.940	0.002
	Within Groups	55.351	54	1.025		
	Total	69.579	56			
Tax incentives should be provided for the purchase of green technologies.	Between Groups	20.167	2	10.084	9.275	0.000
	Within Groups	58.710	54	1.087		
	Total	78.877	56			
Various supports should be provided on e-marketing, e-sales and e-payment for the green entrepreneurs.	Between Groups	15.514	2	7.757	7.778	0.001
	Within Groups	53.854	54	0.997		
	Total	69.368	56			
Technologies should be used that prevent the waste of resources.	Between Groups	32.306	2	16.153	11.814	0.000
	Within Groups	73.834	54	1.367		
	Total	106.140	56			

As can be seen from Table 7, the "Sig." value of all of the options took values less than 0.05. This means that any group separated by business size differs significantly from the overall group average. According to the results of the Tukey post-hoc test, small businesses treated these options more negatively than large and medium businesses (see Appendix A Table A4). Thus, H_{1d} was adopted within these results.

5.2. Verification of Hypothesis 2

The research sought to ascertain whether business approaches had shifted in terms of the business's field of activity. In the second hypothesis, the first ANOVA analysis was conducted in terms of the consumption dimension.

In the ANOVA analysis, it was concluded that the consumption dimension did not differ in terms of the business field, since the "Sig." value was greater than 0.05 (Table 8). Within the scope of this result obtained, H_{2a} was rejected.

Table 8. ANOVA results regarding consumption dimension and business field.

	ANOVA					
		Sum of Squares	df	Mean Square	F	Sig.
The wishes of green consumers must be heeded.	Between Groups	10.186	4	2.547	1.046	0.393
	Within Groups	126.656	52	2.436		
	Total	136.842	56			
Promotions should be carried out to raise consumer awareness of eco-friendly products.	Between Groups	7.069	4	1.767	0.860	0.494
	Within Groups	106.861	52	2.055		
	Total	113.930	56			
Green labeling method must be used.	Between Groups	19.428	4	4.857	2.241	0.077
	Within Groups	112.712	52	2.168		
	Total	132.140	56			
There should be laws regulating environmental awareness.	Between Groups	10.754	4	2.689	1.394	0.249
	Within Groups	100.298	52	1.929		
	Total	111.053	56			

A similar result was observed in the approaches of the groups that differed in terms of business field regarding the production dimension. Namely, in the ANOVA analysis, the "Sig." value was greater than 0.05 (Table 9). These values reflect that the responses of companies to the choices given in terms of business fields do not differ. As a result, H_{2b} was rejected.

Table 9. ANOVA results regarding production dimension and business field.

	ANOVA					
		Sum of Squares	Df	Mean Square	F	Sig.
Cooperation should be made with international and local organizations on eco-production.	Between Groups	2.908	4	0.727	0.418	0.795
	Within Groups	90.461	52	1.740		
	Total	93.368	56			
In the production process, environmental factors must be taken into account.	Between Groups	7.840	4	1.960	1.143	0.347
	Within Groups	89.143	52	1.714		
	Total	96.982	56			
Renewable energy sources must be used in production.	Between Groups	25.926	4	6.482	3.112	0.023
	Within Groups	108.320	52	2.083		
	Total	134.246	56			
Used products must be recycled.	Between Groups	0.849	4	0.212	0.088	0.986
	Within Groups	126.133	52	2.426		
	Total	126.982	56			

The following analysis was performed to test Hypothesis 2c. Table 10 shows the results of the analysis of variance in terms of financial dimension and the business field of companies.

Table 10. ANOVA results regarding financial dimension and business field.

		ANOVA Sum of Squares	Df	Mean Square	F	Sig.
The production of eco-friendly products increases the company's revenues in the long run.	Between Groups	17.307	4	4.327	2.758	0.037
	Within Groups	81.570	52	1.569		
	Total	98.877	56			
The production of eco-friendly friendly products increases the chances of finding new investors.	Between Groups	4.861	4	1.215	0.729	0.576
	Within Groups	86.718	52	1.668		
	Total	91.579	56			
Financial support should be provided by the state for the green entrepreneurs.	Between Groups	12.656	4	3.164	2.041	0.102
	Within Groups	80.607	52	1.550		
	Total	93.263	56			
Green entrepreneurs should receive cheaper funding.	Between Groups	5.691	4	1.423	1.547	0.202
	Within Groups	47.817	52	0.920		
	Total	53.509	56			

The values given in Table 10 reflect that there is no difference in terms of business outcomes across all four approaches. In other words, there was no significant difference between the averages of the answers given by the companies grouped according to the business field. As a result of the ANOVA analysis, H_{2c} is rejected.

The results of the analysis of variance presented in Table 11 show that the attitudes of participants of companies grouped by business field do not differ according to the technological dimension. According to the ANOVA results, H_{2d} was rejected.

Table 11. ANOVA results regarding technology dimension and business field.

		ANOVA Sum of Squares	df	Mean Square	F	Sig.
Technologies that consume less energy should be used in production.	Between Groups	3.203	4	0.801	0.627	0.645
	Within Groups	66.376	52	1.276		
	Total	69.579	56			
Tax incentives should be provided for the purchase of green technologies.	Between Groups	5.188	4	1.297	0.915	0.462
	Within Groups	73.689	52	1.417		
	Total	78.877	56			
Various supports should be provided on e-marketing, e-sales, and e-payment for the green entrepreneurs.	Between Groups	4.953	4	1.238	1.000	0.416
	Within Groups	64.415	52	1.239		
	Total	69.368	56			
Technologies should be used that prevent the waste of resources.	Between Groups	2.456	4	0.614	0.308	0.871
	Within Groups	103.684	52	1.994		
	Total	106.140	56			

5.3. *Verification of Hypothesis 3*

In order to prove Hypothesis 3, different dimensions were handled one by one and analyzed in terms of the operating year of the companies. ANOVA results related to the operating year of the business and consumption dimensions are shown in Table 12.

Table 12. ANOVA results regarding consumption dimension and operating year of the business.

		ANOVA Sum of Squares	Df	Mean Square	F	Sig.
The wishes of green consumers must be heeded.	Between Groups	0.318	2	0.159	0.063	0.939
	Within Groups	136.525	54	2.528		
	Total	136.842	56			
Promotions should be carried out to raise consumer awareness of eco-friendly products.	Between Groups	0.522	2	0.261	0.124	0.883
	Within Groups	113.408	54	2.100		
	Total	113.930	56			
Green labeling method must be used.	Between Groups	6.116	2	3.058	1.310	0.278
	Within Groups	126.025	54	2.334		
	Total	132.140	56			
There should be laws regulating environmental awareness.	Between Groups	2.528	2	1.264	0.629	0.537
	Within Groups	108.525	54	2.010		
	Total	111.053	56			

The companies that participated in the survey were divided into three groups depending on the length of their activities. It can be seen that the average values of the attitudes of the companies in the groups "up to 5 years", "between 5–10 years", and "more than 10 years" in terms of the consumption dimension do not differ from each other. Within the scope of this result obtained, H_{3a} was rejected.

ANOVA results for the production dimension and operating year of the enterprise are given in Table 13. The data in Table 13 shows that there is no difference between the groups divided according to the operating year of the enterprise. Thus, H_{2b} was rejected according to ANOVA results.

Table 13. ANOVA results regarding production dimension and operating year of the business.

		ANOVA Sum of Squares	Df	Mean Square	F	Sig.
Cooperation should be made with international and local organizations on eco-production.	Between Groups	2.424	2	1.212	0.720	0.492
	Within Groups	90.945	54	1.684		
	Total	93.368	56			
In the production process, environmental factors must be taken into account.	Between Groups	0.593	2	0.296	0.166	0.847
	Within Groups	96.390	54	1.785		
	Total	96.982	56			
Renewable energy sources must be used in production.	Between Groups	8.113	2	4.057	1.737	0.186
	Within Groups	126.132	54	2.336		
	Total	134.246	56			
Used products must be recycled.	Between Groups	9.594	2	4.797	2.207	0.120
	Within Groups	117.388	54	2.174		
	Total	126.982	56			

Another ANOVA analysis was performed to prove H_{3c}. The results obtained are given in Table 14.

Table 14. ANOVA results regarding financial dimension and operating year of the business.

		ANOVA Sum of Squares	df	Mean Square	F	Sig.
The production of eco-friendly products increases the company's revenues in the long run.	Between Groups	2.364	2	1.182	0.661	0.520
	Within Groups	96.513	54	1.787		
	Total	98.877	56			
The production of eco-friendly friendly products increases the chances of finding new investors.	Between Groups	3.448	2	1.724	1.056	0.355
	Within Groups	88.131	54	1.632		
	Total	91.579	56			
Financial support should be provided by the state for green entrepreneurs.	Between Groups	7.103	2	3.551	2.226	0.118
	Within Groups	86.161	54	1.596		
	Total	93.263	56			
Green entrepreneurs should receive cheaper funding.	Between Groups	0.709	2	0.354	0.362	0.698
	Within Groups	52.800	54	0.978		
	Total	53.509	56			

In ANOVA analysis, the lowest "Sig." value among the components was "0.118". Since this value is greater than 0.05, it is concluded that the financial dimension does not differ in terms of the operating year of the business. Within the scope of these results, H_{3c} was rejected.

Table 15 presents companies' approaches to technological measurement depending on the year of operation.

Table 15. ANOVA results regarding technology dimension and operating year of the business.

		ANOVA Sum of Squares	Df	Mean Square	F	Sig.
Technologies that consume less energy should be used in production.	Between Groups	7.259	2	3.630	3.145	0.051
	Within Groups	62.320	54	1.154		
	Total	69.579	56			
Tax incentives should be provided for the purchase of green technologies.	Between Groups	5.381	2	2.690	1.977	0.148
	Within Groups	73.496	54	1.361		
	Total	78.877	56			
Various supports should be provided on e-marketing, e-sales, and e-payment for green entrepreneurs.	Between Groups	4.960	2	2.480	2.079	0.135
	Within Groups	64.408	54	1.193		
	Total	69.368	56			
Technologies should be used that prevent the waste of resources.	Between Groups	5.399	2	2.699	1.447	0.244
	Within Groups	100.741	54	1.866		
	Total	106.140	56			

As in other analyses made according to the operating year of the companies, no significant difference among group means was found in the technological dimension either. Therefore, H_{3d} is also rejected.

5.4. Verification of Hypothesis 4

The last thesis put forward to measure the environmental awareness of companies operating in Azerbaijan is Hypothesis 4. Based on this hypothesis, the attitudes of companies towards the environment have been evaluated according to whether they actually implement an environmentally friendly policy.

In Table 16, the companies' approaches to consumption vary depending on their green business practices.

Table 16. ANOVA results regarding consumption dimension and green entrepreneur.

		ANOVA Sum of Squares	Df	Mean Square	F	Sig.
The wishes of green consumers must be heeded.	Between Groups	9.410	2	4.705	1.994	0.146
	Within Groups	127.432	54	2.360		
	Total	136.842	56			
Promotions should be carried out to raise consumer awareness of eco-friendly products.	Between Groups	10.103	2	5.052	2.627	0.081
	Within Groups	103.826	54	1.923		
	Total	113.930	56			
Green labeling method must be used.	Between Groups	21.378	2	10.689	5.211	0.009
	Within Groups	110.762	54	2.051		
	Total	132.140	56			
There should be laws regulating environmental awareness	Between Groups	12.091	2	6.046	3.299	0.044
	Within Groups	98.962	54	1.833		
	Total	111.053	56			

The results of the ANOVA, which are presented in Table 16, show that "Sig." values are higher than 0.05 at all options. So, according to the results of one-way analysis of variance, there is no group that differs significantly from the overall group mean. H_{4a} was rejected.

In the following ANOVA analysis, the production aspects of companies were examined in terms of their level of compliance with environmentally sensitive policies.

Since the data given in Table 17 shows that the "Sig." value is greater than 0.05 over all approaches, it has been concluded that the production size does not differ in terms of whether the companies are green entrepreneurs or not. As a result, hypothesis H_{4b} is rejected.

Table 17. ANOVA results regarding production dimension and green entrepreneur.

		ANOVA Sum of Squares	df	Mean Square	F	Sig.
Cooperation should be made with international and local organizations on eco-production.	Between Groups	12.930	2	6.465	4.340	0.018
	Within Groups	80.438	54	1.490		
	Total	93.368	56			
In the production process, environmental factors must be taken into account.	Between Groups	6.836	2	3.418	2.047	0.139
	Within Groups	90.147	54	1.669		
	Total	96.982	56			
Renewable energy sources must be used in production.	Between Groups	11.594	2	5.797	2.552	0.087
	Within Groups	122.652	54	2.271		
	Total	134.246	56			
Used products must be recycled.	Between Groups	6.638	2	3.319	1.489	0.235
	Within Groups	120.344	54	2.229		
	Total	126.982	56			

The next ANOVA test was performed to prove H4c in terms of the financial dimension. The results are given in Table 18.

Table 18. ANOVA results regarding financial dimension and green entrepreneur.

		ANOVA Sum of Squares	Df	Mean Square	F	Sig.
The production of eco-friendly products increases the company's revenues in the long run.	Between Groups	12.077	2	2.752	1.531	0.225
	Within Groups	86.800	54	1.797		
	Total	98.877	56			
The production of eco-friendly friendly products increases the chances of finding new investors.	Between Groups	11.130	2	2.565	3.736	0.992
	Within Groups	80.449	54	1.490		
	Total	91.579	56			
Financial support should be provided by the state for green entrepreneurs.	Between Groups	11.363	2	5.682	3.746	0.136
	Within Groups	81.900	54	3.511		
	Total	93.263	56			
Green entrepreneurs should receive cheaper funding.	Between Groups	3.294	2	3.224	1.771	0.180
	Within Groups	50.214	54	1.993		
	Total	53.509	56			

As can be seen from Table 18, the "Sig." value exceeded 0.05 in all components. This value, found in the ANOVA analysis, indicates that the financial aspect does not differ depending on the nature of the business following an environmentally responsible policy.

The final ANOVA analysis of the research aimed to determine whether there is a differentiation between the responses given to the technological dimension depending on the green business feature. The connection between these two variables is given in Table 19.

Table 19. ANOVA results regarding technology dimension and green entrepreneur.

		ANOVA Sum of Squares	Df	Mean Square	F	Sig.
Technologies that consume less energy should be used in production.	Between Groups	1.183	2	0.592	0.467	0.629
	Within Groups	68.396	54	1.267		
	Total	69.579	56			
Tax incentives should be provided for the purchase of green technologies.	Between Groups	4.544	2	2.272	1.650	0.201
	Within Groups	74.333	54	1.377		
	Total	78.877	56			
Various supports should be provided on e-marketing, e-sales, and e-payment for green entrepreneurs.	Between Groups	3.732	2	1.866	1.535	0.225
	Within Groups	65.636	54	1.215		
	Total	69.368	56			
Technologies should be used that prevent the waste of resources.	Between Groups	6.554	2	3.277	1.777	0.179
	Within Groups	99.586	54	1.844		
	Total	106.140	56			

The values given in Table 19 reflect that there is no difference between the attitudes of the companies grouped according to the level of implementation of green policies. This result, obtained according to the ANOVA test, rejects H_{4d}.

6. Discussion

As a result of the research, it has been determined that there is no difference between the dimensions of the business according to the fields in which the companies operate. This result obtained from the research differs from the results of the research conducted by Ceyhan and Ada [96]. At the same time, there is no difference between the dimensions of an environmentally responsible business and the period of activity and the level of implementation of environmentally friendly policies.

However, it has been found that large enterprises have more awareness of environmental sensitivity compared to small enterprises. Namely, the relevant hypothesis, which includes only the size of the enterprise, was accepted. This result obtained from the research is similar to the results of the research conducted by Wagner's [91].

That is, the participants of large enterprises argued, in comparison with the participants of other enterprises, that the wishes of green consumers must be heeded; promotions should be carried out to raise consumer awareness of eco-friendly products; green labeling method must be used, and there should be laws regulating environmental awareness. In addition, in the production dimension, cooperation should be made with international and local organizations on eco-production; environmental factors must be taken into account; renewable energy sources must be used, and used products must be recycled.

On the financial dimension, large companies believe that, in the long run, the production of eco-friendly products increases the company's revenues in the long run, that the production of eco-friendly products increases the chances of finding new investors, that green entrepreneurs should receive cheaper funding, and that financial support should be provided by the state for green entrepreneurs. The attitudes of large companies are usually not observed in the approaches of small companies, but they are partially observed in medium-sized companies.

Medium-sized businesses, on the other hand, viewed approaches such as "the production of eco-friendly friendly products increases the chances of finding new investors", "the state should provide financial support for green entrepreneurs", and "green entrepreneurs should receive cheaper funding" favorably.

Large- and medium-sized businesses also welcomed the use of environmentally friendly technologies for the implementation of environmentally sensitive policies. In other words, they argued that technologies that use less energy should be used in production; for tax incentives for the purchase of green technologies; for various supports for e-marketing, e-sales, and e-payment for green entrepreneurs; and that technologies that prevent resource waste should be used.

7. Conclusions

Today, the expansion of environmentally responsible business is one of the most important issues. In this context, the approaches to the basic dimensions of business—consumption, production, finance, and technology—according to some characteristics of companies operating in Azerbaijan were analyzed.

The findings of the research are as follows: (1) the dimensions of an environmentally friendly business differ according to the size of the business, (2) the dimensions of an environmentally friendly business do not differ according to the field of activity of the business, (3) the dimensions of an environmentally friendly business do not differ according to the operating year of the business, and (4) the dimensions of an environmentally responsible business do not differ depending on whether the business is operating environmentally sensitively.

The results of the research show the need for various applications for small- and medium-sized companies in order to increase their environmental responsibility. For example, projects that include various incentives and grants can be developed by the government, NGOs, and other international and local organizations, and the environmental awareness of these companies can be raised.

The results of this article contribute to the development of eco-entrepreneurship in Azerbaijan, providing an empirical basis. Since there is an important gap in the literature on this subject in Azerbaijan, it is thought that this research will be beneficial to the literature. However, we cannot apply the findings to the whole country. With the same analysis, you can get different results because the environmentally friendly practices of companies change over time.

Limitations and future research. There are some limitations regarding the scope of the study. That is, conducting similar studies with a larger sample would give different results. Future research in similar areas may provide useful information on environmentally friendly practices in Azerbaijan by studying the impact on organizational performance.

Funding: This research received no external funding.

Institutional Review Board Statement: The study was conducted in accordance with the Declaration of Helsinki, and approved by the Institutional Review Board (or Ethics Committee) of Economic Think (protocol code 202109 and date of approval 2 September 2021).

Informed Consent Statement: Informed consent was obtained from all subjects involved in the study.

Conflicts of Interest: The authors declare no conflict of interest.

Appendix A

Table A1. Tukey HSD Results Regarding Consumption Dimension and Business Size.

Multiple Comparisons							
Tukey HSD							
Dependent Variable	(I) Business Size	(J) Business Size	Mean Difference (I−J)	Std. Error	Sig.	95% Confidence Interval Lower Bound	Upper Bound
The wishes of green consumers must be heeded.	small enterprise	mid-size enterprise	0.24746	0.39261	0.804	−0.6987	1.1937
		large enterprise	−2.39185 *	0.45514	0.001	−3.4887	−1.2950
	mid-size enterprise	small enterprise	−0.24746	0.39261	0.804	−1.1937	0.6987
		large enterprise	−2.14439 *	0.49735	0.001	−3.3430	−0.9458
	large enterprise	small enterprise	2.39185 *	0.45514	0.001	1.2950	3.4887
		mid-size enterprise	2.14439 *	0.49735	0.001	0.9458	3.3430
Promotions should be carried out to raise consumer awareness of eco-friendly products.	small enterprise	mid-size enterprise	0.01623	0.38652	0.999	−0 9153	0.9477
		large enterprise	−1.76489 *	0.44808	0.001	−2.8448	−0.6850
	mid-size enterprise	small enterprise	−0.01623	0.38652	0.999	−0.9477	0.9153
		large enterprise	−1.74866 *	0.48964	0.002	−2.9287	−0.5686
	large enterprise	small enterprise	1.76489 *	0.44808	0.001	−0.6850	2.8448
		mid-size enterprise	1.74866 *	0.48964	0.002	0.5686	2.9287
Green labeling method must be used.	small enterprise	mid-size enterprise	1.05477	0.40475	0.031	−0.0793	2.0302
		large enterprise	−2.08150 *	0.46921	0.001	−3.2123	−0.9507
	mid-size enterprise	small enterprise	−1.05477	0.40475	0.031	−2.0302	0.0793
		large enterprise	1.02674	0.51273	0.121	−0.2089	2.2624
	large enterprise	small enterprise	2.08150 *	0.46921	0.001	0.9507	3.2123
		mid-size enterprise	−1.02674	0.51273	0.121	−2.2624	0.2089
There should be laws regulating environmental awareness.	small enterprise	mid-size enterprise	0.60446	0.36693	0.235	−0.2798	1.4887
		large enterprise	−2.03762 *	0.42536	0.001	−3.0627	−1.0125
	mid-size enterprise	small enterprise	−0.60446	0.36693	0.235	−1.4887	0.2798
		large enterprise	1.43316	0.46482	0.009	−0.3130	2.5534
	large enterprise	small enterprise	2.03762 *	0.42536	0.001	1.0125	3.0627
		mid-size enterprise	−1.43316	0.46482	0.009	−2.5534	0.3130

* The mean difference is significant at the 0.05 level.

Table A2. Tukey HSD Results Regarding Production Dimension and Business Size.

Multiple Comparisons							
Tukey HSD							
Dependent Variable	(I) Business Size	(J) Business Size	Mean Difference (I−J)	Std. Error	Sig.	95% Confidence Interval Lower Bound	Upper Bound
Cooperation should be made with international and local organizations on eco-production.	small enterprise	mid-size enterprise	0.50304	0.33455	0.297	−0.3032	1.3093
		large enterprise	−1.89342 *	0.38783	0.001	−2.8281	−0.9588
	mid-size enterprise	small enterprise	−0.50304	0.33455	0.297	−1.3093	0.3032
		large enterprise	1.09037	0.42380	0.007	0.3690	2.4117
	large enterprise	small enterprise	1.89342 *	0.38783	0.001	0.9588	2.8281
		mid-size enterprise	−1.09037	0.42380	0.007	−2.4117	−0.3690
In the production process, environmental factors must be taken into account.	small enterprise	mid-size enterprise	−0.28803	0.33325	0.665	−1.0912	0.5151
		large enterprise	−1.78683 *	0.38632	0.001	−2.7179	−0.8558
	mid-size enterprise	small enterprise	0.28803	0.33325	0.665	−0.5151	1.0912
		large enterprise	0.07487	0.42216	0.007	−1.0575	3.0923
	large enterprise	small enterprise	1.78683 *	0.38632	0.001	0.8558	2.7179
		mid-size enterprise	−0.07487	0.42216	0.007	−3.0923	1.0575
Renewable energy sources must be used in production.	small enterprise	mid-size enterprise	−1.86004 *	0.42395	0.003	−2.1617	−0.8817
		large enterprise	−1.90282 *	0.49146	0.001	−3.0872	−0.7184
	mid-size enterprise	small enterprise	1.86004 *	0.42395	0.003	0.8817	2.1617
		large enterprise	1.04278	0.53705	0.137	−0.2515	2.3371
	large enterprise	small enterprise	1.90282 *	0.49146	0.001	0.7184	3.0872
		mid-size enterprise	−1.04278	0.53705	0.137	−2.3371	0.2515
Used products must be recycled.	small enterprise	mid-size enterprise	−1.44828 *	0.42259	0.001	−2.5702	−0.4667
		large enterprise	−1.72100 *	0.48989	0.003	−2.9016	−0.5404
	mid-size enterprise	small enterprise	1.44828 *	0.42259	0.001	0.4667	2.5702
		large enterprise	0.27273	0.53533	0.054	−0.0174	2.5629
	large enterprise	small enterprise	1.72100 *	0.48989	0.003	0.5404	2.9016
		mid-size enterprise	−0.27273	0.53533	0.054	−2.5629	0.0174

* The mean difference is significant at the 0.05 level.

Table A3. Tukey HSD Results Regarding Financial Dimension and Business Size.

Multiple Comparisons							
Tukey HSD							
Dependent Variable	(I) Business Size	(J) Business Size	Mean Difference (I−J)	Std. Error	Sig.	95% Confidence Interval Lower Bound	Upper Bound
The production of eco-friendly products increases the company's revenues in the long run.	small enterprise	mid-size enterprise	0.95538	0.32148	0.010	0.1806	1.7301
		large enterprise	−2.16928 *	0.37268	0.001	−3.0674	−1.2711
	mid-size enterprise	small enterprise	−0.95538	0.32148	0.010	−1.7301	−0.1806
		large enterprise	1.21390	0.40725	0.012	0.2324	2.1954
	large enterprise	small enterprise	2.16928 *	0.37268	0.001	1.2711	3.0674
		mid-size enterprise	−1.21390 *	0.40725	0.012	−2.1954	−0.2324

Table A3. *Cont.*

Multiple Comparisons							
Tukey HSD							
Dependent Variable	(I) Business Size	(J) Business Size	Mean Difference (I−J)	Std. Error	Sig.	95% Confidence Interval Lower Bound	Upper Bound
The production of eco-friendly friendly products increases the chances of finding new investors.	small enterprise	mid-size enterprise	−1.15619 *	0.31923	0.002	−1.9255	−0.3868
		large enterprise	−1.86207 *	0.37007	0.001	−2.7539	−0.9702
	mid-size enterprise	small enterprise	1.15619 *	0.31923	0.002	0.3868	1.9255
		large enterprise	0.70588	0.40439	0.198	−0.2687	1.6805
	large enterprise	small enterprise	1.86207 *	0.37007	0.001	0.9702	2.7539
		mid-size enterprise	−0.70588	0.40439	0.198	−1.6805	0.2687
Financial support should be provided by the state for green entrepreneurs.	small enterprise	mid-size enterprise	−1.19675 *	0.30087	0.001	−1.9218	−0.4717
		large enterprise	−2.13793 *	0.34879	0.001	−2.9785	−1.2974
	mid-size enterprise	small enterprise	1.19675 *	0.30087	0.001	0.4717	1.9218
		large enterprise	0.94118 *	0.38114	0.043	0.0226	1.8597
	large enterprise	small enterprise	2.13793 *	0.34879	0.001	1.2974	2.9785
		mid-size enterprise	−0.94118 *	0.38114	0.043	−1.8597	−0.0226
Green entrepreneurs should receive cheaper funding.	small enterprise	mid-size enterprise	−1.57404 *	0.27234	0.001	−1.0823	−0.2304
		large enterprise	−1.90345 *	0.31571	0.003	−1.8643	−0.3426
	mid-size enterprise	small enterprise	1.57404 *	0.27234	0.001	0.2304	1.0823
		large enterprise	0.52941	0.34500	0.283	0.3020	1.3609
	large enterprise	small enterprise	1.90345 *	0.31571	0.003	0.3426	1.8643
		mid-size enterprise	−0.52941	0.34500	0.283	−1.3609	−0.3020

* The mean difference is significant at the 0.05 level.

Table A4. Tukey HSD Results Regarding Technology Dimension and Business Size.

Multiple Comparisons							
Tukey HSD							
Dependent Variable	(I) Business Size	(J) Business Size	Mean Difference (I−J)	Std. Error	Sig.	95% Confidence Interval Lower Bound	Upper Bound
Technologies that consume less energy should be used in production.	small enterprise	mid-size enterprise	−1.58201 *	0.30314	0.001	−1.0924	−0.5061
		large enterprise	−1.68643 *	0.32631	0.001	−2.4178	−0.7458
	mid-size enterprise	small enterprise	1.58201 *	0 30314	0.001	0.5061	1.0924
		large enterprise	0.5481	0.52846	0.085	1.0977	−0.0972
	large enterprise	small enterprise	1.68643 *	0.32631	0.001	0.7458	2.4178
		mid-size enterprise	−0.05481	0.52846	0.085	0.0972	−1.0977
Tax incentives should be provided for the purchase of green technologies.	small enterprise	mid-size enterprise	−1.02637 *	0.31850	0.001	−1.2084	−0.7927
		large enterprise	−1.37931 *	0.36923	0.001	−1.4895	−0.4691
	mid-size enterprise	small enterprise	1.02637 *	0.31850	0.001	0.7927	1.2084
		large enterprise	0.35294	0.40348	0.658	−0.6194	1.3253
	large enterprise	small enterprise	1.37931 *	0.36923	0.001	0.4691	1.4895
		mid-size enterprise	−0.35294	0.40348	0.658	−1.3253	0.6194

Table A4. *Cont.*

Multiple Comparisons							
Tukey HSD							
Dependent Variable	(I) Business Size	(J) Business Size	Mean Difference (I−J)	Std. Error	Sig.	95% Confidence Interval Lower Bound	Upper Bound
Various supports should be provided on e-marketing, e-sales, and e-payment for green entrepreneurs.	small enterprise	mid-size enterprise	−1.79108 *	0.30505	0.001	−1.0580	−0.2294
		large enterprise	−1.28840 *	0.35363	0.002	−1.4362	−0.6406
	mid-size enterprise	small enterprise	1.79108 *	0.30505	0.001	0.2294	1.0580
		large enterprise	0.49733	0.38643	0.409	−0.4340	1.4286
	large enterprise	small enterprise	1.28840 *	0.35363	0.002	0.6406	1.4362
		mid-size enterprise	−0.49733	0.38643	0.409	−1.4286	0.4340
Technologies should be used that prevent the waste of resources.	small enterprise	mid-size enterprise	−2.01481 *	0.40210	0.001	−1.0953	−0.6595
		large enterprise	−2.01254 *	0.41406	0.001	−1.0147	−0.4045
	mid-size enterprise	small enterprise	2.01481 *	0.40210	0.001	0.6595	1.0953
		large enterprise	1.41850	0.45247	0.007	0.3481	−2.5289
	large enterprise	small enterprise	2.01254 *	0.41406	0.001	0.4045	1.0147
		mid-size enterprise	−1.41850 *	0.45247	0.007	2.5289	−0.3481

* The mean difference is significant at the 0.05 level.

References

1. Lilian, U.O. Are Environmental Rights Human Rights? Issues And Responses. *J. Intern. Hum. Rights Law* **2015**, *1*, 48–64.
2. Zeynalova, Z.; Namazova, N. Revealing Consumer Behavior toward Green Consumption. *Sustainability* **2022**, *14*, 5806. [CrossRef]
3. Dolan, P. The Sustainability of "Sustainable Consumption". *J. Macromark.* **2002**, *22*, 170–181. [CrossRef]
4. Webb, D.J.; Mohr, L.A.; Harris, K.E. A re-examination of socially responsible consumption and its measurement. *J. Bus. Res.* **2008**, *61*, 91–98. [CrossRef]
5. Wise, G.F. Conservation-Based Green Marketing, 2000. Available online: http://yosemite.epa.gov (accessed on 11 December 2021).
6. Sharma, R.; Jha, M. Values influencing sustainable consumption behaviour: Exploring the contextual relationship. *J. Bus. Res.* **2017**, *76*, 77–88. [CrossRef]
7. Sustainable Development Commission. I Will If You Will: Towards Sustainable Consumption, 05/2006. Available online: https://www.sd-commission.org.uk/data/files/publications/I_Will_If_You_Will.pdf (accessed on 2 January 2022).
8. Picazao-Tadeo, A.J.; Beltrán-Esteve, M.; Gómez-Limón, J.A. Assessing eco-efficiency with directional distance functions. *Eur. J. Oper. Res.* **2012**, *220*, 798–809. [CrossRef]
9. Suska, M. Environmental Corporate Social Responsibility (ECSR) on the Example of Polish Champion Oil, Gas and Mining Companies. *Sustainability* **2021**, *13*, 6179. [CrossRef]
10. Antolin-Lopez, R.; Martínez-del-Rio, J.; Céspedes-Lorente, J.J. Environmental entrepreneurship: A review of the current conversation after two decades of research. In Proceedings of the 2014 GRONEN Conference, Helsinki, Finland, 16–18 June 2014.
11. Dobers, P.; Rolf, W. Competing with 'soft' issues—From managing the environment to sustainable business strategies. *Bus. Strategy Environ.* **2000**, *9*, 143–150. [CrossRef]
12. De Clercq, D.; Voronov, M. Sustainability in entrepreneurship: A tale of two logics. *Int. Small Bus. J.* **2011**, *29*, 322–344. [CrossRef]
13. Wagner, M.; Maximilians, J. Ventures for the public good and entrepreneurial intentions: An empirical analysis of sustainability orientation as a determining factor. *J. Small Bus. Entrep.* **2012**, *25*, 519–531. [CrossRef]
14. Cetindamar, D. Corporate Social Responsibility Practices and Environmentally Responsible Behavior: The Case of the United Nations Global Compact. *J. Bus. Ethics* **2007**, *76*, 163–176. Available online: https://link.springer.com/article/10.1007/s10551-006-9265-4 (accessed on 17 November 2021). [CrossRef]
15. Melynk, S.A.; Calantone, R.; Sroufe, C.; Montabon, F.L. Environmentally Conscious Manufacturing: Integrating Environmental Issues into Product Design, Planning and Manufacturing. In Proceedings of the NSF Design and Manufacturing Grantees Conference, Vancouver, BC, Canada, 3–6 January 2000; Volume 13.
16. Matuszewska, A.; Hańderek, A.; Paczuski, M.; Biernat, K. Hydrocarbon Fractions from Thermolysis of Waste Plastics as Components of Engine Fuels. *Energies* **2021**, *14*, 7245. [CrossRef]
17. Schlange, L.E. What drives sustainable entrepreneurs. In Proceedings of the 3rd Applied Business and Entrepreneurship Association International Conference, Kona, HI, USA, 16–20 November 2006.
18. Harbison, F. Entrepreneurial organization as a factor in economic development. *Q. J. Econ.* **1956**, *70*, 364–379. [CrossRef]

19. Rennings, K. Redefining Innovation—Eco-Innovation Research and the Contribution from Ecological Economics. *Ecol. Econ.* **2000**, *32*, 319–332. [CrossRef]
20. Nair, S.R.; Ndubisi, N.O. Evaluating management's environmental commitment and link with frm's environmental orientation. *J. Manag. Res.* **2015**, *15*, 165–178.
21. Uddin, S.F.; Khan, M.N. Exploring green purchasing behaviour of young urban consumers: Empirical evidences from India. *S. Asian J. Glob. Bus. Res.* **2016**, *5*, 85–103. [CrossRef]
22. Fischetti, M. Green entrepreneurs. *Technol. Rev.* **1992**, *95*, 38–45.
23. OECD. Measuring Green Entrepreneurship. In *Entrepreneurship at a Glance 2011*; OECD Publishing: Paris, France, 2011.
24. Czarniewski, S. Small and medium-sized enterprises in the context of innovation and entrepreneurship in the economy. *Pol. J. Manag. Stud.* **2016**, *13*, 30–39. [CrossRef]
25. Dudun, M.; Ivashchenko, N.; Gurinovich, A.; Tolmachev, O.; Sonina, L. Environmental entrepreneurship: Characteristics of organization and development. *Entrep. Sustain. Issues* **2019**, *6*, 1861–1870. [CrossRef]
26. Galkina, T.; Hultman, M. Ecopreneurship: Assessing the feld and outlining research potential. *Small Enterp. Res.* **2016**, *23*, 58–72. [CrossRef]
27. Zemtsov, S.P. Institutions, entrepreneurship, and regional development in Russia. *Zhurnal Novaya Ekon. Assotsiatsiya-J. New Econ. Assoc.* **2020**, *2*, 168–180. [CrossRef]
28. Emami, A.; Dimov, D. Degree of Innovation and the Entrepreneurs' Intention to Create Value: A Comparative Study of Experienced and Novice Entrepreneurs. *Eurasian Econ. Rev.* **2017**, *7*, 161–182. [CrossRef]
29. Pacheco, D.F.; Dean, T.J.; Payne, D.S. Escaping the green prison: Entrepreneurship and the creation of opportunities for sustainable development. *J. Bus. Ventur.* **2010**, *25*, 464–480. [CrossRef]
30. Schaltegger, S. A framework for ecopreneurship: Leading bioneers and environmental managers to ecopreneurship. *Greener Manag. Int.* **2002**, *38*, 45–58. [CrossRef]
31. Dean, T.J.; McMullen, J.S. Toward a theory of sustainable entrepreneurship: Reducing environmental degradation through entrepreneurial action. *J. Bus. Ventur.* **2007**, *22*, 50–76. [CrossRef]
32. Anderson, A.R. Cultivating the garden of Eden: Environmental entrepreneuring. *J. Organ. Change Manag.* **1998**, *11*, 135–144. [CrossRef]
33. Gibbs, D. Sustainability Entrepreneurs, Ecopreneurs and the Development of a Sustainable Economy. *Greener Manag. Int.* **2009**, *55*, 63–78. [CrossRef]
34. Menon, A.; Menon, A. Enviropreneurial marketing strategy: The emergence of corporate environmentalism as marketing strategy. *J. Mark.* **1997**, *61*, 51–67. [CrossRef]
35. Berle, G. *The Green Entrepreneur: Business Opportunities That Can Save the Earth and Make You Money*; Liberty Hall Press: Blue Ridge Summit, PA, USA, 1991.
36. Ndubisi, N.O.; Nair, S.R. Green Entrepreneurship (GE) and Green Value Added (GVA): A conceptual framework. *Int. J. Entrep.* **2009**, *13*, 21–34.
37. Isaak, R. Globalisation and green entrepreneurship. *Greener Manag. Int.* **1997**, *18*, 80–91.
38. O'Neill, K.; Gibbs, D. Rethinking Green Entrepreneurship—Fluid Narratives of the Green Economy. *Environ. Plan. A* **2016**, *48*, 1727–1749. [CrossRef]
39. Dees, J.G.; Khanin, D.; Kruger, N.; Murphy, P.J.; Santos, F.; Scarlata, M.; Walske, J.; Zacharakis, A. Social entrepreneurship and broader theories: Shedding new light on the 'Bigger Picture'. *J. Soc. Entrep.* **2013**, *4*, 88–107.
40. O'Donnell, G. Climate Tech: How ESG Investing is Evolving into 'a Profitable Endeavor'. Available online: https://news.yahoo.com/climate-tech-esg-investing-profitable-endeavor-191604271.html (accessed on 22 October 2021).
41. Anderson, T.L.; Leal, D.R. *Enviro-Capitalists: Doing Good While Doing Well*; Rowman and Littlefield: Lanham, MD, USA, 2008.
42. Tilley, F.; Parrish, B. From poles to wholes: Facilitating an integrated approach to sustainable entrepreneurship. *World Rev. Entrep. Manag. Sustain. Dev.* **2006**, *2*, 281–294. [CrossRef]
43. Thompson, N.; Kiefer, K.; York, J. Distinctions not Dichotomies: Exploring Social, Sustainable, and Environmental Entrepreneurship. In *Social and Sustainable Entrepreneurship (Advances in Entrepreneurship, Firm Emergence and Growth)*; Lumpkin, G.T., Katz, J.A., Eds.; Emerald Group Publishing Limited: Bingley, UK, 2011; Volume 13, pp. 201–220.
44. Jayal, A.D.; Badurdeen, F.; Dillon, O.W., Jr.; Jawahir, I.S. Sustainable manufacturing: Modeling and optimization challenges at the product, process and system levels. *CIRP J. Manuf. Sci. Technol.* **2010**, *2*, 144–152. [CrossRef]
45. Allam, H. Green Entrepreneurship Success Stories in the Arab Countries, Center for Environment & Development in the Arab Region & Europe (CEDARE). Available online: https://www.ilo.org/wcmsp5/groups/public/---dgreports/---exrel/documents/genericdocument/wcms_235612.pdf (accessed on 2 January 2022).
46. Hennemann, J.N.; Draser, B.; Stofkova, K.R. The Green Business and Sustainable Development School—A Case Study for an Innovative Educational Concept to Prevent Big Ideas from Failure. *Sustainability* **2021**, *13*, 1943. [CrossRef]
47. Dixon, S.E.; Clifford, A. Ecopreneurship—A new approach to managing the triple bottom line. *J. Organ. Change Manag.* **2007**, *20*, 326–345. [CrossRef]
48. Schaper, M. The Essence of Ecopreneurship. *Greener Manag. Int.* **2002**, *38*, 26–30. [CrossRef]
49. Schaltegger, S. Research and Markets. In *Making Ecopreneurs: Developing Sustainable Entrepreneurship*; Schaper, M., Ed.; Gower Publishing: Surrey, UK, 2010.

50. Volery, T. Ecopreneurship: Rationale, current issues and future challenges. In Proceedings of the Conference Papers of Swiss Research Institute of Small Business and Entrepreneurship, St. Gallen, Switzerland; 2002; pp. 541–553. Available online: https://www.kmu-hsg.ch/rencontres/IGW_Rencontres/BAND2002_WEB/F_11_Volery.pdf (accessed on 11 December 2021).
51. Mammadli, M.; Sadik-Zada, E.R.; Gatto, A.; Huseynova, R. What Drives Public Debt Growth? A Focus on Natural Resources, Sustainability and Development. *Int. J. Energy Econ. Policy* **2021**, *11*, 614–621. [CrossRef]
52. Basdekidou, V.A. Green Entrepreneurship & Corporate Social Responsibility: Comparative and Correlative Performance Analysis. *Int. J. Econ. Financ.* **2017**, *9*, 1–12.
53. Jones, M.T. Social Responsibility and the "Green" Business Firm. *Ind. Environ. Crisis Q.* **1996**, *9*, 327–345. [CrossRef]
54. Mammadli, M.M. Prospects for the export of agricultural products in Azerbaijan. In Proceedings of the 55th International Scientific Conference on Economic and Social Development, Baku, Azerbaijan, 17–18 June 2020; Volume 4, pp. 496–505.
55. Sadik-Zada, E.R.; Gatto, A.; Blick, N. Rural Electrification and Transition to Clean Cooking: The Case Study of Kanyegaramire and Kyamugarura Solar Mini-Grid Energy Cooperatives in the Kyenjojo District of Uganda. In *Sustainable Policies and Practices in Energy, Environment and Health Research*; Springer: Berlin/Heidelberg, Germany, 2021.
56. Aragón-Correa, J.A.; Sharma, S. A contingent resource-based view of proactive corporate environmental strategy. *Acad. Manag. Rev.* **2003**, *28*, 71–88. [CrossRef]
57. Hart, S.L. A natural-resource-based view of the firm. *Acad. Manag. Rev.* **1995**, *20*, 986–1014. [CrossRef]
58. Lu, I.Y.; Kuo, T.; Lin, T.S.; Tzeng, G.H.; Huang, S.L. Multicriteria decision analysis to develop effective sustainable development strategies for enhancing competitive advantages: Case of the TFT-LCD industry in Taiwan. *Sustainability* **2016**, *18*, 646. [CrossRef]
59. Zeynalova, Z.; Mammadli, M. Analysis of the economic factors affecting household consumption expenditures in Azerbaijan. *J. Crit. Rev.* **2020**, *7*, 241–248. [CrossRef]
60. Davis, K. The Case for and Against Business Assumption of Social Responsibilities. *Acad. Manag. J.* **1973**, *16*, 312–322. Available online: https://57ef850e78feaed47e42-3eada556f2c82b951c467be415f62411.r9.cf2.rackcdn.com/Davis-1973-ForAnd%20Against.pdf (accessed on 5 November 2021).
61. Park, S.R.; Kim, S.T.; Lee, H.-H. Green Supply Chain Management Efforts of First-Tier Suppliers on Economic and Business Performances in the Electronics Industry. *Sustainability* **2022**, *14*, 1836. [CrossRef]
62. Terán-Yépez, E.; Marín-Carrillo, G.; Casado-Belmonte, M.; Capobianco-Uriarte, M. Sustainable entrepreneurship: Review of its evolution and new trends. *J. Clean Prod.* **2020**, *252*, 14–27. [CrossRef]
63. Emami, A.; Khajeheian, D. Social Norms and Entrepreneurial Action: The Mediating Role of Opportunity Confidence. *Sustainability* **2019**, *11*, 158. [CrossRef]
64. Delai, I.; Takahashi, S. Sustainability measurement system: A reference model proposal. *Soc. Responsib. J.* **2011**, *7*, 438–471. [CrossRef]
65. Khan, E.; Dewan, M.; Chowdhury, M. Reflective or formative measurement model of sustainability factor? A three industry comparison. *Corp. Ownersh. Control. J.* **2016**, *13*, 83–92. [CrossRef]
66. Sadik-Zada, E.R. Political Economy of Green Hydrogen Rollout: A Global Perspective. *Sustainability* **2021**, *13*, 13464. [CrossRef]
67. Carlson, J.; Rahman, M.; Voola, R.; Vries, N. De Customer engagement behaviours in social media: Capturing innovation opportunities. *J. Serv. Mark.* **2018**, *32*, 83–94. [CrossRef]
68. Kirkwood, J.; Walton, S. What motivates ecopreneurs to start businesses? *Int. J. Entrep. Behav. Res.* **2010**, *16*, 204–228. [CrossRef]
69. Ballew, M.T.; Omoto, A.M.; Winter, P.L. Using Web 2.0 and Social Media Technologies to Foster Proenvironmental Action. *Sustainability* **2015**, *7*, 10620–10648. [CrossRef]
70. Huang, H. Media use, environmental beliefs, self-efficacy and pro-environmental behavior. *J. Bus. Res.* **2015**, *69*, 2206–2212. [CrossRef]
71. Zeynalova, Z. The applying functions of tax regulation for state development. *Aust. Econ. Pap.* **2020**, *59*, 376–394. [CrossRef]
72. Zahraie, B.; Everett, A.M.; Walton, S.; Kirkwood, J. Environmental entrepreneurs facilitating change toward sustainability: A case study of the wine industry in New Zealand. *Small Enterp. Res.* **2016**, *23*, 39–57. [CrossRef]
73. Khare, A. Antecedents to green buying behaviour: A study on consumers in an emerging economy. *Mark. Intell. Plan.* **2015**, *33*, 309–329. [CrossRef]
74. Laroche, M.; Bergeron, J.; Barbaro-Forleo, G. Targeting consumers who are willing to pay more for environmentally friendly products. *J. Consum. Mark.* **2001**, *18*, 503–520. [CrossRef]
75. Santini, C. Ecopreneurship and Ecopreneurs: Limits, Trends and Characteristics. *Sustainability* **2017**, *9*, 492. [CrossRef]
76. Cleveland, M.; Kalamas, M.; Laroche, M. Shades of green: Linking environmental locus of control and pro-environmental behaviors. *J. Consum. Mark.* **2005**, *22*, 198–212. [CrossRef]
77. Greeno, J.L.; Robinson, S.N. Rethinking Corporate Environmental Management. *Columbia J. World Bus.* **1992**, *27*, 222–232.
78. Polonsky, M.J.; Mintu-Wimsatt, A.T. *Environmental Marketing: Strategies, Practice, Theory, and Research*; Haworth Press: New York, NY, USA, 1995.
79. Welford, R. *Environmental Strategy and Sustainable Development: The Corporate Challenge for the Twent First Century*; Routledge: Abingdon, UK, 1995.
80. Miao, C.; Fang, D.; Sun, L.; Luo, Q. Natural resources utilization efficiency under the influence of green technological innovation. *Resour. Conserv. Recycl.* **2017**, *126*, 153–161. [CrossRef]

81. Spence, M.; Gherib, J.B.B.; Biwolé, V.O. Sustainable entrepreneurship: Is entrepreneurial will enough? A north-south comparison. *J. Bus. Ethics* **2011**, *99*, 335–367. [CrossRef]
82. Zhang, Y.; Berhe, H.M. The Impact of Green Investment and Green Marketing on Business Performance: The Mediation Role of Corporate Social Responsibility in Ethiopia's Chinese Textile Companies. *Sustainability* **2022**, *14*, 3883. [CrossRef]
83. Ambec, S.; Lanoie, P. Does it pay to be green? A systematic overview. *Acad. Manag. Perspect.* **2008**, *22*, 45–62.
84. Kushwaha, G.S.; Sharma, N.K. Factors Influencing Young Entrepreneurial Aspirant's Insight towards Sustainable Entrepreneurship. *Iran. J. Manag. Stud.* **2017**, *10*, 435–466.
85. Silajdžić, I.; Kurtagić, S.M.; Vučijak, B. Green entrepreneurship in transition economies: A case study of Bosnia and Herzegovina. *J. Clean. Prod.* **2015**, *88*, 376–384. [CrossRef]
86. Gliedt, T.; Parker, P. Green community entrepreneurship: Creative destruction in the social economy. *Int. J. Soc. Econ.* **2007**, *34*, 538–553. [CrossRef]
87. Gliedt, T.; Parker, P. Green community entrepreneurship 2.0. *Int. J. Soc. Econ.* **2014**, *41*, 609–625. [CrossRef]
88. Nizaeva, M.; Coskun, A. Determinants of the financing obstacles faced by SMEs: An empirical study of emerging economies. *J. Econ. Soc. Stud.* **2018**, *7*, 81–99. [CrossRef]
89. Noh, H.J. *Financial Strategies to Activate Green Company Eco Systems*; KCMI: Seoul, Korea, 2014.
90. European Union. *Green Budgeting: Towards Common Principles*; Publications Office of the European Union: Luxembourg, 2021.
91. Wagner, M. Eco-entrepreneurship: An empirical perspective based on survey data. In *Frontiers in Eco-Entrepreneurship Research, Advances in the Study of Entrepreneurship, Innovation and Economic Growth*; Libecap, G.D., Ed.; Emerald Group Publishing Limited: Bingley, UK, 2009; Volume 20, pp. 127–152.
92. Dess, G.G.; Keats, B.W. Environmental Assessment and Organizational Performance: An Exploratory Field Study. In *Academy of Management Proceedings*; Academy of Management: Briarcliff Manor, NY, USA, 1987; pp. 21–25.
93. Balatbat, M.C.; Lin, C.Y.; Carmichael, D.G. Management efficiency performance of construction businesses: Australian data. *Eng. Constr. Archit. Manag.* **2011**, *18*, 140–158. [CrossRef]
94. Goodwin, H.; Francis, J. Ethical and responsible tourism: Consumer trends in the UK. *J. Vocat. Mark.* **2003**, *9*, 271–284. [CrossRef]
95. Shin, S.; Cho, M. Green Supply Chain Management Implemented by Suppliers as Drivers for SMEs Environmental Growth with a Focus on the Restaurant Industry. *Sustainability* **2022**, *14*, 3515. [CrossRef]
96. Ceyhan, S.; Ada, S. Environmentally Sensitive Business Administration in Terms of Business Functions. *Int. J. Manag. Econ. Bus.* **2015**, *11*, 26, 115–137.
97. Lerro, M.; Raimondo, M.; Stanco, M.; Nazzaro, C.; Marotta, G. Cause Related Marketing among Millennial Consumers: The Role of Trust and Loyalty in the Food Industry. *Sustainability* **2019**, *11*, 535. [CrossRef]
98. Landy, H. A Wall Street Powerhouse is Finally Matching Its Climate-Change Rhetoric with Action. Available online: https://qz.com/1784949/blackrock-ceo-larry-finks-2020-letter-backs-up-climate-rhetoric-with-action/ (accessed on 10 April 2022).
99. Hair, J.F.; Hult, G.T.M.; Ringle, C.M.; Sarstedt, M. *A Primer on Partial Least Squares Structural Equation Modeling (PLS-SEM)*, 3rd ed.; Sage: Thousand Oaks, CA, USA, 2022.
100. Jolliffe, I.T. Sample Sizes and the Central Limit Theorem: The Poisson Distribution as an Illustration. *Am. Stat.* **1995**, *49*, 3.
101. Everitt, B.; Skrondal, A. *The Cambridge Dictionary of Statistics, 4th ed*; Cambridge University Press: New York, NY, USA, 2010.

sustainability

MDPI

Article

Revealing Consumer Behavior toward Green Consumption

Zivar Zeynalova [1,*] and Natavan Namazova [2,3]

1 School of Public and International Affairs, ADA University, Baku AZ1008, Azerbaijan
2 International Trade, Logistic and Marketing Department, Azerbaijan Technical University, Baku AZ1001, Azerbaijan; natavan.namazova@aztu.edu.az or natavan.namazova@unec.edu.az
3 Centre for Studies on European Economy (AIM), Azerbaijan State University of Economics (UNEC), Baku AZ1001, Azerbaijan
* Correspondence: zzeynalova@ada.edu.az

Abstract: The aim of this study was to determine the relationship between the attitudes and behaviors of the administrative staff of four Baku universities toward the consumption of environmentally friendly products. The study used 536 consumer questionnaires belonging to different socio-economic and demographic consumer groups. The survey evaluated the factors that affected purchasing decisions, purchase intentions, awareness, attitudes and behaviors toward green products of respondents living in an urban area. The results from the survey were analyzed using one-way MANOVA analysis. According to the results of the study, the green consumer behaviors of the survey participants differed according to age, the number of household members, marital status, education level and income. At the same time, the main indicators that consumers paid attention to were price, brand, appearance, advertising, expert opinion and label information. The level of importance of these indicators varied according to the education, income and number of household members of the participants. The results showed that the participants of the survey were partially sensitive to the environment.

Keywords: green product; green consumer behavior; environmental responsibility; environmental concern; green purchasing decisions

Citation: Zeynalova, Z.; Namazova, N. Revealing Consumer Behavior toward Green Consumption. *Sustainability* **2022**, *14*, 5806. https://doi.org/10.3390/su14105806

Academic Editors: Elkhan Richard Sadik-Zada and Andrea Gatto

Received: 7 March 2022
Accepted: 2 May 2022
Published: 11 May 2022

1. Introduction

Global warming and climate change, the consequences of which have become even more serious, have contributed to the growth of many environmental problems, such as the reduction in water resources; the loss of biological diversity; air, water and soil pollution; and the depletion of natural resources [1,2]. The fact that today, environmental problems have reached their peak has led to the growth of environmental concerns [3] and made it necessary to shape consumer behavior more carefully in environmental approaches [4]. Therefore, current consumption patterns need to be shaped correctly in the fight against climate change [5,6]. On the other hand, rapid population growth and advances in technology have led to a further increase in production and consumption. In other words, over the past two centuries, advances in medicine have increased the average human life expectancy, which has led to a rapid increase in the world's population. The world population, which was 2.5 billion in 1950 and increased to 6.1 billion in 2000 and 7.8 billion in 2020. The world population will be between 8.5 and 8.6 billion in 2030, between 9.4 and 10.1 billion in 2050, and between 9.4 and 12.7 billion in 2100 [7]. It is believed that the environmental problems observed with the growth of the population and the level of consumption will negatively affect the living standards of future generations [8]. In other words, it is very important to develop public consciousness on this issue, as the ongoing environmental problems will grow in the future if they are not prevented [9].

In academic studies, special attention is paid to the responsibility and behavior of the consumers [10,11]. One of the reasons this point of view needs to be heard is the power of consumer preferences to influence production [12]. That is, while increasing environmental awareness and concern for the environment is driving most people to use green products in their daily lives, companies are seeing the benefits of green marketing strategies [13–15].

Consumer preferences can have a significant impact on the solution of environmental problems [16,17]. Consumers' preference for products that do not harm the environment and the use of recyclable packaging will make significant contributions to the environment. In other words, consumers play a key role in product policies for environmentalism [18]. Therefore, responsible consumption behaviors in society should be encouraged and supported [19].

A growing population should aim for sustainable food consumption [20,21]. Sustainable food consumption can be achieved by creating more favorable consumption patterns rather than reducing the quantity of goods and services [22]. Determining the attitudes of consumers toward food consumption will contribute to the further development of sustainable food consumption [23]. Meanwhile, reducing food waste plays an important role in sustainable food consumption [24,25]. Data from Eurostat reveals that around 20% of food production in the EU is wasted [26]. More than half of all food waste in the EU is generated by households. A total of 70% of all food waste is produced by households and catering and retail businesses [27].

Reducing waste should be a sense of obligation for consumers and a commitment to environmental action [28]. Considering that food waste includes natural resources, such as water, soil and energy used in food production, it can be easily said that it has negative effects on food safety and the environment. Therefore, it is necessary to reduce food waste and improve global food and nutrition security, both throughout the food supply chain and in the consumption process [29,30].

Another environmental problem is caused by plastic packaging products. Plastic packaging products that are used daily are widely circulated among consumers and these products cause various environmental problems, such as polluting soil and water [31,32]. Therefore, one of the disadvantages of plastics today is related to their management. However, chemical recycling makes it possible to liquefy polymer waste and use them as fuel components, but such technologies are expensive [33]. Despite the fact that the recycling of plastic waste is increasing every year, more than 27% of plastic is thrown away [34]. Knowing the consumers' attitudes toward green products enables the preparation of a more effective information and education policy by both the state and NGOs in society [35]. It would be possible to add a fiscal policy, especially a tax policy, because taxes imposed by the government can affect the behavior of consumers [36,37].

For this purpose, on the basis of national policies, consumers should be encouraged to change their behavior to be more environmentally friendly. In order to transform consumers into environmentally conscious consumers, it is necessary to recognize them, identify their distinctive features and develop appropriate strategies. In this regard, an important research topic is to determine whether there is a significant relationship between the environmental orientation of consumers and their socio-economic characteristics. From this perspective, this study attempted to determine whether there were significant relationships between the general characteristics of consumers and their environmental orientation and green behavior trends. To this end, the thoughts of various consumers on environmental orientations and trends in their environmentally friendly behavior in Azerbaijan were studied.

2. Literature Review

Issues related to the physical environment, such as air, water and soil, can be handled under the name of green, and environmental awareness associated with consumption can also be considered a sign of greenness [38,39].

There are many different and similar definitions in the literature on environmental awareness, which is called the green movement. One of these definitions was stated in the study named *Promoting Environmentalism* made by Zelezny and Schultz in 2000. According to Zelezny and Schult, environmental consciousness represents psychological factors related to individuals' attitudes and evaluations toward the natural environment [18]. Therefore, environmental awareness can be defined as the development of an environmentally sensitive attitude and environmentally beneficial behavior for a sustainable life. The development of environmental awareness of consumers also affects their concerns about the environment [40–44].

2.1. Green Products

Generally, a green product is known as an ecological product or an environmentally friendly product [45]. In a broader sense, a green product refers to a product group that has recycling strategies or recyclable content to reduce its impact on the natural environment, is energy efficient (does the same job but consumes less energy than others), reduces the use of packaging or reduces the use of toxic materials [45–50].

The definitions given to green products show that they are environmentally friendly products. They can be recycled after use. Moisander [51] listed the characteristics of environmentally friendly products and services as follows:

- Should not be dangerous to the health of people and animals;
- Should not harm the environment during its production, use and disposal;
- Should not consume a disproportionate amount of energy and resources during its production, use and disposal;
- Should not cause waste due to excessive packaging.

According to Blair [52], environmentally friendly products have certain characteristics. They are:

- Provide an opportunity to reduce global environmental problems;
- Energy saving;
- Do not create pollution;
- Ease of repair;
- Designed to be reused or recycled;
- Produced with minimal packaging;
- Produced from renewable resources;
- Based on the security principle;
- Produced from local sources to reduce distribution costs;
- Designed to meet genuine and sincere human needs;
- Provide sufficient information on the label;
- Harmless to human health;
- Do not contain harmful substances;
- Not tested on animals.

The characteristics given for green products show that consumers should prefer products with the characteristics shown when they are environmentally friendly and that they should also adopt a holistic approach in order not to ignore one feature while supporting the other [53,54].

2.2. Green Consumers

Green consumers or environmental consumers can be defined as people who aim to protect themselves and the world through the power of their purchasing decisions [55]. In other words, the purchasing, product use and disposal decisions of environmentalist consumers are based on the desire to protect the ecological balance of nature [56–58]. The green consumer contributes to the protection of the environment by refusing to buy products that are harmful to the environment [58,59]. Therefore, green consumers avoid purchasing products that are considered unhealthy; that harm the environment during production, during use and after use; consume excess energy; are repackaged; or contain ingredients from endangered habitats or species [60–63]. This means that a green consumer is generally defined as one who adopts environmentally friendly behaviors and buys green products rather than standard products [49].

It is seen that green consumption is used together with various expressions that are important in terms of an environmental bias, being nature-friendly and engaging in sustainable consumption [48,64]. However, being sensitive to the environment and reflecting on consumption practices are different processes. Therefore, green consumers can be considered in two categories [65]:

- Active green consumers whose purchasing behavior is largely shaped by environmental concerns;
- Passive green consumers whose purchasing behavior is partly shaped by environmental concerns.

In order to ensure sustainable consumption in society, it is necessary to transform sensitivities toward nature into a lifestyle and to cover green consumption behavior practices. Therefore, it is necessary to increase the number of active green consumers in society. On the other hand, it is necessary for consumers to have sufficient information in order to make conscious choices because a lack of information may prevent consumers from making their purchasing decisions toward green consumption.

As mentioned earlier, environmentally conscious consumers who are aware of the fact that the quality of the environment will improve as a result of increased consumption of green products may tend to become more actively involved in green consumption activities. Therefore, an accurate determination of consumer behavior ensures that the direction of consumer needs and desires in society is known correctly.

Since green consumers have a high environmental responsibility, they will not harm the environment or they will try to minimize damage in their production or consumption in their daily lives [66]. Focusing on behavior that minimizes environmental damage [67] is the main goal of environmental consumption behavior.

2.3. *Green Consumption Behavior*

In studies that take into account the ecological concerns of the consumer, this type of behavior was examined within the framework of different concepts, such as "ecological behavior" [68,69], "environmentally friendly behavior" [70,71], "responsible consumer behavior" [7] and "sustainable consumer behavior" [49,53,72]. Even though they are handled in different ways, consumers with environmental awareness are not only interested in the consumption process, but also in the production, scarcity of consumed resources and post-use processes of products [8,73].

Many studies have examined demographic factors, such as age, gender, educational and income levels, and location, as well as psychographic variables that include various values and attitudes, to characterize environmentally sensitive consumers [74–80]. Different results were reached in the studies on the demographic characteristics of environmentally sensitive consumers. Studies that identified age as an important factor in consumer behavior showed that younger people engaged in more responsible environmental behaviors compared to older people. Despite this, a qualitative study by Autio and Heinonen found that although young people aged 16–19 years are aware of the consumption of organic products, the image of a green consumer does not seem attractive to them and they do not buy green products [81,82].

A study by Kreidler and Joseph [83] showed that income and education have a positive effect on environmental behavior. This effect was defined as a linear effect. However, Qasim et al. [84] found that consumer income level did not affect Pakistani consumers' intention to consume organic food.

Roberts and Bacon [44] and Straughan and Roberts [85] determined in their research that environmentally sensitive consumers are relatively younger, better educated, have higher incomes and mostly consist of women. There are also studies in the literature that argue that environmental commitment is higher in women than in men [70,82,86]. Because women perceive environmental health risks more than men, they are more sensitive to the environment [87–89]. The study conducted by Chen and Chai in 2010 revealed that there is no gender-related difference in attitudes toward the environment and green products.

However, studies by Schwartz [90] and Roper Organization [91] found that environmental consumers do not consist only of high-income, young, educated women. Straughan and Robert [85] also found that people living in cities are more environmentally aware than people living in rural areas.

There may be many reasons for differentiating environmental consumption behavior in studies conducted by different researchers at different times. This may include consumer hedonic desires [92], such as planting trees, purchasing organic products and achieving goals, such as saving energy. However, some studies found that there is a weak correlation between consumers' positive attitudes toward environmental issues and their conversion to actual purchasing behavior [93–96].

The complexity of consumer environmental behavior is sometimes related to financial reasons, such as a lack of affordable green products, as well as a loss of comfort and/or loss of time, such as sorting waste for recycling and keeping it at home before sending it to collection points [97–101]. For example, some consumers are reluctant to pay more for green-packed products because their prices are higher and unaffordable [102–104]. A consumer who cannot afford to buy a product in eco-friendly packaging will choose to purchase the product in non-ecological packaging because the main motivation for these actions is need. On the other hand, some consumers are willing to pay more for organic products because they are of better quality [102,105]. However, if prices are cheaper, most customers will buy more organic food [106,107]. Even in the EU, about 43 million people today cannot afford quality food every two days [27].

3. Research Methodology

3.1. Research Technique and Questions

A questionnaire technique was used as the data collection method in the study. The survey was divided into two parts. The 7 items given in the first subsection of the first section were designed to understand the general characteristics of the participants. In the second subsection of the first section, the participants were asked to define green products.

In the second part of the questionnaire, items were given to measure the factors that affected the decisions of consumers while purchasing products, namely, their purchasing intentions, awareness, attitudes and behaviors toward ecological products. The first subsection of the second part of the questionnaire aimed to determine the indicators that consumers paid attention to when purchasing products by using 12 indicators. These indicators included the production and expiry date, being eco-friendly, freshness, ingredients, brand, expert opinions, appearance, price, label information and advertising.

In the second subsection of the second part, ecological consumer behaviors were evaluated. Eighteen statements were presented to measure the ecological consumer behavior of participants in the research. These expressions are given below:

- ✓ I prefer reusable products to disposable ones.
- ✓ I prefer to buy the same product in a larger package.
- ✓ I use the product until it is completely worn out.
- ✓ I buy used goods to reduce unnecessary consumption.
- ✓ I separate waste, such as paper, glass, plastic bottles, batteries, etc.
- ✓ I want to receive documents by e-mail so as not to use extra paper.
- ✓ Instead of buying products that I will use temporarily, I borrow them from my relatives and friends.
- ✓ I don't use a plastic bag and put groceries in my bag when I'm shopping.
- ✓ I donate unused clothes to those who need them.
- ✓ I prefer to buy products that blend quickly with nature.
- ✓ I buy packaged products made from recycled papers.
- ✓ I buy products made from environmentally friendly materials.
- ✓ I pay attention to the type of energy I use so as not to increase air pollution.
- ✓ Although expensive, I buy lamps that consume less electricity.
- ✓ I buy energy-saving electrical appliances.

✓ When the weather gets colder, I prefer to wear warm clothes rather than raise the temperature.
✓ I turn off unnecessary lights.
✓ While brushing my teeth, washing dishes, etc., I turn off the faucet.

Consumers stated whether they agreed with these statements or not. These items are selected from the studies of Barr [108], Pepper et al. [109], Berger and Corbin [110], Gupta and Agrawal [111], Roberts and Bacon [44], Barbarossa and De Pelsmacker [112], Sütterlin et al. [50], Straughan and Roberts [85], and Barr and Gilg [113]. To determine the degree of agreement or disagreement of the participants with a particular statement, a 5-point Likert scale was used.

Thus, the main research questions in the study are given below:

- "What characteristics do survey participants attribute to environmental products?"
- "What factors do survey participants pay attention to when consuming?"
- "What are the main characteristics of the ecological consumer behavior of the survey participants?"

The consumer behaviors of the respondents were compared using one-way MANOVA analysis. The data collected in the survey were subjected to the post hoc data-cleaning technique. IBM SPSS Statistics Version 28 was used for data analysis and evaluation.

3.2. Research Scope

Data were collected from 580 respondents over the age of 18, but due to incorrect and/or incomplete filling of some questionnaires, 536 of them turned out to be suitable. The survey was conducted both face-to-face and online. The questionnaire was prepared using a Google form and distributed via e-mail, Facebook and WhatsApp. In the face-to-face interview, participants were asked to fill out a paper-based questionnaire. The study was conducted among the administrative staff of four Baku universities living in urban areas. Participants from the Azerbaijan State University of Economics, Academy of Public Administration under the President of the Republic of Azerbaijan, Azerbaijan Technical University and Azerbaijan State Pedagogical University were grouped into different categories, such as age, gender, education, marital status, number of households, income level and purchase frequency.

Residents of economically developed regions generally have higher education, higher incomes, etc. [30]. In this context, the respondents were selected from Baku, and universities are one of the most suitable places for the analysis.

The survey was conducted by the authors between November 2021 and January 2022.

4. Data Analysis and Findings

4.1. General Characteristics of Survey Participants

This section presents the data collected on general variables, such as gender, marital status, education level, age, number of household members, monthly income and frequency of purchases.

According to the results presented in Table A1, 47.2% of the study participants were men and 52.8% were women (see Appendix A). When considering the marital status of the survey participants, it can be seen that 40.5% were married and 59.5% were single. Considering the educational status of the participants, it can be seen that 10.8% of them had a high school diploma, 64.9% had a bachelor's degree, 18.3% had a master's degree and 6.0% had a doctoral degree. These results show that all participants were educated and the majority of participants had a high level of education.

Looking at the ages of the respondents, it can be seen that 35.3% of participants were between the ages of 18 and 29, while 18.3% were aged 65 or older. Considering the number of household members of the interviewed participants, it can be seen that 22.2% of them lived alone or with another person. A total of 34.5% of them had a family of 3 or 4 people and 26.7% had a family of 5 or 6 people. Only 16.6% of participants had more than six people in the household. Another difference that members had was to do with their monthly income. While the majority of respondents (37.1%) earned between 301 and

600 manats, only 5.04% of them earned more than 1500 manats. Meanwhile, the lowest salary in Azerbaijan was 250 manats in 2021 and 300 manats in 2022 [114]. Some of the respondents were paid less than 300 manats as they worked half-days. Approximately 30% of the participants noted that they did shopping every day, 24.8% of participants made purchases 2–3 times a week, 17.5% made purchases once a week, and 14.6% of participants made purchases once a month.

4.2. Results for Survey Participants' Statements on Definition of Green Products

As mentioned earlier, green purchasing decisions provide an opportunity to reduce the negative impact on the environment through the consumption of green products [115,116].

Therefore, what is the consumer attitude and behavior toward ecological products in Azerbaijan? To answer this question, consumers were first asked the question, "What is a green product?" in order to identify consumer awareness.

The distribution of consumer responses about what a green product means is shown in Figure 1. More than one answer was selected by participants and the largest share (67%) in the definition of green products was marked as "Environmentally related products". Other highly rated responses were "Naturally produced products" with 46% and "Products that generate less waste" with 39%. The lowest scoring responses were "High value products", "Products that protect natural resources", and "Production of more labor-intensive products". The percentages of these options were 13%, 12% and 10%, respectively. A total of 16% of consumers said that they did not know what green products are.

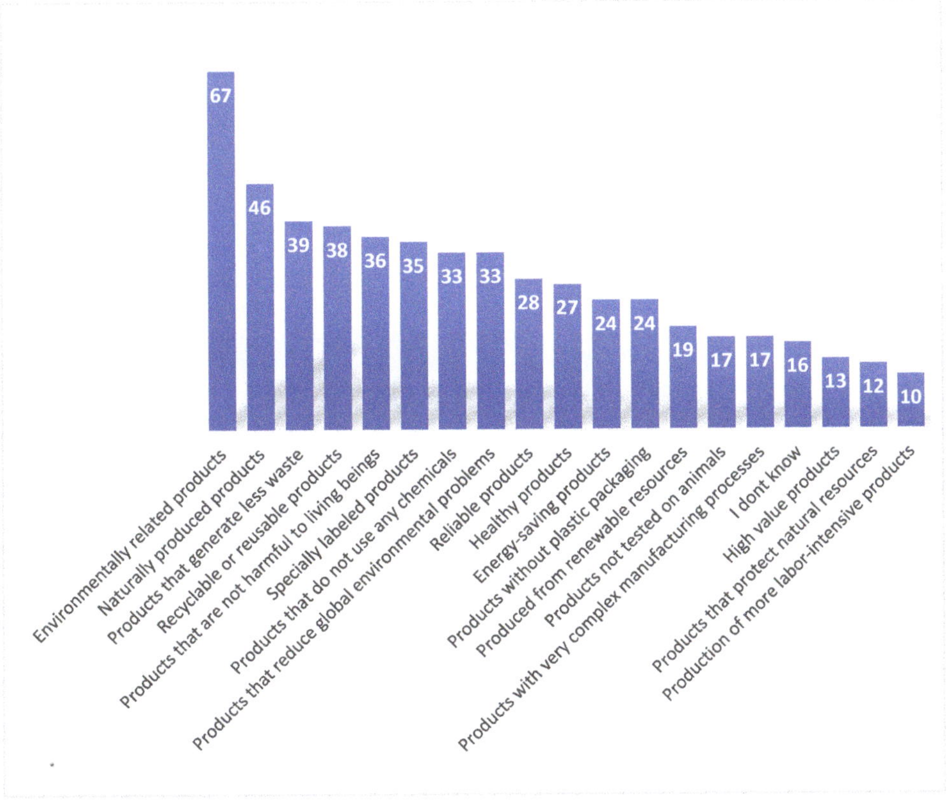

Figure 1. Percentage of Survey Participants' Statements on Definition of Green Products.

4.3. Results for Indicators That Consumers Paid Attention To

This study aimed to determine the indicators that consumers pay attention to when purchasing products by using 12 expressions. Factor analysis was performed on the correlation matrix of these variables.

The factor was the weighted average of the original variables. With factor analysis, we aimed to find the factors from which the original correlation matrix could be formed [117]. As is known, the factor load should be at least 0.30 [118,119]. In addition, the difference between factor loadings given to a variable by more than one factor must be at least 0.1 [118]. However, two indicators—quality and taste—were loaded onto two factors. The factor loads were 0.417 and 0.431, and 0.523 and 0.554, respectively. These indicators were excluded from the analysis because the difference between the factor loadings was less than 0.1.

Before the factors of the variables were given, the KMO and Bartlett's tests were performed. KMO and Bartlett's tests evaluate all data together. A KMO value above 0.5 and a Bartlett's test value below 0.05 indicate a significant correlation in the data. Variable collinearity indicates how strongly a single variable is related to other variables [120].

The KMO and Bartlett's test outputs are shown in Table A2 (see Appendix B). The KMO value obtained was 0.813. This value was good and it meant that the sum of partial correlations was not large in comparison to the sum of the correlations. Therefore, the factor analysis was appropriate in this case. In other words, reliable and different factors were obtained from the factor analysis of these data. The approx. chi-square value of Bartlett's test of sphericity was 10,216.520. The significance value of Bartlett's test of sphericity was less than 0.001. Thus, factor analysis could be applied to the data set of the indicators that consumers paid attention to.

In the next step of the analysis process, factor analysis was performed on 10 items using the principal component calculation method and a direct oblimin rotation. The analysis revealed a three-factor structure that explained 59% of the total variance. It was observed that the factor loads of the items varied between the lowest at 0.402 and the highest at 0.891. The pattern matrix and Cronbach's alpha based on the standardized items for indicators that consumers pay attention to are presented in Table A3 (see Appendix C).

When the factor structure was examined, it was seen that the "Production and expire date, Being eco-friendly, Freshness, Ingredients" were combined under a single factor. "Brand, Expert Opinions, Appearance" were combined under the second factor. Finally, "Price, Label information and Advertising" were combined in the third factor. "Brand" is loaded on both second and third factors. However, since the load of this indicator in the third factor (0.409) was lower than the load in the fourth factor (0.843), it remained in the second factor. "Advertising" was loaded on both the first and third factors. Since the load of the "Advertising" in the third factor (0.891) was higher than the load in the first factor (0.460), it remained in the third factor.

Cronbach's alpha method was used to measure the reliability of the variables used in the study. Since the Cronbach's alpha value for two factors was greater than 0.70, we could accept the analysis of only two factors as reliable. These factors were component 2 (brand, expert opinions, appearance) and component 3 (price, label information and advertising). One-way MANOVA results for the indicators that consumers paid attention to are shown in Table A4 (see Appendix D).

The p-values of Box's test of equality of covariance matrices were higher than 0.05 for income and number of family members in component 2. On the other hand, education, income and number of family members had a value higher than 0.05 in component 3. This meant that the assumption of homogeneity of covariance was not violated and the outputs could be interpreted for these mentioned indicators. Therefore, for the next step, we only examined the values of these indicators.

To test the assumption of homogeneity of variance, we should look at Levene's test of equality of error variances. The p-values for Levene's test were higher than 0.05 for some mentioned indicators. This result allowed us to continue the analysis and interpretation only for indicators with values higher than 0.05. For Pillai's trace, the mentioned indicators had a p-value less than 0.05. This meant that we had evidence of a significant main effect.

Wilks' lambda sig. determines whether the one-way MANOVA is statistically significant. It can be seen from Table A4 that the value of "Sig" was smaller than 0.0005 for the selected indicators. Therefore, we could say that some examined elements of component 2 and component 3 varied according to the demographic characteristics of consumers.

4.4. Data Analysis and Results for Items Oriented toward Measuring Environmental Consumer Behavior

Factor analysis was performed on the correlation matrix of the variables. Four items were excluded from the analysis because the difference between the factor loadings was less than 0.1. These items and their factor loadings are given below:

- "I donate unused clothes to those who need them." (0.725 and 0.748);
- "I don't use a plastic bag and put groceries in my bag when I'm shopping." (0.580 and 0.593);
- "I use the product until it is completely worn out." (0.715 and 0.791);
- "Although expensive, I buy lamps that consume less electricity." (0.814 and 0.863).

Table A5 presents the results of the KMO and Bartlett's tests for items oriented toward measuring environmental consumer behavior (see Appendix E).

The KMO value obtained in this analysis was equal to 0.805. Therefore, factor analysis was appropriate in this case. On the other hand, the approx. chi-square value of Bartlett's test of sphericity was 11,317.065. The significance value of Bartlett's test of sphericity was less than 0.001. Thus, factor analysis could be applied to the data set of the items oriented toward measuring environmental consumer behavior.

As a result of the factor analysis, 14 items were gathered under four factors using the principal components calculation method and direct oblimin rotation. This four-factor structure explained 62% of the total variance. It was observed that the factor loads of the items varied between the lowest at 0.405 and the highest at 0.853. The pattern matrix and Cronbach's alpha based on standardized items for items oriented toward measuring environmental consumer behavior are given in Table A6 (see Appendix F).

In this analysis, three items were combined under component 1 and the Cronbach's alpha value was greater than 0.835. Component 3 and component 4 contained four and three items, respectively, and had a value greater than 0.07. However, the Cronbach's alpha value for component 2 was less than 0.07. This meant that we could only consider the analysis of three components as reliable.

The item "When the weather gets colder, I prefer to wear warm clothes rather than raise the temperature" was loaded on both the second and third factors. Since the load of this item in the second factor (0.827) was higher than the load in the third factor (0.411), it remained in the second factor. One-way MANOVA results for items oriented toward measuring environmental consumer behavior are shown in Table A7 (see Appendix G).

The analysis revealed that the p-values of Box's test of equality of covariance matrices were higher than 0.05 for the income, number of family members and age groups in component 1. On the other hand, demographic characteristics, such as marital status, income and number of family members, had values greater than 0.05 in component 3. Component 4 had significant values for the income and age groups. This meant that outputs could be interpreted for specified groups. Therefore, we only examined the values of these indicators. The p-values for Levene's test were higher than 0.05 for some indicators examined. At the same time, some indicators examined had a p-value that was less than 0.05 for Pillai's trace. This means we had evidence of a significant main effect.

Wilks' lambda sig. was smaller than 0.0005 value for some indicators examined. Therefore, we could say that the examined elements of component 1, component 3 and component 4 varied according to the items oriented toward measuring environmental consumer behavior.

5. Discussion

One of the aims of this research was to determine the meaning that the survey participants attached to environmental products. The distribution of participant responses revealed that 16% of consumers had no idea about green products. As stated in the literature, green products have basic features such as energy savings, benefits for human health, production from renewable sources, being designed for reuse or recycling, and not making waste [45–52]. Despite the fact that the majority of respondents, i.e., 67%, said that green products are related to the environment, they chose only some of the features of these products. For example, only 39% of respondents believed that green products produced less waste, while 33% and 19% believed that they reduce global environmental problems and are made from renewable resources, respectively. Considering the contemporary pollution challenges our Earth is facing, these indicators highlight that even in universities, the level of consciousness on organic products is not at a required level and should be increased. On the other hand, as mentioned earlier, the depletion and waste of natural resources are major human problems [1,2] and can be reduced through the use of environmentally friendly products [29,30].

The survey shows that only 12% of respondents considered green products as products that protect natural resources, and 24% of the survey participants knew that environmentally friendly products had energy-saving features.

One of the main causes of environmental problems is related to the use of plastic packaged products [31,32]. Solving this problem requires less consumption of plastic packaging products. However, 24% of the respondents defined green products as products without plastic packaging. If the interest in ecological products increases among consumers, then their tendency to consume plastic packaged products may decrease.

The price of green products is usually higher than conventional products [102,103]. Therefore, if consumers accept green products as more labor-intensive, chemical-free and healthier, they may be willing to pay more for those products. However, only 10%, 27% and 33% of the respondents considered green products as more labor-intensive, healthy and chemical-free products, respectively.

In general, the characteristics that participants attributed to green products revealed that they did not fully perceive what these products are.

The second aim of this study was to determine the differences between the demographic characteristics of the respondents regarding what they paid attention to when consuming. The study found that there was a significant difference between the averages of income groups that were related to certain characteristics, such as price, brand, appearance, expert opinion and label information. In particular, for consumers in the income group "1001–1500 manats", the price was less important than for other groups. At the same time, low-income groups paid less attention to the brand, appearance, advertising, expert opinion and label information. For example, the "up to 300 manats" and "301–600 manats" income groups were basically not interested in label information, in contrast with other groups.

Another difference lay in the level of education of participants regarding label information and advertising. Participants belonging to the "master" group pay more attention to label information of products than the "high school" and "bachelor" groups. Advertising was more important for the "bachelor" group than for the "master" group.

Regarding the consumption process, price and brand were some of the indicators that affected consumers in large households. Households with up to four people were more nutritionally sensitive to these indicators than others. Participants with five or six people in the household paid less attention to the ingredients of products than other groups. However, expert opinion as an indicator was a more important factor for groups of "more than 6 people" than "up to 2 people".

These data showed that green product sellers can use price, brand, appearance, expert opinion and label information to increase the consumption of eco-friendly products. However, the study found no statistically significant mean differences between the groups in terms of product production and expiry date, ingredients, freshness and being eco-friendly.

The third objective of the study was to determine the main characteristics of the ecological consumer behavior of the survey participants. The results of the one-way MANOVA analysis showed some differences between age groups. Those included in the "18–30 ages" group paid more attention to the type of energy to avoid increasing air pollution than the "65 and over" group. The young group preferred to buy the same product in a larger package more than others, and people in the "46–65" and "65+" groups were generally not interested in receiving documents by e-mail compared to other groups.

The findings showed that in order to become responsible consumers, people over the age of 46 years need to be trained to use less paper and pay attention to the type of energy in order to not increase air pollution. Meanwhile, the younger group should be encouraged to buy products in larger packages to generate less waste.

Another difference in the behavior of the respondents was related to the number of household members. Households with more than six people supported the statement "While brushing my teeth, washing dishes, etc., I turn off the faucet." On the other hand, the "3–4 people" group was more interested in buying energy-saving electrical appliances than the "more than 6 people" group. Households with less than six members should be asked to be more sensitive about water use and appropriate policies should be prepared. Meanwhile, households with more than six members should be motivated to purchase energy-efficient electrical appliances.

Furthermore, there was a difference between the means of the single and married groups. Single people had a higher mean value than married people regarding preferring reusable products to disposable ones. At the same time, single people preferred energy-saving electrical appliances more than married people. The given data showed that married couples should be educated to use disposable products less and increase the use of energy-efficient electrical appliances.

Another difference between variables was observed between the income groups. Most people from the "up to 300 manats" group tended to buy the same product in a larger package. People in the income group "601–1000 manats" tended to separate waste, such as paper, glass, plastic bottles and batteries, more than other groups. Meanwhile, most people in the "1001–1500 manats" income group disagreed with the statement, "Instead of buying products that I will use temporarily, I borrow it from my relatives and friends." This group was also not interested in turning off unnecessary lights. Among the income groups, the lowest income group (up to 300 manats) paid less attention to the type of energy that could increase air pollution. Therefore, in order to maintain personal, environmental and social benefits [121], consumers should become more responsible and reuse an undamaged product instead of buying a new one, as well as prefer reusable products instead of disposables one. Given the low interest in such behavior patterns in the mentioned groups, the benefits of such decisions should be explained to them.

This result indicated that consumers in the middle- and high-income groups should be educated on the use of larger-package products, and consumers in low- and high-income groups should be motivated to separate waste. In addition, people's habits of buying temporary products should be reduced; in contrast, they should increase the habits of turning off unnecessary lights and paying attention to energy use to prevent pollution.

In addition, the study found no significant differences between the groups compared for some statements, such as wearing thick clothing instead of raising the temperature, buying products that biodegrade, buying packaged products made from recycled paper and purchasing products made from environmentally friendly materials.

6. Conclusions

As a result of the study, important factors were identified that increased the propensity of survey participants to choose green products in Azerbaijan. It was established that there was a relationship between the general characteristics of survey participants and product features in the decisions made by participants, where the product features of importance included price, expert opinion, brand, appearance and label information. This conclusion showed that in order to encourage the consumption of green products in society, it is necessary to offer consumers these products with appropriate properties. For example, price is an important factor for some consumer groups. That is, in order for consumers to choose environmentally friendly products, their prices should be more affordable. In addition, experts should provide consumers with detailed information about the importance of environmentally friendly products since some consumers attach great importance to the opinion of experts. Consumers who care about characteristics such as appearance and label information should find this feature in organic products. Moreover, since brands influence consumers, the presence of reliable brands of organic products will positively influence the consumption of these products because consumer confidence in well-known brands is high. If this happens, consumers will seek environmentally friendly products when buying brands they know.

As a result of the data obtained, it is possible to shape the consumption behavior of consumers and orient them toward more environmental characteristics because identifying the characteristics of consumers and assessing their behavior will help to find the shortcomings of consumption policies and increase environmental awareness. Studies show that survey participants in Azerbaijan are partially sensitive to the environment.

The environmental behavior of consumers differed depending on their age, marital status, education level and income, as well as the number of household members. By considering differences in consumer environmental attitudes, the right policies can be identified to increase sensitive consumer behavior.

Groups that do not care about the environment can be offered various aids, such as certain training, more affordable prices and price incentives. The change in consumers' attitudes toward a more balanced consumption of green products is one of the driving forces in ensuring sustainable consumption. Providing information on the importance of green consumption can reduce excessive consumption, reduce water waste and unnecessary energy consumption, or prevent paper waste by increasing skills in the use of technological products because a conscious consumer that is educated will enable the economy and society to change since this education will help consumers to shape their purchasing behavior largely with environmental concerns in mind, that is, to turn them into active green consumers.

In addition to all the above, it should be noted that we cannot generalize these results to all of Azerbaijan since the sample used in the study was not representative. However, some results were identified that may be subject to further verification.

There were some limitations in the present study. As an online resource, the survey was distributed via e-mail, Facebook and WhatsApp. Using other channels, such as Instagram, LinkedIn and Twitter, could lead to additional information being obtained and the results being slightly changed. At the same time, changing the interrogation group area from universities to hospitals, schools, business centers or other workplaces could have produced different results.

In the future, it would be useful to investigate the relationship between technology and consumption more comprehensively since the rapid change in technology affects consumer behaviors and attitudes.

Author Contributions: Conceptualization, N.N.; Data curation, Z.Z.; Formal analysis, Z.Z.; Funding acquisition, N.N.; Investigation, Z.Z.; Methodology, N.N.; Project administration, Z.Z.; Resources, N.N.; Software, N.N.; Supervision, Z.Z.; Validation, Z.Z.; Visualization, Z.Z.; Writing—original draft, Z.Z.; Writing—review & editing, N.N. All authors have read and agreed to the published version of the manuscript.

Funding: This research received no external funding.

Institutional Review Board Statement: The study was conducted in accordance with the Declaration of Helsinki, and approved by the Ethics Committee of Economic Think (protocol code 202107, 8 July 2021).

Informed Consent Statement: Informed consent was obtained from all subjects involved in the study.

Conflicts of Interest: The authors declare no conflict of interest.

Appendix A

Table A1. General Characteristics of the Survey Participants.

	Number	Percent (%)		Number	Percent (%)
	Gender			**Marital Status**	
Male	253	47.2	Married	217	40.5
Female	283	52.8	Single	319	59.5
	Education Level			**Number of Household Members**	
High school	58	10.8	Up to 2 people	119	22.2
Bachelor's	348	64.9	3 or 4 people	185	34.5
Master's	98	18.3	5 or 6 people	143	26.7
Ph.D	32	6	More than 6 people	89	16.6
	Monthly Income			**Shopping Frequency**	
Up to 300 manats	123	22.9	Every day	160	29.9
Between 300–600 manats	199	37.1	2–3 times a week	133	24.8
Between 601–1000 manats	89	16.6	Once a week	94	17.5
Between 1001–1500 manats	98	18.3	2–3 times a month	71	13.2
More than 1500 manats	27	5.04	Once a month	78	14.6
	Age				
18–29 years	189	35.3			
30–45 years	155	28.9			
46–64 years	94	17.5			
65 years and over	98	18.3			
		***N* = 535**			

Appendix B

Table A2. KMO and Bartlett's test for indicators that consumers paid attention to.

Kaiser–Meyer–Olkin Measure of Sampling Adequacy		0.813
Bartlett's Test of Sphericity	Approx. Chi-Square	10,216.520
	Df	351
	Sig.	<0.001

Appendix C

Table A3. Pattern matrix and Cronbach's alpha based on standardized items for indicators that consumers paid attention to.

Pattern Matrix [a]				Cronbach's Alpha Based on Standardized Items
	Component			
	1	2	3	
Production and expiration date	−0.709			0.504
Being eco-friendly	−0.690			
Freshness	−0.612			
Ingredients	−0.402			
Brand		0.843		**0.849**
Expert opinions		0.832		
Appearance		0.755		
Price			0.762	**0.813**
Label information			0.624	
Advertising			0.891	

Extraction method: principal component analysis. Rotation method: oblimin with Kaiser normalization. [a]: rotation converged in 18 iterations.

Appendix D

Table A4. Results of one-way MANOVA for indicators that consumers paid attention to.

	Component 2					
	Box's Test Sig.	Pillai's Trace Sig.	Wilks' Lambda Sig.	Levene's Test (Based on Mean)		
				Brand	Expert Opinions	Appearance
Gender	0.001	0.067	0.000	0.816	0.532	0.970
Marital status	0.005	0.053	0.548	0.589	0.742	0.604
Education level	0.030	0.096	0.513	0.232	0.789	0.563
Income	**0.683**	**0.000**	**0.000**	**0.871**	**0.425**	**0.386**
Number of household members	**0.542**	**0.000**	**0.000**	**0.140**	**0.158**	0.013
Shopping frequency	0.025	0.083	0.263	0.708	0.391	0.663
Age	0.014	0.057	0.384	0.047	0.104	0.393
	Component 3					
	Box's Test Sig.	Pillai's Trace Sig.	Wilks' Lambda Sig.	Levene's Test (Based on Mean)		
				Price	Label Information	Advertising
Gender	0.003	0.084	0.156	0.003	0.089	0.367
Marital Status	0.001	0.091	0.329	0.360	0.419	0.224
Education level	**0.136**	**0.000**	**0.000**	0.037	**0.270**	**0.163**
Income	**0.762**	**0.000**	**0.000**	**0.896**	**0.733**	**0.382**
Number of household members	**0.294**	**0.001**	**0.000**	**0.452**	0.046	0.012
Shopping frequency	0.000	0.065	0.352	0.031	0.014	0.524
Age	0.000	0.043	0.572	0.429	0.603	0.187

Appendix E

Table A5. KMO and Bartlett's test for items oriented toward measuring environmental consumer behavior.

Kaiser–Meyer–Olkin Measure of Sampling Adequacy		**0.805**
Bartlett's Test of Sphericity	Approx. Chi-Square	11,317.065
	Df	453
	Sig.	<0.001

Appendix F

Table A6. Pattern matrix and Cronbach's alpha based on standardized items for items oriented toward measuring environmental consumer behavior.

	Pattern Matrix [a]				**Cronbach's Alpha Based on Standardized Items**
	Component				
	1	**2**	**3**	**4**	
Instead of buying products that I will use temporarily, I borrow them from my relatives and friends.	0.679				
I prefer to buy the same product in a larger package.	0.594				
I want to receive documents by e-mail so as not to use extra paper.	0.707				**0.835**
I buy packaged products made from recycled paper.		−0.853			
I buy products made from environmentally friendly materials.		−0.573			
I prefer to buy products that blend quickly with nature.		−0.680			
When the weather gets colder, I prefer to wear warm clothes rather than raise the temperature.		−0.827			**0.449**
I prefer reusable products to disposable ones.			0.747		
I turn off unnecessary lights.			−0.658		
I buy energy-saving electrical appliances.			−0.635		
While brushing my teeth, washing dishes, etc., I turn off the faucet.			−0.623		**0.769**
I pay attention to the type of energy I use so as not to increase air pollution.				0.745	
I buy used goods to reduce unnecessary consumption.				0.648	
I separate waste such as paper, glass, plastic bottles, batteries, etc.				0.405	**0.803**

Extraction method: principal component analysis. Rotation method: oblimin with Kaiser normalization. [a]: rotation converged in 23 iterations.

Appendix G

Table A7. Results of one-way MANOVA for items oriented toward measuring environmental consumer behavior.

Component 1						
				Levene's Test (Based on Mean)		
	Box's Test Sig.	Pillai's Trace Sig.	Wilks' Lambda Sig.	Instead of buying products that I will use temporarily, I borrow them from my relatives and friends.	I prefer to buy the same product in a larger package.	I want to receive documents by e-mail so as not to use extra paper.
Gender	0.020	0.035	0.305	0.325	0.783	0.453
Marital status	0.001	0.087	0.276	0.275	0.301	0.705
Education level	0.000	0.029	0.330	0.137	0.436	0.273
Income	**0.351**	**0.001**	**0.000**	**0.482**	**0.120**	0.032
Number of household members	0.016	0.001	0.0049	0.307	0.644	0.258
Shopping frequency	0.003	0.169	0.204	0.568	0.432	0.345
Age	**0.487**	**0.000**	**0.000**	0.008	**0.638**	**0.597**

Component 3							
				Levene's Test (Based on Mean)			
	Box's Test Sig.	Pillai's Trace Sig.	Wilks' Lambda Sig.	I prefer reusable products to disposable ones.	I turn off unnecessary lights.	I buy energy-saving electrical appliances.	While brushing my teeth, washing dishes, etc., I turn off the faucet.
Gender	0.001	0.029	0.000	0.240	0.373	0.544	0.743
Marital status	**0.243**	**0.000**	**0.002**	**0.358**	0.011	**0.493**	0.022
Education level	0.002	0.004	0.000	0.518	0.407	0.190	0.673
Income	**0.825**	**0.000**	**0.000**	**0.792**	**0.435**	0.044	**0.763**
Number of household members	**0.810**	**0.000**	**0.001**	**0.027**	0.042	**0.693**	**0.521**
Shopping frequency	0.000	0.010	0.761	0.264	0.534	0.324	0.753
Age	0.001	0.000	0.007	0.621	0.614	0.021	0.279

Component 4						
				Levene's Test (Based on Mean)		
	Box's Test Sig.	Pillai's Trace Sig.	Wilks' Lambda Sig.	I pay attention to the type of energy I use so as not to increase air pollution.	I buy used goods to reduce unnecessary consumption.	I separate waste such as paper, glass, plastic bottles, batteries, etc.
Gender	0.020	0.060	0.000	0.091	0.543	0.274
Marital status	0.000	0.000	0.001	0.282	0.365	0.327
Education level	0.000	0.038	0.000	0.438	0.439	0.193
Income	**0.404**	**0.000**	**0.000**	**0.228**	**0.471**	**0.221**
Number of household members	0.010	0.000	0.000	0.536	0.126	0.437

Table A7. *Cont.*

	Component 4					
				Levene's Test (Based on Mean)		
	Box's Test Sig.	Pillai's Trace Sig.	Wilks' Lambda Sig.	I pay attention to the type of energy I use so as not to increase air pollution.	I buy used goods to reduce unnecessary consumption.	I separate waste such as paper, glass, plastic bottles, batteries, etc.
Shopping frequency	0.028	0.074	0.071	0.197	0.098	0.490
Age	**0.653**	**0.000**	**0.001**	**0.472**	0.019	0.024

References

1. Sharma, B.R.; Sharma, D. Impact of Climate Change on Water Resources and Glacier Melt and Potential Adaptations for Indian Agriculture. In Proceedings of the 33rd Indian Agricultural Universities Association Vice Chancellors Annual Convention on "Climate Change and its Effect on Agriculture", Anand Agricultural University, Anand, India, 4–5 December 2008.
2. Mammadli, M.; Sadik-Zada, E.R.; Gatto, A.; Huseynova, R. What Drives Public Debt Growth? A Focus on Natural Resources, Sustainability and Development. *Int. J. Energy Econ. Policy Econj.* **2021**, *11*, 614–621. [CrossRef]
3. Axelrod, L.J.; Lehman, D.R. Responding to environmental concerns: What factors guide individual action? *J. Environ. Psychol.* **1993**, *13*, 149–159. [CrossRef]
4. Arndt, J.; Solomon, S.; Kasser, T.; Sheldon, K.M. The urge to splurge: A terror management account of materialism and consumer behavior. *J. Consum. Psychol.* **2004**, *14*, 198–212. [CrossRef]
5. European Commission. *Proposal for a Council Decision on the Conclusion on Behalf of the European Union of the Paris Agreement Adopted under the United Nations Framework Convention on Climate Change. Publications Office of the EU*; Document 52016PC0395; European Commission: Brussels, Belgium, 2016.
6. Almond, R.E.A.; Grooten, M.; Petersen, T. *Living Planet Report 2020—Bending the Curve of Biodiversity Loss*; World Wildlife Fund (WWF): Gland, Switzerland, 2020.
7. United Nations. *Department of Economic and Social Affairs, Population Division 2019. World Population Prospects 2019: Highlights*; ST/ESA/SER.A/423; United Nations: New York, NY, USA, 2019.
8. Zinkhan, G.M.; Carlson, L. Green advertising and the reluctant consumer. *J. Advert.* **1995**, *24*, 1–6. [CrossRef]
9. Simãoab, L.; Lisboa, A. Green marketing and green brand—The toyota case. *Procedia Manuf.* **2017**, *12*, 183–194. [CrossRef]
10. Kinnear, T.C.; Taylor, J.R.; Ahmed, S.A. Ecologically Concerned Consumers: Who Are They? *J. Mark.* **1974**, *38*, 20–24. [CrossRef]
11. Sesini, G.; Castiglioni, C.; Lozza, E. New Trends and Patterns in Sustainable Consumption: A Systematic Review and Research Agenda. *Sustainability* **2020**, *12*, 5935. [CrossRef]
12. McMeekin, A.; Southerton, D. Sustainability transitions and final consumption: Practices and socio-technical systems. *Technol. Anal. Strateg.* **2012**, *24*, 345–361. [CrossRef]
13. Hart, S.L. Beyond greening: Strategies for a sustainable world. *Harward Bus. Rev.* **1997**, *75*, 66–76.
14. Luo, X.; Bhattacharya, C.B. Corporate Social Responsibility, Customer Satisfaction, and Market Value. *J. Mark.* **2006**, *70*, 1–18. [CrossRef]
15. Baziana, S.; Tzimitra-Kalogianni, E. Investigation of consumer behavior: A study on organic wine. *Int. J. Soc. Ecol. Sustain. Dev.* **2016**, *7*, 50–61. [CrossRef]
16. Peattie, K. *Environmental Marketing Management: Meeting the Green Challenge*; Pitman: London, UK, 1995.
17. Sadik-Zada, E.R.; Gatto, A.; Blick, N. *Rural Electrification and Transition to Clean Cooking: The Case Study of Kanyegaramire and Kyamugarura Solar Mini-Grid Energy Cooperatives in the Kyenjojo District of Uganda, In Sustainable Policies and Practices in Energy, Environment and Health Research*; Springer: Berlin/Heidelberg, Germany, 2021.
18. Zelezny, L.C.; Schultz, P.W. Promoting environmentalism. *J. Soc. Issues* **2000**, *56*, 365–371. [CrossRef]
19. Zeynalova, Z.; Mammadli, M. Analysis of the economic factors affecting household consumption expenditures in Azerbaijan. *J. Crit. Rev.* **2020**, *7*, 241–248. [CrossRef]
20. Fróna, D.; Szenderák, J.; Harangi-Rákos, M. The Challenge of Feeding the World. *Sustainability* **2019**, *11*, 5816. [CrossRef]
21. Vermeir, I.; Weijters, B.; de Houwer, J.; Geuens, M.; Slabbinck, H.; Spruyt, A.; van Kerckhove, A.; van Lippevelde, W.; de Steur, H.; Verbeke, W. Environmentally Sustainable Food Consumption: A Review and Research Agenda from a Goal-Directed Perspective. *Front. Psychol.* **2020**, *11*, 1603. [CrossRef] [PubMed]
22. Lorek, S.; Spangenberg, J.H. Sustainable consumption within a sustainable economy—Beyond green growth and green economies. *J. Clean. Prod.* **2014**, *63*, 33–44. [CrossRef]
23. Verain, M.C.; Dagevos, H.; Antonides, G. Sustainable food consumption. Product choice or curtailment? *Appetite* **2015**, *91*, 375–384. [CrossRef]
24. Rahman, S.; Luomala, H. A Comparison of Motivational Patterns in Sustainable Food Consumption between Pakistan and Finland: Duties or SelfReliance? *J. Int. Food Agribus. Mark.* **2021**, *33*, 459–486. [CrossRef]

25. Kim, M.J.; Hall, C.M.; Kim, D.-K. Predicting environmentally friendly eating out behavior by value-attitude-behavior theory: Does being vegetarian reduce food waste? *J. Sustain. Tour.* **2019**, *28*, 797–815. [CrossRef]
26. Eurostat. *Sustainable Development in the European Union, Monitoring Report on Progress towards the SDGs in an EU Context*; Eurostat: Luxembourg, 2018.
27. European Commission. *Reflection Paper towards a Sustainable Europe by 2030*; European Commission: Brussels, Belgium, 2019.
28. Han, H.; Yu, J.; Kim, H.C.; Kim, W. Impact of social/personal norms and willingness to sacrifice on young vacationers' pro-environmental intentions for waste reduction and recycling. *J. Sustain. Tour.* **2018**, *26*, 2117–2133. [CrossRef]
29. Food and Agriculture Organization of the United Nations. *Global Initiative on Food Loss and Waste Reduction*; Food and Agriculture Organization of the United Nations: Rome, Italy, 2015; pp. 1–8.
30. Mesterházy, A.; Oláh, J.; Popp, J. Losses in the Grain Supply Chain: Causes and Solutions. *Sustainability* **2020**, *12*, 2342. [CrossRef]
31. Zhang, J.; Wu, X.; Guo, H.; Zheng, X.; Mai, B. Pollution of plastic debris and halogenated flame retardants (HFRs) in soil from an abandoned e-waste recycling site: Do plastics contribute to (HFRs) in soil? *J. Hazard. Mater.* **2020**, *410*, 124649. [CrossRef] [PubMed]
32. Confente, I.; Scarpi, D.; Russo, I. Marketing a new generation of bio-plastics products for a circular economy: The role of green self-identity, self-congruity, and perceived value. *J. Bus. Res.* **2020**, *112*, 431–439. [CrossRef]
33. Matuszewska, A.; Ha´nderek, A.; Paczuski, M.; Biernat, K. Hydrocarbon Fractions from Thermolysis of Waste Plastics as Components of Engine Fuels. *Energies* **2021**, *14*, 7245. [CrossRef]
34. Matuszewska, A.; Owczuk, M.; Biernat, K. Current Trends in Waste Plastics' Liquefaction into Fuel Fraction: A Review. *Energies* **2022**, *15*, 2719. [CrossRef]
35. Gazdecki, M.; Gorynska-Goldmann, E.; Kiss, M.; Szakály, Z. Segmentation of Food Consumers Based on Their Sustainable Attitude. *Energies* **2021**, *14*, 3179. [CrossRef]
36. Zeynalova, Z. The Effect of Tax Revenues on Economic Growth in Azerbaijan. *J. Crit. Rev.* **2020**, *7*, 166–172.
37. Zeynalova, Z. The applying functions of tax regulation for state development. *Aust. Econ. Pap.* **2020**, *59*, 376–394. [CrossRef]
38. Shrum, L.J.; McCarty, J.A.; Lowrey, T.M. Buyer Characteristics of the Green Consumer and Their Implications for Advertising Strategy. *J. Advert.* **1995**, *14*, 71–82. [CrossRef]
39. Sadik-Zada, E.R. Political Economy of Green Hydrogen Rollout: A Global Perspective. *Sustainability* **2021**, *13*, 13464. [CrossRef]
40. Chan, R.Y. Determinants of Chinese consumers' green purchase behavior. *Psychol. Mark.* **2001**, *4*, 389–413. [CrossRef]
41. Ellen, P.S.; Wiener, J.L.; Cobb-Walgren, C. The role of perceived consumer effectiveness in motivating environmentally conscious behaviors. *J. Public Policy Mark.* **1991**, *10*, 102–117. [CrossRef]
42. Minton, A.P.; Rose, R.L. The effects of environmental concern on environmentally friendly consumer behavior: An exploratory study. *J. Bus. Res.* **1997**, *40*, 37–48. [CrossRef]
43. Pinto, D.C.; Nique, W.M.; Anna, E.; Herter, M.M. Green consumer values: How do personal values influence environmentally responsible water consumption? *Int. J. Consum. Stud.* **2011**, *35*, 122–131. [CrossRef]
44. Roberts, J.A.; Bacon, D.R. Explaning the Subtle Relationship Between Environmental Concern and Ecologically Conscious Consumer Behaviour. *J. Bus. Res.* **1997**, *40*, 79–89. [CrossRef]
45. Chen, T.B.; Chai, L.T. Attitude Towards the Environment and Green Products: Consumers' Perspective. *Manag. Sci. Eng.* **2010**, *4*, 27–39.
46. Kim, Y.; Choi, S.M. Antecedents of Green Purchase Behavior: An Examination of Collectivism, Environmental Concern, and PCE. *Adv. Consum. Res.* **2005**, *32*, 592–599.
47. Ottman, J.A.; Stafford, E.R.; Hartman, C.L. Avoiding green marketing myopia: Ways to improve consumer appeal for environmentally preferable products environment. *Sci. Policy Sustain. Dev.* **2006**, *48*, 22–36. [CrossRef]
48. Schlegelmilch, B.B.; Bohlen, G.M.; Diamantopoulos, A. The link between gren purchasing decisions and measures of environmental consciousness. *Eur. J. Mark.* **1996**, *30*, 35–55. [CrossRef]
49. Shamdasani, P.; Chon-Lin, G.O.; Richmond, D. Exploring Green Consumers in an Oriental Culture: Role of Personal and Marketing Mix Factors. *Adv. Consum. Res.* **1993**, *20*, 488–493.
50. Sütterlin, B.; Brunner, T.A.; Siegrist, M. Who puts the most energy into energy conservation? A segmentation of energy consumers based on energy-related behavioral characteristics. *Energy Policy* **2011**, *39*, 8137–8152. [CrossRef]
51. Moisander, J. Motivational Complexity of Green Consumerism. *Int. J. Consum. Stud.* **2007**, *31*, 404–409. [CrossRef]
52. Blair, I. *Greener Marketing: A Responsible Approach to Business*; Charter, M., Ed.; Greenleaf Publishing: Sheffield, UK, 1992.
53. Thøgersen, J. Spillover processes in the development of a sustainable consumption pattern. *J. Econ. Psychol.* **1999**, *20*, 53–81. [CrossRef]
54. Trudel, R. Sustainable consumer behavior. *Consum. Psychol. Rev.* **2019**, *2*, 85–96. [CrossRef]
55. Ali, A.; Khan, A.A.; Ahmed, I. Determinants of Pakistani consumers' green purchase behavior: Some insights from a developing country. *Int. J. Bus. Soc. Sci.* **2011**, *2*, 217–226.
56. Ishaswini, N.; Datta, S.K. Pro-environmental concern influencing green buying: A study on Indian consumers. *Int. J. Bus. Manag.* **2011**, *6*, 124–133.
57. Ölander, F.; Thøgersen, J. Understanding of consumer behaviour as a prerequisite for environmental protection. *J. Consum. Policy* **1995**, *18*, 345–385. [CrossRef]
58. Kumar, P.; Polonsky, M.J. An analysis of the green consumer domain within sustainability research: 1975 to 2014. *Australas. Mark. J.* **2017**, *25*, 85–96. [CrossRef]

59. Bangsa, A.B.; Schlegelmilch, B.B. Linking sustainable product attributes and consumer decision-making: Insights from a systematic review. *J. Clean. Prod.* **2020**, *245*, 118902. [CrossRef]
60. Dursun, İ.; Kabadayı, E.; Tuğer, A. Pro-Environmental Consumption Behavior: Dimensions and Measurement. *Acad. Rev. Econ. Adm. Sci.* **2018**, *11*, 42–66.
61. Gandhi, M.; Sen, K. Environmentally responsive consumption: A study of young consumers in India. *Int. J. Multidiscip. Thought* **2013**, *3*, 439–447.
62. Roberts, J.A. Green consumers in the 1990s: Profile and implications for advertising. *J. Bus. Res.* **1996**, *36*, 217–231. [CrossRef]
63. Shao, J. Sustainable consumption in China: New trends and research interests. *Bus. Strategy Environ.* **2019**, *28*, 1507–1517. [CrossRef]
64. Fu, L.; Sun, Z.; Zha, L.; Liu, F.; He, L.; Sun, X.; Jing, X. Environmental awareness and pro-environmental behavior within China's road freight transportation industry: Moderating role of perceived policy effectiveness. *J. Clean. Prod.* **2019**, *252*, 119796. [CrossRef]
65. D'Souza, C.; Taghian, M. Green Advertising Effects on Attitude and Choice of Advertising Themes. *Asia Pac. J. Mark. Logist.* **2005**, *17*, 51–66. [CrossRef]
66. Roberts, J.A. Profiling Levels of Socially Responsible Consumer Behavior: A Cluster Analytic Approach and Its Implications for Marketing. *J. Mark. Theory Pract.* **1995**, *3*, 97–117. [CrossRef]
67. Lubowiecki-Vikuk, A.; Dąbrowska, A.; Machnik, A. Responsible consumer and lifestyle: Sustainability insights. *Sustain. Prod. Consum.* **2021**, *25*, 91–101. [CrossRef]
68. Kaiser, F.G.; Biel, A. Assessing general ecological behavior: A cross-cultural comparison between switzerland and sweden. *Eur. J. Psychol. Assess.* **2000**, *16*, 44–52. [CrossRef]
69. Kaiser, F.G.; Ranney, M.; Hartig, T.; Bowler, P.A. Ecological Behavior, Environmental Attitude, and Feelings of Responsibility for the Environment. *Eur. Psychol.* **1999**, *4*, 59–74. [CrossRef]
70. Kollmuss, A.; Agyeman, J. Mind the Gap: Why Do People Act Environmentally and What Are the Barriers to Pro-Environmental Behavior? *Environ. Educ. Res.* **2002**, *8*, 239–260. [CrossRef]
71. Zou, J.; Tang, Y.; Qing, P.; Li, H.; Razzaq, A. Donation or Discount: Effect of Promotion Mode on Green Consumption Behavior. *Int. J. Environ. Res. Public Health* **2021**, *18*, 1912. [CrossRef] [PubMed]
72. Prothero, A.; Dobscha, S.; Freund, J.; Kilbourne, W.E.; Luchs, M.G.; Ozanne, L.K.; Thøgersen, J.; Freund, J. Sustainable Consumption: Opportunities for Consumer Research and Public Policy. *J. Public Policy Mark.* **2011**, *30*, 31–38. [CrossRef]
73. Liu, A.; Ma, E.; Qu, H.; Ryan, B. Daily green behavior as an antecedent and a moderator for visitors' pro-environmental behaviors. *J. Sustain. Tour.* **2020**, *28*, 1390–1408. [CrossRef]
74. McCarty, J.A.; Shrum, L.J. A Structural Equation Analysis of the Relationships of Personal Values, Attitudes and Beliefs About Recycling, and the Recycling of Solid Waste Products. *Adv. Consum. Res.* **1993**, *20*, 641–646.
75. Namazova, N. Features of regulating the use of financial resources in Azerbaijan. In Proceedings of the 55th International Scientific Conference on Economic and Social Development, Baku, Azerbaijan, 18–19 June 2020; Volume 4, pp. 213–218.
76. Schahn, J.; Holzer, E. Studies of individual environmental concern: The role of knowledge, gender, and background variables. *Environ. Behav.* **1990**, *22*, 767–786. [CrossRef]
77. Schwepker, C.H.; Cornwell, T.B. An examination of ecologically concerned consumers and their intention to purchase ecologically packaged products. *J. Public Policy Mark.* **1991**, *10*, 77–101. [CrossRef]
78. Stern, P.C. New environmental theories: Toward a coherent theory of environmentally significant behavior. *J. Soc. Issues* **2000**, *56*, 407–424. [CrossRef]
79. Stone, G.; Barnes, J.H.; Montgomery, C. Ecoscale: A scale for the measurement of environmentally responsible consumers. *Psychol. Mark.* **1995**, *12*, 595–612. [CrossRef]
80. Zimmer, M.R.; Stafford, T.F.; Stafford, M.R. Green issues: Dimensions of environmental concern. *J. Bus. Res.* **1994**, *30*, 63–74. [CrossRef]
81. Hines, M.J.; Hungerford, R.H.; Tomera, N.A. Analysis and Synthesis of Research on Responsible Environmental Behavior: A Meta-Analysis. *J. Environ. Educ.* **1987**, *18*, 1–8. [CrossRef]
82. Autio, M.; Heinonen, V. To consume or not to Consume? Young People's Environmentalism in the Affluent Finnish Society. *Young* **2004**, *12*, 137–153. [CrossRef]
83. Kreidler, N.B.; Joseph-Mathews, S. How green should you go? Understanding the role of green atmospherics in service environment evaluations. *Int. J. Cult. Tour. Hosp. Res.* **2009**, *3*, 228–245. [CrossRef]
84. Qasim, H.; Yan, L.; Guo, R.; Saeed, A.; Ashraf, B.N. The defining role of environmental self-identity among consumption values and behavioral intention to consume organic food. *Int. J. Environ. Res. Public Health* **2019**, *16*, 1106. [CrossRef] [PubMed]
85. Straughan, R.D.; Roberts, J.A. Environmental Segmentation Alternatives: A Look at Green Consumer Behaviour in the New Millennium. *J. Consum. Mark.* **1999**, *16*, 558–575. [CrossRef]
86. Fernandez-Feijoo, B.; Romero, S.; Ruiz, S. Effect of stakeholders' pressure on transparency of sustainability reports within the GRI framework. *J. Bus. Ethics* **2014**, *122*, 53–63. [CrossRef]
87. Ciocirlan, C.; Pettersson, C. Does Workforce Diversity Matter in the Fight against Climate Change? An Analysis of Fortune 500 Companies. *Corp. Soc. Responsib. Environ. Manag.* **2012**, *19*, 47–63. [CrossRef]

88. Diamantopoulos, A.; Schlegelmilchb, B.B.; Sinkovicsd, R.R.; Bohlenc, G.M. Can sociodemographics still play a role in profiling green consumers? *J. Bus. Res.* **2003**, *56*, 465–480. [CrossRef]
89. Park, J.; Zheng, W. Human Exposure and Health Effects of Inorganic and Elemental Mercury. *J. Prev. Med. Public Health* **2012**, *45*, 344–352. [CrossRef]
90. Schwartz, J. Earth Day Today. *Am. Demogr.* **1990**, *12*, 40–41.
91. Roper Organization Inc. *Environmental Behaviour*; S.C. Johnson and Son Inc.: New York, NY, USA, 1992.
92. Baudry, J.; Péneau, S.; Allès, B.; Touvier, M.; Hercberg, S.; Galan, P.; Amiot, M.-J.; Lairon, D.; Méjean, C.; Kesse-Guyot, E. Food choice motives when purchasing in organic and conventional consumer clusters: Focus on sustainable concerns (the nutrinet-santé cohort study). *Nutrients* **2017**, *9*, 88. [CrossRef]
93. Kleiner, A. What does it mean to be green? *Harv. Bus. Rev.* **1991**, *3*, 4–11.
94. Roper Starch Worldwide. *Green Gauge Report*; Roper Starch Worldwide Inc.: New York, NY, USA, 1997.
95. Schlossberg, H. Green marketing has been planted—Now watch it grow. *Mark. News* **1991**, *4*, 26–30.
96. Winski, J.M. Green marketing: Big prizes, but no easy answers. *Advert. Age* **1991**, *62*, 26.
97. Barbarossa, C.; Pastore, A. Why environmentally conscious consumers do not purchase green products. *Qual. Mark.Res. Int. J.* **2015**, *18*, 188–209. [CrossRef]
98. Bamberg, S.; Schmidt, P. Incentives, morality, or habit? Predicting students' car use for university routes with the models of Ajzen, Schwartz, and Triandis. *Environ. Behav.* **2003**, *35*, 264–285. [CrossRef]
99. McCarty, J.A.; Shrum, L.J. The recycling of solid wastes: Personal values, value orientations, and attitudes about recycling as antecedents of recycling behavior. *J. Bus. Res.* **1994**, *30*, 53–62. [CrossRef]
100. Tanner, C.; Kast, S.W. Promoting Sustainable Consumption: Determinants of Green Purchases by Swiss Consumers. *Psychol. Mark.* **2003**, *20*, 883–902. [CrossRef]
101. Widayat, W.; Praharjo, A.; Putri, V.P.; Andharini, S.N.; Masudin, I. Responsible Consumer Behavior: Driving Factors of Pro-Environmental Behavior toward Post-Consumption Plastic Packaging. *Sustainability* **2022**, *14*, 425. [CrossRef]
102. Orzan, G.; Cruceru, A.F.; Bălăceanu, C.T.; Chivu, R.-G. Consumers' Behavior Concerning Sustainable Packaging: An Exploratory Study on Romanian Consumers. *Sustainability* **2018**, *10*, 1787. [CrossRef]
103. Yatish, J.; Zillur, R. Factors Affecting Green Purchase Behaviour, and Future Research Directions. *Int. Strateg. Manag. Rev.* **2015**, *3*, 128–143.
104. Royne, M.B.; Martinez, J.; Oakley, J.; Fox, A.K. The Effectiveness of Benefit Type and Price Endings in Green Advertising. *J. Advert.* **2012**, *41*, 85–102. [CrossRef]
105. Choi, S.; Sangno, L. Eco-Packaging and its Market Performance: UPC-level Sales, Brand Spillover Effects, and Curvilinearity. *Sustainability* **2020**, *12*, 9061. [CrossRef]
106. Radman, M. Consumer consumption and perception of organic products in Croatia. *Br. Food J.* **2005**, *107*, 263–273. [CrossRef]
107. Lea, E.; Worsley, T. Australians' organic food beliefs, demographics and values. *Br. Food J.* **2005**, *107*, 855–869. [CrossRef]
108. Barr, S. Factors influencing environmental attitudes and behaviors: A UK case study of household waste management. *Environ. Behav.* **2007**, *39*, 435–473. [CrossRef]
109. Pepper, M.; Jackson, T.; Uzzell, D. An examination of the values that motivate socially conscious and frugal consumer behaviours. *Int. J. Consum. Stud.* **2009**, *33*, 126–136. [CrossRef]
110. Berger, I.E.; Corbin, R.M. Perceived consumer effectiveness and faith in others as moderators of environmentally responsible behaviors. *J. Public Policy Mark.* **1992**, *11*, 79–89. [CrossRef]
111. Gupta, S.; Agrawal, R. Environmentally responsible consumption: Construct definition, scale development, and validation. *Corp. Soc. Responsib. Environ. Manag.* **2017**, *25*, 523–536. [CrossRef]
112. Barbarossa, C.; De Pelsmacker, P. Positive and negative antecedents of purchasing eco-friendly products: A comparison between green and non-green consumers. *J. Bus. Ethics* **2016**, *134*, 229–247. [CrossRef]
113. Barr, S.; Gilg, A.W.; Ford, N. The household energy gap: Examining the divide between habitual-and purchase-related conservation behaviours. *Energy Policy* **2005**, *33*, 1425–1444. [CrossRef]
114. AZERTAG. Minimum Əməkhaqqı 300 Manata Çatdırılıb, Vahid Tarif Cədvəli Üzrə. Available online: https://azertag.az/xeber/Minimum_emekhaqqi_300_manata_chatdirilib_vahid_tarif_cedveli_uzre_isleyenlerin_emekhaqqi_artirilib-1954354. (accessed on 12 January 2022).
115. Leonidou, N.C.; Katsikeas, S.C.; Morgan, A.N. 'Greening' the marketing mix: Do firms do it and does it pay off? *J. Acad. Mark. Sci.* **2013**, *41*, 151–170. [CrossRef]
116. Liobikienė, G.; Grincevičienė, Š.; Bernatonienė, J. Environmentally friendly behaviour and green purchase in Austria and Lithuania. *J. Clean. Prod.* **2017**, *142*, 3789–3797. [CrossRef]
117. Hogarty, K.Y.; Hines, C.V.; Kromrey, J.D.; Ferron, J.M.; Mumford, K.R. The quality of factor solutions in exploratory factor analysis: The influence of sample size, communality and overdetermination. *Educ. Psychol. Meas.* **2005**, *65*, 202–226. [CrossRef]
118. Stevens, J.P. *Applied Multivariate Statistics for the Social Sciences*, 4th ed; Lawrence Erlbaum Associates: Mahwah, NJ, USA, 2002.
119. MacCallum, R.C.; Widaman, K.F.; Zhang, S.; Hong, S. Sample size in factor analysis. *Psychol. Methods* **1999**, *4*, 84–99. [CrossRef]
120. Reddy, L.S.; Kulshrestha, P. Performing the KMO and Bartlett's Test for Factors Estimating the Warehouse Efficiency, Inventory and Customer Contentment for E-retail Supply Chain. *Int. J. Res. Eng. Appl. Manag.* **2019**, *5*, 1–13.
121. Gautam, V. Examining environmental friendly behaviors of tourists towards sustainable development. *J. Environ. Manag.* **2020**, *276*, 111292. [CrossRef] [PubMed]

sustainability

MDPI

Article

Barriers for Prosumers' Open Business Models: A Resource-Based View on Assets and Data-Sharing in Electricity Markets †

Mehdi Montakhabi [1,2,*], Ine Van Zeeland [1] and Pieter Ballon [1]

1 imec-SMIT, Vrije Universiteit Brussel, Pleinlaan 9, 1050 Brussels, Belgium; ine.van.zeeland@vub.be (I.V.Z.); pieter.ballon@vub.be (P.B.)
2 Center of Energy Technologies (CET), Department of Business Development and Technology (BTECH), Aarhus University, 8000 Aarhus, Denmark
* Correspondence: mehdi.montakhabi@vub.be

† This article is an extended version of our paper published in the proceedings of the 16th IEEE International Conference on Distributed Computing in Sensor Systems (DCOSS 2020) (pp. 482–489).

Abstract: This article explores the barriers for open business models in support of sustainability in electricity markets. It puts forward privacy and data protection concerns about sharing prosumers' physical assets as well as data due to to their critical role in decentralized modes of electricity/flexibility trading. In particular, it uses a multiple case study approach to identify actors' resources, examine other interested actors in each resource, define their objectives, and consider privacy and data protection concerns of sharing prosumers' physical assets and data. The findings yield new insights into sharing opportunities beyond electricity/flexibility trading. In doing so, our study contributes to theories of the firm by applying the resource-based view in a new context and to the business model literature by shedding light on barriers in applying open business models.

Keywords: circular business model; open business model; sharing economy; circular economy; peer-to-peer; electricity trading

Citation: Montakhabi, M.; Van Zeeland, I.; Ballon, P. Barriers for Prosumers' Open Business Models: A Resource-Based View on Assets and Data-Sharing in Electricity Markets. *Sustainability* **2022**, *14*, 5705. https://doi.org/10.3390/su14095705

Academic Editors: Elkhan Richard Sadik-Zada and Andrea Gatto

Received: 15 March 2022
Accepted: 4 May 2022
Published: 9 May 2022

1. Introduction

Adverse results of climate change which are increasingly manifesting around the globe [1] have made combating climate change an international mission [2]. The European Green Deal [3] and Paris Agreement [4] are two examples of international efforts to fight climate change. Energy transition as a path from "status quo to the envisioned future" [5] is said to be one of the most effective preventive acts against climate change [6].

While a considerable share of the extracted energy (from any source), before consumption, is transformed into electricity [7], the electricity industry is experiencing a rapid change toward more digital and decentralized forms of commercial interactions [8]. The electricity market is a highly regulated market that has been dominantly controlled by governments, with a clear division of labor and established roles [9]. In this situation, innovations do not frequently happen at the level of firms [10]. Nevertheless, digital and decentralized forms of commercial interactions are said to transform the industry from the bottom up [11]. Technological breakthroughs in the production of high-capacity batteries and solar panels at low prices, and the prevalence of smart devices, have facilitated the emergence of new market models (e.g., peer-to-peer, community self-consumption, and transactive energy models) [12]. Accordingly, electricity markets are witnessing the emergence of new types of multipolar innovation ecosystems around these models [13,14]. Not only do they innovate the paradigms of value generation and capturing, they also transcend firms' boundaries and transform a wide range of established organizations [15].

The wave of the 5D global energy megatrends, namely, decarbonization, decentralization, digital transformation, democratization, and disruption-as-usual, is said to have accelerated the shift from the conventional electricity paradigm to a new era of decentralized, distributed, clean, and smart energy systems [16]. The interactions between the urgent need to tackle climate change, advances in information and communication technologies (ICTs), and the proliferation of distributed energy resources, batteries, and home energy management systems, are, to a large extent, the main antecedents for this change [17]. New market models are said to have the capacity to generate a wide range of economic, social, and environmental values [14,18].

New market models in the electricity market have attracted scholars' and practitioners' attention over the past years [19,20]. Technical challenges of these models, such as their impacts on the power grid [21], different market and pricing mechanisms [22], security and data protection aspects [23], and (other) legal requirements [24] have been extensively studied, generating important insights toward understanding decentralized electricity/flexibility trading.

However, despite their potential value, other opportunities for sharing in the electricity market, beyond electricity and/or flexibility trading, have been overlooked in the literature [25]. Alongside the emergence of new market models, which are based on electricity/flexibility trading [26], the concept of sharing can potentially be applied beyond trading electricity or flexibility between peers [27]. This gap in the literature gains importance because some of these opportunities can already be exploited, whereas market models require several co-innovations and adoptions for emergence [28].

The purpose of this article is to explore the barriers for open business models [29] in support of sustainability in electricity markets. To do so, we narrow our focus to sharing phenomenon as a means for manifestation of open business models [30] in the electricity market and conduct a multiple case study [31] on the P2P, CSC, and TE models. The study builds upon [32]'s perspective on framing prosumers as rational market actors (i.e., firms) and therefore utilizes theories of the firm [33], resource-based theory [34] in particular, to explore opportunities for sharing physical assets as well as data. We winnow our discussions down to prosumers as one of the main emerging actors [35] in electricity markets due to their pivotal role in decentralized models. We investigate drawbacks in implementing prosumers' open business models from the perspective of privacy and data protection [36].

The findings yield insights into sharing opportunities beyond electricity/flexibility trading that have been overlooked so far. In doing so, our study contributes to theories of the firm by applying the resource-based view [37] in a new context (i.e., the electricity market) and to the business model literature by shedding light on prosumers' business models [38] through identifying barriers in applying open business models (i.e., by examining privacy and data protection concerns on sharing prosumers' physical assets and data).

This article extends previous research [39] (the preliminary results of this study were presented at the 16th IEEE International Conference on Distributed Computing in Sensor Systems (DCOSS 2020)) by conducting an extensive literature review on the theoretical and conceptual groundings of the study, improving the methodology by introducing detailed steps for data gathering and data analysis, providing a privacy discussion on the opportunities for sharing around prosumers' physical assets as well as their data, and providing discussions on prosumers' open business models [40]. This article adds two new contributions to our previous research [39] (see the second and third contributions in the next paragraph).

The novel contributions of this article are three-fold:

- First, it contributes to theory by applying resource-based theory to a relatively new context—electricity markets—to identify resources with sharing potential held by different existing and emerging actors. To our knowledge, resource-based theory has not been used before in the context of the electricity trading to discover sharing opportunities [41].

- Second, it introduces new insights on prosumers' business models [42] based on possibilities for sharing beyond trading surplus electricity from distributed resources, or prosumers' flexibility in future electricity markets [43].
- Third, it analyzes the differences between privacy [44] aspects of sharing physical objects versus data protection [45] issues related to sharing prosumers' data in electricity markets. We contend that physical privacy aspects are too often overlooked, though they can present critical barriers to prosumer sharing.

The rest of this manuscript is organized as follows. Section 2 provides a literature review on new market models for electricity and/or flexibility trading, open business models for circularity, and theories of the firm. This is followed by the methodology section, which introduces boundary conditions, case study selection, data gathering, and data analysis (Section 3). Section 4 presents the traditional electricity trading paradigm and introduces three cases: P2P, CSC, and TE. Section 5 presents the findings of the study. Valuable resources of the different actors in the electricity market with a potential for sharing, interested actors, and the benefits they can gain by accessing each resource, are presented in this section. Consequently, opportunities for sharing in the electricity market are introduced. Section 6 analyzes the findings. First, the current status of sharing is compared with future opportunities. Secony, privacy and data protection aspects of prosumers' sharing opportunities are analyzed. Section 7 critically discusses the findings and compares the results with others. Furthermore, we outline how our study contributes to theory. Section 8 discusses the limitations of this study and opportunities for future research. Section 9 concludes the paper.

2. Literature Review

2.1. New Market Models for Electricity or Flexibility Trading

When it comes to new market models for electricity trading, many initiatives can be detected, including peer-to-peer (P2P) electricity trading, community self-consumption (CSC), and transactive energy (TE) models [38]. There is not yet a clear distinction between the three market models, but they promise similar benefits. Although the three models have their own merits, there is no common understanding about them. Knowing that they may be self-contradictory in some situations, it becomes important to highlight the alignments and contradictions between them [22].

There is a rich literature on various market models for electricity transactions. The most cited models are the P2P, the CSC, the TE, or a combination of them [22]. A common element of these models is the encouragement of and incentives for prosumers (and consumers) to take up an active role in electricity markets. This element is usually achieved by allowing prosumers to trade electricity/flexibility with other market participants, such as other prosumers and grid operators, in return for some (financial) incentives. However, the models also have their own specific objectives [32].

2.2. Open Business Models for Circularity

Companies are increasingly exploring and exploiting external resources and knowledge by increasing their openness to innovating their business models, in order to gain a competitive advantage [46–48]. Yet, research on open business models remains scarce [49]. In our study, we define open business models as "a subclass of business models in which collaboration of the focal firm with its ecosystem is a decisive or novel element of value creation and capturing" [49] (p. 175). Through open business models, a company becomes part of a larger innovation ecosystem consisting of individuals, communities, and other organizations, which entails simultaneous competition and cooperation between ecosystem actors [50,51]. Thus, open business models enable an organization to be more effective in both creating and capturing value [52] by leveraging more ideas through inclusion of a variety of external concepts. At the same time, open business models allow greater value capture by utilizing a firm's key asset, resource, or position, not only in that organization's own operations, but also in other companies' businesses [40,52]. This resonates very well with the ideas of the circular and sharing economy and circular business models. In understanding why companies engage in open business models, ref. [49] identified five main antecedents that lead companies to open up their business models: (1) business model inconsistency, (2) a need to create and capture new value, (3) previous experience with collaboration, (4) open business model patterns, and (5) industry convergence. Thus, sharing opportunities for new circular business to create and capture new value are at the core of open business models for circularity.

The concept of the circular economy has gained attention in the literature by transforming the way resources are applied: by shifting from existing open production systems to closed production systems, where resources are reused and kept in a loop of production and consumption [53]. Thus, the circular economy provides large potential for sustainability transformation [54]. Yet, very little is known about the realization of the political goals and scientific principles attached to a circular economy in business practice [55]. Researchers have recognized that in order to facilitate a circular economy, fundamental transitions of established business strategies, value chains, and eventually business models are needed [56]. Hence, the circular economy may be described as a cyclic system that aims to eliminate waste by turning goods that are at the end of their life cycle into resources for new goods [57]. Closing material loops in industrial ecosystems can create a continual use of resources, which is achieved through long-lasting design, proactive maintenance, recycling, repairing, refurbishment, and re-manufacturing [58].

The concept of circular business models has gained increasing focus (e.g., [59–61]) and may be defined as creating, delivering, and capturing value, while implementing circular strategies that can prolong the useful life of products and parts and close material loops [62], p. 187. A circular business should be built on distributed marketplaces or decentralized networks that create a sense of belonging, collective accountability, and mutual benefit through the community they build [63]. The market relations between the circular businesses differ between peer-to-peer (P2P), business-to-consumer (B2C), and business-to-business (B2B), where ref. [64] emphasizes sharing among peers as the most innovative and most interesting variant of the sharing economy. Increasingly, the sharing of products (as a service), specifically through digital sharing platforms, is seen as an enabler for a circular economy (e.g., [65]). Unlocking the potential of the circular economy depends on innovative large and complex dynamic data collection and analysis [66]. Thus, sharing data and opportunities with stakeholders is a critical dimension [67] as the circular economy depends on developing new business models.

2.3. *Theories of the Firm*

Theories of the firm are applied to state why firms exist and how they make decisions to maximize profits, compete, etc. In other words, they are applied to predict behaviors of firms. For more than four decades, economists, sociologists, and organizational scholars have extensively examined the theory of the firm's central question: What determines the boundaries of the firm? Many alternative theories have emerged and are frequently positioned as competing explanations, often with no shortage of critique for one another [68]. By building our study on [32]'s perspective, which considered prosumers as firms, we take the liberty of applying theories of the firm to discuss prosumers' open business models. In the following we provide an overview of three (out of many) theories of the firm—industrial organization theory, resource-based, and dynamic capabilities—that are mostly applied to tackle the increasing importance of transcending organizational boundaries through open business models in the literature [29]. Furthermore, we argue for choosing resource-based theory for this research.

Industrial organization theory considers the strategies that an organization would devise as a means to relate the firm to its environment. In this view, an organization selects a position that provides the best competitive conditions. An organization may decide to maintain this position or, if required, impact rivalry for its own benefit. As an example, we can mention strategies that hinder any new entry into the market in which the organization is competing. In this view, external events dictate a firm's strategies [69].

By contrast, the resource-based view takes an internal approach and considers strategies of organizations beyond responses to the actions of market forces. It prioritizes the resource base of an organization. Opportunities and threats (which are out of the control of organizations) are not the only identifiers of organizations' strategies [70]. Whether an organization continues to exist or maintain its superiority depends on whether it builds idiosyncratic capabilities: capabilities that distinguish it from competitors in changing markets [71]. In [70]'s view, industrial organization theory and resource-based theory complement each other, and can therefore be combined in the theory of industrial organization and resources.

Everything that a firm uses to achieve its planned goals is considered to be part of the firm's "resource base" [72]. Resources thus are "all assets, capabilities, organizational processes, firm attributes, information, knowledge, etc. controlled by a firm that enable the firm to conceive of and implement strategies that improve its efficiency and effectiveness" [73], p. 101. Resource-based theory considers specific types of firms' resources (i.e., VRIN resources) as potential basis for building competitive advantages.

Firms in competition possess various types of resources, which are mostly immobile. The immobility of firms' resources intensifies the variety of their competitive positions in the long run [74,75]. Value, rarity, inimitability, and non-substitutability are the characteristics of special resources of organizations, which impact their market selection and increase their success rate [76]. The possession of special resources is required for the success of a firm. Furthermore, possession of distinguishing capabilities enables a firm to effectively exploit its resources [75].

Resource-based theory has been criticized for being too static and for ignoring the dynamism of markets in recent decades, when competition was fiercer and markets were more turbulent [77]. Resource-based theory assumes that (a) resources are heterogeneously persistent, and (b) benefits originating in a lack of competition in gaining complementary resources are long lasting [74]. Both assumptions lead to a failure of resource-based theory in explaining the effects of turbulent markets and the evolution of firms. Despite the effective influence of resource-based theory on strategic management, its validity is disputed as it cannot answer questions about turbulent market situations, nor transformational mechanisms that use resources to create market advantages for a firm in competition [78]. For instance, ref. [79] calls resource-based theory into question because it fails to identify the mechanisms which allow firms to evolve their resources and capabilities, and transform their resources into competitive advantages.

Subsequently, the concept of "dynamic capabilities" can complement resource-based theory by considering resources' developing characteristics as well as developing capabilities of firms [80]. Dynamic capabilities are defined as the ability to build, develop, or amend the resource base of a firm. They are capabilities that are simultaneously part of the resource base of a firm. Dynamic capabilities can amend and develop themselves; they have the capacity to self-modify [72]. The concept of dynamic capabilities also completes the shortcomings of resource-based theory by justifying the effects of market dynamism [81]. Therefore, the relationship between the concept of dynamic capabilities and resource-based theory is complementary.

The concept of dynamic capabilities has been effectively used as a theoretical framework in abundant empirical research over the past two decades. It has been continually improved and now completes resource-based theory by filling its gaps, specifically to address the natural evolution of organizations' resources and capabilities under conditions of market turbulence. The concept of dynamic capabilities makes it possible to identify processes that are necessary for the evolution of the resources of firms, whether they are industry- or firm-specific [78].

Considering the purpose of this research, which is to identify sharing opportunities and consequently to identify barriers to apply prosumers' open business models, we opt for the resource-based view [37] for two main reasons. First, the sharing concept is tied with resources of others that can be deployed by the firm or resources of a firm that could be deployed by others [82]. Among the three discussed theories, the resource-based view is a better fit with the sharing concept. Second, the electricity market is, arguably, not a really turbulent market [83]. Therefore, between resource-based and dynamic capabilities, the former is a simpler theory to apply in this context.

3. Methodology

To carry out our study, an exploratory approach [84] was deployed. The overarching method for data gathering is multiple case study [31]. In the following subsections, boundary conditions, case study selection, data collection, and analysis are explained in detail.

3.1. Boundary Conditions

In this study, we apply the resource-based view [37]. Hence, each actor is considered as a bundle of resources under a common governance [85]. The resource base of a firm can have various components. Among the common categories of resources are:

1. Physical assets: grid infrastructure, smart meters, batteries, electric vehicles, etc.
2. Digital resources (data): smart meter information, flexibility-related information, supply and demand information, etc.

The resource base of a firm is not limited to these two categories. Other categories of resources have also been mentioned in literature, such as human resources, financial resources, downstream and upstream knowledge, governance-related (administrative) knowledge, and reputation-related resources, to name a few [85].

The sharing economy is concerned with non-VRIN resources [86]. Therefore, we use the two categories mentioned above, physical assets and data, to categorize the non-VRIN resources of different actors in the electricity market to which the sharing concept would apply.

3.2. Case Study Selection

As there is a flurry of initiatives centered on decentralized production and digitally enabled forms of transactions, we decided on a two-step process to narrow the focus. First, the empirical study was set to three cases in order to investigate actors and their resources as well as their motivations. Second, three controversial cases were selected yet with considerable similarities, that is, based on a decentralized nature, and with a diverse range of capacities in value generation [32] that affect electricity trading in similar ways, but at different capacities. The selected cases, arguably, are said to alternate the traditional electricity trading [87].

In doing so, setting market models as cases—thus, P2P, CSC, and TE models—they are not required to be limited to a specific project, pilot test, geographic region, or study [88], allowing a broad range of data for each case. While these are three different decentralized forms of electricity/flexibility trading changing electricity trading in different ways, the conclusions drawn are comparable.

3.3. Data Collection

In the first step of our study, extensive desk research [89] yielded insights into the identification of different existing and emerging actors in the electricity market. In particular, actors' resources, the objectives they pursue in the market, and other interested actors and their motivations are identified. This resulted in an overview of several actors in different market models (traditional, P2P, CSC, and TE models).

Subsequently, semi-structured interviews [90] were planned to validate the findings from the literature review (Appendix A shows the guideline questions for semi-structured interviews). For this purpose, the interviewees represent several stakeholder groups (e.g., existing actors, prosumers, policy makers, and academics) to provide a comprehensive view of the electricity market. Interview questions covered a wide range of aspects related to the current and future electricity markets (actors in the market, their responsibilities, resources, objectives, etc.). Top levels in management hierarchies were chosen for interviews to have a broad view on their companies' business models as well as a good understanding of the electricity market. To identify and select the interviewees, a snowball technique was deployed [13,91]. This helps to approach relevant people that otherwise tend to remain "under the radar"; this continues until data saturation is reached. A total of 26 interviews were conducted between October 2019 and March 2020. Interviews were conducted face-to-face and via Skype. Each interview took forty-five minutes on average.

3.4. Data Analysis

We transcribed interviews, coded the data, synthesized, built narratives, and applied storytelling techniques [92] to present the findings. Furthermore, in order to identify opportunities for sharing in the electricity market, a matrix was built on the findings of the study, which has actors in the electricity markets on its axes. For this study, we consider an opportunity to be a situation in which new services, products, raw materials, markets, and in general business models emerge through the formation of new means-ends combinations [93]. In this article, a sharing opportunity is a combination of a valuable resource of an actor and another actor interested in gaining access to that resource. The interested actor pursues one of several benefits by accessing the resource. In other words, the resource may be utilized for value generation or capturing by the interested actor.

The results of the matrix analysis were validated (i.e., face validity) by two expert members of the Global Observatory on Peer-to-Peer, Community Self-Consumption and Transactive Energy Models. We made a database of expert members in different five sub-tasks of Global Observatory: (i) power systems integration, (ii) hardware, software and data, (iii) transactions and markets, (iv) economic and social value, and (v) policy and regulation. Considering our research design, for the purpose of face validity, we opted for experts from sub-tasks ii and iii.

The next step in our analysis was to focus on the sharing opportunities of prosumers, and examine aspects of privacy and data protection related to the prosumer resources. To this end, we disambiguated concerns over privacy from concerns over personal data, based on a seminal study providing a comprehensive overview of different privacy typologies [94]. Applying these different types of privacy conceptions to the listed prosumer resources serves to clarify that sharing of physical assets can raise other privacy concerns (and merit a different type of protection) than merely concerns related to sharing data. Addressing these concerns will be instrumental to promoting prosumer sharing in electricity markets. Figure 1 shows a summary of our applied methods.

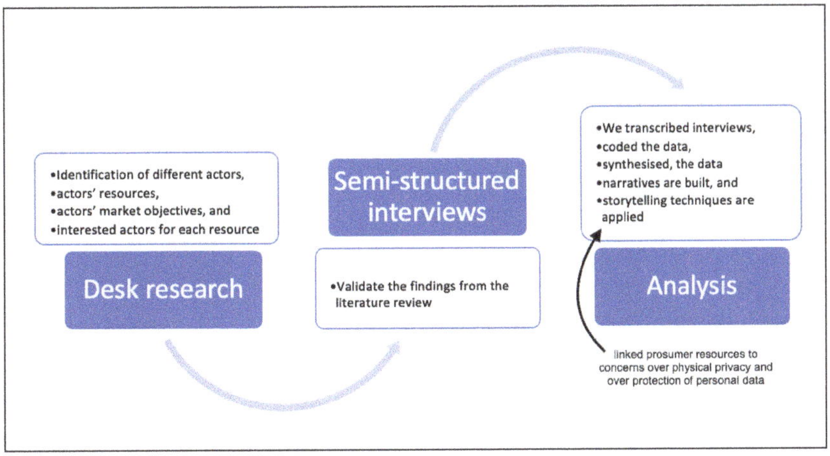

Figure 1. An overview of the applied methods.

4. Three Cases

The traditional electricity trading model refers to central electricity production from non-renewable (fossil-fuel, nuclear, gas, etc.) and renewable (wind, hydroelectric, etc.) resources in power plants. Electricity, in a one-directional flow, passes through the transmission grid, is transformed from high to low voltage, and is delivered to consumers through distribution grids [95]. Figure 2 shows the evolution of power systems during the time. In the following subsections, three cases (P2P, CSC, and TE) are introduced.

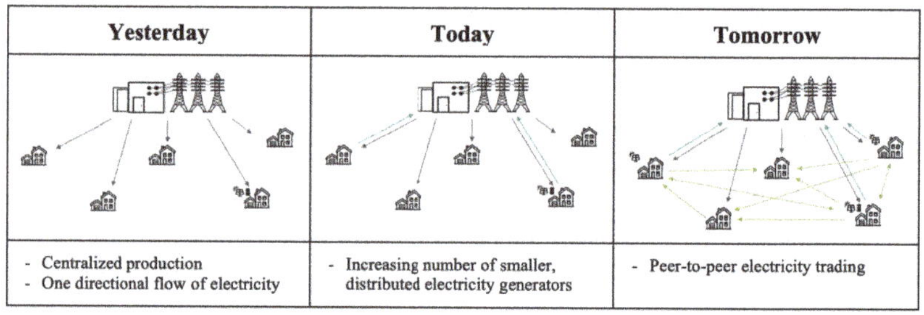

Figure 2. Evolution of power systems [14].

4.1. Case 1: Peer-to-Peer Electricity Trading

P2P market models support trading of electricity between prosumers (directly or through an intermediary) [39]. A possible future scenario of P2P energy trading is shown in Figure 3. Apart from prosumers, the scenario also envisions the presence of representatives who can trade electricity on behalf of citizens and a broker who facilitates and clears the market [35].

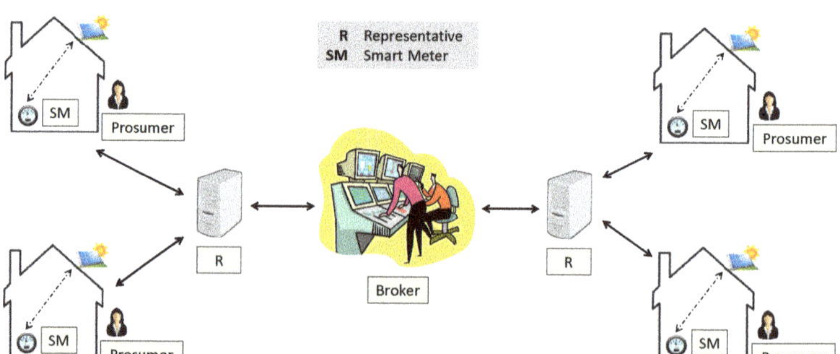

Figure 3. Peer-to-Peer electricity trading [13].

P2P models enable mutual transactions among different entities to trade electricity [96,97]. Energy traders in the P2P market may be of different sizes, i.e., residential houses, neighborhoods, microgrids, or local distribution networks [96,98]. These models are described as ways to allow the grid to take advantage of demand-side, flexible resources, operationally as well as economically [99]. In P2P models, the objective of the market mechanism is to incentivize transactions that prioritize maximizing the benefits of individual prosumers [100]. Such models could involve intermediate parties that facilitate the trades among prosumers, or support fully decentralized trades among them. The market mechanisms used in P2P models are usually set to optimize the trading based on algorithms with objectives of matching the excess supply of prosumers with the demand of consumers.

P2P electricity trading schemes enable consumers and prosumers in the same region to trade electricity they have produced with their own renewable energy resources among each other [101]. The type of sharing that occurs in these schemes is similar to what happens in most other sharing concepts that are based on P2P interaction [102–104], such as Airbnb and Uber.

4.2. Case 2: Community Self-Consumption

Community Self-Consumption (CSC) market models support communities in reducing their dependence from the electricity grid [105] through collectively utilizing their members' available resources [106]. Apart from aiming at grid independence, they could also provide flexibility services to other communities or to grid operators. Typically, a community manager orchestrates the market and assists members to collectively decides how to utilize their resources [107]. The assistance could range from having full responsibility for making decisions on behalf of the community members to only playing the role of a facilitator and strictly following the instructions given by the community's members. The community manager, the rights and privileges he or she holds, the scope of operation he or she has, and the length of his or her term are chosen by the community members via consensus mechanisms. Once these choices are made, the community manager's responsibility is to ensure that every community member receives a fair share of the rewards received according to their contribution to the service provided. The distribution of rewards is typically done via transactions between community members, between the community manager and community members, and the community manager and other communities/grid operators.

In CSC models, the overarching goal is independence from centralized electricity generation through unification of dispersed resources in communities [105]. Community members operate in a collaborative manner to optimize usage of resources [14,107]. CSC incentivizes transactions that prioritize maximizing the benefits of the community. CSC models usually involve a community manager [107] who coordinates transactions within the community as well as the transactions with other communities or the main grid. The market mechanisms used in CSC models are set to optimize the trading based on algorithms with objectives such as minimizing the electricity import by the community from the main grid and maximizing the revenue for the community. They usually involve sharing the individual prosumers' assets among each other or aggregating all the assets within the community in order to maximize the benefits for the entire community.

4.3. Case 3: Transactive Energy Models

Transactive Energy (TE) market models support grid operators in balancing the grid at all times [108]. They achieve this by facilitating demand response services provided by prosumers based on market-based incentives [109]. The key difference between the TE and P2P/CSC market models is that in TE models there is typically a single buyer—the grid operator or aggregator—who demands a certain amount of flexibility, thus offering financial incentives to prosumers to engage and provide flexibility. Prosumers are competing with each other on who would be selected for the service provision. Typically, the flexibility demanded is substantially larger than the flexibility an individual prosumer could provide [110]. Hence, typically, a number of prosumers is selected such that their aggregated flexibility satisfies the needs of the aggregator/grid operator. Similarly to CSC market models, all selected prosumers are rewarded in proportion to their contributed flexibility.

TE models are based on demand response, where end-user loads are automated and engaged through market-based interactions [108,110]. They provide market access to flexibility providers, and a support tool for the grid operators to manage technical complications [111]. It is a distributed control strategy that uses market mechanisms to engage self-interested responsive loads to provide services to the grid [112]. The objective of the market mechanism is to incentive transactions that prioritize and support the stability and reliability of the grid [108]. As the grid operators are the main actors who are responsible for maintaining the stability of the grid, they are usually the primary participants in the market buying electricity/flexibility directly from prosumers or via aggregators [113]. The market mechanisms used in TE models are set to optimize trading based on algorithms with different objectives, such as keeping the grid in balance, reducing grid congestion, and maintaining the voltage and frequency stability.

5. Findings

5.1. Actors' Resources and Interested Parties

In this section, actors (existing and emerging) and their market objectives in the electricity market are briefly introduced. Resources of each of the actors are identified. For the list of actors in the future electricity market, the study considers the future scenarios introduced in [114] and actors introduced based on the future scenarios in [35]. Resources of each actor are categorized under physical and digital groups. An overview of these resources is given in Table 1.

Table 1. Valuable resources of each actor in the future electricity markets and interested actors in each resource.

	Physical Resources ← Interested Parties	Digital Resources (Data)—Interested Parties
Prosumer (Pro)	RES ← Pro, Rep Home Battery ← Pro, Rep EV ← Pro, Rep EV charging station ← Pro, Rep, DSO EV battery ← Pro, Rep HEMS ← Pro, Rep Smart Appliance ←	Smart Meters ← Rep, Ret, DSO, TSO Demand ← Rep, Br, Ret, DSO, TSO Supply ← Rep, Br, Ret, DSO, TSO Flexibility ← Rep, DSO, TSO
Representative (Rep)	Clients' RES ← Pro, Rep Clients' Batteries ← Pro, Rep Clients' EVs ← Pro, Rep Clients' EV Charging Stations ← Pro, Rep, DSO Clients' Batteries of EVs ← Pro, Rep Clients' HMS ← Pro, Rep	Clients' Smart Meters ← Ret, DSO, TSO Clients Demand ← Br, Ret, DSO, TSO Clients' Supply ← Br, Ret, DSO, TSO Clients' Flexibility ← DSO, TSO
Broker (Br)		Sellers'supply information ← Ret, DSO, TSO Sellers'offered price ← Ret Buyers'demand information ← Ret, DSO, TSO Buyers offered price ← Ret Clearance price ← Ret Total traded volume ← Ret, DSO, TSO, Gen
Aggregator (Agg)		Clients'supply capacity ← TSO Clients'demand capacity ← TSO Clients'balancing capacity ← TSO
Retailer (Ret)		Customers'demand information ← DSO, TSO Customers'supply information ← DSO, TSO
DSO	Distribution grid infrastructure ← Pro, Rep, Ret	Smart meters' inflow information ← Rep, Ret Smart meters' outflow information ← Rep, Ret Congestion information ← Rep, Ret
TSO	Transmission grid Infrastructure ← TSO, Gen, Pro	Balancing information ← Agg, Ret, TSO, Gen Congestion information ← Agg, TSO, Gen Demand/supply pred. ← Agg, Ret, DSO, TSO, Gen
Generator (Gen)		Power plants (coal, gas, nuclear, etc.) ← Ret

Note: If the interested actor in a resource is the same as the owner, it indicates other peer actors are interested in that resource.

5.1.1. Prosumers

Prosumers are consumers who can also act as producers. In the electricity market, this means that they can generate electricity and inject it into the grid. To this end, they are in possession of renewable energy sources (e.g., solar panels) and storage devices (e.g., batteries). They will most likely also own smart meters, home energy management systems, and various other smart appliances. Prosumers' main objectives will be to *minimize cost* (specifically, their electricity bills), *maximize profit*, and *mitigate their dependency on the electricity grid*, by maximizing how they use their resources. To summarize, prosumers have the following valuable resources:

- A **renewable energy source (RES)** is a mini-generator located on a prosumer's premises (e.g., a solar panel). Typically, most of the electricity generated by a RES is consumed by its owner, who may inject surplus electricity into the grid.
- **Home batteries** are storage appliances that allow for intentional latency between the provision and the consumption of electricity generated or purchased by prosumers.
- A **smart meter** is an advanced measuring and recording device that keeps track of the electricity flowing in both directions (from the home to the grid, and vice versa) and that can perform two-way communications with other actors or appliances.
- A **home energy management system (HMS)** is a platform that consists of hardware and software to monitor electricity consumption and production. It allows prosumers to manually control and/or automate their household energy consumption.
- **Demand information** is information about the electricity requirements of prosumers, as well as their energy consumption patterns.
- **Supply information** is information about the volume of electricity produced by prosumers' RES, as well as production patterns.
- **Flexibility information** is information about the extent to which prosumers can modify their electricity production or consumption in response to variability, expected or otherwise.
- **Smart appliances** are Internet-connected appliances that connect to each other and/or other intelligent devices in the home. They can often be accessed and controlled remotely.
- An **electric vehicle (EV)** is a vehicle that has an electric motor for propulsion (or two such motors). EV owners typically also possess EV batteries and charging stations.
- An **EV battery** is a battery installed in an EV, which stores (and transports) electricity.
- An **EV charging station** is an appliance that can connect EVs to the grid to (dis)charge electricity.

5.1.2. Representatives (A New Role in Future Electricity Markets)

Representatives are a new type of actor in the peer-to-peer electricity market. Their role is to manage their clients' information and physical assets (RES, home battery, smart appliances, etc.) and to represent them in the peer-to-peer sharing market [35,114]. In doing so, they transform consumers' passive roles into active market participation. The presence of representatives can therefore increase consumer involvement in peer-to-peer electricity markets.

Representatives are expected to facilitate sharing opportunities between prosumers and other, already established market players.

Their main objectives will be to *minimize costs* and *increase profits* for their clients (prosumers). As representatives do not own their clients' assets, their resources mostly fall into categories other than physical and digital resources. Instead, representatives may for instance possess upstream knowledge (sourcing knowledge, and knowledge about product/service- or process-related technologies), downstream knowledge (which is critical to customer interfaces and is related to marketing, sales, distribution, and after-sales services), administrative (governance-related) resources, and reputable resources (brand names, a good reputation for honest business dealings, etc.).

5.1.3. Brokers (A New Role in Future Electricity Markets)

Like representatives, brokers take up a new role in future electricity markets [35]. They are intermediate, (neutral) actors facilitating the trade between peers. Due to their position, they have access to information about all parties participating in peer-to-peer electricity trading, as well as all transactions. Their objectives are to *clear the peer-to-peer market* while *respecting the grid's constraints*, as well as the *prosumers' preferences*.

As for representatives, brokers' special market position is not derived from their own physical or digital resources. Nor do they have upstream or downstream knowledge, or brand and other common resources. Brokers' positions mostly result from legal monopolies or first-mover advantages, which can provide actors with an important share of the market. Essentially, they will have access to all the information contained in a prosumer bid, such as the amount of electricity offered, the price requested, the demand and supply bids, buyer preferences (specific type of energy source, location, etc.), grid access points (location), and electricity source (solar, wind, biomass, etc.).

5.1.4. Aggregators

Aggregators are actors that exist already in current markets. They provide ancillary services to grid operators. They play an important role in balancing the electricity market by aggregating prosumers' flexibility and trading. Their main objective is to *maximize profit* by offering ancillary services to grid operators. To achieve this, they have access to the following information:

- **Clients' supply capacity:** Information about the capacity of their clients' electricity supply to the grid, including supply patterns.
- **Clients' demand capacity:** Information about the capacity of their clients' electricity demand from the grid, including demand patterns.
- **Clients' balancing capacity:** Information about the flexibility of their clients, which can be served to the grid.

5.1.5. Retailers

Retailers provide consumers with electricity. (In the case of prosumers, they also fulfill this role when prosumers' RES do not generate sufficient volumes of electricity). Retailers buy electricity in bulk from generators in the wholesale market and sell it to prosumers in the retail market. In current market settings, retailers are also obliged to buy any of the clients' electricity that is not traded in the P2P electricity market and still injected back to the grid. Their main objective is to *maximize profit*, while ensuring that their clients' demand for electricity is met. Retailers have access to the following information:

- **Customers' demand information:** This indicates how much electricity is consumed by retailers' clients over a specific period of time. This information is valuable to make any estimates about the future demand of the market.
- **Customers' supply information:** This indicates how much electricity is injected into the grid by retailers' clients over a specific period of time. Similarly, this information is valuable to make any estimates about the future demand/supply of the market.

5.1.6. Distribution System Operators (DSO)

DSOs are the operating managers (and sometimes owners) of energy distribution networks [115], operating at low and medium voltage levels. Their main objective is to *avoid congestion*.

- **Distribution grid infrastructure:** Distribution grid refers to the final stage of the electrical grid in which electricity is distributed to homes, industry, and other end-use products. Distribution is the process of reducing power to safe customer-usable levels, and delivering the electric power to the grid.
- **Smart meters' inflow information:** This includes the information regarding the amount and pattern of consumption at any smart meter. The more real-time this

information, the more valuable it is. Modern smart meters make it possible to read the smart meters' information in a real-time pattern. DSO is the sole actor who has access to this information through the smart meters installed at clients' premises.

- **Smart meters' outflow information:** Similar to the smart meters' inflow, this includes the information regarding the amount and pattern of electricity provisioned at any smart meter. The more real-time this information, the more valuable it is.
- **Congestion information:** Congestion can be defined as violations of network constraints (voltage and frequency) due to high electricity demand or excess electricity generation.

5.1.7. Transmission System Operators (TSOs)

A TSO is responsible for maintaining the transmission network, balancing the grid, and charging suppliers transmission network fees based on the electricity consumption/provision data of the suppliers' customers in the grid. A TSO's main objective is to *balance the grid*. A TSO has the following valuable resources:

- **Transmission grid infrastructure:** Electricity transmission is the bulk movement of electricity from a generating site, such as a power plant, to an electrical substation. The interconnected lines that facilitate this movement are known as a transmission grid.
- **Balancing information:** Electricity balancing encompasses all actions and processes, on all timelines, through which TSOs ensure, in a continuous way, the system frequency remains within a predefined stability range, as set forth in the Network Code on System Operation. It complies with the amount of reserves needed with respect to the required quality. This includes deficit, surplus, and reserves for any time period.
- **Congestion information:** Transmission congestion happens when scheduled market transactions (generation and load) result in power flow over a transmission element that exceeds the available capacity for that element.
- **Demand/supply prediction:** This includes all the estimates that TSOs can have based on the comprehensive information they receive for the balancing purposes.

5.1.8. Generators

Generators are the entities who generate electricity to meet the demand for electricity by consumers/prosumers. Their main objective is to *maximize profit* by trading electricity in the wholesale and/or balancing market. They have the following valuable resource:

- **Power plant:** An industrial facility for the generation of electric power. Power plants are connected to electricity grids. Their energy sources vary widely. Most of them burn fossil fuels (e.g., coal, oil, and natural gas). Cleaner energy sources of power plants include nuclear power and, increasingly, renewables (e.g., solar, wind, wave, and hydroelectric).

5.2. Opportunities for Sharing

In this section, the existing (or new) market players who would be interested in utilizing (some of the) resources of other actors, identified in the previous section, and the attainable benefit by accessing the resource are discussed. This part answers two questions: *who is interested in the resource?* and *what benefit can the interested actor gain?*

5.2.1. Prosumers

- **Renewable Electricity Sources (RES):** Other *prosumers/consumers* (or *representatives* on behalf of them), by accessing prosumers' RES, can produce renewable electricity. The owner of the produced electricity is the party with which it is shared. In a sense, rather than trading the electricity generated by the RES, prosumers can lend/rent their RES for a specific time period.
- **Batteries:** Other *prosumers/consumers* (or *representatives* on their behalf) can store electricity by accessing prosumers' batteries. The owner of the stored electricity is the party

with which the battery is shared. Renting out the RES and the battery in combination should be the most beneficial for both—the owner and the renter.

- **Electric vehicles (EV):** Electric vehicles can be used by other *prosumers* in idle time. Car sharing initiatives in a peer-to-peer manner are imaginable for this type of sharing. *Representatives* can represent underutilized capacity of electricity vehicles in a more efficient way on behalf of owners of EVs.
- **EV battery:** Other *prosumers* are interested in EV batteries as portable storage devices. People can, for instance, sell 2kw to users living close to their work place. Rather than feeding this energy to the grid at their home location, they can transport it with their EV and inject it to the grid at their work locations, potentially saving on grid use fees. *Representatives* can represent their clients' EV batteries in a more efficient way.
- **EV charging station:** EV charging stations can be used by other *prosumers* in idle times. This requires bringing the cars to the location of the station to charge them. Other prosumers can be interested in using the EV charging stations for charging their EVs. *Representatives* are also interested in expanding their service basket and in sharing charging stations more efficiently on behalf of the owners. *DSOs* would be interested as a means to help the congestion problem without occupying the distribution grid's capacity.
- **Smart meters:** People cannot share the smart meter, but it generates valuable data to be shared. Considering the frequency of access to the smart meter data, it reveals information that is highly valued by several actors. *DSOs* value this information for billing and solving congestion problems, *TSOs* for balancing, and *retailers* for customer consumption/production estimation purposes. *Representatives* can represent this information on behalf of their clients.
- **Home energy management system (HMS):** As is the case for the smart meter, the data generated by these systems are probably their most valuable output. *Representatives* can use these data for participating in various markets. It is unclear how one can benefit from sharing the system, unless a neighbor (*other prosumer*) could use the system's functionality as well. In a sense, the neighbor (or any user) could send the data of their assets and let the home management system of another user make intelligent decisions for them. It is yet to be seen how practical this solution is. It might well work in apartment buildings, where several flats use only one home management system.
- **Demand information:** *Representatives* are interested to have this information because it enables them to better represent their clients' demand and decide on the way to supply (purchasing from other prosumers, purchasing from grid, using the battery capacity, etc.). *Retailers* can plan their electricity provision based on forecasts based mainly on their clients demand/supply information. *DSOs* can use these data to better handle the distribution grid congestion problem. *TSOs* can better balance the grid by having this information. *Brokers* are interested in the part of this information which would be traded in peer-to-peer electricity markets through their channel.
- **Supply Information:** *Representatives* are interested in this information because it enables them to better represent their clients' supply and decide how to sell/share (selling to other prosumers, selling to the grid, using the battery capacity to store the produced electricity, etc.). *Retailers* can plan their electricity provision based on forecasts based mainly on their clients' demand/supply information. *DSOs* can use these data to better handle the distribution grid congestion problem. *TSOs* can better balance the grid by having this information. *Brokers* are interested in the part of this information which would be traded in peer-to-peer electricity markets through their channel.
- **Flexibility:** *DSOs* are interested in flexibility capacity to overcome the congestion problem. *TSOs* can use this capacity as a means to balance the electricity grid. *Representatives* can represent this capacity on behalf of clients to other interested parties.

Aggregators do the same type of representation for their clients in the current electricity market.

5.2.2. Representatives

As discussed in the presentation of representatives in the previous section, the type of resources representatives own is mostly not relevant to sharing. However, representatives (on behalf of prosumers) have access to the same valuable resources as prosumers and facilitate sharing. What differentiates representatives' role in sharing from prosumers is that representatives have the aggregate capacity of their clients to share. The aggregation of clients' resources in the hands of a representative gives it a service basket and leverages its position to negotiate and enter to markets that is not possible for a single prosumer to access.

- **Sellers' supply information:** This information is about the capacity for supply, and not necessarily all the supply capacity is traded in peer-to-peer markets. The information is useful for planning purposes. *DSOs* are interested in the supply information to have a better congestion management in the distribution grid. *TSOs* can enhance balancing planning by access to this information. *Retailers* can have plan better for their electricity demand from generators, which later they will supply in the retail market.
- **Sellers' offered price:** Access to this information helps *retailers* in their pricing. Retailers' pricing is in direct competition with the supply price in peer-to-peer trading market.
- **Buyers' demand information:** This information is about the capacity for demand, and not necessarily all the demanded amount is traded in the peer-to-peer market. The information is useful for planning purposes. *DSOs* are interested in the demand information to have a better congestion management in the distribution grid. *TSOs* can plan better for balance through access to this information. *Retailers* can plan better for their electricity demand from generators, which they will later supply in the retail market.
- **Buyers' offered price:** Access to this information helps *retailers* in electricity pricing. It indicates potential customers' willingness to pay. It is also worth mentioning that this willingness is for the peer-to-peer market. Considering that retailers nowadays are also offering green electricity to their customers, it would help retailers to tailor their offering in that product segment.
- **Clearance price:** This is the price for complementary/substitute service of *retailers*. It is expected that retailers' pricing is directly impacted by this information.
- **Total traded volume:** *Retailers* and *generators* can adjust their pricing and supply by knowing the actual traded amount of electricity in the peer-to-peer trading market. *DSOs* can better manage the congestion problem in the distribution grid. *TSOs* can better balance the grid by knowing this information.

Consideration: What defines the brokers' position in the market is the type of information that they have access to. It seems some of this information (sellers' supply and price information, and buyers' demand information and offered price) are VRIN resources that, despite the existence of other interested parties, brokers are not supposed to share.

5.2.3. Aggregators

- **Clients' supply capacity:** *TSOs* are interested in this information for prevention of congestion in the transmission grid.
- **Clients' demand capacity:** *TSOs* are interested in this information for prevention of congestion in the transmission grid.
- **Clients' balancing capacity:** *TSOs* are interested because of the use of this capacity in balancing of the grid.

5.2.4. Retailers

- **Customers' demand information:** *DSOs* are interested in this information for distribution grid congestion prevention. *TSOs* are interested because of balancing purposes.

- **Customers' supply information:** As above.

5.2.5. Distribution System Operators (DSOs)

- **Distribution grid infrastructure:** *Prosumers* are interested in the distribution grid to receive their electricity (purchased in the retail or peer-to-peer market) through it. Connectivity to the grid is also necessary for receiving the balancing services which translates into the stability of the electricity stream. *Representatives* have the same dependency/interest as prosumers regarding the distribution grid. The distribution grid makes the existence of *retailers'* services meaningful. This means that they can only deliver what they sell if the client is connected to the grid.
- **Smart meters' inflow information:** *Retailers* are interested in this information because it reveals their customers' consumption behavior. It is valuable for pricing and planning purposes. *Representatives* are also interested in this information because they are the sellers in the peer-to-peer market. Electricity sold in the peer-to-peer market is in competition with retailers' offers. So it has a similar value for representatives. Considering the level of expertise of prosumers, the required expertise to process this information is absent in individual prosumers.
- **Smart meters' outflow information:** As above.
- **Congestion information:** *Representatives* are interested in this information because it shows where the situation is more prone to peer-to-peer trading (considering that peer-to-peer trading of electricity may cause congestion problems in the distribution grid). *Retailers* have similar interests as representatives in this information.

5.2.6. TSOs

- **Transmission grid infrastructure:** *Generators* and *prosumers* are already making use of this infrastructure by using it as a transposition means for their traded electricity.
- **Balancing information:** Active players in the balancing market are the interested parties. This includes *retailers*, *aggregators*, and *generators*. Other *TSOs* (neighboring countries) are interested in these data because of the balancing purposes.
- **Congestion information:** *Aggregators*, other *TSOs*, and *generators* are interested actors in this information. As a source of balancing solutions, this information is meaningful for the interested actors to find where there is a good node to offer their balancing services.
- **Demand/supply prediction:** Active players in the balancing market are the interested parties. This includes *retailers*, *aggregators*, and *generators*. Other *TSOs*, (neighboring) countries, are interested in these data because for balancing purposes. *DSOs* are also interested in these data because of the implications that it could have on the congestion in the distribution grid.

5.2.7. Generators

- **Power plants:** *Retailers* are interested in the production capacity of power plants rather than buying the electricity they produce. Retailers can keep it as a reserve for balancing purposes. This scenario makes sense if sharing the capacity is more beneficial for generators than selling the output electricity.

6. Analysis

6.1. *Current vs. Future Opportunities for Sharing*

This study was initially designed to explore sharing opportunities in future electricity markets. Interestingly, the findings revealed several opportunities already available in the present, which have been overlooked despite their potential for value creation and capturing for several actors. It goes without saying that these are missed opportunities that can be sources of value. This also resonates with the concept of increasing the exploitation of existing resources, which is one of the cornerstones of circularity [116]. To clarify this argument, by following the boundary conditions set for the study (i.e., a "sharing opportunity" is a situation in which at least one actor other than the owner is interested in a resource), we built a matrix based on a combination of actors to highlight sharing opportunities between every two types of actors. We also highlight the presently missed opportunities and future opportunities for sharing. As shown in Table 2, matrix cells present a combination of actors that make up the two axes of the matrix. Each cell lists resources of the actor in the left column that are of interest to the actor in the top row. We underline the items that are currently shared. The remaining items in the list are opportunities for sharing in the future.

One of the implications of the identified sharing opportunities is that we can expect new entrants [35] (e.g., providers of sharing platforms or other supporting services) in electricity markets. Attractive sharing opportunities may attract these actors, which will also lead to facilitation of trading. Among the newly introduced actors, representatives are credible candidates to extend their activities and seize these opportunities [115]. However, privacy concerns may stand in the way of full-fledged, active participation of all prosumers [39]. As a consequence of privacy concerns, which go beyond issues of data protection, prosumers may apply a privacy/sharing calculus, weighing sharing risks against benefits [117]. In that calculus, the compounded physical privacy and data protection concerns may weigh more heavily against the benefits of sharing than data protection concerns alone would do. Those promoting sharing opportunities in electricity markets would do well to consider not only security and data protection measures but also to address the physical privacy concerns that may impede (some) prosumers from reaping the benefits.

Among the existing players in the electricity market, aggregators currently more or less undertake the combined activities of brokers and representatives [13]. A future opportunity imaginable for aggregators is to diversify their activities to offer representative and broker services. An interesting question that remains open is whether a single actor will play representative, broker, and aggregators' roles [115]. It is yet to be seen if specialization and a division of labor, meaning distinct actors undertaking separate roles [13], is the path forward, or if diversification would generate enough return for aggregators.

Table 2. Opportunities for sharing between different actors.

	Pro	Rep	Bro	Agg	Ret	DSO	TSO	Gen
Pro	**Physical assets:** - RES - Batteries - EV - EV charging station - Battery of EV - HMS	**Physical assets:** - RES - Batteries - EV - EV charging station - Battery of EV - HMS **Data:** - SM info - Demand info - Supply info - Flexibility	**Data:** - Demand info - Supply info		**Data:** - SM info - Demand info - Supply info	**Data:** - SM info - Demand info - Supply info - Flexibility	**Data:** - SM info - Demand info - Supply info - Flexibility	
Rep	**Physical assets:** - Clients' RES - Clients' Batteries - Clients' EV - Clients' EV charging station - Clients' Battery of EV - Clients' HMS	**Physical assets:** - Clients' RES - Clients' Batteries - Clients' EV - Clients' EV charging station - Clients' Battery of EV - Clients' HMS	**Data:** - Clients' Demand info - Clients' Supply info		**Data:** - Clients' SM info - Clients' Demand info - Clients' Supply info	**Data:** - Clients' SM info - Clients' Demand info - Clients' Supply info - Clients' Flexibility	**Data:** - Clients' SM info - Clients' Demand info - Clients' Supply info - Clients' Flexibility	
Bro					**Data:** - Sellers'supply info - Sellers'offered price - Buyers'demand info - Buyers offered price - Clearance price - Total traded volume	**Data:** - Sellers'supply info - Buyers'demand info - Total traded volume	**Data:** - Sellers'supply info - Buyers'demand info - Total traded volume	**Data:** - Total traded volume
Agg							**Data:** - Clients' supply capacity - Clients' demand capacity - Clients' balancing capacity	
Ret						**Data:** - Customers' demand info - Customers' supply info	**Data:** - Customers' demand info - Customers' supply info	
DSO	**Physical assets:** - Distribution grid infrastructure	**Physical assets:** - Distribution grid infrastructure **Data:** - SM inflow info - SM outflow info - Congestion info			**Physical assets:** - Distribution grid infrastructure **Data:** - SM inflow info - SM outflow info - Congestion info			
TSO				**Data:** - Balancing info - Congestion info - Demand/supply prediction	**Data:** - Balancing info - Demand/supply prediction	**Data:** - Demand/supply prediction	**Physical assets:** - Transmission grid infrastructure **Data:** - Balancing info - Congestion info	**Data:** - Balancing info - Congestion info - Demand/supply prediction
Gen					**Physical assets:** - Power plants			

Note: Underlined items are the resources which are shared in the current market between the two involved actors.

6.2. Privacy and Data Protection Aspects of Prosumers' Sharing Opportunities

Despite opportunities to generate additional income, promote sustainable consumption, or enjoy the social activity of sharing [118], prosumers may be reluctant to share assets due to privacy concerns [114]. Such concerns may vary, from a desire to limit access to private spaces to discomfort with sharing information about daily behavioral patterns, or distrust against commercial parties whose access to sensitive data may lead to price or other market manipulation attempts. This section presents an analysis of privacy and data protection aspects of the sharing opportunities for prosumers discussed above, as well as a number of potential remedies.

First, a disambiguation between "privacy" and "(personal) data protection" is in order. A comprehensive, systematically developed typology of privacy distinguishes eight basic types: bodily, intellectual, spatial, decisional, communicational, associational, proprietary, and behavioral privacy, with informational privacy as a ninth type that overlaps, but does not coincide with, the eight basic types [94]. This non-exhaustive typology underscores the importance of protecting more than merely personal information when it comes to privacy.

The distinction between different types of privacy is also a legal one. The right to privacy was established in the Universal Declaration of Human Rights (UDHR), which states that "[n]o one shall be subjected to arbitrary interference with his privacy, family, home or correspondence, nor to attacks upon his honour and reputation". The European Convention for Human Rights (ECHR) of the Council of Europe also provides protection for the right to privacy ("respect for private life, family life, home, and correspondence") in its Article 8. This right covers a wide range of issues in the rulings of the European Court of Human Rights (ECtHR): bodily integrity, wiretapping, gender identification, and unwarranted trespassing and protection against environmental nuisances, among other issues [119]. The Treaty on the Functioning of the European Union distinguishes a separate right to the protection of personal data in its Article 16. So does the European Union's Charter of Fundamental Rights, which includes both the right to respect for private and family life in Article 7 and the right to protection of personal data in Article 8. The EU's General Data Protection Regulation (GDPR) regulates this latter right, not the right to private and family life.

In the context of prosumers sharing assets in the electricity market, concerns related to physical assets are discussed here as "privacy concerns", whereas concerns related to digital assets are discussed as "data protection concerns". These labels are chosen for reasons of brevity, not to introduce a strict distinction between informational or data privacy and other privacy concerns. Our aim is to clarify that privacy concerns do not end when (personal) data are secure.

To be clear, this section does not deal with the situation in which prosumers are handling other individuals' data, i.e., the situation in which the prosumer becomes a "controller" in terms of the GDPR. What is discussed are potential personal privacy and data protection concerns of prosumers related to sharing their own assets.

6.3. Privacy Aspects of Sharing Prosumers' Physical Assets

Prosumers experience privacy concerns when they fear an infringement of personal boundaries around their bodies, intimate spaces, or personal assets. Prosumers' physical resources, as listed above, are likely to be found in or around their private homes, and possibly the homes of their families. In "collaborative consumption" schemes [118], privacy concerns about giving strangers access to zones of intimacy (which can include electric vehicles) may affect prosumers' willingness to share. Compared to data protection issues, these physical privacy concerns have been understudied, but they can weigh more heavily than concerns about the protection of personal data [117]. Among the eight basic types of privacy distinguished by [94], five are primarily related to physical aspects: bodily, spatial, behavioral, associational, and proprietary privacy. Each of these types is discussed below in relation to the sharing of physical resources of prosumers in the electricity market.

- **Bodily privacy** refers to the integrity of a person's body, from excluding others from touching one's body to having the freedom to move one's body [94]. Though prosumers may fear infringements of their bodily privacy when allowing strangers into their homes for the purpose of sharing smart appliances or home batteries, privacy would not be their primary concern in this respect, which is why we shall leave bodily privacy concerns aside in this discussion. It is, however, of no small importance to realize that related concerns may dissuade, for instance, female prosumers who live alone from actively participating in collaborative consumption schemes.
- Like bodily privacy, **spatial privacy**, particularly the privacy of the home, is a constitutionally protected right around the world [94]. In some countries, "dwellings" or (in Poland) vehicles are also protected, while the inviolability of property, of computers, or of cell phones sometimes also falls under the protection of the private sphere. Clearly, the need to protect intimate spaces is felt worldwide. This implies that sharing of all physical assets mentioned in Section 5.1—renewable electricity sources, home batteries, electric vehicles, EV charging stations, EV batteries, home energy management systems, and smart appliances—can, to some extent, and depending on circumstances, trigger spatial privacy concerns.
- **Behavioral privacy** has to do with excluding others from observing one's personal actions and behaviors [120], with systemic observation being of particular concern. Visitors to a private space (when sharing renewable electricity sources, electric vehicles, home or EV batteries, EV charging stations, or smart appliances) would undoubtedly learn something of the sharing prosumer's lifestyle, actions, and personal behavior. However, systemic observation is unlikely to occur in the physical environment, but is potentially of graver concern in relation to information-sharing, or information that could be gleaned from the sharing of a home energy management system.
- **Associational privacy** refers to an "individuals' interests in being free to choose who they want to interact with: friends, associations, groups, and communities" [94]. Assuming that prosumers would voluntarily engage in collaborative consumption schemes, associational privacy is unlikely to be affected by the choice to share in itself. In fact, social influence plays an important role in users' sharing decisions: "The more the people in the sharers' environment encourage and support their sharing, the more frequently these users will share" [117]. Collaborative consumption therefore appears to be rather a positive expression of associational privacy preferences.
- **Proprietary privacy** is about property-based interests, meaning that property can be used to shield activity or information from others. Window curtains, locked doors, or closed bags are examples of physical property protections. Naturally, if property is shared, proprietary privacy is affected, but not necessarily infringed upon. Proprietary privacy is mostly a legal issue, in that it mostly applies to situations of unreasonable search and seizure [94], not to voluntary sharing. It is therefore less likely to be a concern specific to collaborative consumption.

For some assets, mitigation of physical privacy concerns can be achieved by introducing "safe spaces", away from intimate zones. For example, lighter smart appliances can be handed over in a public space. For less mobile physical assets, it will be of utmost importance to establish trust between sharing partners, e.g., through privacy-preserving features of sharing platforms or by introducing physical security checks, setting clear boundaries in contracts, or introducing sharing policies [121,122].

For all five types of physical privacy concerns, there are related data protection concerns. Information "leaked" about bodies, spaces, behavior, property (e.g., photos or videos taken surreptitiously), or associations (e.g., contact lists) may be as much a concern, or even more worrisome, to prosumers.

6.4. Data Protection Issues of Sharing Prosumers' Digital Resources

While informational privacy is considered to be a somewhat separate type of privacy that overlaps with the eight basic types, three of those basic types are also closely related to the protection of data: communicational, intellectual, and decisional privacy. These types, and informational privacy, are discussed below in relation to the sharing of digital resources of prosumers in the electricty market.

- **Communicational privacy** is also protected in all constitutions, specifically mediated communications (or "correspondence" in the ECHR) [94] (Unmediated communication, or communication in person, is sometimes protected in the same vein, or sometimes considered as a part of physical privacy). It refers to concerns regarding conscious communication, such as sending messages on sharing platforms or online marketplaces (potential "representatives"). Communicational privacy is affected if platforms are processing communications data [122] or in case of "eavesdropping" [114].
- **Intellectual** and **decisional privacy** are two sides of the same coin: intellectual privacy can be regarded as freedom from intrusion into the functioning of the mind, and decisional privacy is seen as the freedom to exercise one's mind. Prosumers sharing smart meter data, demand and supply data, and/or flexibility information with retailers or system operators may fear that their beliefs about prices and demand may be manipulated by incorrectly presented or skewed information provision by larger players in the market, affecting their ability to make effective, rational decisions in their own best interest.
- Lastly, **informational privacy** concerns around P2P sharing in the electricity market are well documented, including such risks as impersonation, data manipulation, and individual privacy breaches leaking location or trajectory data, payment information, or behavioral patterns to third parties [114,122]. Any kind of privacy also has an informational aspect. At the same time, information always relates to certain aspects of people's physical situations, which have privacy elements beyond information.

Fortunately, there is also ample discussion regarding mitigating data protection issues, partly due to the introduction of the GDPR, which requires that organizations processing personal data take extensive "technical and organizational measures" to protect such data. Technical measures can include secure authentication of parties and messages, encryption, anonymization and pseudonymization, secure channels, prosumer data aggregation, decentralized data storage, unlinkability, multiparty computation, and zero-knowledge proofs [114]. Organizational measures can include access controls, logging, contractual clauses, staff privacy training, confidentiality agreements, and binding policies, to name but a few common approaches. Figure 4 summarizes the privacy and data security aspects of sharing prosumers' physical assets as well as their data.

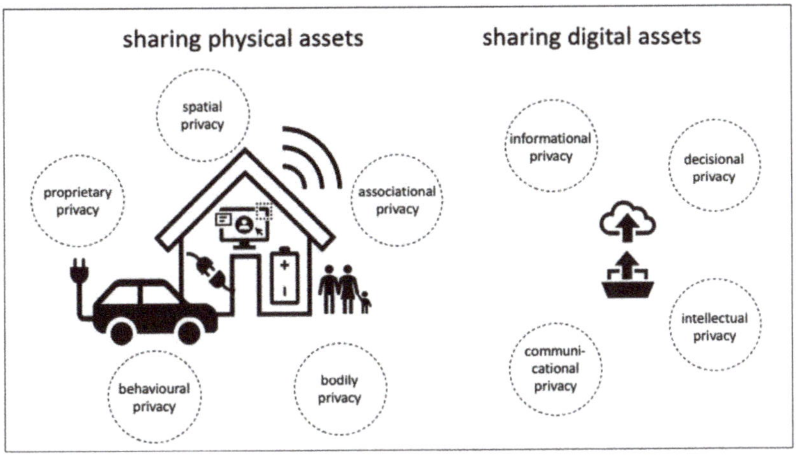

Figure 4. Privacy and data security aspects of sharing prosumers' physical assets as well as their data.

7. Discussion

The findings from this study suggest how prosumers' open business models, in decentralized market models in the electricity market, may be hindered by privacy concerns on sharing physical assets as well as data protection issues with sharing their data.

Our results showed several sharing opportunities in electricity markets exist which are already exploitable by prosumers for value creation and capturing in their business models. Furthermore, all the identified opportunities have a collaborative nature through resource sharing, which emphasizes the openness aspect of prosumers' business models. This is in line with previous studies' claims [46–48] on increasing use of open business models for exploring and exploiting external resources and knowledge. Furthermore, by identifying sharing opportunities between dual combination of actors it provides empirical evidence for [40,52]'s claim that open business models allow greater value capture by utilizing a firm's key asset, resource, or position not only in that organization's own operations, but also in other companies' businesses. Previous studies [29,49] mostly take a positive tone and focus on antecedents of open business models. Nevertheless, our analysis on prosumers' privacy and data protection concerns complements this strand of research by showing barriers for implementing open business models.

Somewhat surprisingly, most of the identified sharing opportunities are already exploitable. In other words, building open business models to make use of these opportunities is not hindered by legal constraints [24] or other structural changes in the electricity grid [123]. This sheds new light on previous studies that suggest the large-scale rollout of P2P, CSC, and TE models, as non-stand-alone innovations, requires several co-innovations and adoptions [13]. Furthermore, the results broaden the scope of sharing, which was mostly focused specific devices (charging stations, batteries, etc.) in previous studies [124], by putting prosumers as the core of discussion rather than a specific device.

Formerly, prosumers were considered passive actors in the electricity market because they could only inject their excess electricity into distribution grids and had no negotiation power on price setting nor on whom to sell [106,125]. As a result of the limited decisions they could take, having a business model for active prosumer participation in the market was deemed unnecessary. However, technology breakthroughs in the production of high-capacity batteries, distributed renewable resources (e.g., solar panels), ease of communication, and prevalence of smart devices (smart meters, home energy management systems, etc.), as shown under actors' resource base in our study, are expanding prosumer control and allowing them to play a more substantial active role in future electricity markets [126,127]. This shift in prosumers' capabilities necessitates better understanding of their options for value generation and capturing through their business models. A clear picture is necessary for prosumers as potential active participants in future electricity markets and for policy makers to better prioritize public resource allocation [32].

By following [128]'s guidelines for making theoretical contributions, we applied the resource-based theory of the firm in relatively new context—the electricity market—to identify resources with sharing potential held by different existing and emerging actors. This is a novel application for the resource-based theory to discover sharing opportunities [41]. Hence, our study contributes to theories of the firm, in particular to the resource-based theory [128]. Furthermore, by highlighting the privacy and data protection aspects on prosumers' sharing of physical assets and data, as potential barriers for open business models, the study contributes to open business models literature. Finally, by analyzing the differences between privacy [44] aspects of sharing physical objects and data protection [45] issues related to sharing prosumers' data in electricity markets, we contend that physical privacy aspects are too often overlooked, though they can present critical barriers to prosumer sharing.

8. Limitations and Opportunities for Future Research

This study has limitations that can be overcome in future research. Furthermore, this study raises important questions for follow-up research. First, it would be fruitful to pursue further research into the dynamics of prosumers' resource transformation in the short run and in the long run. The intersection of open business models [29], digital transformation [129], and dynamic capabilities [130] is another area for further research. The study is partly based on speculations derived from case studies. A follow-up, quantitative study (e.g., a survey) with a large sample size [131] can validate our findings. Among the drawbacks for implementing open business models by prosumers, we focused on privacy as well as data protection concerns. Taking other perspectives (technical, legal, etc.) to study the factors which hinder applying open business models can contribute to theory and practice. Broadening the scope to cover the business models as well as privacy and security aspects of other actors is another path to follow in future studies.

It is valuable to perform an in-depth privacy analysis of identified sharing opportunities that goes beyond concerns over sharing data or the security of IT infrastructure. While there is evidence that prosumers hesitate to participate in particular parts of the sharing economy due to concerns about the privacy of their homes, bodies, or other aspects of their private lives (e.g., in schemes such as Airbnb), there has been no empirical research into prosumers' privacy-related hesitations to participate in sharing schemes for physical assets in the electricity market [132]. This may be partly due to the lack of existing schemes. Although our data collection and validation did not allow for such empirical exploration of potential concerns related to all different types of privacy, there are certainly research strategies that do allow for exploration of future attitudes, e.g., using fictional methods [133] such as "futuristic autobiographies" [134].

Last but not least, the implicit assumption in identifying opportunities for sharing in our study, which is based on scenarios for the future [135,136], is that we can extrapolate and speculate about the future using the knowledge that we have at present. Hence, the possibility of disruption as the result of breakthrough innovations [137] is overlooked in this method. These could alter the whole market. Taking into account possible sources of disruption can be another inspiration for conducting follow-up studies.

9. Conclusions

This article explores the barriers for prosumers' open business models due to their critical role in decentralized modes of electricity/flexibility trading. In particular, it puts forward privacy and data protection concerns on sharing prosumers' physical assets and data. This is an exploratory study that utilizes a multiple case study method to identify existing and emerging actors in electricity markets. Furthermore, resource-based theory is applied to identify resources of the actors under physical assets and data categories. Objectives that different actors pursue in the electricity market are identified, and for each identified resource, interested actors and their motivation are introduced. Accordingly, a matrix is built by dual combination of actors, to identify what opportunities for sharing exist between every two actors. We built on [32]'s perspective and dealt with prosumers as firms in applying the resource-based theory and in our discussions. Identified sharing opportunities, besides participation in the new market models for electricity/flexibility trading, call for developing prosumers' open business model and highlight less attended opportunities. The study contributes to theories of the firm by applying the resource-based view in a new context and to the business model literature by shedding light on barriers in applying open business models.

Author Contributions: Conceptualization, M.M., I.V.Z. and P.B.; methodology, M.M. and I.V.Z.; data curation, M.M.; writing—original draft preparation, M.M. and I.V.Z.; writing—review and editing, M.M. and I.V.Z.; visualization, M.M. and I.V.Z.; supervision, P.B.; project administration, M.M. All authors have read and agreed to the published version of the manuscript.

Funding: This work was supported in part by the Flemish Government through the FWO-SBO project SNIPPET-S007619.

Institutional Review Board Statement: Not applicable.

Informed Consent Statement: Not applicable.

Data Availability Statement: Not applicable.

Acknowledgments: The authors would like to thank the co-authors of the earlier version of this research, which was published in the proceedings of the 16th IEEE International Conference on Distributed Computing in Sensor Systems (DCOSS).

Conflicts of Interest: The authors declare no conflict of interest.

Abbreviations

The following abbreviations are used in this manuscript:

ICTs	Information and Communication Technologies
P2P	Peer-to-Peer
CSC	Community Self-Consumption
TE	Transactive Energy
B2C	Business-to-Consumer
B2B	Business-to-Business
VRIN	Valuable, Rare, Imperfectly imitable and Not substitutable
RES	Renewable Energy Sources
EV	Electric Vehicle
HMS	Home energy Management System
DSO	Distribution System Operator
Pro	Prosumer
Rep	Representative
Br	Broker
Agg	Aggregator
Ret	Retailer
Gen	Generator
TSO	Transmission System Operator
RES	Renewable Electricity Sources
EV	Electric Vehicle
HMS	Home energy Management System
UDHR	Universal Declaration of Human Rights
ECHR	European Convention for Human Rights
ECtHR	European Court of Human Rights
GDPR	General Data Protection Regulation
DCOSS	Distributed Computing in Sensor Systems

Appendix A. Guideline Questions for Semi-Structured Interviews

- Would you please introduce yourself and your organization?
- What are your organization's roles in the electricity market?
- How do you describe the current electricity market (Who are actors and what are their roles)?
- What are the main influencers on the future of electricity market?
- How do you describe the future electricity market (Who are actors and what are their roles in next 5 to 10 years)?
- What are the constraints for P2P electricity trading?
- What are the best and worst scenarios for P2P electricity market?
- Who are the main actors and their roles in the P2P energy trading?
- Do existing trading mechanisms in electricity market need to be changed for P2P electricity trading (How should they be in order to allow for P2P trading)?
- Which non-energy-related market mechanism is applicable for P2P electricity trading?
- What roles would your company have in the future electricity market?
- What are your organization's objectives and business model in P2P electricity market?
- What problem is peer-to-peer is trying to solve?
- Which type of actors are trustworthy to have the critical roles in P2P, CSC, and TE models for electricity/flexibility trading?

References

1. Schellnhuber, H.; Cramer, W.; Nakicenovic, N.; Wigley, T.; Yohe, G. *Avoiding Dangerous Climate Change*; Cambridge University Press: Cambridge, UK, 2006.
2. United Nations Development Programme. Fighting climate change: Human solidarity in a divided world. In *Human Development Report 2007/2008*; Springer: London, UK, 2007.
3. Claeys, G.; Tagliapietra, S.; Zachmann, G. *How to Make the European Green Deal Work*; Bruegel Policy Contribution Issue n° 14 November 2019; Bruegel: Brussels, Belgium, 2019.
4. Streck, C.; Keenlyside, P.; Von Unger, M. The Paris Agreement: A new beginning. *J. Eur. Environ. Plan. Law* **2016**, *13*, 3–29. [CrossRef]
5. Verbong, G.; Geels, F. The ongoing energy transition: Lessons from a socio-technical, multi-level analysis of the Dutch electricity system (1960–2004). *Energy Policy* **2007**, *35*, 1025–1037. [CrossRef]
6. Hoppe, T.; van Bueren, E. Guest editorial: Governing the challenges of climate change and energy transition in cities. *Energy Sustain. Soc.* **2015**, *5*, 19. [CrossRef]
7. Armaroli, N.; Balzani, V. The future of energy supply: Challenges and opportunities. *Angew. Chem. Int. Ed.* **2007**, *46*, 52–66. [CrossRef] [PubMed]
8. Andoni, M.; Robu, V.; Flynn, D.; Abram, S.; Geach, D.; Jenkins, D.; McCallum, P.; Peacock, A. Blockchain technology in the energy sector: A systematic review of challenges and opportunities. *Renew. Sustain. Energy Rev.* **2019**, *100*, 143–174. [CrossRef]
9. Jamasb, T.; Pollitt, M. Electricity market reform in the European Union: Review of progress toward liberalization & integration. *Energy J.* **2005**, *26*, 11–41.
10. Lange, O. A note on innovations. *Rev. Econ. Stat.* **1943**, *25*, 19–25. [CrossRef]
11. LeBaron, B.; Tesfatsion, L. Modeling macroeconomies as open-ended dynamic systems of interacting agents. *Am. Econ. Rev.* **2008**, *98*, 246–250. [CrossRef]
12. Montakhabi, M.; van der Graaf, S.; Madhusudan, A.; Sarenche, R.; Mustafa, M.A. Fostering Energy Transition in Smart Cities: DLTs for Peer-to-Peer Electricity Trading. In Proceedings of the 2021 17th International Conference on Distributed Computing in Sensor Systems (DCOSS), Pafos, Cyprus, 14–16 July 2021; pp. 466–472.
13. Montakhabi, M.; Vanhaverbeke, W. Ecosystem Risks of Peer-to-Peer Electricity Trading. In Proceedings of the DRUID Academy 2022, Aalborg, Denmark, 12–14 January 2022.
14. Montakhabi, M.; Van Der Graaf, S.; Ballon, P.; Walravens, N.; Vanhaverbeke, W. Defining the Business Ecosystem of Peer-to-Peer Electricity Trading. In Proceedings of the 6th International Conference on New Business Models, Halmstad, Sweden, 9–11 June 2021.
15. Chesbrough, H.; Vanhaverbeke, W.; West, J. *New Frontiers in Open Innovation*; Oup Oxford: Oxford, UK, 2014.
16. IEA. *Net Zero by 2050: A Roadmap for the Global Energy Sector*; IEA: Paris, France, 2021.
17. Pratt, A.; Krishnamurthy, D.; Ruth, M.; Wu, H.; Lunacek, M.; Vaynshenk, P. Transactive home energy management systems: The impact of their proliferation on the electric grid. *IEEE Electrif. Mag.* **2016**, *4*, 8–14. [CrossRef]
18. Montakhabi, M.; Vannieuwenhuyze, J.; Ballon, P. Expert Recommendations on Energy Trading Market Models using the AHP model. In Proceedings of the IEEE ENERGYCON 2022, Riga, Latvia, 9–12 May 2022.
19. Jogunola, O.; Ikpehai, A.; Anoh, K.; Adebisi, B.; Hammoudeh, M.; Son, S.Y.; Harris, G. State-of-the-art and prospects for peer-to-peer transaction-based energy system. *Energies* **2017**, *10*, 2106. [CrossRef]
20. Adams, S.; Brown, D.; Cárdenas Álvarez, J.P.; Chitchyan, R.; Fell, M.J.; Hahnel, U.J.; Hojckova, K.; Johnson, C.; Klein, L.; Montakhabi, M.; et al. Social and economic value in emerging decentralized energy business models: A critical review. *Energies* **2021**, *14*, 7864. [CrossRef]
21. Tushar, W.; Saha, T.K.; Yuen, C.; Smith, D.; Poor, H.V. Peer-to-peer trading in electricity networks: An overview. *IEEE Trans. Smart Grid* **2020**, *11*, 3185–3200. [CrossRef]
22. Capper, T.; Gorbatcheva, A.; Mustafa, M.A.; Bahloul, M.; Schwidtal, J.M.; Chitchyan, R.; Andoni, M.; Robu, V.; Montakhabi, M.; Scott, I.; et al. A Systematic Literature Review of Peer-to-Peer, Community Self-Consumption, and Transactive Energy Market Models. *SSRN* **2021**. [CrossRef]
23. Mustafa, M.A.; Cleemput, S.; Abidin, A. A local electricity trading market: Security analysis. In Proceedings of the 2016 IEEE PES Innovative Smart Grid Technologies Conference Europe (ISGT-Europe), Ljubljana, Slovenia, 9–12 October 2016; pp. 1–6.
24. de Almeida, L.; Cappelli, V.; Klausmann, N.; van Soest, H. *Peer-to-Peer Trading and Energy Community in the Electricity Market: Analysing the Literature on Law and Regulation and Looking Ahead to Future Challenges*; Technical Report; International Energy Agency: Paris, France, 2021.
25. Ketter, W.; Collins, J.; Saar-Tsechansky, M.; Marom, O. Information systems for a smart electricity grid: Emerging challenges and opportunities. *ACM Trans. Manag. Inf. Syst. (TMIS)* **2018**, *9*, 1–22. [CrossRef]
26. Xu, Z. *The Electricity Market Design for Decentralized Flexibility Sources*; Oxford Institute for Energy Studies: Oxford, UK, 2019.
27. Schneiders, A.; Fell, M.J.; Nolden, C. Peer-to-peer electricity trading and the sharing economy: Social, markets and regulatory perspectives. *Energy Sources Part B Econ. Plan. Policy* **2022**, 1–17. [CrossRef]
28. Adner, R. *The Wide Lens: A New Strategy for Innovation*; Penguin: London, UK, 2012.

29. Montakhabi, M.; Radziwon, A. Doors wide shut-Towards A Framework for Open-Innovation, Business Model, Strategy and Beyond. In Proceedings of the WOIC 2021: 8th Annual World Open Innovation Conference, Eindhoven, The Netherlands, 8–10 December 2021.
30. Sanasi, S.; Ghezzi, A.; Cavallo, A.; Rangone, A. Making sense of the sharing economy: A business model innovation perspective. *Technol. Anal. Strateg. Manag.* **2020**, *32*, 895–909. [CrossRef]
31. Stake, R.E. *Multiple Case Study Analysis*; Guilford Press: New York, NY, USA, 2013.
32. Montakhabi, M.; van der Graaf, S.; Ballon, P.; Mustafa, M.A. *Prosumers' Business Models in Future Electricity Markets: Peer-to-Peer, Community Self-Consumption, and Transactive Energy Models*; University of Manchester: Manchester, UK, 2021.
33. Teece, D.J.; Kay, N. *The Evolution of the Theory of the Firm*; Edward Elgar Publishing: Cheltenham, UK, 2019.
34. Barney, J.B. The resource-based theory of the firm. *Organ. Sci.* **1996**, *7*, 469–592. [CrossRef]
35. Montakhabi, M.; Zobiri, F.; Van Der Graaf, S.; Deconinck, G.; Orlando, D.; Vanhove, S.; Callaerts, R.; Ballon, P.; Mustafa, M.A. New roles in peer-to-peer electricity markets: Value network analysis. In Proceedings of the 2020 6th IEEE International Energy Conference (ENERGYCon), Gammarth, Tunisia, 28 September–1 October 2020; pp. 389–394.
36. Bygrave, L.A. Privacy and data protection in an international perspective. *Scand. Stud. Law* **2010**, *56*, 165–200.
37. Barney, J.B. Resource-based theories of competitive advantage: A ten-year retrospective on the resource-based view. *J. Manag.* **2001**, *27*, 643–650. [CrossRef]
38. Schwidtal, J.M.; Piccini, P.; Troncia, M.; Chitchyan, R.; Montakhabi, M.; Francis, C.; Gorbatcheva, A.; Capper, T.; Mustafa, M.A.; Andoni, M.; et al. Emerging business models in local energy markets: A systematic review of Peer-to-Peer, Community Self-Consumption, and Transactive Energy models. *SSRN* **2022**, 4032760. [CrossRef]
39. Montakhabi, M.; Van Der Graaf, S.; Ballon, P.; Mustafa, M.A. Sharing beyond peer-to-peer trading: Collaborative (open) business models as a pathway to smart circular economy in electricity markets. In Proceedings of the 2020 16th International Conference on Distributed Computing in Sensor Systems (DCOSS), Marina del Rey, CA, USA, 25–27 May 2020; pp. 482–489.
40. Chesbrough, H. *Open Business Models: How to Thrive in the New Innovation Landscape*; Harvard Business Press: Boston, MA, USA, 2006.
41. Nazir, M.I.; Hussain, I.; Ahmad, A.; Khan, I.; Mallik, A. System Modeling and Reliability Assessment of Microgrids: A Review. *Sustainability* **2021**, *14*, 126. [CrossRef]
42. Muhsen, H.; Allahham, A.; Al-Halhouli, A.; Al-Mahmodi, M.; Alkhraibat, A.; Hamdan, M. Business Model of Peer-to-Peer Energy Trading: A Review of Literature. *Sustainability* **2022**, *14*, 1616. [CrossRef]
43. Neagu, B.C.; Ivanov, O.; Grigoras, G.; Gavrilas, M.; Istrate, D.M. New market model with social and commercial tiers for improved prosumer trading in microgrids. *Sustainability* **2020**, *12*, 7265. [CrossRef]
44. Zhang, D.; Li, Y.; Li, Y.; Shen, Z. Service Failure Risk Assessment and Service Improvement of Self-Service Electric Vehicle. *Sustainability* **2022**, *14*, 3723. [CrossRef]
45. Kousar, S.; Zafar, N.A.; Ali, T.; Alkhammash, E.H.; Hadjouni, M. Formal Modeling of IoT-Based Distribution Management System for Smart Grids. *Sustainability* **2022**, *14*, 4499. [CrossRef]
46. West, J.; Bogers, M. Leveraging external sources of innovation: A review of research on open innovation. *J. Prod. Innov. Manag.* **2014**, *31*, 814–831. [CrossRef]
47. Snihur, Y.; Wiklund, J. Searching for innovation: Product, process, and business model innovations and search behavior in established firms. *Long Range Plan.* **2019**, *52*, 305–325. [CrossRef]
48. Aagaard, A. *Sustainable Business Models*; Springer: Cham, Switzerland, 2019.
49. Frankenberger, K.; Weiblen, T.; Gassmann, O. The antecedents of open business models: An exploratory study of incumbent firms. *R&D Manag.* **2014**, *44*, 173–188.
50. Sarker, S.; Sarker, S.; Sahaym, A.; Bjørn-Andersen, N. Exploring value cocreation in relationships between an ERP vendor and its partners: A revelatory case study. *MIS Q.* **2012**, *36*, 317–338. [CrossRef]
51. Pereira, D.; Leitão, J.; Devezas, T. Coopetition and co-innovation: Do manufacturing and service providers behave differently? In *Industry 4.0*; Springer: Cham, Switzerland, 2017; pp. 397–431.
52. Chesbrough, H.W. Why companies should have open business models. *MIT Sloan Manag. Rev.* **2007**, *48*, 22.
53. Urbinati, A.; Chiaroni, D.; Chiesa, V. Towards a new taxonomy of circular economy business models. *J. Clean. Prod.* **2017**, *168*, 487–498. [CrossRef]
54. MacArthur, E. Towards the circular economy. *J. Ind. Ecol.* **2013**, *2*, 23–44.
55. Ghisellini, P.; Cialani, C.; Ulgiati, S. A review on circular economy: The expected transition to a balanced interplay of environmental and economic systems. *J. Clean. Prod.* **2016**, *114*, 11–32. [CrossRef]
56. Rosa, P.; Sassanelli, C.; Terzi, S. Towards Circular Business Models: A systematic literature review on classification frameworks and archetypes. *J. Clean. Prod.* **2019**, *236*, 117696. [CrossRef]
57. Stahel, W.R. The circular economy. *Nature* **2016**, *531*, 435–438. [CrossRef]
58. Geissdoerfer, M.; Savaget, P.; Bocken, N.M.; Hultink, E.J. The Circular Economy—A new sustainability paradigm? *J. Clean. Prod.* **2017**, *143*, 757–768. [CrossRef]
59. Bocken, N.M.; De Pauw, I.; Bakker, C.; Van Der Grinten, B. Product design and business model strategies for a circular economy. *J. Ind. Prod. Eng.* **2016**, *33*, 308–320. [CrossRef]

60. Lüdeke-Freund, F.; Gold, S.; Bocken, N.M. A review and typology of circular economy business model patterns. *J. Ind. Ecol.* **2019**, *23*, 36–61. [CrossRef]
61. Atasu, A.; Dumas, C.; Van Wassenhove, L.N. The Circular Business Model. Pick a Strategy That Fits Your Resources and Capabilities. Harvard Business Review from the Magazine (July–August 2021). Available online: https://hbr.org/2021/07/the-circular-business-model (accessed on 14 March 2022).
62. Nussholz, J.L. A circular business model mapping tool for creating value from prolonged product lifetime and closed material loops. *J. Clean. Prod.* **2018**, *197*, 185–194. [CrossRef]
63. Botsman, R. Defining the sharing economy: What is collaborative consumption—And what isn't. *Fast Co.* **2015**, *27*, 2015.
64. Schwanholz, J.; Leipold, S. Sharing for a circular economy? An analysis of digital sharing platforms' principles and business models. *J. Clean. Prod.* **2020**, *269*, 122327. [CrossRef]
65. Kathan, W.; Matzler, K.; Veider, V. The sharing economy: Your business model's friend or foe? *Bus. Horiz.* **2016**, *59*, 663–672. [CrossRef]
66. Despeisse, M.; Baumers, M.; Brown, P.; Charnley, F.; Ford, S.J.; Garmulewicz, A.; Knowles, S.; Minshall, T.; Mortara, L.; Reed-Tsochas, F.; et al. Unlocking value for a circular economy through 3D printing: A research agenda. *Technol. Forecast. Soc. Chang.* **2017**, *115*, 75–84. [CrossRef]
67. Genovese, A.; Acquaye, A.A.; Figueroa, A.; Koh, S.L. Sustainable supply chain management and the transition towards a circular economy: Evidence and some applications. *Omega* **2017**, *66*, 344–357. [CrossRef]
68. Zenger, T.R.; Felin, T.; Bigelow, L. Theories of the firm—Market boundary. *Acad. Manag. Ann.* **2011**, *5*, 89–133. [CrossRef]
69. Porter, M.E. Towards a dynamic theory of strategy. *Strateg. Manag. J.* **1991**, *12*, 95–117. [CrossRef]
70. Teece, D.J. Firm organization, industrial structure, and technological innovation. *J. Econ. Behav. Organ.* **1996**, *31*, 193–224. [CrossRef]
71. Dierickx, I.; Cool, K. Asset stock accumulation and sustainability of competitive advantage. *Manag. Sci.* **1989**, *35*, 1504–1511. [CrossRef]
72. Helfat, C.E.; Finkelstein, S.; Mitchell, W.; Peteraf, M.; Singh, H.; Teece, D.; Winter, S.G. *Dynamic Capabilities: Understanding Strategic Change in Organizations*; John Wiley & Sons: Hoboken, NJ, USA, 2009.
73. Barney, J.B.; Hesterly, W.S. *Strategic Management and Competitive Advantage: Concepts*; Prentice Hall: Englewood Cliffs, NJ, USA, 2010.
74. Mahoney, J.T.; Pandian, J.R. The resource-based view within the conversation of strategic management. *Strateg. Manag. J.* **1992**, *13*, 363–380. [CrossRef]
75. Penrose, E.; Penrose, E.T. *The Theory of the Growth of the Firm*; Oxford University Press: Oxford, UK, 2009.
76. Wernerfelt, B. From critical resources to corporate strategy. *J. Gen. Manag.* **1989**, *14*, 4–12. [CrossRef]
77. Eisenhardt, K.M.; Martin, J.A. Dynamic capabilities: What are they? *Strateg. Manag. J.* **2000**, *21*, 1105–1121. [CrossRef]
78. Wang, C.L.; Ahmed, P.K. Dynamic capabilities: A review and research agenda. *Int. J. Manag. Rev.* **2007**, *9*, 31–51. [CrossRef]
79. Williamson, O.E. Strategy research: Governance and competence perspectives. *Strateg. Manag. J.* **1999**, *20*, 1087–1108. [CrossRef]
80. Schilke, O.; Hu, S.; Helfat, C.E. Quo vadis, dynamic capabilities? A content-analytic review of the current state of knowledge and recommendations for future research. *Acad. Manag. Ann.* **2018**, *12*, 390–439. [CrossRef]
81. Zahra, S.A.; George, G. Absorptive capacity: A review, reconceptualization, and extension. *Acad. Manag. Rev.* **2002**, *27*, 185–203. [CrossRef]
82. Allen, D. The sharing economy. *Inst. Public Aff. Rev. Q. Rev. Politics Public Aff.* **2015**, *67*, 24–27.
83. Newbery, D.M. Problems of liberalising the electricity industry. *Eur. Econ. Rev.* **2002**, *46*, 919–927. [CrossRef]
84. Reiter, B. Theory and Methodology of Exploratory Social Science Research. 2017. Available online: https://ttu-ir.tdl.org/handle/2346/86610 (accessed on 14 March 2022).
85. Verbeke, A. *International Business Strategy: Rethinking the Foundations of Global Corporate Success*; Cambridge University Press: Cambridge, UK, 2013.
86. Chen, C.D.; Zhao, Q.; Wang, J.L.; Huang, C.K.; Lee, N.C. Exploring sharing economy success: Resource-based view and the role of resource complementarity in business value co-creation. In Proceedings of the PACIS 2017, Langkawi, Malaysia, 16–20 July 2017.
87. Liu, Y.; Wu, L.; Li, J. Peer-to-peer (P2P) electricity trading in distribution systems of the future. *Electr. J.* **2019**, *32*, 2–6. [CrossRef]
88. Kenyon, A.T.; Marjoribanks, T. Transforming media markets: The cases of Malaysia and Singapore. *Aust. J. Emerg. Technol. Soc.* **2007**, *5*, 103–118.
89. Van Thiel, S. *Research Methods in Public Administration and Public Management: An Introduction*; Routledge: London, UK, 2014.
90. Kallio, H.; Pietilä, A.M.; Johnson, M.; Kangasniemi, M. Systematic methodological review: Developing a framework for a qualitative semi-structured interview guide. *J. Adv. Nurs.* **2016**, *72*, 2954–2965. [CrossRef]
91. Noy, C. Sampling knowledge: The hermeneutics of snowball sampling in qualitative research. *Int. J. Soc. Res. Methodol.* **2008**, *11*, 327–344. [CrossRef]
92. Glassner, A. *Interactive Storytelling: Techniques for 21st Century Fiction*; CRC Press: Boca Raton, FL, USA, 2017.
93. Eckhardt, J.T.; Shane, S.A. Opportunities and entrepreneurship. *J. Manag.* **2003**, *29*, 333–349.
94. Koops, B.J.; Newell, B.; Timan, T.; Škorvánek, I.; Chokrevski, T.; Maša, G. A Typology of Privacy. *Univ. Pa. J. Int. Law* **2016**, *38*, 483.
95. Wood, A.J.; Wollenberg, B.F.; Sheblé, G.B. *Power Generation, Operation, and Control*; John Wiley & Sons: Hoboken, NJ, USA, 2013.

96. Wang, N.; Xu, W.; Xu, Z.; Shao, W. Peer-to-peer energy trading among microgrids with multidimensional willingness. *Energies* **2018**, *11*, 3312. [CrossRef]
97. Liu, W.; Qi, D.; Wen, F. Intraday residential demand response scheme based on peer-to-peer energy trading. *IEEE Trans. Ind. Inform.* **2019**, *16*, 1823–1835. [CrossRef]
98. Long, C.; Wu, J.; Zhou, Y.; Jenkins, N. Peer-to-peer energy sharing through a two-stage aggregated battery control in a community Microgrid. *Appl. Energy* **2018**, *226*, 261–276. [CrossRef]
99. Pisello, A.L.; Petrozzi, A.; Castaldo, V.L.; Cotana, F. On an innovative integrated technique for energy refurbishment of historical buildings: Thermal-energy, economic and environmental analysis of a case study. *Appl. Energy* **2016**, *162*, 1313–1322. [CrossRef]
100. Chakraborty, S.; Baarslag, T.; Kaisers, M. Automated peer-to-peer negotiation for energy contract settlements in residential cooperatives. *Appl. Energy* **2020**, *259*, 114173. [CrossRef]
101. Zhang, C.; Wu, J.; Long, C.; Cheng, M. Review of existing peer-to-peer energy trading projects. *Energy Procedia* **2017**, *105*, 2563–2568. [CrossRef]
102. Sundararajan, A. Peer-to-peer businesses and the sharing (collaborative) economy: Overview, economic effects and regulatory issues. In *Written Testimony for the Hearing Titled the Power of Connection: Peer to Peer Businesses*; 2014. Available online: https://republicans-smallbusiness.house.gov/uploadedfiles/1-15-2014_revised_sundararajan_testimony.pdf (accessed on 3 May 2022).
103. Abidin, A.; Callaerts, R.; Deconinck, G.; Van Der Graaf, S.; Madhusudan, A.; Montakhabi, M.; Mustafa, M.A.; Nikova, S.; Orlando, D.; Schroers, J.; et al. Poster: SNIPPET—Secure and Privacy-Friendly Peer-to-Peer Electricity Trading. In Proceedings of the Network and Distributed System Security Symposium (NDSS 2020), San Diego, CA, USA, 23–26 February 2020.
104. Montakhabi, M.; Madhusudan, A.; van der Graaf, S.; Abidin, A.; Ballon, P.; Mustafa, M.A. Sharing Economy in Future Peer-to-peer Electricity Trading Markets: Security and Privacy Analysis. In Proceedings of the Workshop on Decentralized IoT Systems and Security (DISS), San Diego, CA, USA, 23 February 2020; pp. 1–6.
105. Heinisch, V.; Odenberger, M.; Göransson, L.; Johnsson, F. Organizing prosumers into electricity trading communities: Costs to attain electricity transfer limitations and self-sufficiency goals. *Int. J. Energy Res.* **2019**, *43*, 7021–7039. [CrossRef]
106. Montakhabi, M.; Van Der Graaf, S. Open Business Models' Actionability in Europe; EU Competition Policy Analysis. *J. Bus. Model.* **2021**, *9*, 29–34.
107. Moret, F.; Pinson, P. Energy collectives: A community and fairness based approach to future electricity markets. *IEEE Trans. Power Syst.* **2018**, *34*, 3994–4004. [CrossRef]
108. Nizami, M.S.H.; Hossain, M.J.; Fernandez, E. Multiagent-Based Transactive Energy Management Systems for Residential Buildings with Distributed Energy Resources. *IEEE Trans. Ind. Inform.* **2020**, *16*, 1836–1847. [CrossRef]
109. Lian, J.; Ren, H.; Sun, Y.; Hammerstrom, D.J. Performance Evaluation for Transactive Energy Systems Using Double-Auction Market. *IEEE Trans. Power Syst.* **2019**, *34*, 4128–4137. [CrossRef]
110. Nguyen, H.T.; Battula, S.; Takkala, R.R.; Wang, Z.; Tesfatsion, L. An integrated transmission and distribution test system for evaluation of transactive energy designs. *Appl. Energy* **2019**, *240*, 666–679. [CrossRef]
111. Masood, A.; Hu, J.; Xin, A.; Sayed, A.R.; Yang, G. Transactive energy for aggregated electric vehicles to reduce system peak load considering network constraints. *IEEE Access* **2020**, *8*, 31519–31529. [CrossRef]
112. Hao, H.; Corbin, C.D.; Kalsi, K.; Pratt, R.G. Transactive control of commercial buildings for demand response. *IEEE Trans. Power Syst.* **2016**, *32*, 774–783. [CrossRef]
113. Morstyn, T.; Teytelboym, A.; McCulloch, M.D. Designing decentralized markets for distribution system flexibility. *IEEE Trans. Power Syst.* **2018**, *34*, 2128–2139. [CrossRef]
114. Montakhabi, M.; Madhusudan, A.; van der Graaf, S.; Abidin, A.; Mustafa, M.A. Sharing Economy in Future Electricity Markets: Security and Privacy Analysis. In Proceedings of the Workshop on Decentralized IoT Systems and Security (DISS), in Conjunction with NDSS, San Diego, CA, USA, 23 February 2020.
115. D'Hauwers, R.; Van Der Bank, J.; Montakhabi, M. Trust, Transparency and Security in the Sharing Economy: What is the Government's Role? *Technol. Innov. Manag. Rev.* **2020**, *10*, 6–18. [CrossRef]
116. Bocken, N.; Ritala, P. Six ways to build circular business models. *J. Bus. Strategy* **2021**, *43*, 184–192. [CrossRef]
117. Lutz, C.; Hoffmann, C.P.; Bucher, E.; Fieseler, C. The role of privacy concerns in the sharing economy. *Inf. Commun. Soc.* **2018**, *21*, 1472–1492. [CrossRef]
118. Hamari, J.; Sjöklint, M.; Ukkonen, A. The sharing economy: Why people participate in collaborative consumption. *J. Assoc. Inf. Sci. Technol.* **2016**, *67*, 2047–2059. [CrossRef]
119. Sloot, B.v.d. Privacy as Personality Right: Why the ECtHR's Focus on Ulterior Interests Might Prove Indispensable in the Age of "Big Data". *Utrecht J. Int. Eur. Law* **2015**, *31*, 25–50. [CrossRef]
120. Finn, R.L.; Wright, D.; Friedewald, M. Seven Types of Privacy. In *European Data Protection: Coming of Age*; Gutwirth, S., Leenes, R., de Hert, P., Poullet, Y., Eds.; Springer: Dordrecht, The Netherlands, 2013; pp. 3–32. [CrossRef]
121. Zarifis, A.; Ingham, R.; Kroenung, J. Exploring the language of the sharing economy: Building trust and reducing privacy concern on Airbnb in German and English. *Cogent Bus. Manag.* **2019**, *6*, 1666641. [CrossRef]
122. Yan, K.; Shen, W.; Jin, Q.; Lu, H. Emerging Privacy Issues and Solutions in Cyber-Enabled Sharing Services: From Multiple Perspectives. *IEEE Access* **2019**, *7*, 26031–26059. [CrossRef]
123. Joskow, P.L. Creating a smarter US electricity grid. *J. Econ. Perspect.* **2012**, *26*, 29–48. [CrossRef]

124. Wu, C.; Kalathil, D.; Poolla, K.; Varaiya, P. Sharing electricity storage. In Proceedings of the 2016 IEEE 55th Conference on Decision and Control (CDC), Las Vegas, NV, USA, 12–14 December 2016; pp. 813–820.
125. Toffler, A. Powershift. *Rev. Filos.* **1992**, *39*, 175–178.
126. Parag, Y.; Sovacool, B.K. Electricity market design for the prosumer era. *Nat. Energy* **2016**, *1*, 1–6. [CrossRef]
127. Xu, Y.; Ahokangas, P.; Reuter, E. EaaS: Electricity as a service? *J. Bus. Model.* **2018**, *6*, 1–23.
128. Makadok, R.; Burton, R.; Barney, J. A practical guide for making theory contributions in strategic management. *Strateg. Manag. J.* **2018**, *39*, 1530–1545. [CrossRef]
129. Dabrowska, J. The Good, the Bad, the Ugly, and the Beautiful of Digital Transformation: A Critical Multi-Level Research Agenda. *R&D Manag.* **2021**. [CrossRef]
130. Teece, D.J.; Pisano, G.; Shuen, A. Dynamic capabilities and strategic management. *Strateg. Manag. J.* **1997**, *18*, 509–533. [CrossRef]
131. Trotter, R.T., II. Qualitative research sample design and sample size: Resolving and unresolved issues and inferential imperatives. *Prev. Med.* **2012**, *55*, 398–400. [CrossRef] [PubMed]
132. Garrido, G.M.; Sedlmeir, J.; Uludağ, Ö.; Alaoui, I.S.; Luckow, A.; Matthes, F. Revealing the landscape of privacy-enhancing technologies in the context of data markets for the IoT: A systematic literature review. *arXiv* **2021**, arXiv:2107.11905.
133. Rinehart, R. Fictional methods in ethnography: Believability, specks of glass, and Chekhov. *Qual. Inq.* **1998**, *4*, 200–224. [CrossRef]
134. Cheon, E.; Su, N.M. Futuristic autobiographies: Weaving participant narratives to elicit values around robots. In Proceedings of the 2018 ACM/IEEE International Conference on Human-Robot Interaction, Chicago, IL, USA, 5–8 March 2018; pp. 388–397.
135. Drucker, P. *Managing in a Time of Great Change*; Routledge: London, UK, 2012.
136. De Jouvenel, B. *The Art of Conjecture*; Routledge: London, UK, 2017.
137. Christensen, C.; Raynor, M.E.; McDonald, R. *Disruptive Innovation*; Harvard Business Review: Brighton, MA, USA, 2013.

sustainability

MDPI

Article

Realising the Potential of Renewable Energy as a Tool for Energy Security in Small Island Developing States

Walter Leal Filho [1,2], Abdul-Lateef Balogun [3], Dinesh Surroop [4], Amanda Lange Salvia [5,*], Kapil Narula [6], Chunlan Li [7,8], Julian David Hunt [9], Andrea Gatto [10,11,12], Ayyoob Sharifi [13,14], Haibo Feng [15], Stella Tsani [16] and Hossein Azadi [17,18,19]

1 School of Science and the Environment, Manchester Metropolitan University, All Saints Building, Oxford Road, Manchester M15 6BH, UK; walter.leal2@haw-hamburg.de
2 Research and Transfer Centre "Sustainable Development and Climate Change Management", Hamburg University of Applied Sciences, Ulmenliet 20, D-21033 Hamburg, Germany
3 Geospatial Analysis and Modelling Research (GAMR) Group, Department of Civil and Environmental Engineering, Universiti Teknologi PETRONAS (UTP), Seri Iskandar 32610, Malaysia; alateef.babatunde@utp.edu.my
4 Faculty of Engineering, Department of Chemical and Environmental Engineering, University of Mauritius, Moka 80837, Mauritius; d.surroop@uom.ac.mu
5 Graduate Program in Civil and Environmental Engineering, University of Passo Fundo, Passo Fundo 99052-900, Brazil
6 Climate Change and Natural Resources Sustainability Cluster, United Nations Economic and Social Commission for Western Asia (UN ESCWA), Riad el-Solh Square, Beirut P.O. Box 11-8575, Lebanon; kapilnarula74@gmail.com
7 Center for Geopolitical and Strategic Studies, Institute for Global Innovation and Development, East China Normal University, Shanghai 200062, China; 15598022233@163.com
8 School of Urban and Regional Sciences, East China Normal University, Shanghai 200062, China
9 International Institute for Applied Systems Analysis (IIASA), Schlossplatz 1, A-2361 Laxenburg, Austria; hunt@iiasa.ac.at
10 College of Business & Public Management, Wenzhou-Kean University, Wenzhou 325060, China; a.gatto@greenwich.ac.uk
11 Natural Resources Institute, Livelihoods & Institutions Department, University of Greenwich, Central Avenue, Chatham Maritime, London ME4 4TB, UK
12 Center for Studies on Europe (AIM), Azerbaijan State University of Economics (UNEC), Baku AZ1001, Azerbaijan
13 Graduate School of Humanities, Social Sciences and Network for Education and Research on Peace and Sustainability (NERPS), Hiroshima University, Higashi-Hiroshima 739-8530, Japan; sharifi@hiroshima-u.ac.jp
14 Center for Peaceful and Sustainable Futures (CEPEAS), Hiroshima University, Higashi-Hiroshima 739-8530, Japan
15 Department of Mechanical and Construction Engineering, Northumbria University, Ellison Building, Room 107J, Newcastle upon Tyne NE2 1XE, UK; haibo.feng@northumbria.ac.uk
16 Department of Economics, University of Ioannina, 45110 Ioannina, Greece; stellatsani@uoi.gr
17 Department of Geography, Ghent University, 9000 Ghent, Belgium; hossein.azadi@ugent.be
18 Department of Economics and Rural Development, Gembloux Agro-Bio Tech, University of Liège, 5030 Gembloux, Belgium
19 Faculty of Environmental Science and Engineering, Babeș-Bolyai University, 400000 Cluj-Napoca, Romania
* Correspondence: amandasalvia@gmail.com or 112737@upf.br

Citation: Leal Filho, W.; Balogun, A.-L.; Surroop, D.; Salvia, A.L.; Narula, K.; Li, C.; Hunt, J.D.; Gatto, A.; Sharifi, A.; Feng, H.; et al. Realising the Potential of Renewable Energy as a Tool for Energy Security in Small Island Developing States. *Sustainability* **2022**, *14*, 4965. https://doi.org/10.3390/su14094965

Academic Editor: Donato Morea

Received: 14 February 2022
Accepted: 18 April 2022
Published: 20 April 2022

Abstract: Small Island Developing States (SIDS) are heavily dependent on the use of imported fossil fuels to address their energy needs. This has a negative impact on the environment, SIDS budgets, and energy security. Therefore, this study aimed to investigate the role of renewable energy (RE) as a tool for energy security in SIDS. In this regard, using VOSviewer, a widely known software tool, two bibliometric analyses were performed with a focus on the literature that explores two intertwined issues: (i) the links between RE and energy security; and (ii) the implications of RE and energy security in SIDS. The results from the study show that RE can help SIDS enhance their energy security and assure long-term energy sustainability. In addition, the results show that with the reduction in the cost of batteries in the future, they will eventually replace diesel generators. Moreover, the study showed that renewable energy may assist SIDS in their long-term efforts to achieve food security. The analysis discusses the major obstacles and the potential solutions for the integration of RES into the energy generation of SIDS. The paper concludes with useful recommendations to help island nations reduce their carbon footprint.

Keywords: energy security; energy policy; energy resources; sustainability; bibliometrics

1. Introduction

Energy security is commonly defined in regards to continuous energy availability and efficiency, affordability, accessibility, sustainability, cleanliness, and reliability [1–4], and it is often linked with sustainable policies and innovations [5,6]. Although uninterrupted energy supply is essential for socio-economic growth and development [7], unfortunately, many nations still struggle with the provision of sufficient and uninterrupted energy [8]. This problem is particularly evident in the Small Island Developing States (SIDS).

SIDS possess a set of characteristics that distinguishes them from other developing countries (Table 1) and make them some of the most vulnerable countries in terms of energy needs and security. SIDS have a population of about 61.5 million people, concentrated mainly in the Caribbean (65%), the Pacific, Africa, Indian Ocean, Mediterranean, and South China Sea regions [9].

Table 1. Unique features of Small Island Developing States and their socio-economic and environmental implications.

Features	Implications
Reduced size	Limited natural and financial resources
Geographical isolation	Access to services that are difficult on occasions
Limited economic outputs	Considerable levels of poverty
Vulnerability to climate change	Higher likelihood of damages and casualties due to extreme events and displacement of populations
Food price volatility	Hardships among some groups
Dependency on fossil fuels	High prices of energy

Except for Papua New Guinea and Trinidad and Tobago, most of the SIDS lack conventional energy resources (natural gas, oil, and coal). The energy vulnerability of SIDS relates to their net energy imports, their insularity and remoteness, their peculiar geographic, demographic, environmental and socio-economic conditions [10,11], their limited natural resources, thin economies, remoteness from the main trading centres, and their high exposure to unexpected events, such as extreme weather events [12]. SIDS' economic challenges are exacerbated by the human capital emergence that leaves in search of improved living conditions in more developed countries [13]. Although being blessed with abundant solar, wind, and wave energy resources, among others, RES constitute a small share in their energy mix.

Poor socio-economic conditions and high dependence on conventional energy sources affect the capacity of SIDS to support sustainable living as envisioned in the UN's sustainable development goals (SDG). Most SIDS are unable to meet their energy requirements independently, largely relying on the importation of fossil fuels for survival [4], which puts pressure on SIDS' financial resources [11] (with some exceptions such as Trinidad and Tobago, an example of oil-exporting SIDS [14]). Half of the Pacific SIDS, for example, are considered to be highly vulnerable to oil price shocks [4]. The high transportation cost of conventional fuels means that the SIDS must pay higher energy prices. The costs of electricity and other secondary energy resources are relatively higher than in non-SIDS [4]. Similarly, petroleum products are more expensive in SIDS than in nonisland parts of the world [4]. SIDS are also subject to an irregular supply of resources. In a recent assessment to monitor the sustainability of national energy systems, only 5 SIDS were ranked among 125 countries, and just two of these ranked among the top 50 countries in terms of energy security [3,15].

In a study by Michalena and Hills [12], it was revealed that by renewable energy (RE) improvement, there will be a minor increase in RE penetration, which will lead to putting the region below global rates of RE adoption. In another respect, there was a failure by the governments to build up inward investment through development partners for RE demonstration projects. Their results also show that focus on combating global climate change has not been successful at placing RE in a local context, which contrasts with traditional Pacific communities' strong "feeling of place" and spiritual nature. Lucas et al. [10] claimed that among others, the SIDS are highly vulnerable to climate change impacts. Furthermore, SIDS are among the world's most reliant on imported petroleum goods, and the usage of RE can assist in mitigating the economic risk associated with fossil fuel price fluctuation. According to their study, there are six categories that put RE deployment in islands at risk, such as (i) insufficient data for RE development, (ii) organised regulatory frameworks, (iii) lack of financial supports, (iv) absence of human resources, (v) expensive required facilities, and (vi) socio-cultural barriers.

In another similar study, Surroop et al. [8] stated that a fundamental worldwide development concern is low access to modern energy services, generally known as energy poverty. A collection of SIDS case studies that show efforts to enhance energy availability can serve as a foundation for recommending critical electrification strategies. It is worth mentioning that the establishment of an independent regulatory body with policymaking and adjudicative capabilities, as well as a high degree of commitment from the government, were highlighted as enabling factors. Furthermore, there are some potential contributors, such as the establishment of a cost-reflective pricing structure, to the development of electricity access in the island context. In the global aspect, the most affordable RE are solar energy and wind power, and it is predicted by the experts that the costs will remain low for the up-coming years [16]. Decarbonised electricity could be generated by wind and solar energies. However, it is required to implement other technologies including energy storage, supplemental generation, demand management, and transmission expansion to fulfil the demands reliably.

Food scarcity, low-quality education, poor healthcare facilities and services, discrimination against women, and poor and ineffective governance have been identified as some of the direct implications of energy insecurity [8,17]. Due to incessant power outages and scarcity of clean cooking fuel in many SIDS, households' resort to alternative domestic sources, such as residues from agricultural products, charcoal, manure, and wood. The proliferation of these cheap, low-efficiency fuels triggers chronic respiratory problems caused by household air pollution (HAP) [18,19], which can also lead to death [8,20]. The African, Caribbean, and Pacific (ACP) SIDS host a significant proportion of energy-deprived people who still rely on traditional fuels for basic daily activities including cooking and heating [8]. Water supply systems of many SIDS increasingly require a great deal of energy to power desalination plants. Scarcity of freshwater is a major challenge, particularly in big islands with a larger number of inhabitants and substantial water demand. This increases the demand for energy and further stresses SIDS' limited resources [21].

The high exposure of SIDS to climate-change-induced natural disasters, such as sea-level rise and flooding, further affects the reliability of the electric power supply. Poor power supply is also aggravated by the SIDS' nonconnection to continental grids [10]. Transportation in SIDS is also dependent on energy imports. For instance, in the Pacific SIDS, transport represents approximately 75% of the final use of imported oil [10]. Price instability, interruptions in supply, and inefficient delivery severely affect the transportation sector [22]. The combination of high fuel costs and limited road networks in SIDS pose transportation and sustainable development challenges [23]. Crucial investments in infrastructural development are usually forfeited in the ensuing trade-off between spending on fuel imports and societal development [24].

The electricity and transportation sectors in SIDS are highly reliant on fossil fuels, making these sectors responsible for most of the emissions of greenhouse gases in these countries, which contribute to global warming. Moreover, SIDS have been in the limelight as being faced with a "moral crisis" as their combined greenhouse gases emissions are less than 1%, but they are the first victims of global warming and suffer disproportionately from extreme weather conditions such as storms and floods. Their low increase exposes them to a high risk of submergence of rising sea levels and coastal erosion. Islands such as Tuvalu, Solomon Islands, and the Maldives are currently under threat of submersion [25].

The RE developed in SIDS are mostly hydro, solar, wind, and biomass. The potential of wind resources [26], bioenergy resources, marine energy, and geothermal energy production [4,27] have been variously discussed, but available evidence remains scarce thus far of the actual status of RE in SIDS and how they are being harnessed for energy security. Studies on energy security in SIDS are still in the formative stages and are of high relevance [3].

It is believed that RE can help SIDS significantly in achieving energy security and reducing the import of fossil fuel, but it has been largely ignored by analysts and policymakers. Given the fact that most advances are moving toward a low-carbon future, RE has the potential to accelerate SIDS' energy transformation. Energy intensity trends have been unpredictable over the previous decade, and for SIDS, reductions in energy usage have not been consistent. SIDS have a wide range of initiatives, projects, and policies, and this article provides a comprehensive overview of the current state of energy efficiency in SIDS. Overall, while previous studies have primarily explored policy interventions and recommendations, this study provides a more practical plan by combining institutional arrangements and the link of RE with SIDS.

RE and energy security are widely used concepts, for which an extensive body of literature exists. To this end, bibliometrics and co-occurrence analysis can be of help in sketching and disentangling the current energy policies and sustainability drivers [26,28]. Even though many scholars have examined energy security research from distinct viewpoints and through the application of different approaches, e.g., [29–31], few of them, such as [32–34], have attempted to investigate the bibliometrics and co-occurrence analysis domain of energy security research. Nevertheless, it is vital to employ an effective strategy for analysing the existing state and upcoming trends in these domains because most study areas are dynamic as a society, and the environments change. The results of bibliometrics and co-occurrence analysis are thought to be more practical [35].

Certain valuable information about energy security research is revealed, allowing researchers to continue their research. The goal of this research is to bridge the gap between bibliometrics and co-occurrence analysis, as well as energy security research. Given the increasing number of publications on energy security, it is necessary to provide a comprehensive overview of the current state and future trends in the field. Using bibliometrics and co-occurrence analysis to analyse energy security can help provide a clearer understanding of the contribution of RE to SIDS. Accordingly, in the current study, a bibliometric/co-occurrence analysis of "renewable energy" and "energy security" was performed to provide an overview of the extent to which the use of RE may assist in energy security. Therefore, the main research problem is that SIDS have poor energy security due to their dependence on fossil fuel imports. With this problem in mind, this study aimed to investigate the role of RE as a tool for energy security in SIDS to answer the following questions:

(1) Can RE be used as a tool for energy security in SIDS?
(2) What are the feasible solutions and evidence-based approaches in accelerating the acceptance of RE in order to achieve energy security in SIDS?

Towards this end, the remainder of the paper develops as follows. Section 2 explores the methodology used for bibliometrics and co-occurrence analysis of publications focusing on general and SIDS-focused RE and energy security—it presents a set of case studies. Section 3 presents the results, followed by Section 4, which discusses the obstacles and the means to address them. Section 5 presents the conclusions.

2. Methods

In order to shed light on the prospect and deployment of RE in the SIDS, a twofold approach was used. First, using VOSviewer, a widely known software tool, two bibliometric analyses were performed with a focus on the literature that explores two intertwined issues: (i) the links between RE and energy security; and (ii) RE and energy security in SIDS. This helped with the understanding of the thematic focus of the research in this area and with the identification of the major studies that can inform this work. A bibliometric analysis is performed with a focus on the literature exploring the links between RE and climate change mitigation and energy security. Bibliometrics is a valuable set of methodologies to track and map existing publications on a specific research issue. The outputs are suitable for highlighting major thematic focuses and examining the previous studies' trends [36].

To conduct the first bibliometric analysis using VOSviewer, a search for literature at the nexus of RE and energy security was performed in the Web of Science (using the following search string: TS = ("RE" AND "energy security")) AND LANGUAGE: (English) Timespan: 1900–2021. Data were collected from the following World of Science (WoS) Indexes: SCI-EXPANDED, SSCI, A&HCI, and ESCI. The bibliometric data of 1110 articles were downloaded and used for analysis using VOSviewer. The software can be used to conduct different types of analyses. Here co-citation analysis was used to identify major resources and apply co-occurrence analysis to highlight main thematic focus areas. Co-citation shows connections between two documents that are both cited by a third document [37]. Publications that have stronger connections with other documents have been more influential in the development of the field.

A second bibliometric and co-occurrence analysis was performed. In this case, the following strings were searched: TS = (("RE" AND "energy security") AND ("Small Island Developing State *" OR "sids" OR "Bahrain" OR "Cabo Verde" OR "Comoros" OR "Guinea-Bissau" OR "Maldives" OR "Mauritius" OR "Sao Tomé and Principe" OR "Seychelles" OR "Antigua and Barbuda" OT "Bahamas" OT "Barbados" OR "Belize" OR "Cuba" OR "Dominica" OR "Dominican Republic" OR "Grenada" OR "Guyana" OR "Haiti" OR "Jamaica" OR "Saint Kitts and Nevis" OR "Saint Lucia" OR "Saint Vincent and the Grenadines" OR "Suriname" OR "Trinidad and Tobago" OR "Fiji" OR "Kiribati" OR "Marshall Islands" OR "Micronesia" OR "Nauru" OR "Palau" OR "Papua New Guinea" OR "Samoa" OR "Solomon Islands" OR "Timor-Leste" OR "Tonga" OR "Tuvalu" OR "Vanuatu" OR "American Samoa" OR "Anguilla" OR "Aruba" OR "Bermuda" OR "British Virgin Islands" OR "Cayman Islands" OR "Commonwealth of Northern Marianas" OR "Cook Islands" OR "Curacao" OR "French Polynesia" OR "Guadeloupe" OR "Guam" OR "Martinique" OR "Montserrat" OR "New Caledonia" OR "Niue" OR "Puerto Rico" OR "Saint Maarten" OR "Turks and Caicos Islands" OR "U.S. Virgin Islands")). Secondly, a case study analysis on six SIDS was performed (Figure 1).

The choice was based on: (i) the geographical distribution with a focus on main regions where SIDS are distributed and (ii) the size diversity (very small, small, and middle-sized SIDS). To the best of our knowledge, no studies have covered such a large sample of SIDS, so this study is an important contribution to the literature.

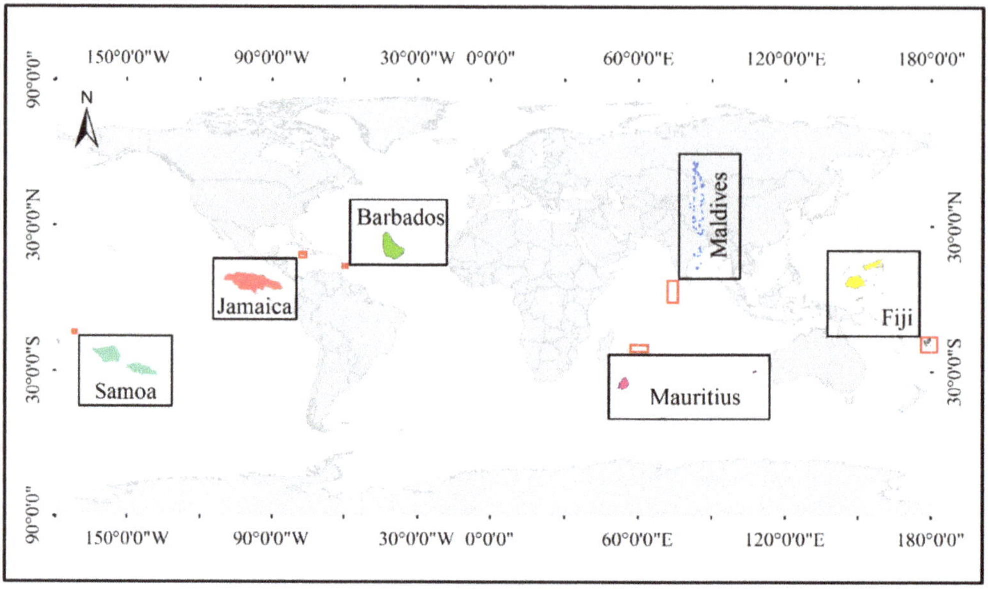

Figure 1. Locations of the sampled SIDS.

3. Results

3.1. Bibliometric Analyses

The present research examined both the overall and SIDS-focused existing scholarship on RE and energy security. The former was investigated to obtain a broader and more solid picture of the phenomenon, considering the larger corpus of literature. The latter was conducted with the aim to better highlight the publications on the energy policy dynamics related to SIDS. The results of the co-citation analysis show that the articles listed in Table 2 were more influential.

Table 2. Some influential publications in the field of energy security.

Reference	Title	Total Link Strength
Kruyt et al. (2009) [38]	Indicators for energy security	357
Winzer (2012) [39]	Conceptualizing energy security	258
Ang et al. (2015) [40]	Energy security: Definitions, dimensions, and indexes	275
Sovacool and Mukherjee (2011) [41]	Conceptualizing and measuring energy security: A synthesized approach	285
Yergin (2006) [42]	Ensuring Energy Security	154
Chester (2010) [43]	Conceptualising energy security and making explicit its polysemic nature	232
Asif and Muneer (2007) [44]	Energy supply, its demand and security issues for developed and emerging economies	108
Sovacool and Brown (2010) [45]	Competing Dimensions of Energy Security: An International Perspective	159
Intharak et al. (2007) [46]	A Quest for Energy Security in the 21st Century	156
Sovacool et al. (2011) [47]	Evaluating energy security performance from 1990 to 2010 for eighteen countries	192

These articles, e.g., [38–42] and some other important publications, such as those by Wang et al. [48] and Ibrahiem and Hanafy [49], have been reviewed to better understand the major issues discussed regarding the links between RE and energy security. The term energy security is extensively used, according to Kruyt et al. [38], but there is no agreement on its precise interpretation. They identified four aspects of energy security, which include energy availability, accessibility, cost, and acceptability, and categorised indicators for energy security using this classification. There is no single perfect indication, according to their findings, because the concept of energy security is largely context based. Applying several signs, on the other hand, leads to a more comprehensive understanding. Due to rising global demand, integrating these indications into a model-based scenario analysis revealed that the currently known fossil fuels are rapidly depleting.

Energy security and RE are top-tier issues in energy policy research. As can be seen in Table 2, the documents are listed regressively, according to their link strength. The best performing papers of this exercise scored a link strength ranging from 357 to 192. Interestingly, all of the studies that appeared in the co-citation analysis have been published in the last 15 years, with the majority of them in the last 10 years, and one work in the last 5 years. The number of yearly publications is as follows: 2015: n = 1; 2012: n = 1; 2011: n = 1; 2010: n = 2; 2009: n = 1; 2007: n = 2; 2006: n = 1.

In terms of number of publications, Energy Policy has 3 publications, followed by Renewable and Sustainable Energy Reviews and Energy, with 2 publications each. Foreign Affairs, Annual Reviews of Environment and Resource, and Asia Pacific Energy Research Centre also appear with 1 document each. In terms of authorship, the most prolific author is Benjamin K. Sovacool, who authored or was listed as the first co-author in three publications. All the other authors obtain one publication each—either as authors or coauthors. None of the authors displayed multiple affiliations.

In order to have more detailed information on the SIDS selected for the case study, a bibliometric analysis was performed on the topics of RE, energy security, and small island developing states. The words relate to all the SIDS chosen for this inquiry (n = 58).

The bibliometric search returned 19 results. All the examined papers had a spatial focus. Three works propose regional investigations: 3 in South Asia, 1 in East Asia, 2 in Southeast Asia, 1 in the Caribbean and Pacific, 1 in the Western Pacific Ocean, and 1 in Gulf Cooperation Countries (GCC). A total of 7 publications concerned domestic case studies from SIDS; Fiji with 2 works and Comoros, Cape Verde, Suriname, Jamaica, and the Maldives scoring 1 paper each. Only one publication explicitly considered SIDS broadly. Two papers did not directly deal with SIDS, but they had language and project implications and secondary analyses on SIDS; one paper focused on Denmark and the other on the EU.

According to the findings, the most trending topics are related to RE and specific sources [50], energy transition [51], energy security [52], mitigation [53], aid [54], efficiency and fuel access topics [55], consumer perception [56], sustainability and environmental preservation [57], ICT and technological innovation [58], specific projects and techniques [59], economic dynamics [60], and energy regulation and policy [61]. In terms of methodology, a wide set of techniques, including case study analysis, impact evaluation, index construction, and principal component analysis by Argentiero and Falcone [62], Karytsas and Choropanitis [63], Li et al. [64], and Luo et al. [65], were used as methods.

The most prolific authors were M. Mohsin and M. Dornan, with two items each. Apart from these authors, the bibliometric analysis did not return any outstanding authors displaying more than one publication. The paper displays the largest number of coauthors; there are six individuals (one publication). Three works were written jointly by five scholars (per each), and four were drafted from four researchers altogether (per researcher). An additional 5 papers were realised by groups made of 3 coauthors, 3 were finalised by 2 scientists each, and 3 articles were published by a single author.

The number of publications on the cross-cutting topics related to RE and energy security in SIDS reveals the newsworthiness of the tackled topic. The first publication dates back to 2009. In 2015, 2016, 2018, and 2019, two papers are recorded each year. In

2020, a publication peak (six records) is noticeable. The year 2021 registered one published article to date. These figures signal an ascending trend in terms of scholarly interest from 2015, which clearly increased in 2020. For instance, Azzuni et al. [66] investigated the impacts of transitioning from fossil fuel to a renewables-dominated energy system on energy security in Jordan. Their results reveal that availability, affordability, environment, health, and employment will be improved, whereas the diversity will remain unchanged. Therefore, Jordan can reach a 100% RE system by 2050, which will improve the country's energy security.

The largest publication outlet was Renewable and Sustainable Energy Reviews, with 6 articles, followed by Utilities Policy, where 2 papers were published. The remaining items (n = 11) were collected in 11 different journals including (i) *RE*; (ii) *Energies*; (iii) *Environmental Science and Pollution Research*; (iv) *The Journal of Cleaner Production*; (v) *Ecological Indicators*; (vi) *Asian Politics & Policy*; (vii) *Energy Economics*; (viii) *Energy Research & Social Science*; (ix) *Energy for Sustainable Development*; (x) *The Journal of the Energy Institute*; and (xi) *Pacific Economic Bulletin*.

3.2. Co-Occurrence Analyses

The output of the analysis, for a minimum occurrence threshold of six keywords, is shown in Figure 2, mapping the co-occurrence analysis on RE and energy security publications. Here, node size is an indication of the keyword's frequency of use, and the proximity of occurrence indicates that two keywords have co-occurred more frequently. It was found that the following are common topics as they have been used more frequently and have higher values of link strength (top 10): sustainability, climate change, GHGs (greenhouse gases), solar energy, wind energy, biofuels, China, biomass energy, efficiency, and bioenergy. Sustainability is strictly intertwined with RE and energy security and is fundamental to fostering connected issues such as energy resilience, vulnerability, and justice [67–69]. This is also due to the polysemic nature of energy security, even though prior scholarship, e.g., [70–72], has attempted to provide definitions and conceptualisations [42]. The achievement of energy efficiency is primarily linked to energy security and RE.

As shown in Figure 2, three major clusters can be identified in red and green—larger clusters—and blue—a smaller cluster. The largest cluster (red colour) is mainly focused on ecological issues—above all, the implications for climate change mitigation and environmental and energy sustainability. This may indicate that different scenarios for achieving energy security have major impacts on the progress towards achieving environmental sustainability, greenhouse gases and climate change mitigation targets, RE consumption and overall energy resilience, and vice-versa [73,74].

Some nonrenewable sources such as coal and natural gas can be spotted, and this may signify their importance for achieving energy security in some regions such as ASEAN and Europe. Additional geographical indications include China, Japan, Germany, Turkey, and North Africa. The first two locations refer to two East Asian ascending markets for RE, which are devoting increasing attention to energy security. Germany is a major world and European player on these themes, whereas Turkey and North Africa are regarded as major channels for both RE pipelines and energy security and supply. Energy security and environmental sustainability, as demonstrated by Murshed [75], have become important to the global policy agenda, with global economic growth strategies being reorganised to assure the reliability of energy supply while also safeguarding environmental well-being. However, one of the biggest obstacles to achieving these broad objectives is technological inefficiency. The findings of the economic analyses show that ICT trade boosts RE consumption, increases RE shares, reduces energy intensity, encourages the use of cleaner fuels, and lowers carbon emissions. Furthermore, ICT reduces carbon emissions indirectly by increasing RE usage, improving energy efficiency, and expanding access to cleaner fuels.

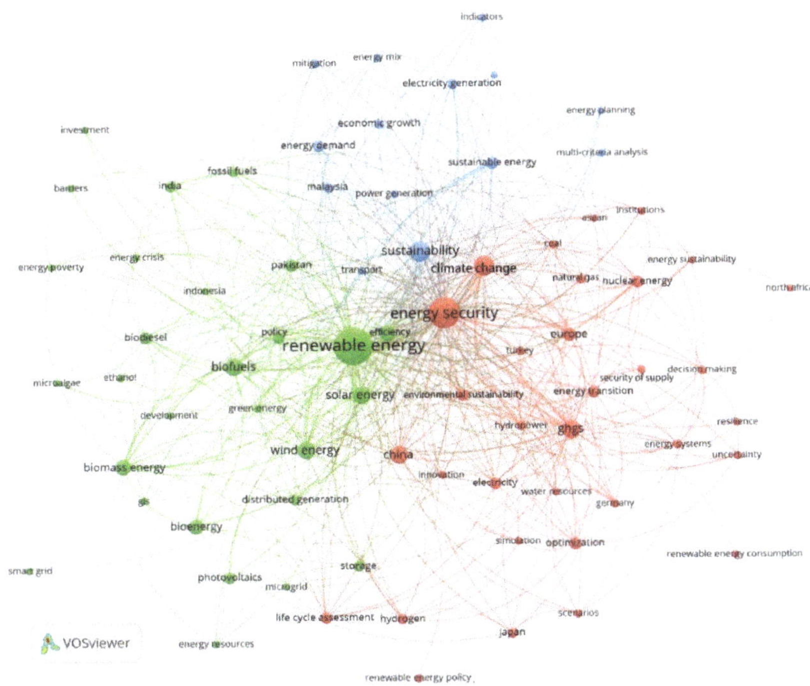

Figure 2. Co-occurrence map: RE and energy security (derived from VOSviewer).

Given the strong nexus between water and energy in some contexts and the water–energy–food nexus and their security, increasing energy security may also enhance water security [76]. The water–energy–food nexus has emerged globally, according to Sarkodie and Owusu [77], since the interrelation of these resources is critical to building conceptual tools for environmental sustainability. As a result, the water–energy–food nexus supports economic development while also enhancing life and well-being. Because of its policy consequences, scholarly study on the water–energy–food nexus is rapidly expanding nowadays. RE technologies have been identified as the cure for achieving environmental sustainability. However, while fossil fuel energy technologies compete with groundwater extraction and usage, certain renewables compete with food for land usage, requiring cost–benefit analysis. Overall, this cluster highlights the significance of optimizing energy systems to simultaneously achieve energy security and environmental sustainability [78]. As suggested by the cluster, institutions and decision-making processes would drive energy system resilience to minimise sectoral uncertainty.

The second major cluster (Figure 3, green colour) can be predominantly connected to environmental and technical topics pursuing social mandates. Smart grids and microgrids are regarded as key energy resilience strategies for overcoming energy security, access, and vulnerability, especially in remote and rural areas of the Global South to target economic, social, and environmental goals [79,80]. According to Daneshvar et al. [81] and Khalil [82], as cities are developed to become smarter, energy systems and access to them consequently becomes an even more integral part of modern living. To become more dependable and resilient, newer energy systems must also undergo transformations. Despite their reliance on natural processes, solar cells and wind energy technologies remain the strongest possibilities for new significant energy sources in smart cities. These energy sources should continue to be the foundation of smart energy systems in the future. In reality, since their emergence a few years ago, solar panels for practical uses in buildings have seen dramatic price drops around the world. Some other important issues can be highlighted based on proximate terms in this cluster. For instance, the term "storage" is closely linked to RE and infrastructure. This may indicate the significance of improvements in battery storage capacity for maximising the benefits accrued from RE [83].

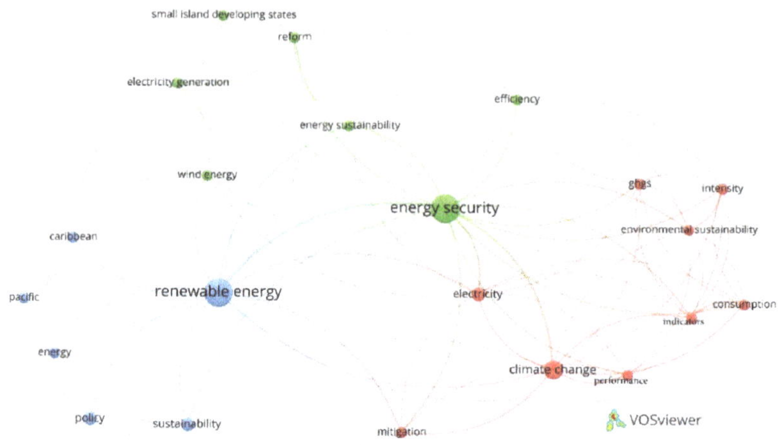

Figure 3. Co-occurrence map: RE and energy security in SIDS (derived from VOSviewer).

"Efficiency" is a frequently used term with strong and close connections to both RE and energy security. This can be interpreted as the significance of efficiency improvements for optimising the performance of RE and their contribution to energy security [84]. Three geographical terms also appear and refer to South and Southeast Asian countries sorting from this study's analyses (i.e., India, Pakistan, and Indonesia). These rank in the lower top 10 countries globally in terms of lack of access to electricity [85]. The robustness of infrastructure investments and development policies to remove energy poverty barriers and to enhance energy efficiency and RE distribution and generation are urgent and on the rise.

Finally, there is a relatively smaller cluster (Figure 3, blue colour) that contains mixed topics—mostly economic and environmental but also technical and methodological. The group includes terms such as sustainability, sustainable energy, mitigation, economic growth, energy demand, energy transport, and energy planning [86]. Another subgroup can be identified with the topics of the energy mix, electricity generation, and power generation. Two widely used techniques can be discerned, which include multicriteria analysis and indicators. Some of the major publications in this area are focused on indicators and multicriteria decision support systems (Table 2). These can be used to guide the development of sustainable and RE systems that can enhance energy security [38,87,88]. Considering the broadness of the approached issues in this third cluster, and the fact that this group only includes a unique spatial location (Malaysia), this last term is explored through specific case studies.

Making use of VOSviewer, a co-occurrence analysis was performed with the goal to corroborate the bibliometric outcomes discussed above, with a focus on SIDS. The term co-occurrence analysis is, indeed, able to render indications within the leading thematic foci and interrelations, as shown in Figure 3, following the same relations as described for general co-occurrence between RE and energy security.

The map illustrates a number of results on the reviewed research items. It is possible to detect two principal bulks of connections, having principal keywords "renewable energy" (in blue) and "energy security" (in green). A third cluster (in red) finds its dominant keyword in "climate change". This illustrates the association between the themes on the one hand and their presence in the literature on the other.

3.3. Case Study Analysis

Table 3 presents some basic characteristics of the sample, as well as the share of RE in final energy consumption. According to data from the World Bank [89], in this sample of SIDS, the percentage of consumed energy that comes from renewable sources corresponds to less than 20%, except for the Pacific Islands, whose values are around 30%.

Table 3. Characteristics of the studied SIDS.

Region	*SIDS*	*Population (Inhab.)*	*Area (km²)*	*GDP (Billion USD)*	*Renewable Energy Consumption (% of Total Final Energy Consumption)*
Caribbean	Barbados	285,719	430	4.8	2.8
	Jamaica	2,890,299	10,990	14.8	16.8
Pacific	Fiji	905,502	18,270	5.1	31.3
	Samoa	196,440	2840	0.9	34.3
African	Mauritius	1,264,613	2040	13.3	11.5
	Maldives	436,330	300	4.6	1.0

Source: based on World Bank [89,90].

The Global Wind Atlas [91] and the Global Solar Atlas [92] are web-based tools to identify potential high-wind areas for wind power generation and to display annual average solar power potential to assist governments and investors in obtaining an initial indication of RE potential before carrying out more detailed analysis (Table 4).

Table 4. Potential for renewable energies in the sample of SIDS.

SIDS	*Wind Energy*		*Solar Energy*	
	Power Density [a] (W/m²)	**Wind Speed [a] (m/s)**	**Photovoltaic Electricity Output (kWh/kWp per Year)**	**Direct Normal Irradiation (kWh/m² per Year)**
Barbados	348	7.56	1663	1777
Jamaica	354	7.09	1498	1482
Fiji	404	7.0	1269	1123
Samoa	319	6.3	1501	1543
Mauritius	418	7.7	1703	1988
Maldives	113	4.6	1599	1715

[a] Average for the 10% windiest areas at height of 100 m. Source: based on Global Wind Atlas [91] and Global Solar Atlas [92].

Among all RE, wind energy is one of the most promising options in many islands [25]. The International Electrotechnical Commission (IEC) International Standard 61400-1 [93] presents wind turbine classes according to the wind conditions of particular sites. Even though more technical details are necessary for a formal classification, the sample of SIDS would be classified as Wind Class III, with annual average wind speeds up to 7.5 m/s. Another important factor presented in Table 4 is the wind power density. This indicator represents the amount of energy available for wind turbines to convert [94]. Therefore, the higher this value is, the better the wind generation potential tends to be.

Due to the average irradiation of approximately 1055 kWh/m² per year, Germany is a renowned investor in RE; therefore, photovoltaic systems are among the most used sources [95–97]. Taking that into consideration, the results for the sample of SIDS are even better. In these conditions, solar energy could be harnessed through solar water heaters (SWH), solar photovoltaic (PV) systems, solar drying, and other systems [25]. According to Kuang et al. [25], the easy implementation of SWH makes it widely used in islands, and the installed capacity of a solar PV system is also flexible, enabling its implementation almost anywhere with low operational costs.

The following sections present additional information regarding RES in the SIDS sample and discuss evidence of the importance of RE for energy security.

3.3.1. Barbados

As with many other countries, Barbados defined targets to increase the use of RE and, consequently, improve its energy efficiency. By 2030, the country is expected to provide 65% of the total electricity demand from renewable sources, and solar energy is considered the most viable resource to achieve this target [98]. Since 2010, the Renewable Energy Rider programme has been contributing to the investment in solar photovoltaic systems, and a rapid increase in solar and wind harnessing is expected for the next decade [98–100]. Following the recommendations from IRENA [101], wind and solar energies are indeed the most favourable sources for this country.

3.3.2. Jamaica

Jamaica is an example of a large country in the Caribbean that has a huge and diversified amount of RE resources, such as geothermal energy, wind and hydropower energy, solar energy, and biomass [101–103]. Although Jamaica is among the islands that developed hydroelectric power more than 60 years ago, the development of its energy system since then has been based on fossil fuels [104]. Jamaica is an exception among other islands because it has a local refinery supplying most of its domestic petroleum needs, but this area also depends on the import of fossil fuels, which are characterised by high energy intensity and low energy efficiency [105]. The island had a large increase (46%) in carbon dioxide emissions between 2000 and 2005, which represented a pressing need to invest in RE technologies and, therefore, to increase national energy security and environmental sustainability [106]. Jamaica's National Energy Policy 2009–2030 acts as an important mechanism towards this need and one of its goals is the recognition of national energy potentials using developing RE, increasing energy security, and decreasing the carbon footprint. In order to ensure that, Jamaica aims to generate 30% of its energy demand from renewable sources by 2030 [103].

3.3.3. Fiji

Like many other Pacific Islands, Fiji considerably depends on imported fossil fuels—100% for transportation and 33% for electricity generation—which is an adverse condition considering increasing fuel costs [107]. As presented by Kumar [108], Fiji has had many historical events of oil supply shortage, including energy crises and war episodes, which have also contributed to high increases in energy prices. Some progress has been observed since the first National Energy Policy was established in 2006, but Fiji continues to fight challenges related to rising expenditures on imported fuels and volatility risk [109]. This represents a huge challenge for the region, given that the fluctuation in fossil fuel supply is directly connected to national energy security and corresponding costs.

3.3.4. Samoa

Like many other SIDS, Samoa also has to deal with extreme weather events, environmental degradation, and rising sea levels [110]. Considering its remoteness from large markets, investments in RE and, consequently, in energy security are even more important. According to information from IRENA [111] and Dornan [112], hydro and solar power have good potentials in Samoa, followed by wind and biomass with moderate potentials.

Samoa meets a third of its power needs through renewable sources, but the rest are met with imported diesel fuels, negatively impacting domestic energy security and prices [113]. Studies from IRENA show that besides helping to achieve a national target of 100% of RE by 2025, increased investments in RE also contribute to being less dependent on fossil fuels [113]. The water availability in Samoa is essential in supporting the high hydropower potential, but solar PV and wind power must contribute as well, accounting for almost 20% of the total electricity demand.

To help meet this local target of achieving 100% RE electricity generation by 2025, a partnership between the government of Samoa and the United Nations Development Programme (UNDP) promoted the project IMPRESS (Improving the Performance and Reliability of Renewable Energy Power System in Samoa). With the aim of supporting national efforts in RE and a cost-effective production, the project also focuses on improving energy efficiency and investing in biomass, biogas, solar, and wind energy systems [114,115].

3.3.5. Mauritius

Mauritius is an island nation that does not have any fossil fuel reserves; therefore, it relies on imports of this resource and local RE sources [116]. For electricity generation, the country imports about 77% of fuel (petroleum products) and coal, and the rest are produced from locally available renewable energies such as biomass, hydro, wind, solar, and landfill gas [117,118].

Mauritius has an urgent need to reduce its fossil fuel dependency, to invest more in RE, for both economic and environmental reasons, and to deal with increased electricity demand [117,118]. Hydroelectric power has been used for many years and corresponds to around 4% of total electricity generated [118]. Despite its advantages, such as low operational costs, this energy resource depends on rainfall for reaching maximum capacity, and the country has been facing some water stress conditions.

On the other hand, solar and wind energy are among the RE sources with the highest potential in Mauritius [119]. The local government is encouraging the use of solar energy and main applications that include SWH and PV systems. In 2014, the first solar farm was launched with a generation capacity of 15 MW, and one year later PV represented almost 1% of total electricity production [118,120]. When it comes to wind energy, Mauritius has potential sites with reliable regimes for investment, and wind power is expected to reach 8% of the contribution to total electricity generation by 2025 [117].

The Mauritius Long-Term Energy Strategy 2009–2025 aims at achieving 35% of RE electricity generation by 2025 [121], 15% more than the value from 2010 (20%). According to Khoodaruth et al. [122], this is a limited scope, and the country has potential for more, as it is feasible to aim for 100% RE by 2050.

3.3.6. The Maldives

The growing energy demand in the Maldives and environmental questions related to carbon emissions are often argued as drivers for more investments in RE technologies. Other important drivers are the financial difficulty caused by the heavy dependency on imported fossil fuels for electricity generation and the significant technoeconomic potentials considering RE [123].

The local government has been investing in sustainable energy and trying to overcome all the barriers that prevent the use of RE. Through the National Energy Policy, which discusses energy efficiency, RE, and energy security, the government set up targets for each aspect and linked them to other national policies and to Sustainable Development Goal 7, which focuses on energy [124]. Nevertheless, according to the Maldives' Renewable Energy Roadmap [125], some of these targets do not meet the most favourable economic conditions to develop RE in the country. New targets should be considered to accelerate the transition to a more RE system and to meet energy security needs.

4. Obstacles and the Means to Address Them

Lucas et al. [10] identified a scarcity of finance, poor human resources, costly infrastructure, socio-cultural impediments, and weak policy and regulatory frameworks as the main challenges for SIDS. Timilsina and Shah [126] highlight the relationships between these factors and argue that the void in technical figures, inadequate assets evaluation, and poor local capacity hamper proper policy resolutions. Dornan [127] argues that the approach of grid extension is not suitable for island countries, and the choice of this model is an obstacle for the deployment of RE. Kaler et al. [128] have identified various shortcomings such as weak institutional frameworks, poor knowledge base, limited financing, poor RE project planning, lack of local capacity, and unreadiness for regional and international cooperation as obstacles to RE deployment. The main obstacles for RE deployment in SIDS were grouped in five categories along with the major obstacles, as shown in Table 5.

Table 5. Main obstacles for RE deployment in SIDS.

Technical	Financial	Policy	Market Barriers	Misc.
Poor resource availability and data gaps in mapping RE resources	High capital costs	Regulatory and policy uncertainty	Inconsistent pricing of energy	Environmental constraints due to the presence of highly sensitive ecosystems
Inadequate infrastructure for evacuation of RE power	Inadequate funding and high cost of capital	Administrative barriers such as slow and opaque decision making	Continued subsidies for fossil fuels	Poor public acceptance and cultural barriers
Inability to absorb RE due to inflexibility and lack of capacity of the electricity grid	Lack of paying capacity for energy services by citizens	Absence of clear goals for RE deployment	Small market size and lack of structured markets	Lack of human capital and skilled workers
Poor scalability of projects	Lack of economies of scale	Poorly defined roles for government departments and weak regulation	Asymmetrical information and market power	Lack of R&D
Lack of accompanying road and transportation infrastructure for accessing remote sites	Poor incentives for RE deployment	Lack of implementing institutions with dedicated responsibilities		Poor integration of stakeholders in project development

The role of international aid for promoting RE in Pacific SIDS was analysed by Keeley [54], who concluded that well-structured action plans, effective regulation, and strong financial status of utilities are essential for attracting funding. Blechinger et al. [129] mentioned the importance of the introduction of proper regulations to accelerate the implementation of the RE potential, while Atteridge and Savvidou [130] called for strengthened local human and institutional capacities to increase the effectiveness of aid efforts. Timilsina and Shah [126] noted that cooperation between public and private sectors could intensify RE expansion. They also highlighted that long-term outlook and policies with incentives are necessary to encourage investment. Lucas et al. [10] identified the technical and financial skills that need to be developed for utilities, the private sector, government, and financial institutions to overcome challenges. Taibi et al. [131] recommended that a framework is important to deploy RE in Pacific Island countries and proposed a focused use of development finance and technical cooperation. Mofor et al. [132] recommended that increased technical conditions to control and preserve RE setups is essential and emphasised capacity-building programs. Furthermore, they support local stakeholder involvement in the development and operation of projects and regional cooperation amongst SIDS for developing RE. Although all the above measures are essential, Michalena and Hills [12] concluded that encouraging RE mechanisms such as energy governance has shown a potential to contribute to the implementation of RE in PSIDS to only a certain extent.

The solution to addressing such issues needs strong, non-government-controlled regulation. In several countries, this has already resulted in improved energy utility performance. However, it cannot tackle investment obstacles such as lack of economies of scale or general investment. Given the fixed costs associated with regulation, the construction of independent regulatory bodies in extremely small countries is also likely to be expensive, considering the possible benefits of such arrangements.

The relationship between political leaders' judgments and technical organisations' decisions in the energy sectors is a larger point. RE at the national level are useful programs because they drive public policy and show where the private sector's investment possibilities exist. Appropriate targets, on the other hand, must be backed up by advice and data from technical institutions, which include both regulatory agencies and electricity producers. This need has not been satisfied by some (but not all) of SIDS's RE objectives. In countries where there are no low-cost RE choices, RE objectives that set a goal of generating 100% of energy from renewables refer to political decision makers acting without (or despite) recommendations from technical bodies in the power sector. This lack of connection between political decision making and advice from the energy sectors is likely to be explained using targets to strengthen SIDS' negotiating position. Political leaders would be completely sensible in adopting such an approach.

5. Conclusions

SIDS are highly dependent on nonrenewable energy (mainly fossil fuels). The fluctuating price of fossil fuels combined with the high transportation cost compromise the energy supply in SIDS and often lead to energy insecurity. This paper considered RE as a medium to improve energy security in SIDS and to ensure energy sustainability. The findings show that SIDS have good potential for exploiting RE, and this offers interesting policy options. This is especially so since some SIDS were compared with Germany, which is a world leader when it comes to integration of RE, and it was concluded that SIDS have a better solar regime.

This implies that SIDS can have a bigger share of RE in the energy mix. However, there are several obstacles to the implementation of these technologies. The different obstacles are categorised into technical aspects (lack of data on resource availability, lack of infrastructure, and lack of smart grid to cope with the intermittency of RE), financial aspects (high capital cost, lack of funding, lack of economies of scale, and lack of incentive), policy aspects (inappropriate regulation and policy, poor governance and slow decision-making process, absence of RE strategy, and lack of institutions), and market size aspects (inconsistent energy price, subsidy on fossil fuels that make the RE cost less competitive, small market size, and lack of market information). The other obstacles are poor public acceptance, cultural barriers, lack of skilled human resources, and lack of research and development. More recently, COVID-19 was added to the list.

Focusing on SIDS, this study aimed to detect and describe key publication characteristics of RE and energy security. To this end, two bibliometric analyses and a co-occurrence analysis on overall trends and SIDS figures were performed. The outcomes returned important information on these energy topics. Despite the low numbers, one can notice an increasing research interest in RE and energy security in SIDS. It is also possible to ascertain selected publication characteristics, including the trendiest topics, journal collection of the publications, author characteristics, and location. Additional key aspects can be acquired from the co-occurrence analyses. In both analyses, several energy- and environment-related topics are recurrent. Some of the most connected terms include climate change and sustainability. These data can be useful for energy policy modelling and upcoming studies.

Based on the case studies analysed and the list of RE projects in SIDS, the authors propose the following additional measures to overcome the existing obstacles and to catalyse RE development in SIDS:

1. Inclusion of economic valuation of climate change mitigation in assessing RE projects;
2. Extensive resource assessment and integrated technoeconomic analysis;
3. Spatial planning for RE deployment along with simultaneous use of available land;
4. Adopting long-term RE targets while maintaining flexibility on technology;
5. Adopting a decentralised model of RE generation and distribution;
6. Development of common standards and guidelines;
7. A coordinated approach to policymaking;
8. Global cooperation for accessing innovative RE financing from private and international funding agencies;
9. Use of fiscal instruments such as tax exemption, import duty waivers on RE equipment, and long-term tax holidays;
10. Fiscal support measures such as feed-in tariffs in the short term;

11. Capacity building through technical inputs;
12. Strengthening of the existing institutional framework.

Moreover, local stakeholder involvement and increased consultative processes are needed to build greater support for the use of RE in SIDS.

In this study, several measures are proposed to overcome different obstacles and challenges to improve the integration of RE. If anything, the COVID-19 pandemic has reiterated the need to further deploy RE to reduce the dependence (and costs) associated with the use of fossil fuels. The authors propose taking the main aspects of sustainability (social, economic, and environmental) into account while assessing RE projects in order to conduct resource assessment, to come up with a strategy and the spatial planning for land use and deployment of RE, to decentralise the energy system, to develop standards, regulations, and guidelines, to come up with a coordinated policy and an institutional framework, to provide financial support through subsidies, and to involve all the stakeholders in different processes.

This paper is not exempt from limitations. Bibliometrics are useful tools to map existing scholarship and dynamics. However, a comprehensive analysis of the publications cannot be ensured. This is the case for grey literature, papers published in nonindexed journals, and papers that are published in languages other than English, which are not indexed in scholarly databases. In order to corroborate the preliminary tests, this work included two bibliometric analyses and two co-occurrence analyses. Future research could expand or differentiate the topics to focus on alternative issues and key bibliometric features or to exploit different techniques.

Author Contributions: Conceptualization, W.L.F.; methodology, W.L.F. and A.G.; writing—original draft preparation, A.-L.B., D.S., A.L.S.; K.N., C.L., J.D.H., A.G., A.S., H.F., S.T. and H.A.; writing—review and editing, W.L.F., A.L.S. and H.A; funding: A.G. All authors have read and agreed to the published version of the manuscript.

Funding: This paper has been funded by the Inter-University Sustainable Development Research Programme (IUSDRP): https://www.haw-hamburg.de/en/ftz-nk/programmes/iusdrp/ (accessed on 13 February 2022).

Institutional Review Board Statement: Not applicable.

Informed Consent Statement: Not applicable.

Data Availability Statement: Not applicable.

Conflicts of Interest: The authors declare no conflict of interest.

References

1. Leal Filho, W.; Voudouris, V. *Global Energy Policy and Security*; Leal Filho, W., Voudouris, V., Eds.; Springer: London, UK, 2013.
2. Pachauri, S. Reaching an international consensus on defining modern energy access. *Curr. Opin. Environ. Sustain.* **2011**, *3*, 235–240. [CrossRef]
3. Raghoo, P.; Surroop, D.; Wolf, F.; Leal Filho, W.; Jeetah, P.; Delakowitz, B. Dimensions of energy security in small island developing states. *Util. Policy* **2018**, *53*, 94–101. [CrossRef]
4. Wolf, F.; Surroop, D.; Singh, A.; Leal, W. Energy access and security strategies in Small Island Developing States. *Energy Policy* **2016**, *98*, 663–673. [CrossRef]
5. Azzuni, A.; Breyer, C. Definitions and dimensions of energy security: A literature review. *Wiley Interdiscip. Rev. Energy Environ.* **2018**, *7*, e268. [CrossRef]
6. Krozer, Y. Valorisation of energy services: Essay on the value addition due to renewable energy. *Energy Sustain. Soc.* **2019**, *9*, 9. [CrossRef]
7. Mulder, P.; Tembe, J. Rural electrification in an imperfect world: A case study from Mozambique. *Energy Policy* **2008**, *36*, 2785–2794. [CrossRef]
8. Surroop, D.; Raghoo, P.; Wolf, F.; Shah, K.U.; Jeetah, P. Energy access in Small Island Developing States: Status, barriers and policy measures. *Environ. Dev.* **2018**, *27*, 58–69. [CrossRef]
9. World Bank. *World Development Indicators—The Little Green Data Book*; World Bank: Washington DC, USA, 2017.
10. Lucas, H.; Fifita, S.; Talab, I.; Marschel, C.; Cabeza, L.F. Critical challenges and capacity building needs for renewable energy deployment in Pacific Small Island Developing States (Pacific SIDS). *Renew. Energy* **2017**, *107*, 42–52. [CrossRef]

11. UNEP. *Global Environment Outlook, Small Island Developing States*; United Nations Environmental Programme: Nairobi, Kenya, 2014.
12. Michalena, E.; Hills, J.M. Paths of renewable energy development in Small Island Developing States of the South Pacific. *Renew. Sustain. Energy Rev.* **2018**, *82*, 343–352. [CrossRef]
13. Weisser, D. On the economics of electricity consumption in small island developing states: A role for renewable energy technologies? *Energy Policy* **2004**, *32*, 127–140. [CrossRef]
14. Scobie, M. Fossil fuel reform in developing states: The case of Trinidad and Tobago, a petroleum-producing small Island developing State. *Energy Policy* **2017**, *104*, 265–273. [CrossRef]
15. World Energy Council. *World Energy Trilemma Index 2017—Monitoring Sustainability of National Energy Systems*; World Energy Council: London, UK, 2017.
16. Szinai, J.K.; Sheppard, C.J.; Abhyankar, N.; Gopal, A.R. Reduced grid operating costs and renewable energy curtailment with electric vehicle charge management. *Energy Policy* **2020**, *136*, 111051. [CrossRef]
17. Oparaocha, S.; Dutta, S. Gender and energy for sustainable development. *Curr. Opin. Environ. Sustain.* **2011**, *3*, 265–271. [CrossRef]
18. Karekezi, S.; McDade, S.; Boardman, B.; Kimani, J. Energy, Poverty, and Development. In *Global Energy Assessment—Towards a Sustainable Future*; Cambridge University Press: Cambridge, UK; New York, NY, USA; International Institute for Applied Systems Analysis: Laxenburg, Austria, 2012; Chapter 2; pp. 151–190.
19. Prasad, G. Improving access to energy in sub-Saharan Africa. *Curr. Opin. Environ. Sustain.* **2011**, *3*, 248–253. [CrossRef]
20. Legros, G.; Havet, I.; Bruce, N.; Bonjour, S. *The Energy Access Situation in Developing Countries*; United Nations Development Programme (UNDP): New York, NY, USA; World Health Organisation (WHO): New York, NY, USA, 2009.
21. Liu, J.; Mei, C.; Wang, H.; Shao, W.; Xiang, C. Powering an island system by renewable energy—A feasibility analysis in the Maldives. *Appl. Energy* **2018**, *227*, 18–27. [CrossRef]
22. Pacific Island Forum. *Framework for Action on Energy Security in the Pacific 2010–2020 (FAESP)*; Pacific Island Forum: Suva, Fiji, 2010.
23. Gay, D.; Rogers, T.; Shirley, R. Small island developing states and their suitability for electric vehicles and vehicle-to-grid services. *Util. Policy* **2018**, *55*, 69–78. [CrossRef]
24. IRENA. A Path to Prosperity: Renewable Energy for Islands. 2015. Available online: https://irena.org/publications/2015/Jun/A-path-to-prosperity-Renewable-energy-for-islands (accessed on 15 November 2019).
25. Kuang, Y.; Zhang, Y.; Zhou, B.; Li, C.; Cao, Y.; Li, L.; Zeng, L. A review of renewable energy utilization in islands. *Renew. Sustain. Energy Rev.* **2016**, *59*, 504–513. [CrossRef]
26. Gatto, A.; Drago, C. A taxonomy of energy resilience. *Energy Policy* **2020**, *136*, 111007. [CrossRef]
27. Hohmeyer, O. 100% Renewable Barbados and Lower Energy Bills—A Plan to Change Barbados' Power Supply to 100% Renewables and its Possible Benefits. 2015. Available online: https://www.uni-flensburg.de/fileadmin/content/abteilungen/industrial/dokumente/downloads/veroeffentlichungen/diskussionsbeitraege/znes-discussionspapers-005-barbados.pdf (accessed on 15 November 2019).
28. Sharifi, A. Urban sustainability assessment: An overview and bibliometric analysis. *Ecol. Indic.* **2020**, *121*, 107102. [CrossRef]
29. Song, Y.; Zhang, M.; Sun, R. Using a new aggregated indicator to evaluate China's energy security. *Energy Policy* **2019**, *132*, 167–174. [CrossRef]
30. Dźwigoł, H.; Dźwigoł-Barosz, M.; Zhyvko, Z.; Miśkiewicz, R.; Pushak, H. Evaluation of the energy security as a component of the national security of the country. *J. Secur. Sustain. Issues* **2019**, *8*, 307–317. [CrossRef]
31. Augutis, J.; Krikštolaitis, R.; Martišauskas, L.; Urbonienė, S.; Urbonas, R.; Ušpurienė, A. Analysis of energy security level in the Baltic States based on indicator approach. *Energy* **2020**, *199*, 117427. [CrossRef]
32. Novikov, V. Bibliometric Analysis of Economic, Social and Information Security Research. *SocioEconomic Chall.* **2021**, *5*, 120–128. [CrossRef]
33. Hou, Y.; Wang, Q. A bibliometric study about energy, environment, and climate change. *Environ. Sci. Pollut. Res.* **2021**, *28*, 34187–34199. [CrossRef]
34. Rosokhata, A.; Minchenko, M.; Khomenko, L.; Chygryn, O. Renewable energy: A bibliometric analysis. *E3S Web Conf.* **2021**, *250*, 03002. [CrossRef]
35. Zhou, W.; Kou, A.; Chen, J.; Ding, B. A retrospective analysis with bibliometric of energy security in 2000–2017. *Energy Rep.* **2018**, *4*, 724–732. [CrossRef]
36. Aria, M.; Cuccurullo, C. bibliometrics: An R-tool for comprehensive science mapping analysis. *J. Informetr.* **2017**, *11*, 959–975. [CrossRef]
37. Van Eck, N.; Waltman, L. Software survey: VOSviewer, a computer program for bibliometric mapping. *Scientometrics* **2009**, *84*, 523–538. [CrossRef]
38. Kruyt, B.; van Vuuren, D.P.; de Vries, H.J.M.; Groenenberg, H. Indicators for energy security. *Energy Policy* **2009**, *37*, 2166–2181. [CrossRef]
39. Winzer, C. Conceptualizing energy security. *Energy Policy* **2012**, *46*, 36–48. [CrossRef]
40. Ang, B.W.; Choong, W.L.; Ng, T.S. Energy security: Definitions, dimensions, and indexes. *Renew. Sustain. Energy Rev.* **2015**, *42*, 1077–1093. [CrossRef]

41. Sovacool, B.K.; Mukherjee, I. Conceptualizing and measuring energy security: A synthesized approach. *Energy* **2011**, *36*, 5343–5355. [CrossRef]
42. Yergin, D. Ensuring energy security. *Foreign Aff.* **2006**, *85*, 69–82. [CrossRef]
43. Chester, L. Conceptualising energy security and making explicit its polysemic nature. *Energy Policy* **2010**, *38*, 887–895. [CrossRef]
44. Asif, M.; Muneer, T. Energy supply, its demand, and security issues for developed and emerging economies. *Renew. Sustain. Energy Rev.* **2007**, *11*, 1388–1413. [CrossRef]
45. Sovacool, B.K.; Brown, M.A. Competing dimensions of energy security: An international perspective. *Annu. Rev. Environ. Resour.* **2010**, *35*, 77–108. [CrossRef]
46. Intharak, N.; Julay, J.H.; Nakanishi, S.; Matsumoto, T.; Sahid, E.; Ormeno, A.A.G.; Aponte, A.A. A Quest for Energy Security in the 21st Century. 2007. Available online: https://aperc.or.jp/file/2010/9/26/APERC_2007_A_Quest_for_Energy_Security.pdf (accessed on 13 February 2022).
47. Sovacool, B.K.; Mukherjee, I.; Drupady, I.M.; D'Agostino, A.L. Evaluating energy security performance from 1990 to 2010 for eighteen countries. *Energy* **2011**, *36*, 5846–5853. [CrossRef]
48. Wang, B.; Wang, Q.; Wei, Y.M.; Li, Z.P. Role of renewable energy in China's energy security and climate change mitigation: An index decomposition analysis. *Renew. Sustain. Energy Rev.* **2018**, *90*, 187–194. [CrossRef]
49. Ibrahiem, D.M.; Hanafy, S.A. Do energy security and environmental quality contribute to renewable energy? The role of trade openness and energy use in North African countries. *Renew. Energy* **2021**, *179*, 667–678. [CrossRef]
50. Zhiznin, S.Z.; Vassilev, S.; Gusev, A.L. Economics of secondary renewable energy sources with hydrogen generation. *Int. J. Hydrog. Energy* **2019**, *44*, 11385–11393. [CrossRef]
51. Cantarero, M.M.V. Of renewable energy, energy democracy, and sustainable development: A roadmap to accelerate the energy transition in developing countries. *Energy Res. Soc. Sci.* **2020**, *70*, 101716. [CrossRef]
52. Hamed, T.A.; Bressler, L. Energy security in Israel and Jordan: The role of renewable energy sources. *Renew. Energy* **2019**, *135*, 378–389. [CrossRef]
53. Acheampong, A.O.; Adams, S.; Boateng, E. Do globalization and renewable energy contribute to carbon emissions mitigation in Sub-Saharan Africa? *Sci. Total Environ.* **2019**, *677*, 436–446. [CrossRef]
54. Keeley, A.R. Renewable Energy in Pacific Small Island Developing States: The role of international aid and the enabling environment from donor's perspectives. *J. Clean. Prod.* **2017**, *146*, 29–36. [CrossRef]
55. Popov, S.P.; Baldynov, O.A. Evaluation of energy efficiency of the long-distance energy transport systems for renewable energy. *E3S Web Conf.* **2019**, *114*, 02003. [CrossRef]
56. Elmustapha, H.; Hoppe, T.; Bressers, H. Consumer renewable energy technology adoption decision-making; comparing models on perceived attributes and attitudinal constructs in the case of solar water heaters in Lebanon. *J. Clean. Prod.* **2018**, *172*, 347–357. [CrossRef]
57. Nathaniel, S.P.; Yalçiner, K.; Bekun, F.V. Assessing the environmental sustainability corridor: Linking natural resources, renewable energy, human capital, and ecological footprint in BRICS. *Resour. Policy* **2021**, *70*, 101924. [CrossRef]
58. Anser, M.K.; Khan, M.A.; Nassani, A.A.; Aldakhil, A.M.; Hinh Voo, X.; Zaman, K. Relationship of the environment with technological innovation, carbon pricing, renewable energy, and global food production. *Econ. Innov. New Technol.* **2020**, 1–36. [CrossRef]
59. Anoune, K.; Bouya, M.; Astito, A.; Abdellah, A.B. Sizing methods and optimization techniques for PV-wind based hybrid renewable energy system: A review. *Renew. Sustain. Energy Rev.* **2018**, *93*, 652–673. [CrossRef]
60. Paramati, S.R.; Apergis, N.; Ummalla, M. Dynamics of renewable energy consumption and economic activities across the agriculture, industry, and service sectors: Evidence in the perspective of sustainable development. *Environ. Sci. Pollut. Res.* **2018**, *25*, 1375–1387. [CrossRef]
61. Nicolli, F.; Vona, F. Energy market liberalization and renewable energy policies in OECD countries. *Energy Policy* **2019**, *128*, 853–867. [CrossRef]
62. Argentiero, M.; Falcone, P.M. The role of Earth observation satellites in maximizing renewable energy production: Case studies analysis for renewable power plants. *Sustainability* **2020**, *12*, 2062. [CrossRef]
63. Karytsas, S.; Choropanitis, I. Barriers against and actions towards renewable energy technologies diffusion: A Principal Component Analysis for residential ground source heat pump (GSHP) systems. *Renew. Sustain. Energy Rev.* **2017**, *78*, 252–271. [CrossRef]
64. Li, R.; Duan, N.; Zhang, Y.; Liu, Z.; Li, B.; Zhang, D.; Dong, T. Anaerobic co-digestion of chicken manure and microalgae Chlorella sp.: Methane potential, microbial diversity, and synergistic impact evaluation. *Waste Manag.* **2017**, *68*, 120–127. [CrossRef] [PubMed]
65. Luo, L.; Yang, L.; Hanafiah, M.M. Construction of renewable energy supply chain model based on LCA. *Open Phys.* **2018**, *16*, 1118–1126. [CrossRef]
66. Azzuni, A.; Aghahosseini, A.; Ram, M.; Bogdanov, D.; Caldera, U.; Breyer, C. Energy security analysis for a 100% renewable energy transition in Jordan by 2050. *Sustainability* **2020**, *12*, 4921. [CrossRef]
67. Valentine, S.V. Emerging symbiosis: Renewable energy and energy security. *Renew. Sustain. Energy Rev.* **2011**, *15*, 4572–4578. [CrossRef]

68. Sovacool, B.K. Energy policymaking in Denmark: Implications for global energy security and sustainability. *Energy Policy* **2013**, *61*, 829–839. [CrossRef]
69. Radovanović, M.; Filipović, S.; Pavlović, D. Energy security measurement—A sustainable approach. *Renew. Sustain. Energy Rev.* **2017**, *68*, 1020–1032. [CrossRef]
70. Santos, T. Regional energy security goes South: Examining energy integration in South America. *Energy Res. Soc. Sci.* **2021**, *76*, 102050. [CrossRef]
71. Hafezi, R.; Alipour, M. Energy Security and Sustainable Development. In *Affordable and Clean Energy: Encyclopedia of the UN Sustainable Development Goals*; Springer: Cham, Switzerland, 2020.
72. Nepal, R.; Musibau, H.; Taghizadeh-Hesary, F. Does renewable energy promote energy security and economic growth in the Association of Southeast Asian Nations? In *Energy Sustainability and Development in ASEAN and East Asia*; Routledge: London, UK, 2020; pp. 81–105.
73. McCollum, D.L.; Krey, V.; Riahi, K.; Kolp, P.; Grubler, A.; Makowski, M.; Nakicenovic, N. Climate policies can help resolve energy security and air pollution challenges. *Clim. Chang.* **2013**, *119*, 479–494. [CrossRef]
74. Child, M.; Koskinen, O.; Linnanen, L.; Breyer, C. Sustainability guardrails for energy scenarios of the global energy transition. *Renew. Sustain. Energy Rev.* **2018**, *91*, 321–334. [CrossRef]
75. Murshed, M. An empirical analysis of the non-linear impacts of ICT-trade openness on renewable energy transition, energy efficiency, clean cooking fuel access, and environmental sustainability in South Asia. *Environ. Sci. Pollut. Res.* **2020**, *27*, 36254–36281. [CrossRef] [PubMed]
76. Sharifi, A.; Yamagata, Y. Principles and criteria for assessing urban energy resilience: A literature review. *Renew. Sustain. Energy Rev.* **2016**, *60*, 1654–1677. [CrossRef]
77. Sarkodie, S.A.; Owusu, P.A. Bibliometric analysis of water–energy–food nexus: Sustainability assessment of renewable energy. *Curr. Opin. Environ. Sci. Health* **2020**, *13*, 29–34. [CrossRef]
78. Gopalakrishnan, K.; Peeta, S. (Eds.) *Sustainable and Resilient Critical Infrastructure Systems: Simulation, Modeling, and Intelligent Engineering*; Springer: Berlin/Heidelberg, Germany, 2010.
79. Clastres, C. Smart grids: Another step towards competition, energy security, and climate change objectives. *Energy Policy* **2011**, *39*, 5399–5408. [CrossRef]
80. Bhattacharyya, S.C. Financing energy access and off-grid electrification: A review of status, options, and challenges. *Renew. Sustain. Energy Rev.* **2013**, *20*, 462–472. [CrossRef]
81. Daneshvar, M.; Asadi, S.; Mohammadi-Ivatloo, B. Overview of the Grid Modernization and Smart Grids. In *Grid Modernization—Future Energy Network Infrastructure*; Springer: Cham, Switzerland, 2021; pp. 1–31.
82. Khalil, E.E. Introduction to energy management in smart grids. In *Solving Urban Infrastructure Problems Using Smart City Technologies*; Elsevier: Amsterdam, The Netherlands, 2021; pp. 399–410.
83. Banos, R.; Manzano-Agugliaro, F.; Montoya, F.G.; Gil, C.; Alcayde, A.; Gómez, J. Optimization methods applied to renewable and sustainable energy: A review. *Renew. Sustain. Energy Rev.* **2011**, *15*, 1753–1766. [CrossRef]
84. Selvakkumaran, S.; Limmeechokchai, B. Energy security and co-benefits of energy efficiency improvement in three Asian countries. *Renew. Sustain. Energy Rev.* **2013**, *20*, 491–503. [CrossRef]
85. González-Eguino, M. Energy poverty: An overview. *Renew. Sustain. Energy Rev.* **2015**, *47*, 377–385. [CrossRef]
86. International Energy Agency; International Renewable Energy Agency; United Nations; World Bank Group; World Health Organization. *Tracking SDG7: The Energy Progress Report 2018*; World Bank: Washington, DC, USA, 2018.
87. OECD & Joint Research Centre. *Handbook on Constructing Composite Indicators: Methodology and User Guide*; OECD: Paris, France, 2008.
88. Wang, J.J.; Jing, Y.Y.; Zhang, C.F.; Zhao, J.H. Review on multi-criteria decision analysis aid in sustainable energy decision-making. *Renew. Sustain. Energy Rev.* **2009**, *13*, 2263–2278. [CrossRef]
89. World Bank. Renewable Energy Consumption (% of Total Final Energy Consumption). World Bank Open Data. 2015. Available online: https://data.worldbank.org/indicator/EG.FEC.RNEW.ZS (accessed on 10 December 2019).
90. World Bank. Population, Area and GDP. World Bank Open Data. 2017. Available online: https://data.worldbank.org/ (accessed on 10 December 2019).
91. Global Wind Atlas. World Bank Group. 2018. Available online: https://globalwindatlas.info/ (accessed on 15 January 2020).
92. Global Solar Atlas. World Bank Group. 2018. Available online: https://globalsolaratlas.info/ (accessed on 15 January 2020).
93. International Electrotechnical Commission. *Wind Turbines. Part I: Design Requirements*; IEC 61400-1 Edition 3; International Electrotechnical Commission: Geneva, Switzerland, 2005.
94. Ghosh, T.K.; Prelas, M.A. *Energy Resources and Systems: Volume 2: Renewable Resources*; Springer: Berlin/Heidelberg, Germany, 2011; Volume 2.
95. Bayer, B.; Matschoss, P.; Thomas, H.; Marian, A. The German experience with integrating photovoltaic systems into the low-voltage grids. *Renew. Energy* **2018**, *119*, 129–141. [CrossRef]
96. Bogdanov, D.; Farfan, J.; Sadovskaia, K.; Aghahosseini, A.; Child, M.; Gulagi, A.; Oyewo, A.S.; Barbosa, L.D.; Breyer, C. Radical transformation pathway towards sustainable electricity via evolutionary steps. *Nat. Commun.* **2019**, *10*, 1077. [CrossRef] [PubMed]
97. Fraunhofer ISE. Recent Facts about Photovoltaics in Germany. 2018. Available online: https://www.ise.fraunhofer.de/content/dam/ise/en/documents/publications/studies/recent-facts-about-photovoltaics-in-germany.pdf (accessed on 15 January 2020).

98. Wyllie, J.O.; Essah, E.A.; Ofetotse, E.L. Barriers of solar energy uptake and the potential for mitigation solutions in Barbados. *Renew. Sustain. Energy Rev.* **2018**, *91*, 935–949. [CrossRef]
99. Energy Division. Barbados National Energy Policy (2017–2037). 2018. Available online: https://www.energy.gov.bb/web/component/docman/doc_download/76-final-draft-of-national-energy-policy (accessed on 15 January 2020).
100. Taibi, E.; Gualberti, G.; Bazilian, M.; Gielen, D. A framework for technology cooperation to accelerate the deployment of renewable energy in Pacific Island Countries. *Energy Policy* **2016**, *98*, 778–790. [CrossRef]
101. IRENA. Renewable Energy Country Profiles: Caribbean. 2012. Available online: https://www.irena.org/documentdownloads/publications/_caribbeancomplete.pdf (accessed on 20 January 2020).
102. Dornan, M.; Shah, K.U. Energy policy, aid, and the development of renewable energy resources in Small Island Developing States. *Energy Policy* **2016**, *98*, 759–767. [CrossRef]
103. Jamaica. Jamaica's National Energy Policy 2009–2030. The Ministry of Energy and Mining. 2009. Available online: https://www.mset.gov.jm/sites/default/files/National%20Energy%20Policy_0.pdf (accessed on 20 January 2020).
104. Harrison, C.; Popke, J. Geographies of renewable energy transition in the Caribbean: Reshaping the island energy metabolism. *Energy Res. Soc. Sci.* **2018**, *36*, 165–174. [CrossRef]
105. Contreras-Lisperguer, R.; Batuecas, E.; Mayo, C.; Díaz, R.; Pérez, F.J.; Springer, C. Sustainability assessment of electricity cogeneration from sugarcane bagasse in Jamaica. *J. Clean. Prod.* **2018**, *200*, 390–401. [CrossRef]
106. United Nations. The Second National Communication of Jamaica to the United Nations Framework Convention on Climate Change. 2011. Available online: https://unfccc.int/resource/docs/natc/jamnc2.pdf (accessed on 20 January 2020).
107. Prasad, R.D.; Bansal, R.C.; Raturi, A. A review of Fiji's energy situation: Challenges and strategies as a small island developing state. *Renew. Sustain. Energy Rev.* **2017**, *75*, 278–292. [CrossRef]
108. Kumar, S. Co-integration and the Demand for Energy in Fiji. *Glob. Energy Issues* **2011**, *35*, 85–97. [CrossRef]
109. IRENA. Fiji Renewables Readiness Assessment. 2015. Available online: https://www.irena.org/publications/2015/Jul/Renewables-Readiness-Assessment-Fiji (accessed on 20 January 2020).
110. Stock, P. Island Innovations—UNDP and GEF: Leveraging the Environment for the Sustainable Development of Small Island Developing States. 2014. Available online: https://www.ws.undp.org/content/dam/samoa/docs/UNDP_WS_IslandInnovations_UNDP_GEF_Leveraging_the_Env.pdf?download (accessed on 20 January 2020).
111. IRENA. Renewable Energy Country Profiles: Pacific. 2012. Available online: https://www.irena.org/-/media/Files/IRENA/Agency/Publication/2012/_PacificComplete.pdf (accessed on 20 January 2020).
112. Dornan, M. Renewable energy development in small island developing states of the Pacific. *Resources* **2015**, *4*, 490–506. [CrossRef]
113. IRENA. Renewables Can Supply Nearly 100% of Samoa's Electricity Needs. 2016. Available online: https://www.irena.org/newsroom/articles/2016/Jun/Renewables-Can-Supply-Nearly-100-of-Samoas-Electricity-Needs (accessed on 20 January 2020).
114. UNDP. Samoa is Ready to IMPRESS with the Launch of a Large-Scale Renewable Energy Project. 2017. Available online: https://www.ws.undp.org/content/samoa/en/home/presscenter/pressreleases/2017/10/31/samoa-is-ready-to-impress-with-launch-of-large-scale-renewable-energy-project.html (accessed on 20 January 2020).
115. MNRE. Impress Project. Ministry of Natural Resources and Environment. 2018. Available online: https://www.mnre.gov.ws/impress-project/ (accessed on 20 January 2020).
116. Timmons, D.; Dhunny, A.Z.; Elahee, K.; Havumaki, B.; Howells, M.; Khoodaruth, A.; Lema-Driscoll, A.K.; Lollchund, M.R.; Ramgolam, Y.K.; Rughooputh, S.D.; et al. Cost minimization for fully renewable electricity systems: A Mauritius case study. *Energy Policy* **2019**, *133*, 110895. [CrossRef]
117. Surroop, D.; Raghoo, P. Energy landscape in Mauritius. *Renew. Sustain. Energy Rev.* **2017**, *73*, 688–694. [CrossRef]
118. Surroop, D.; Raghoo, P. Renewable energy to improve the energy situation in African island states. *Renew. Sustain. Energy Rev.* **2018**, *88*, 176–183. [CrossRef]
119. IRENA. Renewable Energy Country Profiles: Africa. 2011. Available online: https://www.irena.org/-/media/Files/IRENA/Agency/Publication/2011/CountryProfiles_Africa_WEB.pdf (accessed on 20 January 2020).
120. Bundhoo, Z.M. Renewable energy exploitation in the small island developing state of Mauritius: Current practice and future potential. *Renew. Sustain. Energy Rev.* **2018**, *82*, 2029–2038. [CrossRef]
121. MREPU. Republic of Mauritius Long-Term Energy Strategy 2009–2025. 2009. Ministry of Renewable Energy and Public Utilities. Available online: https://sustainabledevelopment.un.org/content/documents/1245mauritiusEnergy%20Strategy.pdf (accessed on 20 January 2020).
122. Khoodaruth, A.; Oree, V.; Elahee, M.K.; Clark, I.I.W.W. Exploring options for a 100% renewable energy system in Mauritius by 2050. *Util. Policy* **2017**, *44*, 38–49. [CrossRef]
123. van Alphen, K.; Hekkert, M.P.; van Sark, W.G. Renewable energy technologies in the Maldives—Realizing the potential. *Renew. Sustain. Energy Rev.* **2008**, *12*, 162–180. [CrossRef]
124. Republic of Maldives. Maldives Energy Policy & Strategy 2016. Ministry of Environment and Energy. 2016. Available online: https://www.environment.gov.mv/v2/en/download/4295 (accessed on 20 January 2020).
125. IRENA. Renewable Energy Roadmap: The Republic of Maldives. Background Report. 2015. Available online: https://www.irena.org/EventDocs/Maldives/Maldivesroadmapbackgroundreport.pdf (accessed on 20 January 2020).
126. Timilsina, G.R.; Shah, K.U. Filling the gaps: Policy supports and interventions for scaling up renewable energy development in Small Island Developing States. *Energy Policy* **2016**, *98*, 653–662. [CrossRef]

127. Dornan, M. Access to electricity in Small Island Developing States of the Pacific: Issues and challenges. *Renew. Sustain. Energy Rev.* **2014**, *31*, 726–735. [CrossRef]
128. Kaler, P.J.; Taibi, E.; Roesch, R.; Benmarraze, S.; Cioci, M. *SIDS Lighthouses Quickscan Interim Report*; IRENA: Abu Dhabi, United Arab Emirates, 2017.
129. Blechinger, P.; Cader, C.; Bertheau, P.; Huyskens, H.; Seguin, R.; Breyer, C. Global analysis of the techno-economic potential of renewable energy hybrid systems on small islands. *Energy Policy* **2016**, *98*, 674–687. [CrossRef]
130. Atteridge, A.; Savvidou, G. Development aid for energy in Small Island Developing States. *Energy Sustain. Soc.* **2019**, *9*, 10. [CrossRef]
131. Taibi, E.; del Valle, C.F.; Howells, M. Strategies for solar and wind integration by leveraging flexibility from electric vehicles: The Barbados case study. *Energy* **2018**, *164*, 65–78. [CrossRef]
132. Mofor, L.; Isaka, M.; Wade, H.; Soakai, A. *Pacific Lighthouses Report: Renewable Energy Road-Mapping for Islands*; IRENA: Abu Dhabi, United Arab Emirates, 2013.

sustainability

Article

E-Commerce Engagement: A Prerequisite for Economic Sustainability—An Empirical Examination of Influencing Factors

Sarmann I. Kennedyd [1,*], Rob Kim Marjerison [2], Yuequn Yu [2], Qian Zi [2], Xinyi Tang [3] and Ze Yang [4]

1 Northeastern University, Boston, MA 02115, USA
2 Wenzhou-Kean University, Wenzhou 325015, China; rmarjeri@kean.edu (R.K.M.); yuyueq@kean.edu (Y.Y.); ziq@kean.edu (Q.Z.)
3 University of Sydney, Sydney, NSW 2006, Australia; xtan2843@uni.sydney.edu.au
4 University of Warwick, Coventry CV4 7AL, UK; ze.yang@warwick.ac.uk
* Correspondence: s.kennedyd@northeastern.edu

Abstract: Economic sustainability for firms of all sizes and sectors is likely to depend on some type of online commercial activity. While technical barriers to e-commerce are not high, adaptability to new online markets is a critical part of sustainable economic growth for many firms. The Chinese e-commerce market has increased dramatically to become larger than that of the United States, Europe, and Japan combined. This study explores the underlying factors that influence Chinese online consumers' acceptance and patronage of the online shopping platforms where those firms must operate. Firm competition in the e-commerce platform in China is highly competitive, making exploration of the factors that influence consumer purchase behaviour more valuable. After an extensive qualitative focus group study, a quantitative online survey of 691 savvy Chinese online shoppers was completed. When the data was subjected to structural equation modelling (SEM) for analysis, it was found that a model of three factor constructs explains whether an online shopping platform would be the preferred online shopping platform of choice. E-commerce platform preference (EPP) can predict purchase intention (PI) and site commitment (SC). The results explain why e-commerce platforms should address important EPP factors such as: order fulfilment and delivery process, company image enhancers, the variety of products offered, the design of the online shopping platform, trust of its recommendation system, and finally, awareness of the online shopping platform itself. These findings may be of interest to e-commerce practitioners as well as those whose research interests include e-commerce and consumer behaviour.

Keywords: economic sustainability; consumer experience; sustainable commerce; online consumer behaviour; e-commerce platform; online purchase intention; online customer engagement

Citation: Kennedyd, S.I.; Marjerison, R.K.; Yu, Y.; Zi, Q.; Tang, X.; Yang, Z. E-Commerce Engagement: A Prerequisite for Economic Sustainability—An Empirical Examination of Influencing Factors. *Sustainability* **2022**, *14*, 4554. https://doi.org/10.3390/su14084554

Academic Editor: Elkhan Richard Sadik-Zada

Received: 20 February 2022
Accepted: 24 March 2022
Published: 11 April 2022

1. Introduction

1.1. Sustainability

Consideration of sustainability as a concept is not a recent phenomenon, over time evolving into a fairly well developed theoretical paradigm that is most frequently conceptualized by the use of the three pillars model [1], each with a separate assessment, Triple Bottom Line, or 3BT [2]. Separate discussions on ecological, social, and economic areas is a simplistic view [3] and presents a challenge for those seeking empirical analysis of the three pillars due to the inseparable and interdependent nature of factors that comprise them [4,5]. Further, there exists a lack of consensus around a precise definition of each of the pillars; indeed, some have proposed a 4th or even more additional pillars to encompass other areas of concern including cultural sustainability [6]. Consensus on how the model is segmented and described may not be necessary and there is, however, a consensus on the relevance of concern for sustainability in all three areas. Examples include ecological sustainability as manifested by the discussion of climate change [7], social sustainability viewed through the lenses of inequality of global wealth distribution and the problems that arise from increasing poverty [8–10], and finally, economic sustainability, which is a necessary precursor to addressing both social and ecological sustainability related concerns [11–13].

1.2. Economic Sustainability

Economic sustainability is complicated by the shifting backdrop of global economic trends caused in part by changing social factors [14], and ecological factors which in turn influence business decisions [15], demonstrating the interconnectedness of the three pillars. Another important factor in the study of economic sustainability is the rapidly changing technologies that both drive and are driven by commercial activity. Advances in technology as key drivers in both globalization and e-commerce have changed the commercial landscape dramatically within a relatively few years, creating opportunities for great successes as well as dismal failures of established firms [16,17]. Economic sustainability for existing firms has depended in many cases upon their ability to transition to online commerce [17,18]. Likewise, sustainable commercial growth for existing firms has often depended on firms ability to adapt to new, and often international, markets [16,19] and the challengers that come with the intercultural aspects of international commerce [20,21]. It is the nexus of these two trends that this study will examine, specifically e-commerce acceptance and global expansion by an established firm.

1.3. E-Commerce

As e-commerce has matured, the direct B2C via company website model has largely disappeared, replaced by the model where sellers and buyers connect via an online platform such as e-bay, Craig's List, or Amazon.com in the US, and Taobao or JD in China [22]. The platform model provides opportunities for micro, small and medium sized firms to engage in commerce on a competitive basis, thus creating a viable avenue for social and economic sustainability [23].

It is within this dynamic and complex context that this study will explore consumer preferences and behaviours in use of e-commerce platforms and EPP.

China currently has the largest e-commerce markets in the world, mostly boosted by small and medium sized businesses (SMEs) [24,25]. In 2017, e-commerce market revenue in China was USD 558.9 billion, and it is projected to grow to USD 1262.5 billion by 2023, almost doubling in a six-year period [26]. E-commerce represents more than 35% of China's retail sales, the highest proportion in the world [27].

Against this backdrop, Weise [28] reported that Amazon China had closed its Chinese e-commerce business. Amazon International originally entered the Chinese e-commerce market by purchasing local online book retailer, Joyo.com, in 2004 and later changing its name to Amazon China. As an online retailer, Amazon China was selling its own inventory and using its own logistics network, unlike its major competitor Taobao that is owned and operated by Alibaba. These online business models rely heavily on individual SMEs to sell a wide variety of products and use local delivery companies to lower prices and create delivery efficiency, which appeals to Chinese consumers' loyalty. The Chinese market can be complicated and it is not an easy market to operate in, due to its uncertain legal and regulatory environment; however, Walmart China has been thriving and expanding its business into the online domain by partnering with the second most dominant online retailer in China, JD.com, and is now offering speedy home deliveries within hours [28].

1.4. Gap in the Literature

The objective of this paper is threefold: The first is to determine the factors that contributed to Amazon China's withdrawal from the Chinese market from an online consumer's perspective. Specifically, it is proposed that certain factors constitute and lead to e-commerce platform preference (EPP). Second, an analysis of the effect of EPP factors on consumers' purchase intention (PI) and site commitment (SC) is undertaken, and third, whether there is a relationship between purchase intention and site commitment is explored.

This study provides an empirical rationalization which states that the Amazon China's pull out and its failure to compete for online customers in China were related to fundamental business strategy failures in marketing, advertising, e-commerce basics, cultural adaptation, and supply chain logistics, rather than other macro-economic or administrative factors. To investigate why Chinese online consumers did not embrace the Amazon China platform—unlike its counterparts in the United States, Japan, or Europe—a survey of previous e-commerce and marketing literature was done, and we used existing scales to review these factors. New scales and constructs for a qualitative perspective employing a focus group were developed, and the factors affecting the development of online shoppers' EPPs and the relationship of these preferences with online store purchase intentions, continued site commitment for repeat business, and customer loyalty was explored and discussed. Influencing factors include not offering a free and efficient delivery process for minimum purchase orders and failure to launch an Amazon mobile application (app) in its most basic design at an early stage to compete with local competitors like Taobao. The necessity of an app is based on Chinese consumers being very mobile-oriented, with 66% of their online purchases done on a mobile phone app [29].

The possible underlying reasons for Amazon China pulling out from the Chinese market can be rationalized by EPP factors. These include the availability of a wide variety of products for online purchase, awareness campaigns to draw future customers and retain existing ones, an efficient order fulfilment and delivery process, the image of the company's effect on consumer purchase, the design of the online shopping platform site, and trust in the online recommendations system of the shopping platform. Surprisingly, the price of the products sold on the shopping platform does not contribute to EPP behaviour development in online shoppers.

There is a gap in the literature where there is no existing EPP model that ties together all the different variables that could potentially influence consumers' motivation to buy online from a specific e-commerce platform. In addition, there is little existing research that investigates the effect of the possible motivating variables on intention to purchase from an online platform and how these factors influence site commitment for repeat online customers specifically, within the context of China.

2. Theoretical Framework and Hypotheses

Many studies have asserted that online retail success may be dependent on factors such as providing excellent merchandising, retailing, ease of navigation, convenience of finding products online, ease of paying for those purchases, delivery efficiency, and enhanced security to protect online customers [30]. This review of current literature discusses these dimensions as well as the proposed relationships among them in relation to price, product variety, the online shop's site design, company image, order fulfilment and delivery process, trust in product recommendation credibility, and promotional awareness campaigns aimed at attracting new online customers and retaining existing ones. How these factors relate to purchase intention (PI) from an online shopping platform and site commitment (SC) generated for future loyal customers to the online shopping platform is explored.

2.1. Price

Price is the amount of money given in exchange for tangible or intangible products; it also forms part of the marketing mix aimed at attracting more customers. Price influences how the product is perceived by intended consumers [31]. E-commerce growth is affected by several factors including the proliferation of technology, perceived security and value proposition [32]. However, assuming such factors are equal, and that the consumers are rational, a lower price indicates a higher quantity demand [33]. Customizing prices is becoming more popular and in the era of e-business, this strategy is ever more important, as it is both profitable and easy to implement [34]. Price has a direct effect on the benefits the consumer perceives when comparing the same product and its price among competitors [35]. According to Gefen and Devine [36], the price, and the perceived fairness of price [31] of products may affect consumer purchase behaviour. Online consumers are more influenced by the price of the products than offline consumers [37]. Moreover, due to the average living standard in China, most consumers are sensitive to the price of products [38]. For example, price consciousness is positively correlated with consumers' trust for purchasing certain product types, such as electronic products [39]. Further, Zhao and Jin [40] found that some of the challenges Amazon China faced in the Chinese online market is the price competition from its direct competitors like Taobao and JD.com. The price war in the Chinese online marketplace placed enormous pressure on e-commerce companies' profits [40]. Further, the authors state that Amazon lost its competitiveness in the Chinese online marketplace because of its excessive focus on the quality of its products, rather than the price attractiveness aimed at its major Chinese customer base, which led to a narrow customer segment, compared to its main competitor, Taobao [40]. Based on the above, we propose the following hypothesis:

H1: *The price of products sold on an online shopping platform has a positive effect on e-commerce platform preference development.*

2.2. Product Variety

Product variety in an online shopping context means that consumers can compare many choices from a wide range of products sold online, enabling them to make a better purchase decision [41]. Product variety has also been described as the availability of a wide range of commodities available for sale in a store or online that helps customers to view its diversity and select high-quality goods on an online shopping platform [42]. Moe [43] reported that, by its nature, online shopping gives consumers better decision-making opportunities, because of the availability of information on a wide variety of different products. Thus, the availability of a variety of products to choose from motivates customers to purchase from a specific online platform [44]. Consumers usually prefer to have more information to make a choice on various products available for purchase on an e-commerce site [45]. Moreover, the frequency of customers' online shopping is positively correlated with the variety of products offered in the online shopping store [46]. Meanwhile, Pandian, et al. [47] found that—compared with shopping in physical retail stores—online shoppers have a better product evaluation capability due to the variety available to them online.

The positive influence of product variety on the consumer to improve product selection in an online shopping store can possibly contribute to explaining Amazon China's withdrawal from the Chinese market. According to Liang et al. [48], product variety is very important to online shoppers and it is considered to be the foremost reason to shop online. Taobao, the Alibaba online shopping platform, has significant advantages in product variety categories offered for sale on its site, as it connects various SMEs to their end users [49]. It can therefore be said that if Amazon China's product variety was of the same scope as that of their major competitor Taobao, it would have had a strong competitive advantage in the Chinese online shopping market.

H2: *The product variety offering on an online shopping platform has a positive effect on e-commerce platform preference development.*

2.3. Site Awareness

Site awareness is the extent to which a site is recognized by potential customers [50]. It also refers to consumers' ability to remember a certain website when they need a specific product or service from that site [51]. Site awareness contributes greatly to the competitive advantage of an e-commerce platform within the marketplace and is a vital aspect of Chinese consumers' selection process of an e-commerce platform when planning to purchase online [52]. Furthermore, Yoon [53] found that high consumer site awareness would influence online shoppers' perceived trust and satisfaction with an e-commerce website. Additionally, site awareness is an essential purchase motive that a consumer associates with a specific online platform [54]. A company's site awareness is among one of the most important factors that encourage consumers to select and evaluate the products or services an online shopping platform offers [55]. It is for this purpose that many online platforms provide additional external informational streams—rather than pure advertising—to attract additional customers and to retain current consumers [56].

The selection process of a suitable online seller from among many competing ones has become a time-consuming process for consumers; therefore, generating effective awareness campaigns is becoming increasingly important for customer acquisition and retention [57]. Online site awareness is the process of generating useful information online in the form of advertisement to attract new customers and retain current customers [56]. Competition among different online shopping platforms in China's e-commerce sector is very intense and a few major e-commerce platforms—such as Taobao and JD.com—dominate the market; in comparison, Amazon China had the smallest scale of this vast Chinese e-commerce market [58]. The first big challenge for any businesses in the e-commerce domain is raising awareness to attract customers [59]. Amazon China failed in this aspect, as it had a low proportion of awareness and usage from Chinese online patrons [60].

H3: *The site awareness of an online shopping platform has a positive effect on e-commerce platform preference development.*

2.4. Site Commitment

Commitment is the intention to establish and preserve a relationship [61]. Consumer commitment refers to establishing a relationship with a brand and keeping this relationship stable with short-term sacrifices but long term advantages [62]. Site commitment refers to shoppers establishing and maintaining a constant valued relationship with online retailers to fulfil their purchase needs [63]. Site commitment plays an essential role in forming and maintaining a consumer's ongoing loyalty and obtaining their repeat business, which further contributes to the site's continuance in operations and future growth [64–66]. Site commitment can be influenced by many factors including post-purchase peer reviews [67,68]. Site commitment is therefore a necessary factor that guides customers to establishing and maintaining a relationship with an online store [69]. High site commitment means that an online shopper can be introduced to a platform's new products or services with very minimal effort and with potential higher benefits [70]. For example, Park and Kim [30] showed that site commitment can affect purchase behaviour, as there is a positive relationship between site commitment and purchase behaviour.

H4: *Site commitment to an online shopping platform has a positive effect on e-commerce platform preference development.*

2.5. Intention to Purchase

Intention refers to an unsolidified act of will or a goal that a person is working toward to achieve [71] and purchase intention is the potential or likelihood of a user buying a product or service [72]. Measuring the real transactional behaviour of a buyer directly is difficult; hence, measuring purchase intention is a good substitute for real purchasing behaviour [73]. The theory of reasoned action argues that behavioural intentions are antecedents to specific behaviours of an individual [74]. Intention to purchase products or services can be influenced by related factors such peer influence via social media interactions [75], perceived brand value and community [76], word-of-mouth communication with others of shared ethnicity [21], as well as internalized personal behaviours [77], all of which combined can demonstrate consumers' willingness and plans to buy a particular product or service [78]. Therefore, intention to purchase can directly forecast the behaviour of a future buyer [79]. Further, the intention to purchase can directly predict the actual purchase behaviours made by consumers [79].

H5: *E-commerce platform preference for an online shopping platform has a positive effect on intention to purchase from that platform.*

H6: *Intention to purchase from an online shopping platform has a positive effect on site commitment (customer loyalty development).*

2.6. Trust in Online Product Recommendations

Consumers' intention to purchase products is based on the recommendations made by others in the same online shopping platforms of the intended product to purchase [80]. An online recommendation to increase sales volumes is regarded as word-of-mouth marketing [81]. However, Chinese online shoppers are sometimes overwhelmed with information and their purchase behaviour can easily be influenced by any of several sources [75] and, in particular, by an informative and fair recommendation system in the shopping platform's system [82]. Trust in online recommendations influences positively site commitment if consumers have higher trust in the credibility of the online product recommendations [83].

H7: *Trust in online recommendations credibility of an online shopping platform has a positive effect on e-commerce platform preference development.*

2.7. Company Image

Consumers are affected by many factors during the shopping process, including store image, store reputation, store awareness, and product features. These factors not only affect consumers' perception of the quality of goods or services, but also their evaluation of the goods or services [84]. Brand image as defined by [85] is "the set of beliefs, ideas, and impression that a person holds regarding an object" (p. 273). Consumers acquire a series of values related to both the product and company image; therefore, the company image is the "sum of value" of the enterprise [86]. In fact, many consider company image as a guarantee of quality assurance that minimizes underperformance issues in certain products [87].

Online shopping platforms' image will affect consumers' purchase intention to varying degrees. For instance, a study [88] on gender role in brand image and purchase intention found that brand image positively influenced purchase intention specially more on female e-shoppers as it lessens the risk perception. Brand image and its associated company image is often used by consumers as a standard to evaluate the quality of goods or services available on the platform. Thus, company image becomes a relevant basis of judgment for product or service selection and generates intention to purchase, eventually establishing site commitment [89]. Further, Treacy and Wiersema [90] confirm that positive experiences will eventually lead to a positive image.

E-commerce platform image can create loyalty in customers when they are exposed to a particular product or service they like and having an enhanced brand image could be a consequence of the product or service itself. The more positive the store image is, the easier it is to gain and attract more customers by affecting their purchase intention. The relationship between the retailer's image and the consumer's willingness to purchase from that retailer is greatly intertwined relationship. For instance, if the customer has a positive perception of the image of the online store, it will have a favourable effect on the consumer's online purchase intention [91].

H8: *The company image of an online shopping platform has a positive effect on e-commerce platform preference development.*

2.8. Site Design

An online store's success is very often determined by the quality of the website's design [92]. The relationship between website design and e-commerce performance has been thoroughly investigated. Website design reflects users' preferences for the website's user interfaces, design, and navigation [93]. Moreover, creating useful data for online content requires a full understanding of several disciplines, such as customer habits, particularly those users who are lead users, those who may provide valuable input into the evolution of the customer interaction [94], as well as information and data warehouse [95]. For example, Cho and Park [96] found that e-commerce customer satisfaction was influenced by the website design. Additionally, how content is arranged on the website is highly indicative of how well-designed the website is [92]. When purchasing online, customers prefer to interact through a technology interface, rather than with employees. Thus, website design aspects are useful influencing factors of online retailers' ability to guarantee customer quality, satisfaction, and loyalty in the long term [97].

Similarly, efficiency and ease of order fulfilment are also crucial parts of the website's service quality; ease of use and a well-organized design of the different webpages contribute to meeting customers' needs for quick service, providing an excellent interaction experience [98]. An e-commerce platform must also use seamless integrated web designs across multiple connection platforms—such as desktops, smartphones, and tablets—where elements such as colours, size of fonts, pictures, and other design elements should be similar across all media to create a reconciled style in e-commerce for branding competitiveness [99]. In the case of Amazon China, they used to focus more on primary functions such as displaying the products and its reviews with less colours, which looked simplistic [100]. Chinese consumers, however, want to know the current trends among their friends [75] and other influencers [75] so they can browse using the vogue button in Taobao. Amazon could have enriched its online platform similarly to make it more attractive for Chinese customers [101].

Compared with the Chinese mainstream online shopping platforms like Taobao and JD.com, Amazon China's website design was not customized for the tastes of Chinese online shoppers. In fact, the Amazon China platform seemed very similar to the Amazon US platform, with no features catering to the local culture and customs that is necessary to attract Chinese online shoppers [100].

H9: *The website design of an online shopping platform has a positive effect on e-commerce platform preference development.*

2.9. Order Fulfilment and Delivery

Online customers assign a lot of importance to how easily they can find products when shopping online. Features such as a live platform comprising short videos provide online shoppers with more accurate information about their purchases. Furthermore, providing video content to online shoppers enables e-commerce platforms to convert browsing customers into buying customers [102,103]. Similarly, ease of payment when making purchases contributes to convenience, which online shoppers often name as the reason they shop mostly online. After-sales service is another significant component that adds to whether the company's order fulfilment is adequate and convenient [104]. Offering after-sales service to customers shows the company's commitment to customers and represents an aspect of its quality offerings to customers [105]. However, many consumers think that after-sales service can hardly be ensured in the depersonalized world of e-commerce, which may affect whether they prefer to purchase online [106]. Order fulfilment considerations such as shipping costs need to be considered, as the shipping cost can increase the total cost of the final products, which, in turn, can affect customer site commitment [107,108]. Therefore, order fulfilment comprising how customers find, evaluate, and pay for the products and the ease of delivery and after-sales support may contribute to customers developing a preference for a certain e-commerce website. Fulfilment was also found to be one of the key ingredients in e-service quality adoption considerations in the e-tail industry [109].

H10: *The order fulfilment and delivery of an online shopping platform has a positive effect on e-commerce platform preference development.*

2.10. Summary of the Hypotheses

Based on our review of the literature above, we propose to test the following hypotheses in the research model below. We aim to determine if the seven factors of price, product variety, website design, company image, order fulfilment and delivery, trust in online recommendation, and website awareness lead to the development of EPP for an e-commerce website. We also propose that EPP could predict online purchase intention (PI) as well as site commitment (SC) to a certain online shopping platform. Moreover, we propose that online PI could predict the development of site commitment, which refers to a customer buying mostly from one online store or developing customer loyalty. Table 1 presents all nine variables in the research model.

Table 1. Research Variables—Definitions.

Variable	Definition
Price	Price is the amount of money required for tangible or intangible transactions, and it is also part of the marketing mix to attract and gain more customers.
Variety	An improved ability to compare a mix of choices to a wide range of products, and eventually the possibility to make a better purchase decision to select a product for purchase.
Purchase intention	The likelihood of a user actually buying a product or a service.
Site awareness	Similar to brand awareness, site awareness means the extent to which potential customers recognize a website.
Site commitment	Creating a lasting desire in future and current online buyer to maintain a valuable relationship with an online seller or to prevent the tendency of that potential online buyer to change or move to another online retailer, thus disrupting long-term engagement of loyalty and repeat sales with customers for the online portal store.
Trust in site recommendation	The willingness of a consumer to trust the product recommendations of shoppers.
Company image	How consumers perceive company image is related to branding, public relationship work, journalism, staff, and consumers' advocacy group. To establish a company image that meets public expectations is crucial for market competitiveness of enterprises through spending a large amount of money on advertising and marketing.
Site design	It is a process about platform development for generating a website through various tools and applications in order to achieve a satisfying look that focuses on figurative elements. Furthermore, the website designer has to consider more on their stakeholders, the target of the platform, and attractive appeal of the design.
Order fulfilment and delivery	How buyers process their online orders, viewing, selecting, comparing, and feeling through the different types of product through live demos, available information, then paying for it, and ultimately getting delivery of their purchases.

Hypothesis summary:

H1: *The price of products sold on an online shopping platform has a positive effect on e-commerce platform preference development.*

H2: *The product variety offering of an online shopping platform has a positive effect on e-commerce platform preference development.*

H3: *The site awareness of an online shopping platform has a positive effect on e-commerce platform preference development.*

H4: *Site commitment to an online shopping platform has a positive effect on e-commerce platform preference development.*

H5: *E-commerce platform preference for an online shopping platform has a positive effect on intention to purchase from that platform.*

H6: *Intention to purchase from an online shopping platform has a positive effect on site commitment or customer loyalty development.*

H7: *Trust in online recommendations of an online shopping platform has a positive effect on e-commerce platform preference development.*

H8: *The company image of an online shopping platform has a positive effect on e-commerce platform preference development.*

H9: *The website design of an online shopping platform has a positive effect on e-commerce platform preference development.*

H10: *The order fulfilment and delivery of an online shopping platform has a positive effect on e-commerce platform preference development.*

3. Methodology

This study uses several scales that were designed, tested, and validated in previous studies to construct an online survey that explored online shoppers' perceptions regarding EPP, as well as its relationship to developing intention to purchase from an e-commerce platform and to generating continued site commitment for repeat business and continued patronage. Three new constructs were designed to determine the outcome of focus group feedback on factors such as order fulfilment and order processing in an online shopping environment. The new constructs were tested to ascertain how these variables contribute to the development of online shoppers' preference for buying from an online platform, a factor which we named e-commerce platform preference or EPP. Moreover, the relationship between EPP and PI as well as the relationship between PI and SC were tested using structural equation modelling.

3.1. *Qualitative Analysis—Focus Group*

A group of twenty Chinese students from an undergraduate American school of business in China were selected as a focus group to share their preferences and experiences when shopping online. Students were informed of the purpose of the focus group meeting and were encouraged to share their experiences and preferences (or lack thereof) when they shopped on the Amazon China website, compared with other Chinese online shopping platforms. Their feedback mainly indicated that even though they fancied shopping on Amazon China because of the implication that this company sells foreign and "authentic" brands, they felt that essential matters that promote site commitment and customer loyalty were not addressed on the platform, such as a wide array of products to choose from. They found the website design to not be customized according to Chinese design styles and colours, which affected transaction completion. Furthermore, there were more critical issues, such as free delivery—as provided by its competitors for most purchases—not being offered based on minimum purchases as Amazon China did.

Focus group participants also reported that Amazon China did not initially offer a mobile app to easily complete purchases or use debit and credit cards for transactions, whereas Taobao is a pioneer in the mobile payment app area and caused the increased dependence of Chinese society on mobile payment. Moreover, issues such as after-sales service or the resolution of disputes were not easily handled by Amazon China, unlike the time-based complaint resolution process on Taobao. Live product demo videos that could showpiece products' key features and benefits to give the buyer the guarantee and accuracy of their purchases were not incorporated for many products. This is also in contrast with Taobao, who increasingly use short video product demonstrations that enhance customers' exactitude of their purchases. Most participants in the focus group were also concerned about the image of the company "Amazon China" as a foreign entity that was competing with a valued local Chinese vanguard brand with excellent company image association. We incorporated the focus group's findings into an online survey to obtain a complete picture of the Chinese online consumers' expectations from online shopping platforms. The scale items in the fulfilment and delivery process construct, company image, and website design were mainly adopted from this focus group's feedback.

Some of the focus group statements described their opinion on factors they considered to be important when shopping online during a lengthy class engagement. The following factors were shown to have influenced their online purchase behaviour and customer loyalty: "I buy products online because I am loyal to the company's culture and orientation" and "I feel proud if it is a Chinese company". Another statement was "I buy products from Taobao because the company Alibaba is engaged in a lot of social responsibility and environmental programs that help Chinese society". Another participant stated that "good customer service and a good company reputation are important factors that influence my online purchase decision making process". One student clarified this statement, further adding: "JD.com is a famous Chinese brand but had a bad reputation due to its CEO [allegedly] misbehaving towards a female, which led me not to buy from the JD platform". It can therefore be said that it is not an issue of pure nationalism, but a perceived overall company image or corporate social responsibility. For example, as a company image related response, they chose a certain online shopping platform "because it resonated with them due to its high social responsibility" and because of "good value and service as well as a good company reputation".

Considering order processing and delivery, participants stated that "comparing the website designs of Amazon China and of Taobao, the latter seems more appealing to me and the orientation of the website is more logical". These statements were also made: "It is easier to complete the transaction", "I couldn't understand the logic of Amazon China website, as I was used to Taobao", "payments were not mobile app payments", and "Amazon China used debit and credit cards, not very efficient, as we no longer carry cards". These comments indicate that the design, logical flow of the website, and ease of paying for purchases are important measurements when shopping online. In terms of customer service, these statements were provided: "It is very difficult to get the company to resolve issues after the sale on Amazon China" and "returning products was not seamless and shipping back was also cumbersome". Regarding product evaluations, the following statement was provided: "I could not use the customized feature of live product demos on Amazon China, so it's difficult to evaluate products for quick purchase". The focus group described payment choices, after-sales services, and product demonstrations as important components of the online shopping experience that influence consumers' choices and online purchase behaviour.

3.2. Survey Questionnaire Design

To collect research data for this study, a survey was designed using constructs from previous research as well as the focus group findings. We designed nine latent variables which are measured by 31 observed variables. For instance, price is measured by three questions: "Buying products on Amazon China may be more expensive than on another online platform". All of the variables considered were measured on a 6-point Likert scale (1 = strongly disagree, to 6 = strongly agree). The survey was then distributed to students from an American university in China as well as their friends and family members via WeChat, a popular social media app in China, to reach more diverse demographics.

All of the following scales are detailed in Appendix A. Price measure scale items were adopted from the work of [36]; product variety scale items were adopted from Clemes et al. [110]; site awareness and site commitment items were both adopted from Park and Kim Park and Kim [30] study; intention to purchase items were adapted from Pavlou and Gefen [111]; and trust in product recommendation scale items were adopted from the scale of [80], which was originally adapted from Gefen and Devine [36]. The company image scale (four observed variables), order process and delivery scale (four observed variables) and website design scale (three observed variables) items were all derived from the focus group. For instance, "I chose Amazon China because it has high company social responsibility" is one of the questions of company image scale.

3.3. Response Rate and Sample Size Determination

Using an a priori sample size determination with an online calculator [112–114], it was determined that the minimum sample size needed to detect effect was 226. The minimum sample size for model structure was 144, after calculating the anticipated effect size at 0.3, desired statistical power level at 0.8, a probability level of 0.01, and factoring in nine latent variables and 31 observed variables. The collected sample size of 691 was justified for the structural equation modelling analysis in this study. After eight weeks of email follow up with the target participants, 691 final usable samples were collected with the desired demographic distribution from almost every province in China with a response rate of 76.8%.

3.4. Data Collection Process

Before the survey was distributed, it was emailed to 50 participants as a pre-test survey. The initial data collection goal was to find participants from different regions that represent Chinese consumers' behaviours when purchasing from an online platform. The questionnaire was distributed to students on a university campus via WeChat. In addition, to create snowball sampling for survey participation, we also asked them to send the survey link to their friends and family members to further ensure good representation. After eight weeks, 691 usable responses were received to the questionnaires from different age groups, education, income, gender, and different provinces in China.

3.5. Demographic Data

The essential demographic survey information is presented in Table 2. The questionnaire instrument collected data from individual respondents with high online shopping experience. Of the participants, 41% were male and 59% female, and almost 95% of the respondents were from urban areas. Of these, 26% of participants were from a metropolis, 41% were from large cities, 28% were from small cities, and 5% were from a small towns and rural areas. We can therefore confidently claim that the survey is geographically representative of the Chinese population. Further, the participants of the three age groups 25–31, 32–38, and 39–55 were evenly distributed with 10–17% in each age group and together representing 43% of the total study population. The millennial generation (age group 18–24) comprised the remaining 57% of the population. Therefore, the age ranges between 18 and 55 comprise more than 99% of the survey responses and covers the majority of Chinese internet users.

Table 2. Socio-demographic breakdown of online shoppers in this study.

	Frequencies	Valid %
Gender		
Male	285	41
Female	406	59
Age		
18–24	391	57
25–31	116	17
32–38	70	10
39–55	112	16
Above 55	2	0
Average monthly salary		
¥0–¥2999	292	42
¥3000–¥4499	102	15
¥4500–¥5999	97	14
¥6000–¥7999	75	11

Table 2. *Cont.*

	Frequencies	Valid %
¥8000–¥9999	42	6
¥10,000–¥14,999	46	7
¥15,000–¥19,999	12	2
More than ¥20,000	25	4
Education level		
None	7	1
Middle school	13	2
High school	68	10
Technical school	11	2
College	118	17
Bachelor's degree	416	60
Master's degree	46	7
PhD degree	12	2
Online Shopping Frequency		
More than one time a day	46	7
Daily	25	4
2–3 times a week	220	32
Weekly	187	27
Monthly	157	23
Rarely	56	8
Geographic residence		
Metropolis	183	26
Large city	280	41
City	193	28
Town	21	3
Rural	14	2

Additionally, for more than half of the respondents, their monthly income is less than CNY 3000 which is consistent with the fact that almost all participants are students. Furthermore, the proportion of respondents gradually decreased to 7% as the monthly income increase to CNY 15,000, and still, 6% of participants had a monthly income higher than CNY 15,000. Approximately 92% of the participants shop online at least once a month, 27% shop online once a week, and nearly 7% shop more than once a day. This data indicates that our participants have enough knowledge of shopping online.

4. Results

To test the direct and indirect relationships between the constructs, this study employed structural equation modelling [90,115]. This analysis method enabled the study of the direct relationship among the study's constructs (latent variables) while simultaneously measuring the effects of the observed variables on those latent constructs. AMOS v24 software was used to allow multiple estimation methods and selected a maximum likelihood as the choice of estimation method for this study. Additionally, we verified the issue of multicollinearity, when one or more of the independent variables are highly intercorrelated [116]. Thus, this would affect the estimates of the regression coefficients to yield statistically significant results [117]. More importantly, multicollinearity leads to distortion in the predictive ability of the independent variables on the dependent variables [118]. In this study, the Tolerance and Variance Inflation Factor (VIF) were assessed to determine the existence of significant multicollinearity issues in the independent variables. According to [117] multicollinearity would be a concern if the VIF value is higher than 5 and tolerance value is <0.20; notably though, other studies also cite a VIF value of 10 and above. Table 3 indicates that multicollinearity for the most variables is within the acceptable limits of VIF values <5 and tolerance values exceeded 0.20, as suggested by [116], even though two variables approach close to the limit. Thus, significant multicollinearity issues are not found in the present study data. Furthermore, we examined the possibility of a Common Method

Bias (CMB) in the research variables. To test that we followed the Harman single-factor test in factor analysis using SPSS v25 with nine factors were extracted using the Principal Axis Factoring, and the first factor Total Variance Explained showed 50.8% of variance, which is slightly more than the 50% marker. Hence, taking into consideration the somewhat presence of a common method variance, we acknowledge that this is one of the limitations of our study.

Table 3. Multicollinearity test.

Independent Variables	Tolerance	VIF
Product Price	0.93	1.07
Product Variety	0.42	2.37
Site Design	0.25	4.04
Company Image	0.20	5.00
Order Fulfilment & Delivery	0.23	5.00
Site Recommendation Trust	0.30	3.39
Site Awareness	0.38	2.62

VIF: Variance Inflation Factor.

4.1. *Confirmatory Factor Analysis*

A nine-factor measurement model with reflective scales was used to estimate the model with a confirmatory factor analysis (CFA) to determine if the observed variables and their latent factor structures fit the hypothesized model. In particular, this study employed CFA for the next step analysis of using causal models in SEM [106]. Table 4, the CFA first order model fit indices showed the following results in: $\chi^2 = 1516.078$, DF = 364, $p = 0.001$ and the Chi-squared goodness of fit is significant, which indicates poor fit to the data; however, due to the large sample size ($n = 691$) it is not expected to fit. Table 4 shows the other fit indices with an acceptable or excellent level of fit: CFI = 0.940, TLI = 0.929, SRMR = 0.040, RMSEA = 0.068 [119]. We evaluated two CFA models. The first-order CFA consisted of nine latent factors that showed excellent convergent and discriminant validity measures. Table 5 lists the variables that are correlated under each latent factor and explain only their unique latent variable and discriminant validity, which measures if the variables correlate with other factors other than their latent variable. The average variance extracted is a measure of convergent validity for all nine constructs, ranging between 0.706 and 0.894. These values exceed 0.5 and are indicative of the high reliability of convergent validity among the latent factors. Further, composite or construct reliability [120] measures the internal consistency within a scale items ranged between 0.878 and 0.962, which exceeds 0.7 and is indicative of good inter-item reliability for the model scales [117,121]. The second-order CFA model consisted of three first order and seven first order representative of EPP indicators. Similarly, the second-order fit indices in Table 6 show acceptable or excellent levels of fit: CFI = 0.920, TLI = 0.911, SRMR = 0.046, RMSEA = 0.076 [120]. Figure 1 presents the second-order factor inter-correlations for the ten factors, with statistical significance set at $p < 0.001$.

The EPP (E-commerce Platform Preference) is based on a second-order factor confirmatory factor analysis (CFA) where the EPP is an endogenous variable and not directly observed but emergent as a latent variable and inferred from the measured first-order variables. The EPP second-order factor CFA was used based on theoretical model assumptions. We found that the second-order factor models allowed the advantage of a more parsimonious model that higher-order factors underlie the data. Moreover, given the researchers' investigation of various factors found in prior literature (product price, variety of products, site design, company image, order fulfilment, site trust, and site awareness) as the underlying dimensions that may influence E-commerce Engagement, there exists the need for empirical testing for a second-order factor which is both theoretically and statistically acceptable and [116,122]. The researchers theorized that, individually first-order measured indicators alone cannot explain the phenomena under investigation in the study to explain the influencing factors of e-commerce engagement. Therefore, the need for the inclusion of a second-order factor which can be used to test the assumption that the correlations among the set of the first-order factors is accounted for one or more higher order factors to further adequately explain a priori theoretical model [123].

Table 4. First Order Model Fit Indices for Confirmatory Factor Analysis.

Model	χ^2	*df*	χ^2/df	CFI	TLI	RMSEA	SRMR
First order	1516.078	364	4.165	0.940	0.929	0.068	0.040

Table 5. Factor Loadings and Reliability Assessment.

Constructs/Indicators	Factor Loadings	CR	AVE	CA	Mean	SD
Product Price	-	0.900	0.749	0.833	3.429	[0.94]
Price1	0.865					
Price2	0.855					
Price3	0.877					
Product variety	-	0.878	0.706	0.787	3.178	[0.87]
Variety1	0.852					
Variety2	0.848					
Variety3	0.822					
Site Design	-	0.940	0.887	0.871	3.624	[1.02]
Design1	0.942					
Design2	0.942					
Design3	0.941					
Company image	-	0.925	0.754	0.888	3.807	[1.13]
Image1	0.910					
Image2	0.908					
Image3	0.840					
Image4	0.812					
Order Fulfilment	-	0.942	0.765	0.922	3.843	[1.10]
Process2	0.905					
Process1	0.895					
Process4	0.875					
Process3	0.875					
Process5	0.820					
Site Recommendation Trust	-	0.962	0.894	0.941	3.837	[1.09]
Trust1	0.947					
Trust2	0.947					
Trust3	0.942					
Intention to purchase	-	0.957	0.881	0.933	3.816	[1.07]
Intention2	0.945					
Intention3	0.936					
Intention1	0.934					
Site commitment	-	0.936	0.785	0.906	4.005	[1.2]
Commitment2	0.904					
Commitment3	0.898					
Commitment4	0.885					
Commitment1	0.856					
Site awareness	-	0.903	0.756	0.834	3.651	[1.10]
Awareness2	0.895					
Awareness3	0.979					
Awareness1	0.836					

Note: AVE: Average variance extracted; CR: Composite Reliability; CA: Cronbach's Alpha.

Table 6. Second Order Model Fit Indices for Confirmatory Factor Analysis.

Model	χ^2	*df*	χ^2/df	CFI	TLI	RMSEA	SRMR
Second order	1936.105	390	4.964	0.920	0.911	0.076	0.046

		Mean	1	2	3	4	5	6	7	8	9	10
1	Intention	3.82	[1.07]									
2	Commitment	4.00	.901**	[1.2]								
3	EcomPref	3.62	.934**	.968**	[1.01]							
4	Awareness	3.65	.864**	.925**	.942**	[1.10]						
5	Trust	3.84	.904**	.868**	.931**	.855**	[1.09]					
6	Process	3.84	.919**	.957**	.996**	.931**	.921**	[1.10]				
7	Corportate Image	3.81	.885**	.934**	.977**	.904**	.889**	.969**	[1.13]			
8	Design	3.62	.769**	.811**	.861**	.788**	.767**	.851**	.852**	[1.02]		
9	Variety	3.18	.807**	.855**	.895**	.822**	.797**	.883**	.885**	.918**	[0.87]	
10	Price	3.43	.136**	.104**	.126**	.132**	.161**	.120**	.093**	.150**	.165**	[0.94]

**. Correlation is significant at the 0.01 level (1-tailed). [ ] means SD.

Figure 1. Factor Inter-correlations for the Second-Order Model.

4.2. Structural Model

The structural equation model was tested using AMOS 24 with latent factors. We used SEM to analyse both the structural model between the latent variables and the measurement relationships between the latent variables and their observable indicators. We used CFA to evaluate the validity of the manifest indicators associated with the underlying latent factors. Next, we used a multivariate analysis of the structural relationships among the latent variables to draw conclusion [124]. Table 4 shows the model fit used to assess the structural model to the absolute, comparative, and parsimonious fit indices to test if the research data fits the proposed model. To test the absolute fit model, the X^2 test for the model is not significant (28) 66.44, $p < 0.001$, even though CHI2/DF 2.37 is less than five. The comparative fit model GFI of 0.984 is greater than the cut-off of 0.90, AGFI 0.969, CFI 0.981, compared to the cut-off of 0.95, SRMR 0.028, and RMSEA 0.041. This shows that the overall model fits the results of the research data.

Table 7 shows the resulting values of the hypothesis testing. This is also indicated in a figure form in Appendix B.

Table 7. Hypothesis Test Results.

			β	*t*-Values	*p*
EPP	→	PI	0.890	25.696	***
EPP	→	SC	0.849	14.175	***
EPP	→	Awareness	0.821	28.247	***
EPP	→	Trust	0.863	28.247	***
EPP	→	Fulfilment	0.937	32.477	***
EPP	→	Image	0.910	30.876	***
EPP	→	Design	0.867	28.469	***
EPP	→	Variety	0.744	22.677	***
PI	→	SC	0.099	1.8970	0.058
Intention	→	Intention1	0.905	38.466	***
Intention	→	Intention2	0.916	38.466	***
Intention	→	Intention3	0.901	36.960	***
Commitment	→	Commitment4	0.822	28.130	***
Commitment	→	Commitment3	0.870	28.131	***
Commitment	→	Commitment2	0.885	28.868	***
Commitment	→	Commitment1	0.794	24.464	***

Note: *** $p < 0.001$. β = Standardized path coefficients, *t*-values = critical ratios; EPP = E-commerce Platform Preference, PI = Purchase Intention, SC = Site Commitment.

Table 8 shows the relationships among the study constructs and variables. The results of the structural equation model analysis show that eight of the ten hypotheses are significantly supported by the final model. E-commerce preference of a platform is influenced by the variety of the products sold on that platform; thus, H2 is supported ($\beta = 0.744$, t-value = 22.677, $p < 0.001$). The online shopping website's aesthetic and adaptive cultural design was also validated, supporting H3 ($\beta = 0.867$, t-value = 28.469, $p < 0.001$). Trust of the recommendation system of online e-commerce platform was also validated, supporting H6 ($\beta = 0.863$, t-value = 28.247, $p < 0.001$), as was awareness of the online shopping platform, and hence, H7 is accepted ($\beta = 0.821$, t-value = 28.247, $p < 0.001$), while order fulfilment and delivery processes and company image significantly affect more than four factors of developing EPP ($\beta = 0.937$, t-value = 32.477, $p < 0.001$ and $\beta = 0.910$, t-value = 30.876, $p < 0.001$) respectively, thus supporting both H5 and H4. However, the price construct was not statistically significant to contribute to developing EPP ($\beta = 0.099$, t-value = 1.897, $p = 0.058$) because the t-value falls below $z < 1.96$; thus, H1—stating that price contributes to EPP, especially when it comes to the Chinese online shoppers' preference or lack thereof—is not accepted. Table 8 summarizes the study hypotheses and its outcomes, and Figure 2 shows the SEM, the path coefficients, and the associated p-values.

Table 8. Hypothesis relationship with the structural model fit results.

Hypothesis	Support
H1. The price of products sold on AC is positively associated with e-commerce platform preference of AC	Not Supported
H2. The variety of products sold on AC is positively associated with e-commerce platform preference of AC	Supported
H3. The design of the AC website is positively associated with e-commerce platform preference of AC	Supported
H4. The image of AC is positively associated with e-commerce platform preference of AC	Supported
H5. The ease of the fulfilment process on AC is positively associated with e-commerce platform preference of AC	Supported
H6. The trust of AC is positively associated with e-commerce platform preference of AC	Supported
H7. The awareness of AC is positively associated with e-commerce platform preference of AC	Supported
H8. E-commerce preference is positively associated with AC site commitment	Supported
H9. E-commerce preference is positively associated with AC intention to purchase	Supported
H10. Intention to purchase on AC is positively associated with AC site commitment	Not Supported

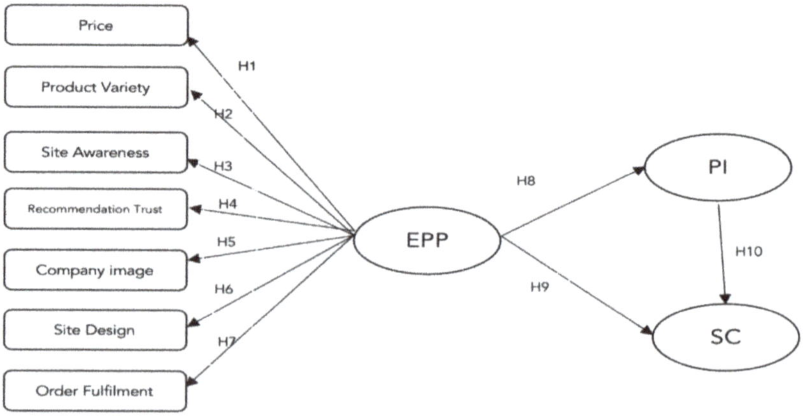

Figure 2. Conceptual Framework. Note: EPP: E-commerce Platform Preference; PI: Purchase Intention; SC: Site Commitment.

5. Discussion

This study makes both theoretical and practical contributions to the topic of the factors that most impact online commerce platform preference sustainability from the customer acceptance perspective.

The purpose of this study was to discover through empirical study the factors that affect consumers' online shopping platform preferences. This is investigated within the context of the maturity and vastness of the Chinese online marketplace. This study contributes to the literature in how the consumers' online shopping platform preferences develop, in addition to understanding whether these EPP (E-commerce Platform Preference) attitudes led to purchase intentions on an e-commerce platform, and if their EPP had the strong influence necessary to achieve site commitment for continued customer loyalty necessary for sustained economic activity, which is imperative for an e-commerce site's future business operations and existence, and more importantly, to the sustainable commercial activity of the countless small and medium sized firms that use the platform to connect with consumers. In contrast, previous studies in this field [125] studied the internal organizational incentives such as e-commerce adoption influenced by top management support, HR IT competence, and financial resources available that create e-business value that leads to e-commerce adoption. Another study, Wang and Ahmed [126], concluded from a macro view that e-commerce adoption is incentivized more from an external pressure, perceived benefits, as well as organizational readiness by itself. Moreover, Thaw, et al. [127] studied the factors that influence e-commerce adoption from the perspective of trust in perceived privacy, security, and the trustworthiness of the vendors. Further, Abubakar, Ilkan and Sahin [88] study looked into the effects that e-referral, eWOM (word of mouth), and gender have on brand image and its related positive impact on purchase intentions. While there is considerable research on e-commerce adoption, to the best of the authors' knowledge, no research has conceptualized and studied e-commerce platform preference motivations in association with purchase intention on the site and finally site commitment for continuous and sustainable patronage. This study adds more to our understanding on how e-commerce platform preference attitudes develop and its close positive associations with purchase intention and site commitment in one single parsimonious model. The findings in this study contribute to the understanding of what factors and characteristics impact consumer acceptance of the e-commerce experience determined by the platform in a sustainable site commitment.

Practical Contribution

These findings contribute to close a gap in the existing literature by establishing that online shopping site (platform) preference is strongly and positively determined by six of the seven essential characteristics investigated: (a) the variety of diverse items on sale, in line with previous research [41,42]; (b) the online shopping portal's site design in terms of layout, colour, font, and design, similar to previous findings [30,92,93]; (c) developing trust in the online recommendation system of the e-commerce portal similarly in line with previous studies [80–82]; (d) ease of order fulfilment and low-cost delivery processes in agreement with previous studies [107,108]; (e) the company's image or brand image in terms of its reputation within society [89–91]; and finally (f) high awareness of the online shopping platform through the media, word of mouth, or being generally well known by most people [51,52,56,58]. These findings are also based upon statements made during the focus group investigation: " ... I didn't even know there was an Amazon in China ... " or "we have Taobao for all our online needs, why venture into another company?"

Specifically, this study found that EPP emergent factor also has a significant impact on consumers' intention to purchase from an online shopping platform, in this case Amazon China, as well as site commitment for repeat purchases and customer loyalty to the same online shopping website. Meanwhile, site commitment and online customer loyalty are not affected by the intention to purchase alone. Customers may have the intention to purchase from an online store, but that intention alone does not create a loyal customer. These results show that the competitiveness and development of customer retention of an e-commerce platform from the online shoppers' perspective depends on those six essential factors that generate EPP attitude (variety of products, website design, company image, order fulfilment and delivery process, trust in the recommendation system, and raising awareness of the e-commerce platform). Interestingly, the price factor in our original theoretical model was found to have no such significant influence on EPP development. This finding was surprising and contradictory to what Zhao and Jin [40] reported, that Amazon China's main challenges were related to the untenable price competition with other online platforms such as Taobao and JD.com. In fact, it was expected that, as Amazon China was mostly offering more foreign expensive products than the local e-commerce sites, that the price element would be a strong determinant factor for EPP. According to Kim, et al. [35], the price of products could have a direct influence as a means of comparison among different competitors; however, in this case, price was not a major contributor in developing EPP behaviour that leads to PI (purchase intention) or SC (site commitment).

The effects of different variables on e-commerce preference, including product variety, site design, company image, order fulfilment and delivery process, trust in the recommendation system, and the awareness of the e-commerce platform were investigated and found to significantly impact EPP behaviour development. Among those factors, order fulfilment and delivery process ($\beta = 0.937$, $t = 32.477$, $p < 0.001$) was the most influential in EPP behaviour development, compared with the other factors [38]. Company image was the second most influential factor ($\beta = 0.910$, $t = 30.876$, $p < 0.001$). The other four variables were also shown to have a great influence on EPP, with β-values ranging between 0.7 to 0.9, significant t-values between 22 to 33, and p-values under 0.001. Remarkably, price was not found to be an influential factor for Chinese consumers' preference in terms of their interaction with the Amazon China online platform ($\beta = 0.099$, t-value = 1.897, p-value = 0.058). Prior research indicates that Chinese consumers' price sensitiveness might be due to the average living standards in the country [38]. These contradictory findings could be explained by changes in China since the previous research was completed, specifically that China lately witnessed increasing living standards and income levels.

From the examination of the relationship between EPP and intention to purchase products from an online shopping platform, it was found that intention to purchase is strongly affected by EPP ($\beta = 0.890$, $t = 25.696$, $p < 0.001$). Further, it was found that site commitment is also strongly influenced by EPP ($\beta = 0.849$, $t = 14.175$, $p < 0.001$). This shows that EPP can predict both the intention to purchase products from an online shopping platform and site commitment for sustained customer loyalty.

In terms of the relationship between intention to purchase products and site commitment, it was found that site commitment is not influenced by the behaviour of developing intention to purchase products from an online shopping platform ($\beta = 0.099$, $t = 1.897$, $p = 0.058$). This result is quite interesting, as it can be said that if one has a PI to an online shopping website, one could also develop the habit to repeatedly buy from the site to become committed as a repeat customer; however, that is not necessarily the case, as the Chinese e-commerce market is very competitive and diverse with many e-commerce sites in existence. Thus, the case of Amazon China abandoning this market indicates that the various reasons that affected the Chinese consumers' lack of purchase intention and site commitment to its online platform can be explained by a failure to address related important factors, such as order fulfilment and delivery process, company image, variety of products offered, site design of the online shopping platform, trust of its recommendation system, and finally, a lack of awareness of the shopping platform itself. All of these factors are highly related to developing strong EPP behaviour and later strong purchase intentions and sustained site commitment in the form of customer loyalty.

6. Conclusions

This paper has attempted to make an advance in the understanding of the concept of how an e-commerce platform can be sustainable—specifically in the Chinese e-commerce market context. The emerging theme from this study is that the e-commerce platform company's focus should be on improving their EPP (E-commerce Platform Preference) factors such as order fulfilment and delivery process, creating a more favourable company image, offering a large variety of products, tailoring the website design to local cultural tastes and habits, building a trustworthy online recommendation system, and increasing awareness of the e-commerce website. Furthermore, this study highlights two important insights. First, developing PI (Purchase Intention) behaviour alone does not lead to online shopping site commitment or customer loyalty and repeat purchasing. Second, online site commitment is generated by EPP factors; among these factors, order fulfilment and delivery process and the company image of the online shopping platform contribute the most, indicating that the prioritization of these six factors should be a starting point when practitioners design online shopping platforms.

This study is not without limitations. Although the data used in this study are based on the consumers' perspective, it would be more informative if combined with direct company information, even though we acknowledge that publicly available data on Amazon China—such as its annual report (2018), which only reports a consolidated financial statement for the whole company—are limited. Additionally, the Chinese government regulates and protects its burgeoning local e-commerce market heavily through licensing requirements and other measures that may significantly limit foreign investments; thus, the extent of government intervention was not investigated in this study. Another limitation in the study is the presence of Common Method Bias (CMB) of 53% in the survey data, slightly above the oft cited threshold of 50% using Harman's Single-Factor Test. Future research should determine how government interferences that protect the local Chinese e-commerce platforms may tip the scales against foreign ventures. Another important angle to study especially in the Chinese e-commerce market is the effect of company image (corporate image) on EPP factors and its eventual site commitment aspect.

Author Contributions: Conceptualization, S.I.K., R.K.M., Y.Y., Q.Z., X.T., Z.Y.; methodology, S.I.K.; software, S.I.K., Y.Y., Q.Z., X.T., Z.Y., X.T.; validation, Y.Y., Q.Z., X.T., Z.Y.; formal analysis, S.I.K.; investigation, S.I.K.; resources, Y.Y., Q.Z., X.T., Z.Y., R.K.M.; writing—original draft preparation, Y.Y., Q.Z., X.T., Z.Y.; writing—review and editing, R.K.M.; visualization, S.I.K.; supervision, S.I.K.; project administration, S.I.K. All authors have read and agreed to the published version of the manuscript.

Funding: This research received no external funding.

Institutional Review Board Statement: For data collected anonymously, that poses no risk, received no funding and was gathered previously, no IRB statement is required by University policy.

Informed Consent Statement: Not applicable.

Data Availability Statement: Not applicable.

Conflicts of Interest: The authors declare no conflict of interest.

Appendix A

Constructs	Items (Anchors: Strongly Disagree/Strongly Agree)	References
Price	1. Buying products on Amazon China may be more expensive than at another online platform 2. I will probably save more money buying products at another online shopping platform than at Amazon China 3. It may be possible to get a better discount from another online platform than from Amazon China 4. It may be cheaper to buy products at Amazon China than at another online platform	[36]
Product variety	1. Shopping on Amazon China offers a wide variety of products 2. I always purchase the types of products I want from Amazon China 3. I can buy the products that are not available in another online shopping platform through Amazon China	[110]
Site commitment	1. I will not change my online shopping on Amazon China in the future. 2. I will continuously purchase products on Amazon China in the future. 3. I will recommend Amazon China to other people. 4. I will visit Amazon China first when I want to buy products.	[30]

Constructs	Items (Anchors: Strongly Disagree/Strongly Agree)	References
Purchase Intention	1. With regard to the products that Amazon China sells, I would consider buying them. 2. With regard to the products that Amazon China sells, I am likely to buy them. 3. With regard to the products that Amazon China sells, I am willing to buy them.	[111]
Trust in Site recommendations	1. I think the product recommendations on Amazon China are credible. 2. I trust the product recommendations on Amazon China. 3. I believe the product recommendations on Amazon China are trustworthy.	[36,80]
Company image	1. I am a loyal customer of Amazon China. 2. I chose Amazon China because it has high company social responsibility. 3. I think Amazon China gives good value and service. 4. I think Amazon China has a good reputation.	These scales are derived from this study's focus group
Website design	1. I really like the page design (layout, style, colour matching, etc.) of Amazon China. 2. I think the design of Amazon China platform is logical. 3. I think it is quick and easy to complete a transaction on Amazon China.	These scales are derived from this study's focus group
Site awareness	1. My neighbours know Amazon China very well. 2. Amazon China is very famous as an Internet shopping platform. 3. Amazon China is known through the advertising media (TV, newspaper, Internet, etc.)	[30]
Fulfilment and delivery processes	1. I am satisfied with the payment choices on Amazon China. 2. I am satisfied with the after-sales services (such as returns) provided by Amazon China. 3. I am satisfied with the shipping cost of the purchases from Amazon China. 4. I would like to use the Live platform on Amazon China.	These scales are derived from this study's focus group

Appendix B

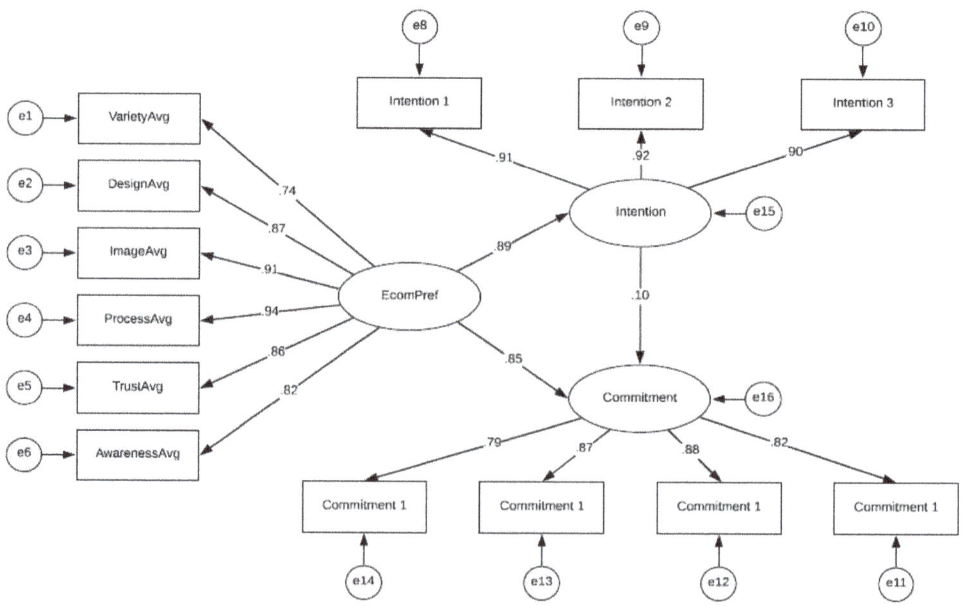

Figure A1. Hypothesized structural model with all relationships at the $p < 0.001$ except intention to commitment = ns.

References

1. Purvis, B.; Mao, Y.; Robinson, D. Three Pillars of Sustainability: In Search of Conceptual Origins. *Sustain. Sci.* **2019**, *14*, 681–695. [CrossRef]
2. Elkington, J. Enter the Triple Bottom Line. In *The Triple Bottom Line*; Routledge: London, UK, 2013; pp. 23–38.
3. Kuhlman, T.; Farrington, J. What Is Sustainability? *Sustainability* **2010**, *2*, 3436–3448. [CrossRef]
4. Hansmann, R.; Mieg, H.A.; Frischknecht, P. Principal Sustainability Components: Empirical Analysis of Synergies Between the Three Pillars of Sustainability. *Int. J. Sustain. Dev. World Ecol.* **2012**, *19*, 451–459. [CrossRef]
5. Komeily, A.; Srinivasan, R.S. A Need for Balanced Approach to Neighborhood Sustainability Assessments: A Critical Review and Analysis. *Sustain. Cities Soc.* **2015**, *18*, 32–43. [CrossRef]
6. Soini, K.; Birkeland, I. Exploring the Scientific Discourse On Cultural Sustainability. *Geoforum* **2014**, *51*, 213–223. [CrossRef]
7. Oreskes, N. The Scientific Consensus On Climate Change: How Do We Know We're Not Wrong? In *Climate Modelling*; Lloyd, A., Winsberg, E., Eds.; Springer International Publishing: Cham, Switzerland, 2018; pp. 31–64.
8. Abbass, K.; Begum, H.; Alam, A.S.A.F.; Awang, A.H.; Abdelsalam, M.K.; Egdair, I.M.M.; Wahid, R. Fresh Insight Through A Keynesian Theory Approach to Investigate the Economic Impact of the Covid-19 Pandemic In Pakistan. *Sustainability* **2022**, *14*, 1054. [CrossRef]
9. Mehta, A.K.; Bhide, S. Poverty Trends and Measures. In *Poverty, Chronic Poverty and Poverty Dynamics*; Mehta, A.K., Bhide, S., Kumar, A., Shah, A., Eds.; Springer: Singapore, 2018; pp. 9–36.
10. Zucman, G. Global Wealth Inequality. *Annu. Rev. Econ.* **2019**, *11*, 109–138. [CrossRef]
11. Gatto, A. A Pluralistic Approach to Economic and Business Sustainability: A Critical Meta-Synthesis of Foundations, Metrics, and Evidence of Human and Local Development. *Corp. Soc. Responsib. Environ. Manag.* **2020**, *27*, 1525–1539. [CrossRef]
12. Goerner, S.J.; Lietaer, B.; Ulanowicz, R.E. Quantifying Economic Sustainability: Implications for Free-Enterprise Theory, Policy and Practice. *Ecol. Econ.* **2009**, *69*, 76–81. [CrossRef]
13. Klaassen, G.A.J.; Opschoor, J.B. Economics of Sustainability or the Sustainability of Economics: Different Paradigms. *Ecol. Econ.* **1991**, *4*, 93–115. [CrossRef]
14. Civero, G.; Rusciano, V.; Scarpato, D. Consumer Behaviour and Corporate Social Responsibility: An Empirical Study of Expo 2015. *Br. Food J.* **2017**, *119*, 1826–1838. [CrossRef]
15. Risitano, M.; Romano, R.; Rusciano, V.; Civero, G.; Scarpato, D. The Impact of Sustainability on Marketing Strategy and Business Performance: The Case of Italian Fisheries. *Bus. Strategy Environ.* **2021**, *11*, 2968. [CrossRef]
16. Woohyoung, K.; Kim, H.; Hwang, J. Transnational Corporation's Failure in China: Focus On Tesco. *Sustainability* **2020**, *12*, 7170. [CrossRef]
17. Rosário, A.; Raimundo, R. Consumer Marketing Strategy and E-Commerce in the Last Decade: A Literature Review. *J. Theor. Appl. Electron. Commer. Res.* **2021**, *16*, 3003–3024. [CrossRef]
18. Barari, M.; Ross, M.; Surachartkumtonkun, J. Negative and Positive Customer Shopping Experience in an Online Context. *J. Retail. Consum. Serv.* **2020**, *53*, 101985. [CrossRef]
19. Ma, S.; Guo, X.; Zhang, H. New Driving Force for China's Import Growth: Assessing the Role of Cross-Border E-Commerce. *World Econ.* **2021**, *44*, 3674–3706. [CrossRef]
20. Gatto, A.; Sadik-Zada, E.R. Revisiting the East Asian Financial Crises: Lessons From Ethics and Development Patterns. In *Economic Growth and Financial Development*; Shahbaz, M., Soliman, A., Ullah, S., Eds.; Springer International Publishing: Cham, Switzerland, 2021; pp. 23–31.
21. Levy, S.; Gvili, Y.; Hino, H. Engagement of Ethnic-Minority Consumers with Electronic Word of Mouth (Ewom) On Social Media: The Pivotal Role of Intercultural Factors. *J. Theor. Appl. Electron. Commer. Res.* **2021**, *16*, 2608–2632. [CrossRef]
22. Huynh, T.K.; Le, H.-D.; Van Nguyen, S.; Tran, H.M. Applying Peer-to-Peer Networks for Decentralized Customer-to-Customer Ecommerce Model. In *Future Data and Security Engineering. Big Data, Security and Privacy, Smart City and Industry 4.0 Applications*; Dang, T.K., Küng, J., Takizawa, M., Chung, T.M., Eds.; Springer: Singapore, 2020; Volume 1306, pp. 21–34.
23. Cao, K.; Xu, Y.; Cao, J.; Xu, B.; Wang, J. Whether A Retailer Should Enter An E-Commerce Platform Taking into Account Consumer Returns. *Int. Trans. Oper. Res.* **2020**, *27*, 2878–2898. [CrossRef]
24. Linnhoff-Popien, C.; Schneider, R.; Zaddach, M. *Digital Marketplaces Unleashed*; Springer: Berlin/Heidelberg, Germany, 2018.
25. Kennedyd, S.I.; Yunzhi, G.; Ziyuan, F.; Liu, K. The Cashless Society Has Arrived: How Mobile Phone Payment Dominance Emerged in China. *Int. J. Electron. Gov. Res.* **2020**, *16*, 94–112. [CrossRef]
26. Striapunina, K. E-Commerce Arpu in United States 2024. Available online: https://www.statista.com/Forecasts/280703/E-Commerce-Arpu-In-United-States (accessed on 19 February 2022).
27. Angelovska, N. Top 5 Online Retailers: 'Electronics and Media' Is the Star of E-Commerce Worldwide. *Forbes*, 20 May 2019.
28. Weise, K. Amazon Gives up on Chinese Domestic Shopping Business. Available online: https://www.nytimes.com/2019/04/18/technology/amazon-china.html%20website (accessed on 19 February 2022).
29. Keyes, D. Amazon Is Struggling to Find Its Place China. *Bus. Insider* **2017**, *11*, 112.
30. Park, C.H.; Kim, Y.G. Identifying Key Factors Affecting Consumer Purchase Behavior in an Online Shopping Context. *Int. J. Retail Distrib. Manag.* **2003**, *31*, 16–29. [CrossRef]
31. Anderson, P.M.; He, X. Price Influence and Age Segments of Beijing Consumers. *J. Consum. Mark.* **1998**, *15*, 152–169. [CrossRef]

32. Rodríguez-Ardura, I.; Meseguer-Artola, A.; Vilaseca-Requena, J. Factors Influencing the Evolution of Electronic Commerce: An Empirical Analysis in a Developed Market Economy. *J. Theor. Appl. Electron. Commer. Res.* **2008**, *3*, 18–29. [CrossRef]
33. Lancaster, G.; Massingham, L. *Essentials of Marketing Management*; Routledge: London, UK, 2010.
34. Shapiro, C.; Varian, H.R. The Art of Standards Wars. *Calif. Manag. Rev.* **1999**, *41*, 8–32. [CrossRef]
35. Kim, H.-W.; Xu, Y.; Gupta, S. Which Is More Important in Internet Shopping, Perceived Price Or Trust? *Electron. Commer. Res. Appl.* **2012**, *11*, 241–252. [CrossRef]
36. Gefen, D.; Devine, P. Customer Loyalty to An Online Store: The Meaning of Online Service Quality. *ICIS* **2001**, *12*, 80.
37. Degeratu, A.M.; Rangaswamy, A.; Wu, J. Consumer Choice Behavior in Online and Traditional Supermarkets: The Effects of Brand Name, Price, and Other Search Attributes. *Int. J. Res. Mark.* **2000**, *17*, 55–78. [CrossRef]
38. Ghauri, P.N.; Usunier, J.-C. *International Business Negotiations*; Emerald Group Publishing: Bentley, UK, 2003.
39. Stieler, M. *Creating Marketing Magic and Innovative Future Marketing Trends*; Springer: Berlin/Heidelberg, Germany, 2017.
40. Zhao, L.; Jin, S. *China's Dilemma of Cross-Border E-Commerce Company-Take Amazon China as an Example*; Atlantis Press: Paris, France, 2016.
41. Keeney, R.L. The Value of Internet Commerce to the Customer. *Manag. Sci.* **1999**, *45*, 533–542. [CrossRef]
42. Ahn, T.; Ryu, S.; Han, I. The impact of the online and offline features on the user acceptance of Internet shopping malls. *Electron. Commer. Res. Appl.* **2004**, *3*, 405–420. [CrossRef]
43. Moe, W.W. Buying, searching, or browsing: Differentiating between online shoppers using in-store navigational clickstream. *J. Consum. Psychol.* **2003**, *13*, 29–39. [CrossRef]
44. Hoskyn, J.; Hill, S.; Collier, M. *eBay.co.uk for Dummies*; John Wiley & Sons: Hoboken, NJ, USA, 2011; ISBN 1-119-99664-3.
45. Rohm, A.J.; Swaminathan, V. A typology of online shoppers based on shopping motivations. *J. Bus. Res.* **2004**, *57*, 748–757. [CrossRef]
46. Smith, A.; Smith, T. *SharePoint 2013 User's Guide: Learning Microsoft's Business Collaboration Platform*; Apress: New York, NY, USA, 2013; ISBN 1-4302-4833-5.
47. Pandian, D.; Fernando, X.; Baig, Z.; Shi, F. *ISMAC in Computational Vision and Bio-Engineering 2018 (ISMAC-CVB)*; Springer: Berlin/Heidelberg, Germany, 2019; Volume 30, ISBN 3-030-00665-4.
48. Liang, B.; Tu, Y.; Cline, T.; Ma, Z. China's E-Tailing Blossom: A Case Study. In *E-Retailing Challenges and Opportunities in the Global Marketplace*; IGI Global: Hershey, PA, USA, 2016; pp. 72–98.
49. Hong, Y.; Pavlou, P.A. Fit does matter! An empirical study on product fit uncertainty in online marketplaces. In *An Empirical Study on Product Fit Uncertainty in Online Marketplaces*; Springer: Berlin/Heidelberg, Germany, 2010.
50. Huang, R.; Sarigöllü, E. How brand awareness relates to market outcome, brand equity, and the marketing mix. In *Fashion Branding and Consumer Behaviors*; Springer: Berlin/Heidelberg, Germany, 2014; pp. 113–132.
51. Aaker, D.A. *Managing Brand Equity*; Free Press: New York, NY, USA, 1991; ISBN 978-0-02-900101-1.
52. Zheng, L.; Favier, M.; Huang, P.; Coat, F. Chinese consumer perceived risk and risk relievers in e-shopping for clothing. *J. Electron. Commer. Res.* **2012**, *13*, 255.
53. Yoon, S.-J. The antecedents and consequences of trust in online-purchase decisions. *J. Interact. Mark.* **2002**, *16*, 47–63. [CrossRef]
54. Greenfield Online. Cybershoppers Research Report #13197. Online Marketing Research Conducted on Behalf of Better Business Bureau. Available online: http://greenfieldcentral.com/newsroom.htm (accessed on 3 September 2020).
55. Lovelock, C.; Patterson, P. *Services Marketing*; Pearson Australia: Melbourne, Australia, 2015; ISBN 1-4860-0476-8.
56. andreassen, T.W.; Lindestad, B. Customer loyalty and complex services: The impact of corporate image on quality, customer satisfaction and loyalty for customers with varying degrees of service expertise. *Int. J. Serv. Ind. Manag.* **1998**, *9*, 7–23. [CrossRef]
57. Board, O.E. *Oswaal CBSE Question Bank Class 10 Social Science Chapterwise & Topicwise (for March 2020 Exam)*; Oswaal Books: Agra, India, 2019; ISBN 978-93-89067-39-2.
58. Leung, Y. Top 5 China Shopping Apps: China Ecommerce Trends. Available online: https://www.dragonsocial.net/blog/top-china-shopping-apps-china-ecommerce/ (accessed on 4 August 2021).
59. Grant, S. E-commerce for small businesses. In Proceedings of the 2nd International Conference IeC'99, Manchester, UK, 1–3 November 1999.
60. Wang, X. Foreign direct investment and innovation in China's e-commerce sector. *J. Asian Econ.* **2012**, *23*, 288–301. [CrossRef]
61. Gursoy, D. *Routledge Handbook of Hospitality Marketing*; Routledge: London, UK, 2017; ISBN 1-315-44550-6.
62. Batat, W. *Experiential Marketing: Consumer Behavior, Customer Experience and the 7Es*; Routledge: London, UK, 2019; ISBN 1-315-23220-0.
63. Palmatier, R.W.; Steinhoff, L. *Relationship Marketing in the Digital Age*; Taylor & Francis: Abingdon, UK, 2019; ISBN 978-1-351-38823-8.
64. Morgan, R.M.; Hunt, S.D. The commitment-trust theory of relationship marketing. *J. Mark.* **1994**, *58*, 20–38. [CrossRef]
65. Pritchard, M.P.; Havitz, M.E.; Howard, D.R. Analyzing the commitment-loyalty link in service contexts. *J. Acad. Mark. Sci.* **1999**, *27*, 333–348. [CrossRef]
66. Garbarino, E.; Johnson, M.S. The different roles of satisfaction, trust, and commitment in customer relationships. *J. Mark.* **1999**, *63*, 70–87. [CrossRef]
67. Kaur, K.; Singh, T. Impact of Online Consumer Reviews on Amazon Books Sales: Empirical Evidence from India. *J. Theor. Appl. Electron. Commer. Res.* **2021**, *16*, 2793–2807. [CrossRef]

68. Kim, H. Use of Mobile Grocery Shopping Application: Motivation and Decision-Making Process among South Korean Consumers. *J. Theor. Appl. Electron. Commer. Res.* **2021**, *16*, 2672–2693. [CrossRef]
69. Gbadamosi, A. *Young Consumer Behaviour: A Research Companion*; Routledge: London, UK, 2017; ISBN 1-351-81905-4.
70. Reiners, T.; Wood, L.C. *Gamification in Education and Business*; Springer International Publishing: Berlin/Heidelberg, Germany, 2014; ISBN 978-3-319-10208-5.
71. Schwarz, L.; Corrigan, F.; Hull, A.; Raju, R. *The Comprehensive Resource Model: Effective Therapeutic Techniques for the Healing of Complex Trauma*; Taylor & Francis: Abingdon, UK, 2016; ISBN 1-317-42554-5.
72. Khosrowpour, M. *Managing Information Technology in a Global Economy*; IGI Global: Hershey, PA, USA, 2001; ISBN 1-930708-07-6.
73. Lin, H.-F. Understanding Behavioral Intention to Participate in Virtual Communities. *Cyberpsychol. Behav.* **2006**, *9*, 540–547. [CrossRef] [PubMed]
74. Liu, C.; Marchewka, J.T.; Lu, J.; Yu, C.-S. Beyond concern—A privacy-trust-behavioral intention model of electronic commerce. *Inf. Manag.* **2005**, *42*, 289–304. [CrossRef]
75. Marjerison, R.K.; Hu, Y.A. Exploring the Impact of Peer Influence on Online Shopping: The Case of Chinese Millennials. In *Quality Management for Competitive Advantage in Global Markets*; IGI Global: Hershey, PA, USA, 2021; pp. 196–210.
76. Rodríguez-Ardura, I.; Martínez-López, F.J.; Gázquez-Abad, J.C.; Ammetller, G. A Review of Online Consumer Behaviour Research: Main Themes and Insights. In *Customer Is NOT Always Right? Marketing Orientations in a Dynamic Business World*; Campbell, C.L., Ed.; Springer International Publishing: Cham, Switzerland, 2017; p. 599. [CrossRef]
77. Cavazos-Arroyo, J.; Máynez-Guaderrama, A.I. Antecedents of Online Impulse Buying: An Analysis of Gender and Centennials' and Millennials' Perspectives. *J. Theor. Appl. Electron. Commer. Res.* **2022**, *17*, 122–137. [CrossRef]
78. Carrillat, F.A.; Jaramillo, F.; Mulki, J.P. Examining the impact of service quality: A meta-analysis of empirical evidence. *J. Mark. Theory Pract.* **2009**, *17*, 95–110. [CrossRef]
79. Moutinho, L.; Chien, C.S. *Problems in Marketing: Applying Key Concepts and Techniques*; Sage: Wuhan, China, 2007; ISBN 1-84920-262-1.
80. Hsiao, K.; Lin, J.C.; Wang, X.; Lu, H.; Yu, H. Antecedents and consequences of trust in online product recommendations: An empirical study in social shopping. *Online Inf. Rev.* **2010**, *34*, 935–953. [CrossRef]
81. Harrison-Walker, L.J. The measurement of word-of-mouth communication and an investigation of service quality and customer commitment as potential antecedents. *J. Serv. Res.* **2001**, *4*, 60–75. [CrossRef]
82. Marjerison, R.K.; Yipei, Y.; Chen, R. The Impact of Social Media Influencers on Purchase Intention towards Cosmetic Products in China. *J. Behav. Stud. Bus.* **2019**, *12*, 1–18.
83. Chen, H.; Li, Z. The Factors Influencing Chinese Online Shopper's Satisfaction in Web2. 0 Era. In Proceedings of the 2008 International Symposium on Electronic Commerce and Security, Guangzhou, China, 3–5 August 2018; IEEE: Piscataway, NJ, USA, 2008; pp. 86–90.
84. Abedin, E.; Mendoza, A.; Karunasekera, S. Exploring the Moderating Role of Readers' Perspective in Evaluations of Online Consumer Reviews. *J. Theor. Appl. Electron. Commer. Res.* **2021**, *16*, 3406–3424. [CrossRef]
85. Bueno, S.; Gallego, M.D. eWOM in C2C platforms: Combining IAM and customer satisfaction to examine the impact on purchase intention. *J. Theor. Appl. Electron. Commer. Res.* **2021**, *16*, 1612–1630. [CrossRef]
86. Ind, N. The corporate brand. In *The Corporate Brand*; Springer: Berlin/Heidelberg, Germany, 1997; pp. 1–13.
87. Balmer, J.M.; Gray, E.R. Corporate brands: What are they? What of them? *Eur. J. Mark.* **2003**, *37*, 972–997. [CrossRef]
88. Abubakar, A.M.; Ilkan, M.; Sahin, P. eWOM, eReferral and gender in the virtual community. *Mark. Intell. Plan.* **2016**, *34*, 692–710. [CrossRef]
89. Richardson, P.S.; Dick, A.S.; Jain, A.K. Extrinsic and intrinsic cue effects on perceptions of store brand quality. *J. Mark.* **1994**, *58*, 28–36. [CrossRef]
90. Treacy, M.; Wiersema, F. Customer intimacy and other value disciplines. *Harv. Bus. Rev.* **1993**, *71*, 84–93.
91. Fiore, A.M.; Kim, J.; Lee, H.-H. Effect of image interactivity technology on consumer responses toward the online retailer. *J. Interact. Mark.* **2005**, *19*, 38–53. [CrossRef]
92. Ranganathan, C.; Grandon, E. An exploratory examination of factors affecting online sales. *J. Comput. Inf. Syst.* **2002**, *42*, 87–93.
93. Kim, J.; Lee, J. Critical design factors for successful e-commerce systems. *Behav. Inf. Technol.* **2002**, *21*, 185–199. [CrossRef]
94. Su, J.; Chen, X.; Zhang, F.; Zhang, N.; Li, F. An Intelligent Method for Lead User Identification in Customer Collaborative Product Innovation. *J. Theor. Appl. Electron. Commer. Res.* **2021**, *16*, 1571–1583. [CrossRef]
95. Jennings, M. *Theory and Models for Creating Engaging and Immersive Ecommerce Websites*; ACM: New York, NY, USA, 2000; pp. 77–85.
96. Cho, N.; Park, S. Development of Electronic Commerce User-Consumer Satisfaction Index (Ecusi) for Internet Shopping. *Ind. Manag. Data Syst.* **2001**, *101*, 400–406. [CrossRef]
97. Wolfinbarger, M.; Gilly, M.C. eTailQ: Dimensionalizing, measuring and predicting etail quality. *J. Retail.* **2003**, *79*, 183–198. [CrossRef]
98. Sethuraman, R.; Parasuraman, A. Succeeding in the Big Middle through technology. *J. Retail.* **2005**, *81*, 107–111. [CrossRef]
99. Bilgihan, A.; Kandampully, J.; Zhang, T.C. Towards a unified customer experience in online shopping environments: Antecedents and outcomes. *Int. J. Qual. Serv. Sci.* **2016**, *8*, 102–119. [CrossRef]
100. Li, K. The Impact of the Context Culture On Shopping Website Design: For Example, Taobao and Amazon China. *J. Educ. Inst. Jilin Prov.* **2015**, *11*, 120–122.

101. Liu, C.; Hong, J. Strategies and Service Innovations of Haitao Business in the Chinese Market: A Comparative Case Study of Amazon.cn vs. Gmarket.co.kr. *Asia Pac. J. Innov. Entrep.* **2016**, *10*, 101–121. [CrossRef]
102. Depamphilis, D. *Mergers, Acquisitions, and Other Restructuring Activities*; Elsevier Science & Technology: Amsterdam, The Netherlands, 2019.
103. Eagle, W. *YouTube Marketing for Dummies*; John Wiley & Sons: Hoboken, NJ, USA, 2019.
104. Chen, S.-J.; Chang, T.-Z. A Descriptive Model of Online Shopping Process: Some Empirical Results. *Int. J. Serv. Ind. Manag.* **2003**, *14*, 556–569. [CrossRef]
105. Bauer, H.H.; Falk, T.; Hammerschmidt, M. Etransqual: A Transaction Process-Based Approach for Capturing Service Quality in Online Shopping. *J. Bus. Res.* **2006**, *59*, 866–875. [CrossRef]
106. Teo, T.S.; Wang, P.; Leong, C.H. Understanding Online Shopping Behaviour Using A Transaction Cost Economics Approach. *Int. J. Internet Mark. Advert.* **2004**, *1*, 62–84. [CrossRef]
107. Karol, R.; Nelson, B.; Nicholson, G. *New Product Development for Dummies*; Wiley: Hoboken, NJ, USA, 2011; ISBN 978-1-118-05128-3.
108. Manning, G.A. *Financial Investigation and Forensic Accounting*; CRC Press: Boca Raton, FL, USA, 2016.
109. Kaya, B.; Behravesh, E.; Abubakar, A.M.; Kaya, O.S.; Orús, C. The Moderating Role of Website Familiarity in the Relationships Between E-Service Quality, E-Satisfaction and E-Loyalty. *J. Internet Commer.* **2019**, *18*, 369–394. [CrossRef]
110. Clemes, M.D.; Gan, C.; Zhang, J. An Empirical Analysis of Online Shopping Adoption in Beijing, China. *J. Retail. Consum. Serv.* **2014**, *21*, 364–375. [CrossRef]
111. Pavlou, P.A.; Gefen, D. Building Effective Online Marketplaces with institution-Based Trust. *Inf. Syst. Res.* **2004**, *15*, 37–59. [CrossRef]
112. Cohen, J. Set correlation and multivariate methods. In *Statistical Power Analysis for the Behavioral Sciences*, 2nd ed.; Lawrence Erlbaum Assoc. Inc.: Hillsdale, NJ, USA, 1988; pp. 467–530.
113. Soper, D.S. A-Priori Sample Size Calculator for Structural Equation Models [Software]. Available online: https://www.danielsoper.com/statcalc/calculator.aspx?id=89 (accessed on 7 January 2020).
114. Westland, J.C. Lower Bounds On Sample Size in Structural Equation Modeling. *Electron. Commer. Res. Appl.* **2010**, *9*, 476–487. [CrossRef]
115. Jöreskog, K.G.; Sörbom, D. *LISREL 8: Structural Equation Modeling with the SIMPLIS Command Language*; Scientific Software International, Wiley: Hoboken, NJ, USA, 1993; ISBN 0-89498-033-5.
116. Hair, J.F.; Ringle, C.M.; Sarstedt, M. Pls-Sem: Indeed A Silver Bullet. *J. Mark. Theory Pract.* **2011**, *19*, 139–152. [CrossRef]
117. Hair, J.F.; anderson, R.E.; Babin, B.J.; Black, W.C. Multivariate Data Analysis: A Global Perspective. In *The Great Facilitator*; Springer: Cham, Switzerland, 2010; Volume 7, pp. 121–130.
118. Tabachnick, B.G.; Fidell, L.S.; Ullman, J.B. *Using Multivariate Statistics*; Pearson: Boston, MA, USA, 2007; Volume 5.
119. Hu, L.T.; Bentler, P.M. Cutoff Criteria for Fit Indexes in Covariance Structure Analysis: Conventional Criteria versus New Alternatives. *Struct. Equ. Modeling Multidiscip. J.* **1999**, *6*, 1–55. [CrossRef]
120. Maclennan, J.; Tang, Z.; Crivat, B. *Data Mining with Microsoft Sql Server 2008*; John Wiley & Sons: Hoboken, NJ, USA, 2011.
121. Malhotra, N.K.; Dash, S. *Marketing Research an Applied Orientation (Paperback)*; Pearson Publishing: London, UK, 2011.
122. Chen, F.F.; Sousa, K.H.; West, S.G. Teacher's Corner: Testing Measurement Invariance of Second-Order Factor Models. *Struct. Equ. Modeling* **2005**, *12*, 471–492. [CrossRef]
123. Byrne, B.M. Structural Equation Modeling: The Basics. In *Structural Equation Modeling with Amos*; Routledge: London, UK, 2016; pp. 25–37.
124. Fiedler, K.; Kutzner, F.; Krueger, J.I. The Long Way From A-Error Control to Validity Proper: Problems with A Short-Sighted False-Positive Debate. *Perspect. Psychol. Sci.* **2012**, *7*, 661–669. [CrossRef]
125. Otoo, A.A.; Otoo, C.; Antwi, M. Influence of Organisational Factors On E-Business Value and E-Commerce Adoption. *Int. J. Sci. Res. Sci. Eng. Technol.* **2019**, *6*, 264–276.
126. Wang, Y.; Ahmed, P.K. The Moderating Effect of the Business Strategic Orientation on Ecommerce Adoption: Evidence from UK Family Run Smes. *J. Strateg. Inf. Syst.* **2009**, *18*, 16–30. [CrossRef]
127. Thaw, Y.Y.; Mahmood, A.K.; Dominic, P. A Study on the factors that influence the consumers trust on ecommerce adoption. *arXiv* **2009**, arXiv:0909.1145.

sustainability

Article

Creating Sustainable Organizations through Knowledge Sharing and Organizational Agility: Empirical Evidence from China

Rob Kim Marjerison [1,*], Matthew Andrews [2] and George Kuan [2]

[1] Global Business, College of Business and Public Management, Wenzhou-Kean University, Wenzhou 325060, China
[2] International Business, International School of Management, 17 Boulevard Raspail, 75007 Paris, France; matthew.andrews@ism.edu (M.A.); glk205@nyu.edu (G.K.)
* Correspondence: rmarjeri@kean.edu

Abstract: This study explores the relationships among knowledge sharing, adaptability, and performance, and furthermore seeks to address a gap in the existing literature on how those relationships may vary between organizational sectors. Economic sustainability for firms of all sizes and sectors is likely to depend to a large extent on the creation of a sustainable organizational culture built on collaboration, innovation, and adaptability. The importance of knowledge management in developing sustainable and higher functioning organizations is well accepted in the literature. Likewise, the ability of organizations to realize competitive advantage by adapting and responding in a timely manner to changes in the landscape is well supported. Building on previous research, this study further examines how organizations in different sectors may experience that interaction differently. Based on data gathered through 720 online surveys and subjected to empirical analysis, the findings suggest that work groups that are more agile can more readily realize the benefits of a knowledge sharing organization culture. Further, in contrast to the main body of existing literature, the findings indicate that there is little difference in these benefits among organizations operating in different sectors, notably, within the context of mainland China. These findings may be of interest to those with an interest in knowledge sharing, organizational agility, organizational behavior, sustainable organizations, collectivistic cultures, to practitioners with an interest in developing higher functioning organizations, and to social scientists in related research areas such as cultural studies and psychology.

Keywords: organizational sustainability; social and economic sustainability; knowledge sharing; organizational agility; organizational behavior; sustainable organizations

Citation: Marjerison, R.K.; Andrews, M.; Kuan, G. Creating Sustainable Organizations through Knowledge Sharing and Organizational Agility: Empirical Evidence from China. *Sustainability* **2022**, *14*, 4531. https://doi.org/10.3390/su14084531

Academic Editors: Elkhan Richard Sadik-Zada and Andrea Gatto

Received: 7 March 2022
Accepted: 7 April 2022
Published: 11 April 2022

1. Introduction

There is consensus in the existing literature that economic sustainability is a precursor to social sustainability and, in its absence, a constraint on ecological sustainability [1]. Likewise, there is consensus that among the factors that contribute to the success of organizations, knowledge sharing and the ability to adapt quickly to changes in circumstances are increasingly seen as essential areas of interest [2,3]. How those factors may affect performance differently across organizational sectors is not well explored in the existing literature. Likewise, research on knowledge sharing and organizational adaptability is China is underrepresented.

Increasingly, work groups and teams are used by companies to perform vital tasks, making group effectiveness strategically vital to organizations and essential for enterprises to survive and prosper in turbulent environments [4–8]. The existing literature on the relationships between knowledge sharing and organizational agility and organizational performance indicates that organizations in different sectors are likely to be affected differently by these factors. The examination of how knowledge sharing and adaptability can affect work group performance and how that effect may vary across organizational sectors is the aim of this study.

As researchers have sought to better understand factors which impact work group performance, some studies have focused on factors related to group dynamics such as cohesion [9], the influence of member behavior [10], basic processes and group composition [11], or the influence of perceived power imbalances [6]; while other studies have paid attention to the organizational context and climate [12]. Other studies meanwhile have focused on the influence of knowledge management systems [8], or on the impact of transactive memory systems [13,14].

There are immense variations among the different contexts that groups operate in, and a great diversity of functions that groups fulfill. This may partly explain the lack of a clear conceptual definition of work group performance in the literature and a lack of agreement among scholars about how to measure it. Some approaches taken to measure group work performance include having stakeholders evaluate their own team's performance [10,13]; having managers, external to the teams, evaluate the performance of teams under their supervision [6]; having stakeholders (such as customers and clients) assess team performance [10]; and having a combination of team members, managers, and leaders evaluate the team's performance [15]. Still other scholars contend that more objective data are required, such as sales commission [5,8].

Organizational agility (OA) has also been receiving increased attention from scholars. There is broad consensus in the literature that OA is considered a dynamic capability or a set of capabilities which allow organizations to respond rapidly to change and to realize superior performance in dynamic, turbulent, and competitive contexts [16–19]. The concept is broad and touches on multiple aspects of a firm's strategy and operations. Various definitions of OA are complex, overlapping in some areas and diverging in others, which makes it challenging to establish the shared meaning of terms among different scholars and to identify clear common themes [19].

Recent studies have paid increasing attention to mechanisms to help managers evaluate needed levels of OA [18], practical methodologies to implement OA [20], the role of OA as a moderator impacting market performance [21], the influence of talent management on OA [22], the relationship between IT capability and OA [23], and factors that either enable or hinder agility [16,24]. Moreover, there is a significant amount of research on agile manufacturing [25–32] and enterprise agility [33]. The impact of OA on firm performance is a critical concern in much of the research; however, there appears to be a gap in the literature regarding assessing the impact of OA on work groups or teams within larger organizations.

It is well established among scholars that knowledge is a critical resource for organizations [34,35]. Recent research has also focused on knowledge as a crucial element to successfully formulating and implementing a firm's purpose [36,37]. For decades, knowledge management has gained traction as a scholarly discipline and as an array of management activities to effectively manage and extract value from knowledge creation and integration [38–41]. Knowledge sharing processes are an important concern in knowledge management, and several recent studies have attempted to define the term according to specific knowledge sharing behaviors (KSBs) that can be measured and influenced [42–46]. One stream of research has attempted to deepen our understanding of how reward systems impact KSBs [42,43,47,48]. Another stream of research has looked more broadly at factors such as organizational culture [44], social norms [49–51], organizational climate [52], paths among organizational subunits and path dependency [53,54], and perceived enjoyment of participants [55].

Despite the growing interest in organizational agility, knowledge sharing, and group performance, little research has attempted to integrate these streams of research to examine the impact on group performance. More specifically, there appears to be a gap in the existing literature on how knowledge sharing behaviors are moderated by organizational agility, and how that effect may vary among organizational sectors. Based on this gap, the research questions put forward are:

RQ1: What is the relationship between knowledge sharing behavior and work group performance?

RQ2: Does organizational agility have an effect on the relationship between knowledge sharing behavior and work group performance?

RQ3: Does the mediating effect of organizational agility between knowledge sharing behavior and work group performance vary by the sector of the organization?

Knowledge sharing and adaptability are important drivers of performance. How they may affect organizations in different operating sectors differently, particularly in China, could have substantial implications both for foreign organizations that operate in China, and for domestic Chinese organizations.

This study contributes to the existing literature by investigating the links between organizational agility, knowledge sharing, and work group performance and how they may vary among organizations in different sectors.

There exists a gap in the existing literature regarding how those factors may manifest different across organizational sectors. Additionally, research on knowledge sharing and organizational adaptability is China is underrepresented. This study seeks to address both of those gaps.

2. Literature Review

2.1. Economic Sustainability

Consideration of sustainability as a concept is not a recent phenomenon, but it has, in recent years, become increasingly relevant and has evolved into a fairly well developed theoretical paradigm, most often conceptualized by the use of the three pillars model [56] of ecological, economic, and social sustainability. Separate discussions on ecological, social, and economic areas is a simplistic view [57] and presents a challenge for those seeking empirical analysis of the three pillars due to the inseparable and interdependent nature of factors that comprise them [58,59]. There is however, a consensus on the relevance of concern for sustainability in all three areas. Examples include ecological sustainability as manifested by the discussion of climate change [60]; social sustainability viewed through the lens of the inequality of global wealth distribution and the problems that arise from poverty [61,62]; and, finally, economic sustainability, which is a necessary precursor to addressing both social and ecological sustainability related concerns [63–66].

Economic sustainability is complicated by the shifting backdrop of global economic interconnectivity resulting in increasingly competitive conditions for organizations operating in all organizational sectors. Economic sustainability for many firms has depended upon their ability to leverage their intellectual capital [2], both formal such as patents and copyrighted materials, and informal such as the knowledge and expertise accumulated by members of the organizations [38]. Likewise, sustainable operations for many firms has depended on their ability to adapt to changes in circumstances both internal and external to their organizations [21,67].

It is the nexus of these two topics, knowledge sharing and adaptability, and how they affect organizational performance and enable the sustainable operation of organizations in different operational sectors that this study will examine.

2.2. Knowledge Sharing Behavior

Knowledge sharing behavior (KSB) is a crucial element of knowledge management systems [42]. The origins of the concept of knowledge sharing can be found in the knowledge management literature [40,41] which is related to the knowledge-based theory of the firm [34]. Bartol and Srivastava [42] define knowledge sharing as "individuals sharing organizationally relevant information, ideas, suggestions, and expertise with one another" (p. 65), and Wilson et al. [68] consider sharing knowledge as one of the basic processes of group learning along with the storage and retrieval of knowledge. Drawing on the prior work of Nonaka [69] and Polyanyi [70], Bartol and Srivastava explain that knowledge can be shared explicitly or tacitly, and requires effort from the knowledge sharer. Because it involves exchanges among individuals, knowledge sharing is considered distinct from knowledge transfer, which generally describes intra-organizational exchanges of knowledge between entities such as departments, or describes inter organizational movements of knowledge [42]. Yang and Chen [46] provide one of the broadest definitions of knowledge sharing and emphasize the behavioral component, describing it as a "set of behaviors about knowledge exchange which involve the actors, knowledge content, organizational content, appropriate media, and societal environment" (p. 96). Due to its voluntary nature and its contribution to the effectiveness of the organization, KSB is related to organizational citizenship behavior [45].

Methods and approaches to operationalizing and measuring KSB can vary from study to study. In their investigation, Cabrera et al. [49] operationalized KSB according to two types: voluntarily seeking ideas and information from co-workers, and providing insights and ideas to co-workers. Citing the work of Davenport and Prusak [41] who proposed evaluating KSB based on knowledge sharing activities during meetings, Hung et al. [43] measured this behavior according to participant outcomes, which included the number of ideas generated, the usefulness and creativity of the ideas generated, and the perceived meeting satisfaction.

One recent stream of research has attempted to provide insights about organizational reward systems and motivational drivers to achieve effective knowledge sharing. Conclusions and findings, however, have been inconsistent, particularly when it comes to the role of extrinsic rewards. Bartol and Srivastava [42] for example, contended that monetary rewards could be effective for encouraging KSB and predicted that rewards based on collective performance would be an appropriate and effective way to encourage cooperation and engagement among employees. Lee and Ahn [48], however, investigated reward systems to encourage knowledge sharing within organizations and concluded that individual-based reward systems were more efficient than group-based ones. Hung et al. [43] evaluated the impact of intrinsic versus extrinsic motivation on knowledge sharing, which, as explained above, was defined as participant outcomes during meetings. They concluded that economic rewards did not lead to an increase in knowledge sharing, whereas reputational feedback, which results in employees feeling that knowledge sharing enhances their reputation, had a significant positive effect.

Bock et al. [47] studied motivational drivers which determined individual KSB and developed a framework that considered a number of factors impacting the intention to share knowledge, such as "anticipated extrinsic rewards, anticipated reciprocal relationships, sense of self-worth, and three facets of organizational climate" [47]. Their findings suggested that these factors, applied as antecedents to attitude and subjective norms, positively contributed to KSB with the exception of extrinsic rewards which appeared to negatively impact attitudes towards knowledge sharing. For example, Ipe [44] described four factors that influence KSB: (a) the nature of the organizational knowledge; (b) the motivation of the actors to share knowledge; (c) opportunities to share the knowledge; and (d) the culture of the work environment. The fourth factor, organizational culture, is the most critical because the other three factors are embedded in it. Ipe's work built on the prior work of Schulz [51] who examined the relationship between the production and distribution of organizational knowledge among subunits and concluded that different learning processes were adopted according to the nature of the knowledge, and that this, in turn, affected how the knowledge was shared. Cabrera et al. [49] came to similar conclusions when they evaluated determinants of individual engagement in knowledge sharing between organizational subunits. Among the organizational variables assessed, normative pressures, described as "perceptions of support from colleagues and supervisors towards knowledge sharing" (p. 259), showed the greatest impact. This concurs with the findings of Ryu et al. [50] who attempted to better understand factors which impacted the knowledge sharing of hospital physicians and concluded that subjective norms (the internalization of outside influences) had the greatest influence on their intention to share knowledge followed by attitude.

Chen et al. [71] set out to study the impact of knowledge management systems, organizational climate, and attitude on the intention of employees to share knowledge. Their findings showed that attitude was the most significant factor but that knowledge management systems self-efficacy, and organizational climate, by positively contributing to attitude, indirectly affected knowledge sharing. Witherspoon et al. [72] assessed the antecedents of organizational knowledge sharing, including the intentions and attitude of the knowledge sharer, rewards for knowledge sharing, and the organizational culture. Their findings provided support for a positive relationship between all three areas studied and KSB; furthermore, their findings suggested it is easier to motivate employees to share knowledge in collectivist cultures than in individualist ones.

From a different perspective, the findings of Ton et al. [73] suggest that knowledge hiding, as opposed to knowledge sharing, also affects group performance, but in the opposite manner. Groups in which the members are prone to hoarding behavior when it comes to knowledge tend to perform at a lower level.

To conclude, many studies have been carried out which explore antecedents of KSB and assess factors which positively impact KSB, but few studies have attempted to measure the impact of KSB on other aspects of the organization or on work group performance. There seems to be an underlying assumption in much of the research that KSB is desirable and linked to positive organizational outcomes; however, there is little empirical research which provides support for this assumption. Therefore, this paper sets out to explore the extent to which organizations with a higher level of knowledge sharing behavior, are likely to experience a higher level of performance at the work group level. Thus, hypothesis one is put forward as

Hypothesis 1 (H1): *KSB has an effect on WGP.*

Likewise, based on the prior literature, it seems likely that organizations that have heightened levels of KSB are likely to be more readily able to adapt to changes in circumstances. Thus, hypothesis two is put forward as

Hypothesis 2 (H2): *KSB has an effect on OA.*

2.3. *Organizational Agility*

Interest in the topic of organizational agility has been gaining attention in recent scholarly research. While distinct from the concept of agile management practices well known in software development [74], OA does share many aspects in common with agile practices, including an emphasis on continuous, iterative improvement cycles; effectively meeting clients' needs; rapid product development; and flexibility [26,74].

Based on her analysis of 75 scholarly papers published between 1994 and 2018, Walter [19] provides the following operational definition of OA: "Organizational Agility is a learned, permanently-available dynamic capability that can be performed to a necessary degree in a quick and efficient fashion, and whenever needed in order to increase business performance in a volatile market environment" (Walter, 2021, p. 379). Walter, furthermore, argues that OA should be viewed as a continuum, integrated into the context of the organization and its business environment, and "independent of the industry" (p. 381). Walter identified and described the following four categories of agility: drivers (environmental changes impacting the organization), capabilities (an organization's ability to handle change), enablers (tools, practices, and technology), and dimensions (parts of the organization that must be agile to achieve OA such as management, technology, and the workforce).

As stated in the Introduction, a substantial amount of research exists that deals with agility in the manufacturing sector [13,33] Moreover, there have been studies which put forward practical management tools or de-scribe mechanism for achieving OA cost effectively and in a manner coherent with the organization's strategy [18,26,30].

Another stream of OA research has attempted to evaluate the factors which influence OA such as company culture [16], talent management [22], and IT capability [23]. Van Oosterhout [24] analyzed change factors and assessed "agility gaps" (p. 132) which companies faced in four different industry sectors; they highlighted "the existence of inflexible legacy systems" (p. 132) as a significant perceived barrier to increased agility.

Finally, there is support in the literature for the impact of agility, including agile manufacturing on the performance of enterprises [75,76]. Vickery et al. [77], examining the roles of supply chain information technologies and supply chain operational initiatives in creating agility and encouraging performance, concluded that agility acted as a mediator related to firm performance. Another study [78] similarly showed that, particularly in turbulent contexts, agile manufacturing increased the competitiveness of firms and led to improved operational, financial, and market performance. Akkaya and Qaisar [21] studied OA and its influence on market performance related to dynamic capabilities and concluded that OA played an important moderating role.

To conclude, there is robust support in the scholarly literature for the influence of factors such as organizational culture and/or IT capabilities on OA, as well as support for the positive impact of OA on firm performance, including as a mediator of firm performance. There are few studies, however, which address the role that OA plays in influencing KSB or its impact on work group performance. Therefore, hypotheses three and four are put forward as

Hypothesis 3 (H3): *OA has an effect on WGP.*

Hypothesis 4 (H4): *OA has a moderating effect on the relationship between KSB and WGP.*

2.4. Work Group Performance

While researchers have identified many different types of work groups, Sundstrom et al. [79] identified basic defining characteristics of work groups which include "shared duties in an organization and interdependence in carrying them out" (p. 49). This is similar to Edmondson's definition of work teams [80,81]. In this paper, we do not make a distinction between work group performance and team performance. The ability of teams to effectively communicate both their explicit and tacit knowledge to develop shared mental models is a key factor in determining their success [68]. As stated in the Introduction, there does not appear to be a shared conceptual understanding among different scholars as to exactly what work group performance (WGP) consists of; and, similar to KSB, researchers have adopted various approaches for measuring it.

Drawing on the earlier work of Janz et al. [10], which emphasizes that team performance is specific to tasks, Choi et al. [13] evaluated team performance of knowledge workers based on the perceptions of multiple stakeholders (such as clients and customers) of the quality of their deliverables, their effective time management, and their ability to meet deadlines. In their study, Chung et al. [6] measured group performance by having managers rate groups under their supervision according to four items: work quality, work quantity, group initiative, and overall performance; Frazier and Bowler [12] similarly used a survey tool completed by the work group supervisors who rated the overall performance of the group. Hoegle and Gemuenden [15] measured team performance using a survey instrument completed by team members, leaders, and managers who supervised the teams. Iyengar et al. [8] on the other hand, citing the advice of Argote and Miron-Spektor [82], used sales commissions as a measure of group performance, considering these data to be more objective than self-reported measures. This study assesses the perceptions of team members of the creativity, efficiency, effectiveness, and initiative of their own teams, as well as the quality of the work produced.

A significant number of studies have investigated factors that influence work group performance, such as characteristics of the group [9,11]. Janz et al. [10] asserted that process behaviors, such as helping, sharing, and innovating, positively impacted effectiveness but that the relationship was affected by contextual factors such as goals, feedback, and time pressure. Hoegle and Gemuenden [15], who developed a team work quality construct, analyzed six factors that they predicted would contribute to effective team performance such as communication, coordination, balanced contributions, mutual support, efforts made by members, and group cohesion. Their findings suggested that these factors were positively associated with team performance.

Other scholars have examined the extent to which transactive memory systems and IT support positively impacted team performance [13,14].

Still other studies have assessed the role of voice climate [12], the impact of power structures [6], and the effects of repository knowledge management systems [8].

The literature on this topic provides support for the impact of factors such as process behaviors positively impacting team performance and hence contributing to organizational sustainability. This provides a firm basis for H1 which posits a positive relationship between KSB and WGP.

2.5. Organizational Sectors

There are several ways to categorize organizations into different sectors, depending on the context of the discussion; categories can be based on size as measured either by revenue, sales, or the number of employees [83]. Other categorical systems may include the profit-nonprofit; Social Enterprise spectrum [84]; the public vs. private sector view; or the ownership structure view of sole proprietor, partnership, corporation [85–88]. Organizational sectors can also be differentiated based on the tenure of the organization, or start-up vs. established firms [89].

Previous research on organizational sectors has found differences between sectors in several areas, including organizational commitment [84,90] and how organizational effectiveness and performance are measured [91]. Other research has shown a difference in how organizations in different sectors are affected by their approach to intellectual capital, knowledge, and how it is used and shared [92,93]. Differences have also been reported between conflict management styles of organizations in different sectors [94]. Of particular interest are findings in the prior research that show differences in how organizations in different sectors can be transformed through innovation [87] as innovation based changes in organizational behavior are closely related to knowledge sharing driven organizational agility.

For purposes of this study, to examine the relationship between KSB, WGP, and the effect of OA on that relationship, organizations were divided into broad sectors based on the findings of previous research which indicated that they are likely some similarities within a country/culture, across certain sectors [86,95,96]. It follows, then, that a comparison of the KSB-WGP and OA relationships would likely be more meaningful between those broad sectors of public and private sectors [90], specifically, government, manufacturing-based private enterprise, services-based private enterprise, agriculture, healthcare, and education.

Private enterprise is divided into manufacturing and services sectors due to differing aspects that agility is factored into in these business models [97]. The time it takes a manufacturing firm to adapt to changes in market conditions is expected to be considerably greater than the expectation for services firms [98].

Likewise, the agriculture sector is of interest as a separate sector because it is a unique, large, and important part of the economic system in China [99]. Furthermore, the production cycle is considerably different from manufacturing or services based firms [100], meaning that both knowledge sharing, and particularly OA, may differ from other sectors of the economy [101,102]. This is of particular relevance to this study when considering the recent focus on the sustainability of the agriculture sector in China [103].

In the public sector, prior research indicates that the performance of public sector organizations tend to be affected differently than profit-making organizations by information sharing [104], and by employee learning practices which can be either formal, as structured training, or informal as knowledge sharing behavior between employees [105].

Within the broad category of the public sector, education is unique and of special interest as a critical factor of socioeconomic sustainability and its importance in poverty reduction [106] and socioeconomic mobility [107–109]. Education has also been found to play a key role as a driver of innovation-based economic development [110–113]. Education may shape the views of future business and public leaders and thereby provide insights into trends in leadership styles [114]. Further, education as a sector is unique because it often straddles the line between private and public sectors [115–117]. As such, it has been found to respond differently than other organizations to situations requiring OA [118,119], and is therefore appropriate to consider education as separate sector.

Healthcare is of special interest because of the importance of OA in the healthcare systems for social, and ultimately, economic sustainability during periods of health related-crises such as the COVID pandemic that began in 2020 [120,121]. Further, similar to the education sector, the healthcare sector may be in the public, or private sectors, and because there is evidence that OA in the healthcare sector faces unique challenges, unlike that of other organizational sectors [122,123].

Having established both the importance of education and healthcare, and the basis for consideration of them as separate factors, and with these five sectors as reference points based on the findings of prior research on related topics, it seems probable that one or more of the constructs of KSM, WGP, and OA, as well as the interactive relationships between them, are likely to differ between sectors thus becoming the Conceptual Framework as indicated in Figure 1. Therefore, hypotheses five and six are put forward for testing:

Hypothesis 5 (H5): *The effect of KSB on WGP will differ between OSs.*

Hypothesis 6 (H6): *The moderating effect of OA between KSB and WGP will differ between OSs.*

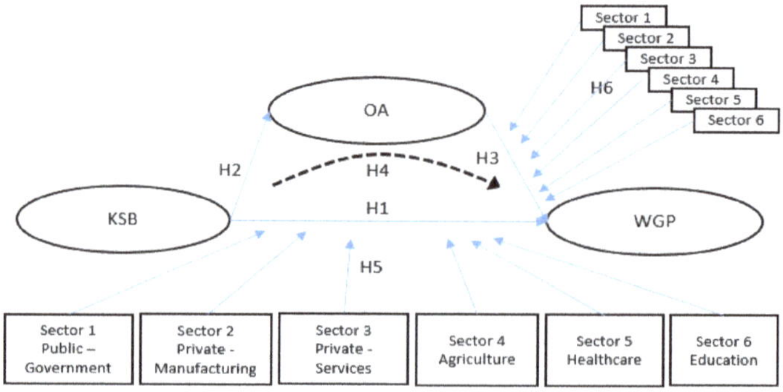

Figure 1. Conceptual Model.

2.6. Conceptual Framework

To address the research questions, this study will test the hypotheses.

3. Methodology

The detailed methodology of empirical research, quantitative methods, including Cronbach's alpha, Principal Component Analysis, and linear regression analysis were adopted to analyze the data. from an online survey. For this study, a regression analysis was employed as part of the methodology for the following reasons: firstly, means of factors were calculated to observe the performance of individual factors compared to the full scale, and also to see the overall level, to ultimately better identify the gap and where progress can be made in the future; secondly, Cronbach's alpha was applied to measure internal consistency between a set of items, and to further investigate how closely they are related to each other as a group, as well as to measure the scale reliability. Researchers apply this method to assure the credibility of the data among multiple items measuring a single factor, to finally confirm the reliability of the dataset [124].

The rationale for considering the usage of linear regression is that, compared to some machine algorithms, it has a considerably lower time complexity [125]. Considering the causal effect, this research requires distinct methods to conduct analysis, thus producing precise and explicable results. An evident advantage of linear regression models is linearity, which makes the estimation procedure simple and able to be interpreted and understood in an accessible manner [126]. In this way, linear and similar models are widespread in academic and quantitative research fields, including psychology, social science, and medicine [127,128].

The three constructs of Knowledge Sharing Behavior, Organizational Agility, and Work Group Performance were tested for validity using Cronbach's Alpha test, hypotheses 1–3 were tested separately using simple regression analysis for R^2, the unstandardized regression coefficients (beta), t-statistics, and associated p-values.

To test H1-H3, a simple regression analysis was performed to investigate the relationships between Organizational Agility (OA) and the relationship between Knowledge Sharing Behavior (KSB) and Work Group Performance (WGP). The dependent variable was WGP. The independent variable was KSB. The mediator variable for the analysis was OA.

Survey Instrument and Data Collection

The instrument was adapted from surveys previously used and validated in related research. The construct of Organizational Agility was measured through use of five questions adapted from the research of Lu and Ramamurthy [23]. The construct of Work Group Performance was measured with five questions adapted from an instrument previously used by Choi et al. [13], and Knowledge Sharing was measured with seven questions from a study carried out by Oliveira et al. [129]. All construct questions in the survey were presented with a 7-point Likert scale response.

An online survey as a data collection tool offers a variety of benefits, including wide geographic coverage, anonymity for respondents, a decreased bias level compared to the pressure given by in-person interviews, and, importantly, cost-saving [130]. Easily under-stood and neutral language was used during the survey translation to facilitate the participants' comprehension of the material. Items in the survey from which the data were derived for this study were, in turn, adapted from previous studies and used with minor modification for context. The survey items used in the previous study were initially writ-ten in English, so the survey was translated from English to Chinese. To validate the question translation, the survey translation was checked by two native Mandarin speakers with PhDs from Western universities and then translated back into English. This translation and back-translation approach is a widely applied technique in cross-cultural social research to achieve translation accuracy and credibility with minimized differences [131]. Additionally, the survey was distributed to a pilot group of nine people for trial and adjusted according to their feedback to assure the translation accuracy, proper length and time taken, as well as the ease of understanding.

Previously, the survey link was shared through the social media platform WeChat, the most widely used social network application in China. The survey was distributed via the software application Wen Juan Xing where the data were compiled and downloaded. As is typical in reposting behavior in social media in China [132], a snowball sampling effect was observed.

Citizens from Zhejiang province, the emerging economic hub of southeast China, contributed most to the sample of the study. The Wechat social media platform is widely used in China for online surveys, customer feedback, and for posting product and service reviews. Due to the significant market penetration by the application, nearly all smartphone users are familiar with the app and were therefore qualified for inclusion in the survey. A total of 720 completed surveys were obtained through convenience sampling. The surveys were completed over a 21-day period and participation in the survey was voluntary and anonymous. After inspection, all surveys were determined to be valid. The survey instrument is Appendix A.

4. Results

4.1. Sampling

The demographic detail of the sample is shown in Table 1. Among the total number of respondents, more than half are female (n = 333, 46.3%) compared to the male respondents (n = 292, 40.6%) while N = 95, 13% preferred not to disclose their gender or chose "other."

Table 1. Demographics of the sample.

Variables	Subcategory	Frequency	Percent
Gender	Male	292	40.6
	Female	333	46.3
	Not disclosed	95	13.2
Age	Under 20	75	10.4
	20–29	120	16.7
	30–35	140	19.4
	36–40	131	18.2
	41–45	123	17.1
	45–50	65	9
	51–55	44	6.1
	Order than 55	22	3.1
Education	Under middle school diploma	81	11.3
	Middle school diploma	178	24.7
	High school diploma	117	16.3
	2–3 years vocational college	125	17.4
	4-year college degree	172	23.9
	Graduate college degree	47	6.5
Sector	Government	124	17.2
	Manufacturing	167	23.2
	Services	199	27.6
	Agriculture	123	17.1
	Healthcare	80	11.1
Position	Middle Manager	109	15.1
	Senior Manager	120	16.7
	Clerical	161	22.4
	Entry-level worker	109	15.1
	Agriculture worker	97	13.5
	Factory worker	88	12.2
	Teacher	36	5
Experience	Less than 1 year	104	14.4
	1–3 year	175	24.3
	4–7 year	184	25.6
	8–12 year	217	30.1
	More than 12 years	40	5.6

The most represented age group is 30–35 (n = 140, 19.4%) but the respondents are mostly between 20–45 which seems consistent with the working age demographic. The lowest number of participants belong to the age group older than 55 (n = 22, 3.1%), which makes sense as they are most likely to be retired, or not a heavily adopted group for smartphone use.

A breakdown of the educational level, while seemingly unremarkable, serves to confirm the applicability of the sample as representative of membership in organizations, and subsequently in work groups as defined for the purposes of this study. The vast majority of respondents fall into the middle school, high school, vocational, and four-year college categories. Most of the participants work in the services sector (n = 199, 27.6%), and 23.2% of participants work in the manufacturing sector.

The highest number of the participants work in the organization as clerical workers (n = 161, 22.4%), and most of the respondents reported their work experience is 8~12 years (n = 217, 30.1%). All in all, the demographic breakdown of the respondents seems as expected, unremarkable, and also well positioned to represent the target population of those who are part of work groups. Clarification of the work group as part of a larger organization, or perhaps as all members of a small organization was included in the introduction to the online survey. See Appendix A.

Additional demographic information was included in the survey but was not included in the hypothesis development and was discarded.

4.2. Reliability Analysis

The Cronbach alpha values shown in Table 2 indicate that all the three constructs are greater than 0.9, suggesting very good internal consistency reliability. According to Pallant [133], Cronbach alpha values above 0.7 are considered acceptable; however, values above 0.8 are preferable. This means that the constructs used in the research are reliable for further analysis.

Table 2. Cronbach Results.

Constructs	Cronbach Alpha	N of Items
Agility	0.97	5
Performance	0.98	5
KSB	0.98	7

The Principal Component Analysis with Promax rotation was used to explore the principal components of the KSB, WGB, and OA scales. The results of the initial analysis revealed three components with Eigenvalues over 1, explaining 72%, 11.12%, and 8.44% of the variance respectively. Seven items loaded on component 1(KSB), five items loaded on component 2 (Agility), and five items loaded on component 3 (Performance) as shown in Table 3.

Table 3. PCA Analysis.

Items	Components		
	Agility	Performance	KSB
Agility 1	0.931		
Agility 2	0.936		
Agility 3	0.924		
Agility 4	0.970		
Agility 5	0.922		
Performance 1		0.947	
Performance 2		0.958	
Performance 3		0.847	
Performance 4		0.966	

Table 3. *Cont.*

Items	Components		
	Agility	Performance	KSB
Performance 5		0.948	
KSB 1			0.822
KSB 2			0.962
KSB 3			0.953
KSB 4			0.951
KSB 5			0.955
KSB 6			0.961
KSB 7			0.944

Each of three main components or constructs (Agility, Performance, and KSB) are reliable. In other words, the questions used for each of the three constructs are measuring the same thing. For example, agility has five questions. All of the five questions are measuring agility. KSB has seven questions. All of the seven questions measure KSB. PCA analysis also confirmed that there are three main factors. It identified three main themes or components in the data set. Component loading table shows which items or questions are correlated with each of the three components. For example, each of the five questions measuring agility are highly correlated with the same component which is Agility.

4.3. *Hypothesis Testing—Regression*

The regression analysis was conducted for testing the formulated hypotheses H1–H3. Table 4 shows the results of the three separate simple linear regression analyses including the value of R^2, the unstandardized regression coefficients (beta), t statistics, and associated *p*-value.

Table 4. Regression Analysis.

Independent Variables	Dependent Variable	R^2	Unstandardized Coefficients	t	Sig.	Impact
KSB	OA	0.42	0.59	22.56	0.000	Significant
KSB	WGP	0.50	0.69	26.95	0.000	Significant
OA	WGP	0.45	0.71	24.15	0.000	Significant

As shown in Table 4, the impact of KSB on WGP was significant, predicting 50% variance in the OGP. The coefficient of KSB indicates that a 1 unit increase in KSB is associated with a 0.69-unit increase in WGP.

Thus, H1 is supported.

Confirmation of H1 indicates that those organizations with a higher level of knowledge sharing behavior are likely to be more successful, at least to the extent that the individuals in those organizations perceive success, and at the workgroup level in the organizations.

For H2, the analysis indicates that there was a significant impact of KSB on OA. KSB accounted for 42% of the variability in OA. The coefficient of KSB indicates that, if KSB increases by 1 unit, the OA score will be increased by 0.59 units.

Thus, H2 is supported.

Confirmation of H2 indicates that those organizations that are more readily able to adapt to changes in circumstances and are more flexible in ways to react to unexpected events are likely to be more successful at the work group level.

There is a significant impact of OA on WGP. OA accounted for 45% of the variance in the WGP. The beta coefficient of OA suggests that a 1 unit increase in OA is associated with a 0.71 increase in WGP.

Thus, H3 is supported.

Confirmation of H3 indicates that those organizations that are more flexible and more readily able to adapt to changing circumstances are more like to experience a higher level of work group performance.

Thus, the hypotheses H1, H2, and H3 were supported and together served to combine, strengthen, and validate prior research on the individual topics of KSB, OA, and WGP and to present a foundation on which to build the analysis for H4 and H5.

4.4. Hypothesis Testing—Mediation

To investigate the possible mediation effect of Organizational Agility (OA) on the relationship between Knowledge Sharing Behavior (KSB) and Work Group Performance (WGP) as was put forth in hypothesis 4 (H4), a simple mediation analysis was performed. The dependent variable was WGP. The independent variable was KSB. The mediator variable for the analysis was OA. The indirect effect of KSB on WGP was statistically significant, $E = 0.23$, 95% CI (0.18, 0.28). Furthermore, 67% of the relationship operates directly and 33% of the relationship operates indirectly via OA, as shown in Table 4.

Thus, hypothesis H4 is supported.

4.5. Hypothesis Testing—Moderation

To test hypothesis 5 (H5), the possible difference in the effect of OA on the KSB-WGP relationship between different organizational sectors, or industries, a moderation analysis was performed. The outcome variable for the analysis was WGP. The predictor variable for the analysis was KSB. The moderator variable evaluated for the analysis was organizational sectors (OS).

As indicated in Table 5, the different interactions between KSB and OA were not statistically significant. In other words, organizations in different business sectors do not show significant differences when it comes to the relationship between KSB and WGP which was tested by H1. As indicated in Table 5, all the p values are greater than 0.05. This relationship does not show significant variation among different organizational sectors.

Table 5. Moderation Analysis.

Interaction Terms	Coefficients	SE	T	p
KSB-Manufacturing	−0.01	0.08	−0.16	0.87
KSB-Service	−0.08	0.08	−1.06	0.29
KSB-Agriculture	0.03	0.09	0.43	0.66
KSB-Healthcare	−0.13	0.10	−1.28	0.19
KSB-Education	−0.05	0.15	−0.36	0.71

Note: "Government" was used as reference category for comparison.

Thus, hypothesis 5 (H5), the effect of KSB on WGP will differ between OSs, is not supported.

4.6. Hypothesis Testing—Moderated Mediation

To test H6, whether the mediation effect of OA on the relationships between KSB and WGP was different among organizations in different sectors, a moderated mediation analysis was performed. KSB was the predictor variable, agility as the mediator. The outcome variable was WGP and OS was the proposed moderator.

The result shown in Table 6 indicates that the mediating effect of agility on the relationship between KSB and OGP does not change significantly based on different organizational sectors (as shown in Table 6; zero is within the confidence interval).

Table 6. Moderation Mediation Analysis.

	Index	BootSE	BootLLCI	BootULCI
Manufacturing	0.04	0.05	−0.06	0.15
Service	−0.02	0.05	−0.01	0.08
Agriculture	0.09	0.06	−0.03	0.21
Healthcare	−0.03	0.06	−0.16	0.09
Education	−0.002	0.10	−0.21	0.19

Thus, the hypothesis (H6) that mediating effect of Agility between KSB and WGP will differ between OSs is not supported.

4.7. Hypothesis Summary

In summary, as shown in Table 7, H1–H4 are supported by the data, while H5 and H6 are not supported. The implications and possible factors that may be instrumental in these findings are discussed in Section 5.

Table 7. Hypothesis Summary.

Hypothesis	Status
H1: KSB has an effect on WGP	Accepted
H2: KSB has an effect on OA	Accepted
H3: OA has an effect on WGP.	Accepted
H4: OA has a mediating effect on the relationship between KSB and WGP.	Accepted
H5: The effect of KSB on WGP will differ between OSs.	Rejected
H6: The mediating effect of OA between KSB and WGP will differ between OSs.	Rejected

5. Discussion

The results of this study contribute to the body of literature in two areas: the relationships between KSB, OA, and WGP, which serve to solidify, coalesce, and build on the findings of prior research by validating the effect of KSB on WGP and by demonstrating the mediating effect of OA on the relationship between KSB and WGP. Interestingly, this was achieved by providing evidence in contrast to the findings of much of the earlier work on the dissimilarities in organizational behavior between organizations of different sectors. It is important to note that nearly all of the preponderance of research done in these areas was not done in China, but rather, with few exceptions, in Western countries. It seems to follow then, that there is a greater similarity in organizational behaviors in different organizational sectors in China than in most other contexts. This is a major contribution of this study.

5.1. Theoretical Contribution

The contributions of this study are threefold. Firstly, by extending the previous work on connecting and validating the relationship between KSB and WGP, and addressing the gap in the existing literature by examining this relationship within the context of China.

Secondly, by demonstrating, empirically, the mediating effect of OA on the relationship between KSP and WGP. While there exists, as discussed above, some related literature, this study contributes with specificity how the effects of KSB on WGP can be increased when there exists a simultaneous ability of the organization to respond and adapt quickly to changes in the operational environment.

Thirdly, the rejection of H5 and H6 indicate a difference in the sample population of this study from the data sets of the majority of the previous literature. As part of the theoretical contribution of this study is the context of Chinese organizations, the difference in the results can most likely be found in the context of the study. If in fact, as seems likely, the basis of this difference is the sample, then this study presents empirical evidence that the influence of culture is the cause. It seems that cultural collectivity, as described by Hofstede [134], within China, is greater than expected and is a sufficiently powerful factor that it will overcome the differences in the relationships between KSB, WGP, and OA that have been established in the past research conducted in a variety of other contexts and cultures.

5.2. Managerial Implications

The practical, or managerial implications are straightforward. Those in managerial and leadership positions within organizations located in, or operating in China, can benefit from the implications of the findings of this study. For example, organizations can more readily apply benchmarking in certain areas of organizational operations, to organizations in different sectors than the one in which they operate, with confidence that the results are generalizable across organizational sectors. Another managerial implication would be recognizing the strength of the cultural influence on workers across all operational sectors.

Secondly, given the diverse sectors that some large organizations operate in China, a firm may wish to apply the same processes for KS and OA across subsidiaries and operations in different sectors, with the expectation of similar outcomes. This could add considerable insights for both Chinese and foreign managers and leaders in those organizations within China that have foreign managers and for Chinese organizations that operate outside of China.

5.3. Limitations and Future Directions

Although this study examined relevant and well validated constructs, there are some limitations that should be noted. Firstly, there was a need to examine work group performance rather than financial performance at the macro organizational level, which may not accurately reflect how the organizational behaviors influence differing views of performance. The strength of this study, within the context of China, is also a weakness as the findings on the different sectors are in contrast to much of the existing literature, which was not based in China. Therefore, the contribution of this study's specific findings, within the context of China, could also be seen as a limitation as the results lack generalizability, while calling for additional research on the topic, both within China and in other contexts.

As with most survey data, there are ways to improve the reliability of the data, both with increased sample size and greater granularity in the sample demographics. Given the high likelihood that there is a collectivistic, cultural component that is influential in the findings, and given the consensus regarding the cultural changes that China has undergone in the last two generations, it seems that a similar study that included specific examination of the age of the respondents may reveal further insights into the effect of culture on the findings.

6. Conclusions

Development of the hypotheses H1–H6 is well grounded in the findings of the existing literature. As H1–H4 are supported by the data, this study adds to the literature connecting knowledge sharing behavior to performance, and provides new evidence of how organizational agility can increase the positive influence of knowledge sharing. Additionally, this study provides a new contribution with regard to the limitations of how knowledge sharing and agility affect performance across various organizational sectors. The rejection of H5 and H6 indicates the presence of a factor present in this study that is in contrast with most of the existing literature. The findings of this study present a compelling argument that the collectivist nature of organizational behavior in China is a more powerful influential factor than in other cultural settings, resulting in less variance in the way that organizations are affected by knowledge sharing behaviors and organizational agility than in other cultural settings.

Author Contributions: Conceptualization, G.K., R.K.M. and M.A.; methodology, R.K.M. and G.K.; validation, R.K.M. and M.A.; formal analysis, R.K.M.; investigation, M.A. and G.K.; resources, R.K.M.; writing—original draft preparation, R.K.M. and M.A.; writing—review and editing, R.K.M., M.A. and G.K.; visualization, M.A. and G.K. All authors have read and agreed to the published version of the manuscript.

Funding: This research received no external funding.

Institutional Review Board Statement: For data collected anonymously, that poses no risk, received no funding, and was gathered previously, no IRB statement is required by university policy.

Informed Consent Statement: Not applicable.

Conflicts of Interest: The authors declare no conflict of interest.

Appendix A

This survey is used for a research study about agile management, work performance, and knowledge sharing.

Participation in this survey is 100% voluntary. Your personal information and participation information will not be disclosed, anonymity and confidentiality of your information are guaranteed. The results of this survey will not be used for any purpose other than this study.

This survey will take about 6 min of your time. We appreciate your participation. If you are willing to help us, please continue to complete the survey.

Please answer the following questions about your work, or job. "Our Work Team" refers to the group of people that you work with. It could be a department of a large company, or it could be everyone that works at a small company.

Demographics:

1. Gender
   - ○ Male
   - ○ Female
   - ○ Not say
2. Age
   - ○ Under 20
   - ○ 20~29 21–29
   - ○ 30~35 30–35
   - ○ 36~40 36–40
   - ○ 41~45 41–45
   - ○ 45~50 45–50
   - ○ 51~55 51–55
   - ○ Older than 55
3. What is your education level?
   - ○ Under middle school diploma degree
   - ○ Middle school diploma degree
   - ○ High school diploma degree
   - ○ 2~3 years Vocational College
   - ○ 4 year college degree
   - ○ Graduate College degree (Master degree or higher)
4. What business sector is your job/company in?
   - ○ Government
   - ○ Manufacturing
   - ○ Services/Banking
   - ○ Agriculture
   - ○ Hospitality/Tourism
   - ○ Healthcare

- ○ Education
- ○ Retail

5. What is your job in your organization?
   - ○ Middle Manager
   - ○ Senior Manager
   - ○ Clerical
   - ○ Entry Level worker
   - ○ Agriculture worker
   - ○ Factory worker
   - ○ Teacher
6. How many years have you worked for the organization?
   - ○ Less than 1 year
   - ○ 1~3 years
   - ○ 4~7 years
   - ○ 8~12 years
   - ○ More than 12 years

For the next section, you will be asked about different variables.

Noted: These are seven-point scale questions. The higher score means the more important or the stronger your agreement is. Your opinion is measured on the scales from 1 to 7. From Rank 1 (Strongly Disagree) to Rank 7 (Strongly Agree), your degree of agreement increases, and rank 4 indicates neutral.

Organizational Agility

7. Our work team fulfills demands for rapid-response and special requests needed to reach our goals.

   1 strongly disagree
   7 strongly agree
8. Our work team can quickly scale up or scale down our work level to support unexpected changes.

   1 strongly disagree
   7 strongly agree
9. Whenever there is a disruption in supply chain from our suppliers, our work team can quickly make necessary alternative arrangements.

   1 strongly disagree
   7 strongly agree
10. Our work team constantly looks for ways to improve our ability to reach our goals.

    1 strongly disagree
    7 strongly agree
11. Our work team treats chaos and changes in market conditions and as opportunities.

    1 strongly disagree
    7 strongly agree

Performance

12. Our Work Team comes up with new, original ideas for handling work.

    1 strongly disagree
    7 strongly agree
13. Our Work Team is redesigning job tasks for greater effectiveness and efficiency, even if it isn't required.

    1 strongly disagree
    7 strongly agree

14. Our Work Team is taking initiative and doing whatever is necessary to be successful.
    1 strongly disagree
    7 strongly agree
15. Our Work Team is looking for better solutions.
    1 strongly disagree
    7 strongly agree
16. Our Work Team is achieving a high quality of work.
    1 strongly disagree
    7 strongly agree

    Knowledge Sharing Behavior
17. My work team usually shares knowledge about our work with each other.
    1 strongly disagree
    7 strongly agree
18. My work team spends a lot of time-sharing knowledge.
    1 strongly disagree
    7 strongly agree
19. I usually share my knowledge with the other members of my team.
    1 strongly disagree
    7 strongly agree
20. I often share the reports and official documents from my work with the members of my team.
    1 strongly disagree
    7 strongly agree
21. I believe that other members of my work team share their knowledge with me.
    1 strongly disagree
    7 strongly agree
22. I believe that other members of my work team share information about our work with me.
    1 strongly disagree
    7 strongly agree
23. I always share my knowledge when asked by the members of my team.
    1 strongly disagree
    7 strongly agree

References

1. Alhaddi, H. Triple Bottom Line and Sustainability: A Literature Review. *Bus. Manag. Stud.* **2015**, *1*, 6–10. [CrossRef]
2. Ahmad, F.; Karim, M. Impacts of Knowledge Sharing: A Review and Directions for Future Research. *J. Workplace Learn.* **2019**, *31*, 207–230. [CrossRef]
3. Elali, W. The Importance of Strategic Agility to Business Survival during Corona Crisis and Beyond. *Int. J. Bus. Ethics Gov.* **2021**, *4*, 1–8. [CrossRef]
4. Andrews, M.; Smits, S. Using Tacit Knowledge Exchanges to Improve Teamwork. *ISM J. Int. Bus.* **2019**, *3*, 15–23.
5. Argote, L.; Miron-Spektor, E. Organizational Learning: From Experience to Knowledge. *Organ. Sci.* **2011**, *22*, 1123–1137. [CrossRef]
6. Chung, M.-H.; Ko, Y.; Kim, J.-Y. Group Power Structure, Inter-Subgroup Cross-Dependency, and Work Group Performance. *Asia Pac. J. Manag.* **2020**, *37*, 297–323. [CrossRef]
7. Edmondson, A.C. Strategies for Learning from Failure. *Harv. Bus. Rev.* **2008**, *89*, 48–55.
8. Iyengar, K.; Sweeney, J.; Montealegre, R. Repository KMS Use and Work Group Performance: The Contingent Role of Group Composition. *E-Serv. J.* **2020**, *11*, 1–33. [CrossRef]
9. Beal, D.J.; Cohen, R.R.; Burke, M.M.; McLendon, C.L. Cohesion and Performance in Groups: A Meta-Analytic Clarification of Construct Relations. PsycNET. Available online: https://doi.apa.org/doiLanding?doi=10.1037%2F0021-9010.88.6.989 (accessed on 28 January 2022).

10. Janz, B.D.; Colquitt, J.A.; Noe, R.A. Knowledge Worker Team Effectiveness: The Role of Autonomy, Interdependence, Team Development, and Contextual Support Variables. *Pers. Psychol.* **1997**, *50*, 877–904. [CrossRef]
11. Guzzo, R.A.; Dickson, M.W. Teams in Organizations: Recent Research on Performance and Effectiveness. *Annu. Rev. Psychol.* **1996**, *47*, 307. [CrossRef]
12. Frazier, M.L.; Bowler, W.M. Voice Climate, Supervisor Undermining, and Work Outcomes: A Group-Level Examination. *J. Manag.* **2015**, *41*, 841–863. [CrossRef]
13. Choi, S.; Lee, H.; Yoo, Y. The Impact of Information Technology and Transactive Memory Systems on Knowledge Sharing, Application, and Team Performance: A Field Study. *MIS Q.* **2010**, *34*, 855–870. [CrossRef]
14. Kanawattanachai, P.; Yoo, Y. The Impact of Knowledge Coordination on Virtual Team Performance Over Time. *MIS Q.* **2007**, *31*, 783–808. [CrossRef]
15. Hoegl, M.; Gemuenden, H.G. Teamwork Quality and the Success of Innovative Projects: A Theoretical Concept and Empirical Evidence. *Organ. Sci.* **2001**, *12*, 435–449. [CrossRef]
16. Cetindamar, D.; Katic, M.; Burdon, S.; Gunsel, A. The Interplay among Organisational Learning Culture, Agility, Growth, and Big Data Capabilities. *Sustainability* **2021**, *13*, 13024. [CrossRef]
17. Mrugalska, B.; Ahmed, J. Organizational Agility in Industry 4.0: A Systematic Literature Review. *Sustainability* **2021**, *13*, 8272. [CrossRef]
18. Teece, D.; Peteraf, M.; Leih, S. Dynamic Capabilities and Organizational Agility: Risk, Uncertainty, and Strategy in the Innovation Economy. *Calif. Manag. Rev.* **2016**, *58*, 13–35. [CrossRef]
19. Walter, A.-T. Organizational Agility: Ill-Defined and Somewhat Confusing? A Systematic Literature Review and Conceptualization. *Manag. Rev. Q.* **2021**, *71*, 343–391. [CrossRef]
20. Nejatian, M.; Zarei, M.H.; Nejati, M.; Zanjirchi, S.M. A Hybrid Approach to Achieve Organizational Agility: An Empirical Study of a Food Company. *Benchmarking Int. J.* **2018**, *25*, 201–234. [CrossRef]
21. Akkaya, B.; Qaisar, I. Linking Dynamic Capabilities and Market Performance of SMEs: The Moderating Role of Organizational Agility. *Istanb. Bus. Res.* **2021**, *50*, 197–214. [CrossRef]
22. Harsch, K.; Festing, M. Dynamic Talent Management Capabilities and Organizational Agility—A Qualitative Exploration. *Hum. Resour. Manage.* **2020**, *59*, 43–61. [CrossRef]
23. Lu, Y.; Ramamurthy, K. Understanding the Link Between Information Technology Capability and Organizational Agility: An Empirical Examination. *MIS Q.* **2011**, *35*, 931–954. [CrossRef]
24. van Oosterhout, M.; Waarts, E.; van Hillegersberg, J. Change Factors Requiring Agility and Implications for IT. *Eur. J. Inf. Syst.* **2006**, *15*, 132–145. [CrossRef]
25. Cho, H.; Jung, M.; Kim, M. Enabling Technologies of Agile Manufacturing and Its Related Activities in Korea. *Comput. Ind. Eng.* **1996**, *30*, 323–334. [CrossRef]
26. Gunasekaran, A.; Yusuf, Y.Y. Agile Manufacturing: A Taxonomy of Strategic and Technological Imperatives. *Int. J. Prod. Res.* **2002**, *40*, 1357–1385. [CrossRef]
27. Sharifi, H.; Zhang, Z. A Methodology for Achieving Agility in Manufacturing Organisations: An Introduction. *Int. J. Prod. Econ.* **1999**, *62*, 7–22. [CrossRef]
28. Sharifi, H.; Zhang, Z. Agile Manufacturing in Practice—Application of a Methodology. *Int. J. Oper. Prod. Manag.* **2001**, *21*, 772–794. [CrossRef]
29. Sindhwani, R.; Malhotra, V. A Framework to Enhance Agile Manufacturing System: A Total Interpretive Structural Modelling (TISM) Approach. *Benchmarking Int. J.* **2017**, *24*, 467–487. [CrossRef]
30. Vinodh, S.; Aravindraj, S.; Pushkar, B.; Kishore, S. Estimation of Reliability and Validity of Agility Constructs Using Structural Equation Modelling. *Int. J. Prod. Res.* **2012**, *50*, 6737–6745. [CrossRef]
31. Vinodh, S.; Madhyasta, U.R.; Praveen, T. Scoring and Multi-Grade Fuzzy Assessment of Agility in an Indian Electric Automotive Car Manufacturing Organisation. *Int. J. Prod. Res.* **2012**, *50*, 647–660. [CrossRef]
32. Zhang, Z.; Sharifi, H. Towards Theory Building in Agile Manufacturing Strategy—A Taxonomical Approach. *IEEE Trans. Eng. Manag.* **2007**, *54*, 351–370. [CrossRef]
33. Lin, C.-T.; Chiu, H.; Tseng, Y.-H. Agility Evaluation Using Fuzzy Logic. *Int. J. Prod. Econ.* **2006**, *101*, 353–368. [CrossRef]
34. Grant, R.M. Toward a Knowledge-Based Theory of the Firm: Knowledge-Based Theory of the Firm. *Strateg. Manag. J.* **1996**, *17*, 109–122. [CrossRef]
35. Grant, R.M. *Contemporary Strategy Analysis: Text and Cases*, 8th ed.; John Wiley & Sons Ltd.: Chichester, UK, 2015; ISBN 978-81-265-5450-8.
36. Hollensbe, E.; Wookey, C.; Hickey, L.; George, G.; Nichols, C.V. Organizations with Purpose. *Acad. Manag. J.* **2014**, *57*, 1227–1234. [CrossRef]
37. Lleo, A.; Bastons, M.; Rey, C.; Ruiz-Perez, F. Purpose Implementation: Conceptualization and Measurement. *Sustainability* **2021**, *13*, 1921. [CrossRef]
38. Agrawal, A.; Mukti, S.K. Knowledge Management & It's Origin, Success Factors, Planning, Tools, Applications, Barriers and Enablers: A Review. *Int. J. Knowl. Manag. IJKM* **2020**, *16*, 43–82. [CrossRef]
39. Hansen, M.T.; Birkinshaw, J. The Innovation Value Chain. *Harv. Bus. Rev.* **2007**, *85*, 121.

40. McIver, D.; Lengnick-Hall, C.A.; Lengnick-Hall, M.L.; Ramachandran, I. Understanding Work and Knowledge Management from a Knowledge-in-Practice Perspective. *Acad. Manag. Rev.* **2013**, *38*, 597–620. [CrossRef]
41. Prusak, L.; Davenport, T.H. Working Knowledge: How Organizations Manage What They Know. In *Harvard Business School Press Books*; Harvard Business Press: Brighton, MA, USA, 2000; p. 1.
42. Bartol, K.M.; Srivastava, A. Encouraging Knowledge Sharing: The Role of Organizational Reward Systems. *J. Leadersh. Organ. Stud.* **2002**, *9*, 64–76. [CrossRef]
43. Hung, S.-Y.; Durcikova, A.; Lai, H.-M.; Lin, W.-M. The Influence of Intrinsic and Extrinsic Motivation on Individuals' Knowledge Sharing Behavior. *Int. J. Hum.-Comput. Stud.* **2011**, *69*, 415–427. [CrossRef]
44. Ipe, M. Knowledge Sharing in Organizations: A Conceptual Framework. *Hum. Resour. Dev. Rev.* **2003**, *2*, 337–359. [CrossRef]
45. Yun, S.; Takeuchi, R.; Liu, W. Employee Self-Enhancement Motives and Job Performance Behaviors: Investigating the Moderating Effects of Employee Role Ambiguity and Managerial Perceptions of Employee Commitment. *J. Appl. Psychol.* **2007**, *92*, 745–756. [CrossRef]
46. Yang, C.; Chen, L.-C. Can Organizational Knowledge Capabilities Affect Knowledge Sharing Behavior? *J. Inf. Sci.* **2007**, *33*, 95–109. [CrossRef]
47. Bock, G.W.; Zmud, R.W.; Kim, Y.G.; Lee, J.N. Behavioral Intention Formation in Knowledge Sharing: Examining the Roles of Extrinsic Motivators, Social-Psychological Forces, and Organizational Climate. *MIS Q.* **2005**, *29*, 87–111. [CrossRef]
48. Lee, D.-J.; Ahn, J.-H. Reward Systems for Intra-Organizational Knowledge Sharing. *Eur. J. Oper. Res.* **2007**, *2*, 938–956. [CrossRef]
49. Cabrera, Á.; Collins, W.C.; Salgado, J.F. Determinants of Individual Engagement in Knowledge Sharing. *Int. J. Hum. Resour. Manag.* **2006**, *17*, 245–264. [CrossRef]
50. Ryu, S.; Ho, S.H.; Han, I. Knowledge Sharing Behavior of Physicians in Hospitals. *Expert Syst. Appl.* **2003**, *25*, 113–122. [CrossRef]
51. Schulz, M. The Uncertain Relevance of Newness: Organizational Learning and Knowledge Flows. *Acad. Manag. J.* **2001**, *44*, 661–681. [CrossRef]
52. Yu, T.-K.; Lu, L.-C.; Liu, T.-F. Exploring Factors That Influence Knowledge Sharing Behavior via Weblogs. *Comput. Hum. Behav.* **2010**, *26*, 32–41. [CrossRef]
53. Hansen, M.T. Knowledge Networks: Explaining Effective Knowledge Sharing in Multiunit Companies. *Organ. Sci.* **2002**, *13*, 232–248. [CrossRef]
54. Howell, K.E.; Annansingh, F. Knowledge Generation and Sharing in UK Universities: A Tale of Two Cultures? *Int. J. Inf. Manag.* **2013**, *33*, 32–39. [CrossRef]
55. Papadopoulos, T.; Stamati, T.; Nopparuch, P. Exploring the Determinants of Knowledge Sharing via Employee Weblogs. *Int. J. Inf. Manag.* **2013**, *33*, 133–146. [CrossRef]
56. Purvis, B.; Mao, Y.; Robinson, D. Three Pillars of Sustainability: In Search of Conceptual Origins. *Sustain. Sci.* **2019**, *14*, 681–695. [CrossRef]
57. Kuhlman, T.; Farrington, J. What Is Sustainability? *Sustainability* **2010**, *2*, 3436–3448. [CrossRef]
58. Hansmann, R.; Mieg, H.A.; Frischknecht, P. Principal Sustainability Components: Empirical Analysis of Synergies between the Three Pillars of Sustainability. *Int. J. Sustain. Dev. World Ecol.* **2012**, *19*, 451–459. [CrossRef]
59. Komeily, A.; Srinivasan, R.S. A Need for Balanced Approach to Neighborhood Sustainability Assessments: A Critical Review and Analysis. *Sustain. Cities Soc.* **2015**, *18*, 32–43. [CrossRef]
60. Oreskes, N. The Scientific Consensus on Climate Change: How Do We Know We're Not Wrong? In *Climate Modelling: Philosophical and Conceptual Issues*; Lloyd, E.A., Winsberg, E., Eds.; Springer International Publishing: Cham, Switzerland, 2018; pp. 31–64. ISBN 978-3-319-65058-6.
61. Mehta, A.K.; Bhide, S. Poverty Trends and Measures. In *Poverty, Chronic Poverty and Poverty Dynamics*; Springer: New York, NY, USA, 2018; pp. 9–36.
62. Zucman, G. Global Wealth Inequality. *Annu. Rev. Econ.* **2019**, *11*, 109–138. [CrossRef]
63. Gatto, A.; Sadik-Zada, E.R. Revisiting the East Asian Financial Crises: Lessons from Ethics and Development Patterns. In *Economic Growth and Financial Development*; Springer: New York, NY, USA, 2021; pp. 23–31.
64. Goerner, S.J.; Lietaer, B.; Ulanowicz, R.E. Quantifying Economic Sustainability: Implications for Free-Enterprise Theory, Policy and Practice. *Ecol. Econ.* **2009**, *69*, 76–81. [CrossRef]
65. Klaassen, G.A.; Opschoor, J.B. Economics of Sustainability or the Sustainability of Economics: Different Paradigms. *Ecol. Econ.* **1991**, *4*, 93–115. [CrossRef]
66. Marjerison, R.K.; Chae, C.; Li, S. Investor Activity in Chinese Financial Institutions: A Precursor to Economic Sustainability. *Sustainability* **2021**, *13*, 12267. [CrossRef]
67. Brown, J.L.; Agnew, N.M. Corporate Agility. *Bus. Horiz.* **1982**, *25*, 29–33. [CrossRef]
68. Wilson, J.M.; Goodman, P.S.; Cronin, M.A. Group Learning. *Acad. Manag. Rev.* **2007**, *32*, 1041–1059. [CrossRef]
69. Nonaka, I. A Dynamic Theory of Organizational Knowledge Creation. *Organ. Sci.* **1994**, *5*, 14–37. [CrossRef]
70. Polanyi, M. *Tacit Dimension*, 1st ed.; University of Chicago Press: Chicago, IL, USA, 1966.
71. Chen, S.-S.; Chuang, Y.-W.; Chen, P.-Y. Behavioral Intention Formation in Knowledge Sharing: Examining the Roles of KMS Quality, KMS Self-Efficacy, and Organizational Climate. *Knowl.-Based Syst.* **2012**, *31*, 106–118. [CrossRef]
72. Witherspoon, C.L.; Bergner, J.; Cockrell, C.; Stone, D.N. Antecedents of Organizational Knowledge Sharing: A Meta-analysis and Critique. *J. Knowl. Manag.* **2013**, *17*, 250–277. [CrossRef]

73. Ton, A.D.; Szabó-Szentgróti, G.; Hammerl, L. Competition within Cross-Functional Teams: A Structural Equation Model on Knowledge Hiding. *Soc. Sci.* **2022**, *11*, 30. [CrossRef]
74. Andrews, M.; Smits, S.J. The Role of Knowledge Management in Balancing Exploration and Exploitation in E-Commerce Firms. *J. Organ. Psychol.* **2021**, *27*, 96–108.
75. Hazen, B.T.; Bradley, R.V.; Bell, J.E.; In, J.; Byrd, T.A. Enterprise Architecture: A Competence-Based Approach to Achieving Agility and Firm Performance. *Int. J. Prod. Econ.* **2017**, *193*, 566–577. [CrossRef]
76. Tallon, P.P.; Pinsonneault, A. Competing Perspectives on the Link between Strategic Information Technology Alignment and Organizational Agility: Insights from a Mediation Model. *MIS Q.* **2011**, *35*, 463–486. [CrossRef]
77. Vickery, S.K.; Droge, C.; Setia, P.; Sambamurthy, V. Supply Chain Information Technologies and Organisational Initiatives: Complementary versus Independent Effects on Agility and Firm Performance. *Int. J. Prod. Res.* **2010**, *48*, 7025–7042. [CrossRef]
78. Vázquez-Bustelo, D.; Avella, L.; Fernández, E. Agility Drivers, Enablers and Outcomes: Empirical Test of an Integrated Agile Manufacturing Model. *Int. J. Oper. Prod. Manag.* **2007**, *27*, 1303–1332. [CrossRef]
79. Sundstrom, E.; McIntyre, M.; Halfhill, T.; Richards, H. Work Groups: From the Hawthorne Studies to Work Teams of the 1990s and beyond. *Group Dyn. Theory Res. Pract.* **2000**, *4*, 44–67. [CrossRef]
80. Edmondson, A. Psychological Safety and Learning Behavior in Work Teams. *Adm. Sci. Q.* **1999**, *44*, 350–383. [CrossRef]
81. Edmondson, A.C. The Local and Variegated Nature of Learning in Organizations: A Group-Level Perspective. *Organ. Sci.* **2002**, *13*, 128–146. [CrossRef]
82. Argote, L.; Ingram, P.; Levine, J.M.; Moreland, R.L. Knowledge Transfer in Organizations: Learning from the Experience of Others. *Organ. Behav. Hum. Decis. Process.* **2000**, *82*, 1–8. [CrossRef]
83. Medina, C.C.; Lavado, A.C.; Cabrera, R.V. Characteristics of Innovative Companies: A Case Study of Companies in Different Sectors. *Creat. Innov. Manag.* **2005**, *14*, 272–287. [CrossRef]
84. Goulet, L.R.; Frank, M.L. Organizational Commitment across Three Sectors: Public, Non-Profit, and For-Profit. *Public Pers. Manag.* **2002**, *31*, 201–210. [CrossRef]
85. Chen, C.-A.; Bozeman, B. Organizational Risk Aversion: Comparing the Public and Non-Profit Sectors. *Public Manag. Rev.* **2012**, *14*, 377–402. [CrossRef]
86. Cun, X. Public Service Motivation and Job Satisfaction, Organizational Citizenship Behavior: An Empirical Study Based on the Sample of Employees in Guangzhou Public Sectors. *Chin. Manag. Stud.* **2012**, *6*, 330–340. [CrossRef]
87. Wise, L.R. The Use of Innovative Practices in the Public and Private Sectors: The Role of Organizational and Individual Factors. *Public Prod. Manag. Rev.* **1999**, *23*, 150–168. [CrossRef]
88. Yang, S. Organizational Sectors and the Institutionalization of Job-Training Programs: Evidence from a Longitudinal National Organizations Survey. *Sociol. Perspect.* **2006**, *49*, 325–342. [CrossRef]
89. Peng, H.; Zhou, C.; Liu, Y. Entrepreneurial Experience and Performance: From the Aspect of Sustainable Growth of Enterprises. *Sustainability* **2020**, *12*, 7351. [CrossRef]
90. Grego-Planer, D. The Relationship between Organizational Commitment and Organizational Citizenship Behaviors in the Public and Private Sectors. *Sustainability* **2019**, *11*, 6395. [CrossRef]
91. Baruch, Y.; Ramalho, N. Communalities and Distinctions in the Measurement of Organizational Performance and Effectiveness Across For-Profit and Nonprofit Sectors. *Nonprofit Volunt. Sect. Q.* **2006**, *35*, 39–65. [CrossRef]
92. Ferreira, A.I. Competing Values Framework and Its Impact on the Intellectual Capital Dimensions: Evidence from Different Portuguese Organizational Sectors. *Knowl. Manag. Res. Pract.* **2014**, *12*, 86–96. [CrossRef]
93. Verma, T.; Dhar, S. The Impact of Intellectual Capital on Organizational Effectiveness: A Comparative Study of Public and Private Sectors in India. *IUP J. Knowl. Manag.* **2016**, *14*, 7–27.
94. Ahmed, K.A.G.; Ahmed, G. The Relationships between Conflict Management Styles, Job Satisfaction and Organizational Commitment among Workers in Public and Private Sectors. *Univers. J. Psychol.* **2015**, *3*, 41–50. [CrossRef]
95. Winley, G.; Arjpru, C.; Wongwuttiwat, J. National Information Technology Policy in Thailand: A Comparison among Organizational Sectors. *Electron. J. Inf. Syst. Dev. Ctries.* **2007**, *31*, 1–28. [CrossRef]
96. Winley, G.K.; Wongwuttiwat, J. The Structure of the Information Technology Profession: A Comparison Among Organizational Sectors in Thailand. *Electron. J. Inf. Syst. Dev. Ctries.* **2012**, *51*, 1–30. [CrossRef]
97. Sahoo, C.K.; Jena, S. Organizational Performance Management System: Exploring the Manufacturing Sectors. *Ind. Commer. Train.* **2012**, *44*, 296–302. [CrossRef]
98. Zehra, N. Training & Development Barometer for Effective Transformation of Organizational Commitment and Overall Performance in Banking Sectors of KPK, Pakistan: Qualitative Study of Workforce of Bank of Khyber. *Int. J. Acad. Res. Bus. Soc. Sci.* **2016**, *6*, 246–267.
99. Tso, T. Agriculture of the Future. *Nature* **2004**, *428*, 215–217. [CrossRef] [PubMed]
100. Cook, M.; Barry, P. Organizational Economics in the Food, Agribusiness, and Agricultural Sectors. *Am. J. Agric. Econ.* **2004**, *86*, 740–743. [CrossRef]
101. Dong, H.; Dacre, N.; Bailey, A. Sustainability in Complex Agriculture Projects: A Study of Agile Agricultural Co-Operative Institutions. *Br. Acad. Manag.* **2021**, 1–10. [CrossRef]
102. Li, L.; Lin, J.; Turel, O.; Liu, P.; Luo, X.R. The Impact of E-Commerce Capabilities on Agricultural Firms' Performance Gains: The Mediating Role of Organizational Agility. *Ind. Manag. Data Syst.* **2020**, *120*, 1265–1286. [CrossRef]

103. Jiao, X.; Mongol, N.; Zhang, F. The Transformation of Agriculture in China: Looking Back and Looking Forward. *J. Integr. Agric.* **2018**, *17*, 755–764. [CrossRef]
104. Thatcher, J.B.; Brower, R.S.; Mason, R.M. Organizational Fields and the Diffusion of Information Technologies Within and Across the Nonprofit and Public Sectors: A Preliminary Theory. *Am. Rev. Public Adm.* **2006**, *36*, 437–454. [CrossRef]
105. Birdi, K.S.; Patterson, M.G.; Wood, S.J. Learning to Perform? A Comparison of Learning Practices and Organizational Performance in Profit- and Non-Profit-Making Sectors in the UK. *Int. J. Train. Dev.* **2007**, *11*, 265–281. [CrossRef]
106. Curtis, B. Education as Poverty Reduction. In *Understanding Global Poverty*; Routledge: Oxfordshire, UK, 2021; pp. 246–266. ISBN 1-00-304382-8.
107. Abaidoo, A. The Nexus between Education and Poverty Reduction in Ghana from 2013 to 2017. *Cogent. Soc. Sci.* **2021**, *7*, 1986933. [CrossRef]
108. Liu, F.; Li, L.; Zhang, Y.; Ngo, Q.-T.; Iqbal, W. Role of Education in Poverty Reduction: Macroeconomic and Social Determinants Form Developing Economies. *Environ. Sci. Pollut. Res.* **2021**, *28*, 63163–63177. [CrossRef] [PubMed]
109. Notanubun, Z. The Effect of Organizational Citizenship Behavior and Leadership Effectiveness on Public Sectors Organizational Performance: Study in the Department of Education, Youth and Sports in Maluku Province, Indonesia. *Public Organ. Rev.* **2021**, *21*, 1–18. [CrossRef]
110. Cascavilla, I.; Hahn, D.; Minola, T. How You Teach Matters! An Exploratory Study on the Relationship between Teaching Models and Learning Outcomes in Entrepreneurship Education. *Adm. Sci.* **2022**, *12*, 12. [CrossRef]
111. Colombelli, A.; Loccisano, S.; Panelli, A.; Pennisi, O.A.M.; Serraino, F. Entrepreneurship Education: The Effects of Challenge-Based Learning on the Entrepreneurial Mindset of University Students. *Adm. Sci.* **2022**, *12*, 10. [CrossRef]
112. Joensuu-Salo, S.; Viljamaa, A.; Varamäki, E. Sustainable Entrepreneurs of the Future: The Interplay between Educational Context, Sustainable Entrepreneurship Competence, and Entrepreneurial Intentions. *Adm. Sci.* **2022**, *12*, 23. [CrossRef]
113. Reyes, A. Connecting Higher Education and Innovation to Local Development. *Futures* **2018**, *103*, 73–83. [CrossRef]
114. Marjerison, R.K.; Pan, J. Decision-Making Styles of the Next Generation of Chinese Business Leaders. In *Handbook of Research on Emerging Business Models and the New World Economic Order*; IGI Global: Hershey, PA, USA, 2022; pp. 365–386.
115. Carlsen, M.H. Familiar Strangers–Managing Engagements in Public-Private Partnerships in Education. *Nord. J. Stud. Educ. Policy* **2021**, 1–14. [CrossRef]
116. Ishaq, K.; Azan, N.; Zin, M.; Rosdi, F.; Abid, A.; Ijaz, M. The Impact of ICT on Students' Academic Performance in Public Private Sector Universities of Pakistan. *Int. J. Innov. Technol. Explor. Eng. IJITEE* **2020**, *9*, 1117–1121.
117. Marginson, S. Public/Private in Higher Education: A Synthesis of Economic and Political Approaches. *Stud. High. Educ.* **2018**, *43*, 322–337. [CrossRef]
118. McCormack, T.J.; Lemoine, P.A.; Waller, R.E.; Richardson, M.D. Global Higher Education: Examining Response to the COVID-19 Pandemic Using Agility and Adaptability. *J. Educ. Dev.* **2021**, *5*, 10. [CrossRef]
119. Menon, S.; Suresh, M. Factors Influencing Organizational Agility in Higher Education. *Benchmarking Int. J.* **2020**, *28*, 307–332. [CrossRef]
120. Jha, M.K. An Integrated Framework of Leadership for Healthcare Organizations to Navigate through COVID-19 Crisis. *Asia Pac. J. Health Manag.* **2021**, *16*, 16–20. [CrossRef]
121. Suresh, M.; Roobaswathiny, A.; Lakshmi Priyadarsini, S. A Study on the Factors That Influence the Agility of COVID-19 Hospitals. *Int. J. Healthc. Manag.* **2021**, *14*, 290–299. [CrossRef]
122. Gordon, A.; Pollack, J. Managing Healthcare Integration: Adapting Project Management to the Needs of Organizational Change. *Proj. Manag. J.* **2018**, *49*, 5–21. [CrossRef]
123. Li, S.A.; Jeffs, L.; Barwick, M.; Stevens, B. Organizational Contextual Features That Influence the Implementation of Evidence-Based Practices across Healthcare Settings: A Systematic Integrative Review. *Syst. Rev.* **2018**, *7*, 72. [CrossRef] [PubMed]
124. Tavakol, M.; Dennick, R. Making Sense of Cronbach's Alpha. *Int. J. Med. Educ.* **2011**, *2*, 53–55. [CrossRef] [PubMed]
125. Meer, P.; Mintz, D.; Rosenfeld, A.; Kim, D.Y. Robust Regression Methods for Computer Vision: A Review. *Int. J. Comput. Vis.* **1991**, *6*, 59–70. [CrossRef]
126. Vieira, S.; Pinaya, W.H.L.; Mechelli, A. Introduction to Machine Learning. In *Machine Learning*; Elsevier: Amsterdam, The Netherlands, 2020; pp. 1–20.
127. Goldstein, R. Regression Methods in Biostatistics: Linear, Logistic, Survival and Repeated Measures Models. *Technometrics* **2006**, *48*, 149–150. [CrossRef]
128. Vittinghoff, E.; Glidden, D.V.; Shiboski, S.C.; McCulloch, C.E. *Regression Methods in Biostatistics: Linear, Logistic, Survival, and Repeated Measures Models*; Springer: New York, NY, USA, 2006.
129. Oliveira, M.; Curado, C.; Balle, A.R.; Kianto, A. Knowledge Sharing, Intellectual Capital and Organizational Results in SMES: Are They Related? *J. Intellect. Cap.* **2020**, *21*, 893–911. [CrossRef]
130. Gosling, S.D.; Vazire, S.; Srivastava, S.; John, O.P. Should We Trust Web-Based Studies? A Comparative Analysis of Six Preconceptions about Internet Questionnaires. *Am. Psychol.* **2004**, *59*, 93–104. [CrossRef]
131. Brislin, R.W. Back-Translation for Cross-Cultural Research. *J. Cross-Cult. Psychol.* **1970**, *1*, 185–216. [CrossRef]
132. Lin, Y.; Marjerison, R.K.; Kennedyd, S. Reposting Inclination of Chinese Millennials on Social Media: Consideration of Gender, Motivation, Content and Form. *J. Int. Bus. Cult. Stud.* **2019**, *12*, 20.

133. Pallant, J. *SPSS Survival Manual: A Step by Step Guide to Data Analysis Using IBM SPSS*; Routledge: Oxfordshire, UK, 2020; ISBN 1-00-311745-7.
134. Hofstede, G. *Culture's Consequences: International Differences in Work-Related Values*; Sage: Thousand Oaks, CA, USA, 1984; Volume 5, ISBN 0-8039-1306-0.

sustainability

Article

A Novel Multi-Factor Three-Step Feature Selection and Deep Learning Framework for Regional GDP Prediction: Evidence from China

Qingwen Li [1], Guangxi Yan [2,*] and Chengming Yu [2]

1 College of Business and Trade, Hunan Industry Polytechnic, Changsha 410036, China; lqw0911123@163.com
2 School of Traffic and Transportation Engineering, Central South University, Changsha 410075, China; 1109150133@csu.edu.cn
* Correspondence: yanguangxi@csu.edu.cn

Abstract: Gross domestic product (GDP) is an important index reflecting the economic development of a region. Accurate GDP prediction of developing regions can provide technical support for sustainable urban development and economic policy formulation. In this paper, a novel multi-factor three-step feature selection and deep learning framework are proposed for regional GDP prediction. The core modeling process is mainly composed of the following three steps: In Step I, the feature crossing algorithm is used to deeply excavate hidden feature information of original datasets and fully extract key information. In Step II, BorutaRF and Q-learning algorithms analyze the deep correlation between extracted features and targets from two different perspectives and determine the features with the highest quality. In Step III, selected features are used as the input of TCN (Temporal convolutional network) to build a GDP prediction model and obtain final prediction results. Based on the experimental analysis of three datasets, the following conclusions can be drawn: (1) The proposed three-stage feature selection method effectively improves the prediction accuracy of TCN by more than 10%. (2) The proposed GDP prediction framework proposed in the paper has achieved better forecasting performance than 14 benchmark models. In addition, the MAPE values of the models are lower than 5% in all cases.

Keywords: GDP prediction; feature selection; deep learning; temporal convolutional network

Citation: Li, Q.; Yan, G.; Yu, C. A Novel Multi-Factor Three-Step Feature Selection and Deep Learning Framework for Regional GDP Prediction: Evidence from China. *Sustainability* **2022**, *14*, 4408. https://doi.org/10.3390/su14084408

Academic Editors: Jungho Baek, Elkhan Richard Sadik-Zada and Andrea Gatto

Received: 27 January 2022
Accepted: 6 April 2022
Published: 7 April 2022

1. Introduction

Regional gross domestic product (GDP) can fully reflect basic economic indicators such as a region's economic growth rate and changes in economic scale, which is equal to the sum of the added value of various industries in the region [1]. It is widely used all over the world and has become a general macroeconomic indicator to measure regional economic conditions [2]. The effective forecasting of regional GDP in economic operation and development can not only determine a certain degree of macroeconomic trend and guide the healthy development of macroeconomics but also provide a crucial basis for sustainable urban development [3]. The research on regional GDP can explore the internal driving force of local economic growth and promote the optimization and upgrading of local industrial structure [4]. Besides this, by predicting the regional GDP, local governments can make more comprehensive scientific and economic choices [4]. The government can forecast and prospect the development of the market economy so that the development plans could be formulated according to the forecast results and decisions that are beneficial to the local economy can be conducted [5]. The formulation of macro-control economic policies and the adjustment of corporate development strategies all depend on accurate forecasting of regional GDP [6].

Especially at the moment, the sudden outbreak of COVID-19 since the beginning of 2020 had such a great impact on the operation of the world economy that the major economic indicators have declined significantly [5–7]. However, with the continuous progress of production resumption, the major economic indicators (especially in China) have shown a rebounding momentum and continual improvement [8,9]. Given the complex world economic situation at this stage, monitoring the current state of economic operation and forecasting the future GDP trend plays a paramount role in the overall control of the macroeconomy and are of practical significance for the formulation of economic policies and macro-control in the next step [10–12]. GDP prediction technology can guide the direction of regional sustainable development in the future. Therefore, the research on this direction can promote regional industrial upgrading and form the direction of green development.

At present, the mainstream GDP forecasting framework mainly includes statistical models, machine learning models, and hybrid models [13]. Statistical models mainly use multiple regression and time series modeling to construct mathematical formulas for GDP changes. Machine learning models mainly include support vector machines, decision trees, and so on, which establish a nonlinear mapping of input and output. Compared with these two kinds of models, the hybrid model can effectively improve the prediction performance of the model from data analysis, feature extraction, and nonlinear modeling by effectively combining various components [14]. Considering the complexity of GDP prediction feature categories, feature engineering, and deep learning are adopted in the paper to establish an accurate multi-factor GDP prediction framework.

2. Related Works

The government forecasts the development of the market economy formulates development plans based on the forecasting results and then makes decisions that are beneficial to regional development [15,16]. Economic forecasting is based on historical statistics or survey data, using scientific methods to predict the prospects of economic phenomena [17]. The GDP forecasting is based on actual data, using scientific methods and data models to predict the future GDP value [18]. The time series forecasting methods (TSFM) are widely used in economic forecasting, which refers to a quantitative forecasting method that sorts the historical data of the prediction target according to the time series [19]. TSFM can analyze the changing trend of data and establish a mathematical model for extrapolation. Forecasting based on mathematical models and computing platforms is a commonly used time series forecasting method. Elkhan Richard et al. proposed a novel data-driven nonlinear regression model that can effectively analyze the deep correlation between national income, oil production capacity, and the environment [20].

In a recent GDP forecasting study, researchers proposed many models for forecasting analysis methods. With the development of machine learning and Nonlinear modeling technology, data-driven methods have become one of the mainstream models in GDP forecasting [21]. Abonazel and Abd-Elftah utilized the autoregressive integrated moving average (ARIMA) model to predict the Egyptian GDP data [22]. Wu and Chen studied the application of a support vector machine (SVM) in a financial time series prediction of the debt to GDP ratio index [23]. The experiment results proved the effectiveness of the evaluation indexes. Ghanem et al. proposed a functional link artificial neural network (FLANN) to predict electricity prices under the impact of COVID-19, which showed significant improvement in the forecasting accuracy [24]. Elkhan Richard et al. proposed a new nonlinear panel ARDL approach, which can effectively analyze the correlation between economic indicators and alcohol consumption and achieve better results than traditional algorithms [25]. Britta et al. proposed a new Nonlinear autoregressive distributed lag framework to analyze the core correlations between income and quality wine imports, which effectively solved the problems existing in traditional algorithms [26].

These forecasting methods have their specific characteristics. At the current stage of GDP forecast analysis, a single predictive analysis model only reflects a part of the information of the analysis object. To effectively improve the adaptability of single prediction algorithms and the comprehensive analysis ability of GDP data, scholars use an optimization algorithm to optimize the input features and parameters of the single neural network model. Yusof et al. combined artificial bee colony (ABC) with the least squares support vector machine (LSSVM) in gold prices forecasting [27]. Long et al., utilized the genetic algorithm (GA) to optimize the key information of SVM and predicted the total GDP data of Anhui province using the GDP data from 1989 to 2007 [28]. Guleryuz et al. employed particle swarm optimization (PSO) with an adaptive neuro-Fuzzy inference system (ANFIS) to predict industrial energy demand, which is affected by many economic and social parameters [29]. The above research works proved that the heuristic algorithm-based models outperformed the single predictors.

Although the above optimization algorithms have proved the availability in the forecasting application, more improvements can be added to achieve better modeling performance. The traditional time series method requires stable time series data in the forecast model, but it has poor fitting ability to complex nonlinear systems, and the forecast accuracy of GDP growth is not accurate enough [30]. To further improve the forecasting accuracy for GDP growth trends, novel deep learning models are gradually investigated and applied, which are getting better at fitting complex systems. Sa'adah and Wibowo used two learning models for the prediction of GDP in Indonesia: the long short-term memory (LSTM) and recurrent neural network (RNN), resulting in an accuracy of over 80% [31]. Liu et al. applied the gated recurrent unit (GRU) network in the prediction of Chinese energy consumption [32]. In the comparison with the multiple linear regression (MLR) and the support vector regression (SVR) models, the GRU proved the superiority of complex nonlinear sequence data processing with lower prediction errors. In the research work of Wang et al., the temporal convolutional network (TCN) is employed for the short-term prediction of power system load, the TCN displayed faster speed and less storage demand by the specific structure, which leads to the best performance compared with SVR and LSTM [33]. It is meaningful to analyze the performance of TCN in the prediction study. Therefore, the TCN is applied as the main predictor in the paper.

In economic forecasting, a reasonable forecast should be selected according to the characteristics of the data and application scenarios. In the analysis of various influencing factors of GDP, the feature cross-validation can accurately measure the effect of the proposed model used on the actual datasets, which helps to select the suitable parameters [34,35]. The multi-feature modeling can sort the importance of variables that affect GDP, then complete the steps of feature extraction and feature filtering, and select important variables for calculation [36]. Ortega-Bastida et al. used an autoencoder (AE) to simultaneously reduce the reconstruction errors and data redundancy of the impacts with the corresponding GDP values to guarantee precision in forecasting [37]. Nahil and Lyhyaoui proposed the kernel principal component analysis (KPCA) to raise generalization performance and provide effective input variables for the predictor SVR [38]. The hybrid framework performed significantly better than the single SVR. Wang and Li utilized several feature extraction methods, which contained variational mode decomposition, Kullback-Leibler divergence (KLD), energy measure (EM), and sample entropy (SE) to extract the best features of raw time-series data [39]. The experimental results showed that the feature extraction method can optimize the predictor. Kosana et al. applied Q-learning as the selection method for time series prediction, which could dynamically select the best approach to increase the total model accuracy [40]. The hybrid model was regarded as the online model selection with Q-learning (OMS-QL). Xu et al. also chose Q-learning to conduct feature selection that outperformed other feature selection algorithms in the comparative experiments [41]. Maeda-Gutiérrez et al. combined the Boruta algorithm with a random forest algorithm as a hybrid BorutaRF framework that contains the function of feature selection and classification using the cross-validation strategy [42].

Based on the above literature survey, the existing excellent feature selection and prediction framework can be summarized in Table 1.

Based on the above literature research, it is sinnvoll to research the application of multi-feature analysis methods with the validation process to optimize the parameters in the forecasting model. Considering the excellent feature analysis ability of feature crossing,

BorutaRF, and Q-learning algorithm, this study proposed a new three-stage feature selection to optimize the input of TCN and establish a multi-data-driven GDP prediction model. The innovation and contribution of the paper are presented as follows:

(1) To comprehensively analyze various influencing factors and improve the prediction accuracy of the traditional single-variable GDP time series prediction framework, a multi-factor data-driven GDP prediction framework is proposed to process the multivariate economic data. This multi-feature prediction and feature engineering is significantly efficient for general prediction structures.
(2) With the excellent forecasting effect compared to the traditional shallow neural network and recursive neural network, the TCN neural network adopted in this paper is firstly applied in the field of GDP prediction, which can conduct analysis for complex nonlinear data and increase the GDP prediction accuracy.
(3) A new three-stage feature selection framework is proposed in the paper. The feature crossing algorithm could select to explore the potential features and the deep information of the raw data. The Q-learning and BorutarRF algorithm could select features from different aspects and guarantee the quality of TCN input. The hybrid three-stage feature selection structure is applied firstly in GDP forecasting to improve the main predictor TCN.

Table 1. Basic information about these GDP forecasting models.

References	Published Year	Feature Selection
[35]	2020	AE
[38]	2018	KPCA
[39]	2018	KLD, EM, SE
[40]	2022	Q-learning
[41]	2021	Q-learning
[42]	2021	BorutaRF

The structure of this study is listed as follows: Section 2 introduces the technical details of deep learning and feature engineering commonly used in current mainstream GDP forecasting models. Section 3 mainly introduces in detail the applied data information, the proposed methods, and the total framework of the paper. In Section 4, comparative experiments are conducted to verify the performance improvement of the proposed model. Section 5 concludes the main contributions of this paper and prospects the research direction of GDP prediction application.

3. Methodology

3.1. Framework of the Proposed Regional GDP Forecasting Model

The influencing factors of regional GDP are complex, and the prediction accuracy of simple single series is not satisfactory enough under the circumstance. This paper proposes a regional GDP forecasting method based on the integration of economic, educational, employment, and industrial data. The accuracy of the prediction model can be improved greatly through feature analysis of the multivariate data. Moreover, a three-step feature selection method is proposed to better obtain the features that are helpful to GDP prediction. TCN network can deeply explore the nonlinear relationship between features and target prediction. In this paper, the TCN model is used to predict regional GDP in combination with selected features. The specific model framework is shown in Figure 1.

3.2. Multivariate Economic Characteristic Data

Regional economic forecasting is a typical time series forecasting problem, but it is different from traditional single time series forecasting. The regional economy is affected by education, industrial structure, transportation and logistics, geographical location, and other factors. Therefore, multivariate data analysis is indispensable to realizing accurate regional economic forecasts. In this paper, education, industry, historical economic data, population, and other factors that have a great impact on regional economic forecast are taken as the prediction features. Historical information is the first place to be considered. The historical economic information like real GDP, real consumption, and so on can reflect quite a few laws in economic development, which is instructive for GDP forecasting [43]. Employment and population decide the purchasing power of the public and further influence the market circulation and the improvement of GDP [44]. As for education and technology, positive education can promote employment, and science and technology is the basement of productivity-increasing [45]. Thus, education data is also in need to improve the accuracy of GDP prediction. Finally, industrial information is directly associated with GDP. For example, industrial output, energy consumption, the transportation efficiency are all crucial parts of the economy [46]. The stability and accuracy of the regional economic forecasting model can be greatly improved through multivariate data fusion. As shown in Figure 2, there are 20 features, which are mainly divided into four categories: historical economic indicators, employment and population, educational scale, and industrial structure.

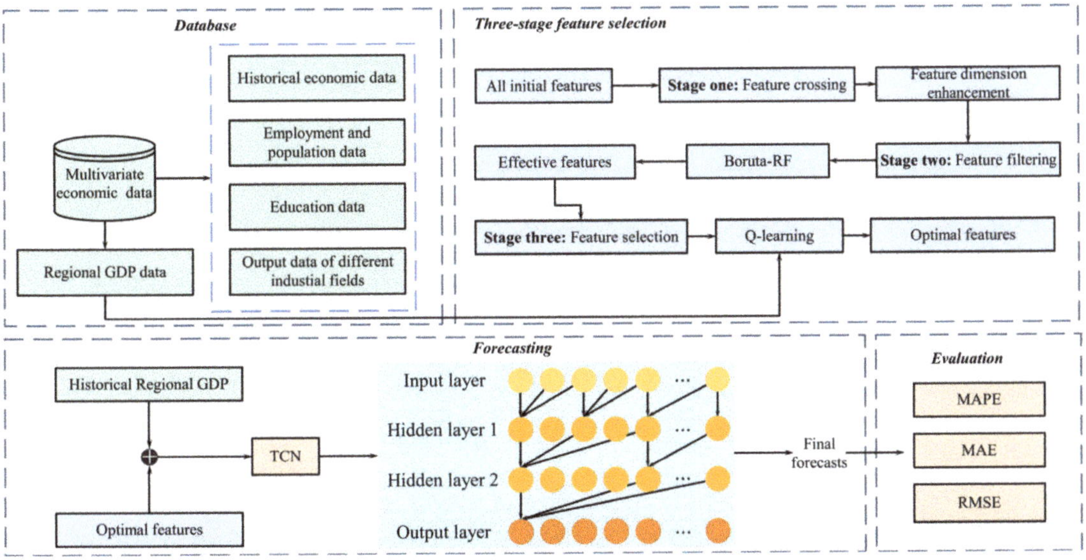

Figure 1. The core structure of the proposed GDP prediction model.

3.3. *Three-Stage Feature Selection*

3.3.1. Stage I: Feature Crossing

The 20-dimensional initial features used are divided into four groups according to different domains. The degree of correlation of features within and between groups is different, and the amount of useful information contained in the features is also different. Directly applying feature selection is easy to cause the loss of useful information, and the data cannot be exploited to the full [47]. Thus, feature crossing is utilized to solve the problems. On the one hand, the different information contained in the feature can be combined effectively by applying feature cross, which is able to improve the performance of feature selection. On the other hand, it also allows models to learn more complex nonlinear features. In this part, four feature crossing schemes are proposed. According to whether features belong to the same category, different statistical aggregation or simple calculating methods are carried out on features. Then, the new features are obtained. As shown in Figure 3, distinguishing the feature class by color and the detailed feature crossing method are illustrated below.

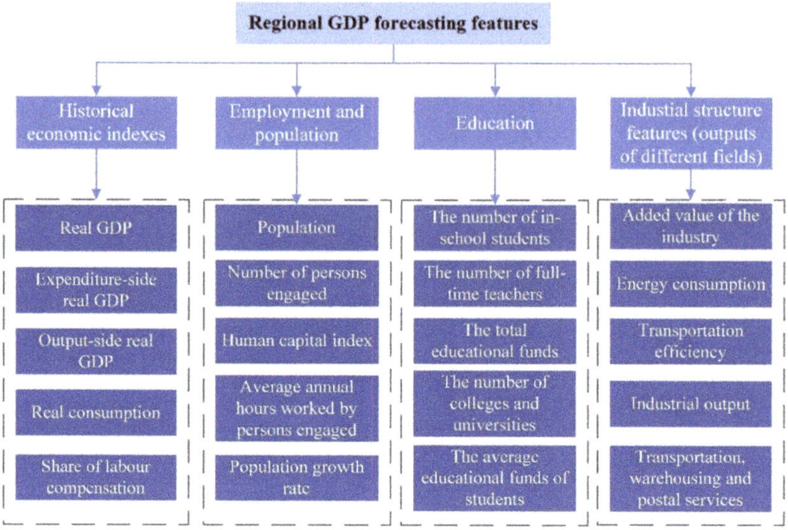

Figure 2. Initial features for regional GDP forecasting.

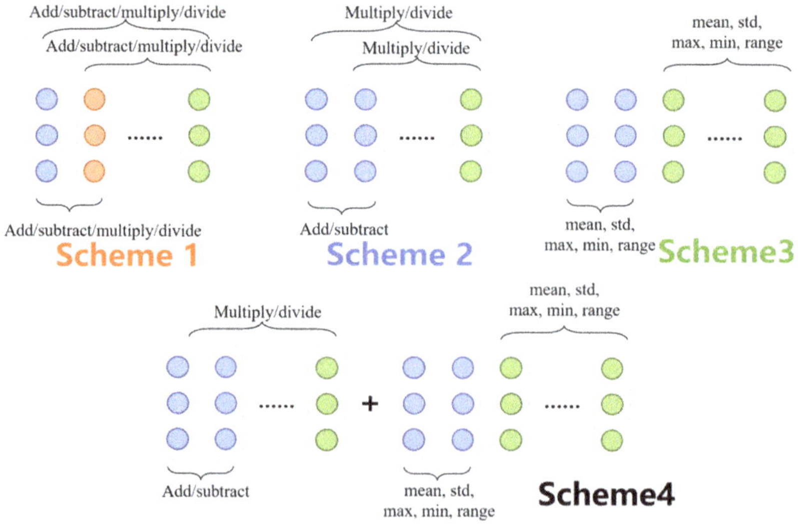

Figure 3. Feature crossing strategy.

Scheme 1: Traverse features. The 20-dimensional features are added, subtracted, multiplied, and divided.

Scheme 2: Intra-group features. If two sets of features belong to the same class, addition and subtraction are carried on each pairwise feature; otherwise, multiply and divide.

Scheme 3: Group aggregation features. For all features that belong to the same class, calculate the mean, standard deviation, maximum, minimum, and range of them to generate new features.

Scheme 4: Intra-group feature and group aggregation features. Namely, the new feature splicing obtained by Schemes 2 and 3.

3.3.2. Stage II: Feature Filtering Based on Boruta-RF

Through feature crossing, the input feature dimension is greatly increased. Among all features, some of them have no correction with the dependent variables. If all the crossed features were taken as the inputs of the feature selection part, the efficiency would be impacted. As a result, a feature filter method is adopted before feature selection. Boruta is an important feature filtering algorithm proposed by M. Kursa and R. Rudnicki [48]. The application of this method as a feature selection method belongs to filter feature selection. Different from most feature screening methods, which aim at the maximization of the evaluation function index or the optimization of the model loss function, this method aims to filter out features unrelated to dependent variables and get all feature sets related to dependent variables. Generally, it is believed that if a feature is added or deleted and the model performance does not change, then the feature is not important. However, it is not completely true. When the feature only has no or little effect on the improvement of model performance, it is not necessarily irrelevant to the dependent variable. Boruta can retain all the features related to the target value, so it is of great significance to use this algorithm as a preliminary feature filtering algorithm for feature selection.

The core idea of Boruta is to Shuffle each feature to generate its shadow feature. The shadow features are chaotic and do not correspond to the original samples. All the features and shadows are utilized to train models. Then, the shadow features with the highest importance scores are taken as the baseline, and the feature sets related to the dependent variable are selected from the crossed features according to the baseline. The specific steps are clarified as follows.

Step 1: shuffle each feature Xi in the feature matrix X and combine the shuffled features and original features [49].

$$\mathrm{Xs} = \mathrm{shuffle}(\mathrm{X}) \tag{1}$$

$$\mathrm{Xn} = [\mathrm{X}; \mathrm{Xs}] \tag{2}$$

where Xs and Xn are the shadow feature matrix and newly generated feature matrix respectively.

Step 2: Train the Random Forest model using the new generated features and calculate the average relative entropy of each feature. Then, calculate Zscore [50], which is given below for all features including the shadow features, and take the highest Z score value of shadow features as the baseline, named Zbase.

$$\mathrm{Zscore} = \frac{\overline{\mathrm{G}}}{\sigma_{\mathrm{G}}} \tag{3}$$

where G is the related entropy of features, $\overline{\mathrm{G}}$ is the mean value of G, and σ_{G} is the standard deviation.

Step 3: Set a percentage parameter perc, compare Zscore values of the crossed features with perc•Zbase and get rid of the features whose Zscore value is lower than perc•Zbase. In the process, Benjamini Hochberg FDR and Bonferroni are adopted to guarantee the stability of the algorithm.

Step 4: Delete all shadow features and repeat the above steps until all features are marked as important or unimportant.

3.3.3. Stage III: Feature Selection Based on Reinforcement Learning

Reinforcement learning is a real-time learning method that focuses on online learning and environmental feedback [51]. It is an algorithm used to describe and solve the problem that agents maximize returns or achieve specific goals through Learning strategies during their interaction with the environment. Different from the traditional evolutionary algorithms, it can realize dynamic optimization and is inclusive of the error path optimization [52]. As a classic reinforcement learning algorithm with excellent decision-making, Q-learning is widely used in decision-making and optimization problems. In this part, a feature selection method based on Q-learning is proposed. The filtered features are further screened to avoid over-fitting the model due to excessively high feature dimensions. The specific steps are given as follows:

Step 1: Initialize the core parameters of Q-learning (the state matrix S and the action matrix a). The state matrix S represents the selection of these features. The action matrix a is the action to keep or leave these features [53].

$$S = [s_1, s_2, \ldots, s_m] \tag{4}$$

$$a = [\Delta s_1, \Delta s_2, \Delta s_3, \ldots, \Delta s_m] \tag{5}$$

where s_m represents the selection of these features, and s_m is 0 or 1 (0 represents that the feature is not required, and 1 represents that the feature is retained). Δs_m is the action of adding or deleting the m-th feature.

Action a: Select an action strategy according to ε-greedy.

$$a_n = \begin{cases} \textit{Action based on } max\mathrm{Q}(\mathrm{S},\mathrm{a})(\textit{probability of } 1-\varepsilon) \\ \textit{Random action}(\textit{probability of } \varepsilon) \end{cases} \tag{6}$$

$$\varepsilon \in (0, 1) \tag{7}$$

where ε is the exploration probability.

Step 2: Establish the reward R, which will affect the agent's action. In this part, the MAPE of the TCN is taken as a reward.

Step 3: The agent performs an action based on a comprehensive analysis of the current environment and the state S.

Step 4: Calculate the evaluation function Q and update the Q table. Based on the reward R received from the environment, the agent updates the state and Q table by adjusting the action of input feature changes. The calculation formula of the Q value is shown as follows [54]:

$$Q_{n+1}(S_n, a_n) = Q_n(S_n, a_n) + \beta_n(R(S_n, a_n) + \gamma \max Q_n(S_{n+1}, a_{n+1}) - Q_n(S_n, a_n)) \tag{8}$$

where a represents the agent's behavior; S stands for the current status of an agent; R is the immediate return; γ is the discount parameter; β represents learning speed.

Step 5: When the termination condition is met, the agent stops its action. At this point, the state matrix S is the final selection result of model input features. Otherwise, repeat steps 3 to 4.

3.4. *TCN for Regional GDP Forecasting*

TCN is a neural network model which integrates extended causal convolution and residual connection and can be used for time series prediction [55]. TCN is composed of multiple TCN residual blocks stacked. Each TCN residual block has an important parameter pair (k, d) which represents the convolution kernel size and expansion coefficient respectively [56]. The final output of the TCN residual block is the sum of the outputs of the two paths. One path takes the input values through two levels of the same DCC and outputs them. Firstly, the input value enters the DCC after the weight initialization of layer 1. Then, the output is nonlinear transformed by the ReLU activation function. Finally, the nonlinear outputs are regularized to reduce the overfitting of the model and are input to layer 2 DCC for the same transformation again. The other path is for the input value to reach the output directly through the one-dimensional convolution layer. The path is RC, which is derived from the residual neural network. It can alleviate the problems of gradient disappearance and gradient explosion existing in the deep neural network and contribute to the construction of the deep neural network.

The core component of TCN is DCC. DCC increases the value of expansion coefficient d based on causal convolution, thus expanding the receptive field of the network, that is, accepting longer historical data [57]. The first is the application of causal convolution, which means that there is no leakage of information in the past. In the network applied in this paper, the convolution kernel is 2, the expansion coefficient is 1, and the receptive field is 3. And the ŷt GDP sequences are calculated from input sequences [xt − 2, xt − 1, xt] and have nothing to do with the input sequences [xt + 1, xt + 2, ...]. Therefore, the application of causal convolution in TCN will not give rise to information leakage. However, causal convolution has the problem of a small receptive field. Therefore, DCC expands the network receptive field by increasing the expansion coefficient. The receptive field of DCC in the same layer can be expanded to 4. The extended convolution operation can be obtained by the following equation [58]:

$$\mathrm{TCN}(t) = \sum_{l=0}^{p-1} f(l)X_{t-dl} \tag{9}$$

where TCN(t) is the extended convolution operation, X represents the time series data, f is the filter function, p is the length of the data, l is the element in X.

4. Case Study

4.1. GDP Dataset

The case study is the key to evaluating the performance of different GDP prediction frameworks. In order to select valuable regional GDP data sets, based on the analysis of GDP data in the references [59], this paper adopts the data of three Provinces in China to construct the experimental analysis. The data comes from the National Statistical Yearbook, which contains GDP data and other features data for each quarter from 2005 to 2021. Table 2 gives the basic information and input features of these three data. Figures 4–6 show the temporal fluctuation characteristics of three sets of GDP data. It is necessary to ensure the stability and robustness of the proposed model. To fully prove the stability and validity of the GDP prediction model, the proposed model and other benchmark models are evaluated by the ten-fold cross-validation method. In addition, ten repeated tests were used to evaluate the performance of the model. The average value of the evaluation index of the ten predicted results was used to analyze the effect of the model. This paper mainly constructs the single-step forecasting model, that is, the model predicts the GDP of the next quarter through the current moment and historical data. The key software platform used for experiments and modeling in this paper is the Python 3.8.5 platform, mainly using TensorFlow 2.3 to build the neural network. The Python was designed by Guido van Rossum, who works in Google. The version of python used in this paper is 3.8.5. The TensorFlow was created by Google open source, the version used in this paper is 2.3.

4.2. Performance Evaluation Indexes

The regression analysis index is the key to evaluating the performance of the model proposed in this paper. To fully analyze the modeling performance of each model, three classic indexes, which are the MAE (Mean Absolute Error), the MAPE (Mean Absolute Percentage Error), and the RMSE (Root Mean Square Error), are used in all case studies. These indexes can be obtained by the following Equation (10) [60]:

$$\begin{cases} \mathrm{MAE} = \left(\sum_{T=1}^{n} \left|Y(T) - \widehat{Y}(T)\right|\right)/n \\ \mathrm{MAPE} = \left(\sum_{T=1}^{n} \left|(Y(T) - \widehat{Y}(T))/y(T)\right|\right)/n \\ \mathrm{RMSE} = \sqrt{\left(\sum_{T=1}^{n} \left[Y(T) - \widehat{Y}(T)\right]^2\right)/n} \end{cases} \tag{10}$$

where Y (T) represents true GDP data. $\widehat{Y}(T)$ represents the GDP data calculated by the proposed model. N means the number of samples.

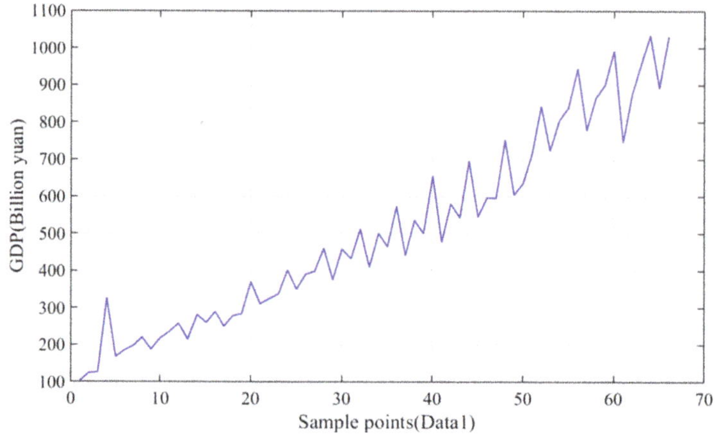

Figure 4. Raw GDP Data 1.

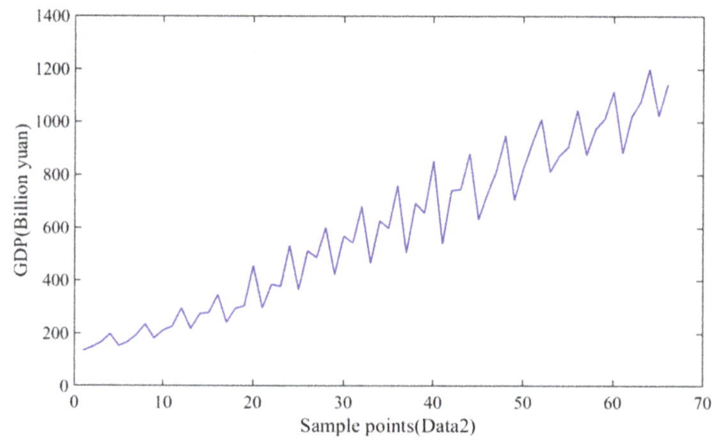

Figure 5. Raw GDP Data 2.

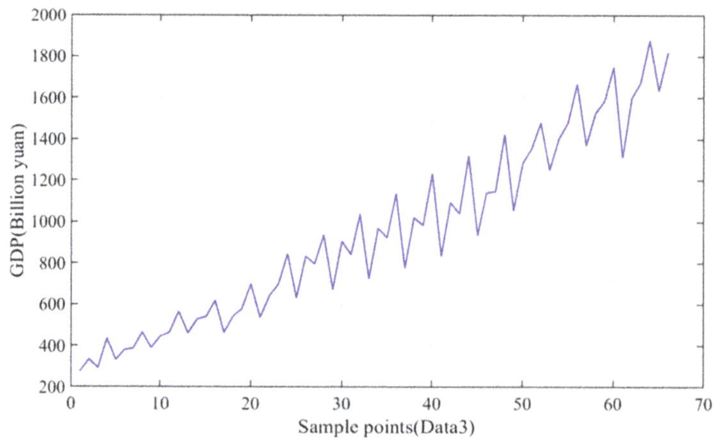

Figure 6. Raw GDP Data 3.

Table 2. Basic information about this GDP dataset.

GDP Data	Data 1	Data 2	Data 3
Province	Beijing	Hunan	Zhejiang
Minimum	102.3870	134.9570	274.6090
Mean	506.0497	592.2015	945.7440
Maximum	1034.3040	1200.0900	1878.7420
Standard derivation	256.8505	309.3011	441.7130

At the same time, it is necessary to select appropriate indicators to evaluate the performance differences between different models. This study utilized the Promoting percentages of the MAE (PMAE), the Promoting percentages of the MAPE (PMAPE), and the Promoting percentages of the RMSE (PRMSE) to evaluate the performance differences between different algorithms. These indexes can be obtained by the following Equation (11) [61]:

$$\begin{cases} P_{MAE} = \frac{(MAE_a - MAE_b)}{MAE_a} \\ P_{MAPE} = \frac{(MAPE_a - MAPE_b)}{MAPE_a} \\ P_{RMSE} = \frac{(RMSE_a - RMSE_b)}{RMSE_a} \end{cases} \quad (11)$$

4.3. Contrast Experiment with Benchmark Algorithms

4.3.1. Experimental Results and Analysis of Different Predictors

To fully compare and analyze the modeling effects of different predictors and prove the superiority of the TCN algorithm, this paper adopts TCN, GRU, LSTM, RNN, ELM, and RBF models to construct comparative experiments. The experiment includes a classical deep learning model and a traditional shallow neural network. Table 3 gives the regression analysis indexes of the prediction results of these algorithms. From Table 3, the following conclusions can be drawn:

(1) Compared with the traditional RBF and ELM algorithms, the neural network model based on deep learning can obtain fewer error prediction results. The experimental cases fully prove that the deep learning method can achieve satisfactory modeling results in this field. The feasible reason is that the multi-layer deep network structure has certain advantages in mining the deep feature information of data.

(2) Compared with the traditional RNN prediction model, other deep learning models can achieve more satisfactory prediction results. This proves that other deep neural networks with special structures can better resume an excellent GDP forecasting framework. The possible reason is that the RNN model has problems such as gradient descent and gradient disappearance, which to some extent limits the training effect of the model and reduces the overall accuracy.

(3) Compared with GRU and LSTM, the TCN model adopted in this paper can achieve smaller prediction errors in all cases. This fully proves the practical value and modeling ability of the TCN algorithm in GDP forecasting. The feasible reason is that the TCN algorithm fully combines the characteristics of CNN and RNN. Therefore, TCN improves the parallel computing capability of the model while maintaining the advantages of timing modeling, which further improves the performance of the model.

4.3.2. Experimental Results and Analysis of Different Hybrid Models

In order to fully verify the application value of the GDP prediction model proposed in this paper, two parts of comparative experiments are set up in this section.

Part I: To prove that the three-stage feature selection method adopted in this paper can effectively optimize the prediction performance of the TCN algorithm, the proposed FC-BorutaRF-Q-TCN algorithm is compared with FC-BorutaRF-TCN, FC-Q-TCN, and TCN respectively.

Part II: To fully prove that the feature selection model based on reinforcement learning adopted in this paper has excellent feature selection ability, the Q-learning algorithm is compared with classical GA and PSO.

Table 3. The regression analysis indexes of several predictors.

Series	Forecasting Models	MAE (Billion Yuan)	MAPE (%)	RMSE (Billion Yuan)
Data 1	TCN	25.2155	5.2263	34.0967
	GRU	26.3096	5.4412	34.9834
	LSTM	26.6932	5.4694	34.9907
	RNN	27.1426	5.5086	35.0814
	ELM	27.7391	5.6319	36.0643
	RBF	27.6943	5.6393	35.8998
Data 2	TCN	21.2653	3.9465	28.4004
	GRU	22.3954	4.0541	29.6895
	LSTM	22.5898	4.1824	29.9655
	RNN	23.2437	4.2517	30.4454
	ELM	25.7957	5.1268	33.1268
	RBF	25.3667	5.1499	32.6133
Data 3	TCN	33.9287	3.7871	50.4280
	GRU	34.7236	3.9526	50.6098
	LSTM	34.1358	3.9577	51.6374
	RNN	35.5990	3.9983	52.2030
	ELM	36.0811	4.0071	52.6383
	RBF	35.8042	4.0245	52.8738

Table 4 gives the regression analysis indexes of the prediction results of these algorithms. Tables 5 and 6 show the promoting percentages of FC-Borutarf-Q-TCN by other models. Figure 7 gives the loss of different feature selection algorithms during iteration. Table 7 shows the results of feature selection. From Tables 4–7 and Figure 7, the following conclusions can be drawn:

(1) Compared with the single TCN model, all the hybrid models can achieve better prediction accuracy. The experimental results fully prove the ability of the feature engineering algorithm to optimize the prediction results of the predictor. The possible reason is that the feature engineering algorithm deeply excavates the deep correlation between GDP and other feature historical data and labels from two perspectives and selects the best quality features, which effectively optimizes the modeling ability of TCN.

(2) The prediction results of FC- BorutaRF-Q-TCN are obviously better than those of FC-Q-TCN and FC-BorutaRF-TCN. This fully proves that the three-stage feature selection algorithm adopted in this paper can deeply mine the feature information of original data and achieve better results than the single feature selection algorithm. The feasible reason is that the BorutaRF algorithm and Q-Learning algorithm fully analyze the characteristic information obtained from the original data of the FC method and optimize it from two different perspectives. Therefore, TCN can obtain the best input features and establish the optimal GDP prediction model.

(3) Compared with PSO and GA, the feature selection algorithm based on reinforcement learning adopted in this paper can obtain the best results. This fully proves the ability of the Q-learning algorithm to analyze feature quality and make a selection. The possible reason is that, compared with other heuristic algorithms, the reinforcement learning algorithm improves the intelligence of the model by constantly training agents. Therefore, Q-learning can effectively evaluate the quality of features and select the optimal input for the TCN algorithm.

(4) Based on feature selection results, it can be found that the largest number of retained features are historical GDP data and industrial features. In addition, the data on

educational features and population features are less reserved. It proves that historical GDP data and industrial features play a paramount role in the composition of regional GDP. Based on the feature selection results, a more accurate prediction framework can be constructed in the process of establishing a GDP prediction model in the future.

Table 4. The indexes evaluation results of several forecasting models.

Series	Forecasting Models	MAE (Billion Yuan)	MAPE (%)	RMSE (Billion Yuan)
Data 1	FC-BorutaRF-Q-TCN	22.1614	4.8334	30.3021
	FC-BorutaRF-PSO-TCN	23.0066	4.9145	31.6693
	FC-BorutaRF-GA-TCN	22.9586	4.9517	30.5457
	FC-BorutaRF-TCN	23.6087	5.1034	32.1499
	FC-Q-TCN	24.1884	5.1061	33.7000
Data 2	FC-BorutaRF-Q-TCN	18.0778	3.3787	25.3888
	FC-BorutaRF-PSO-TCN	19.8081	3.5706	26.8640
	FC-BorutaRF-GA-TCN	19.8178	3.5513	26.4723
	FC-BorutaRF-TCN	20.4341	3.8091	27.5156
	FC-Q-TCN	20.8454	3.8084	27.0660
Data 3	FC-BorutaRF-Q-TCN	28.4412	3.2037	37.9007
	FC-BorutaRF-PSO-TCN	30.4008	3.5108	39.4640
	FC-BorutaRF-GA-TCN	30.6192	3.5527	40.7590
	FC-BorutaRF-TCN	33.3320	3.6871	49.9555
	FC-Q-TCN	33.3321	3.7136	49.7054

Table 5. The promoting percentages of FC-BorutaRF-Q-TCN by other models.

Methods	Indexes	Data 1	Data 2	Data 3
FC-BorutaRF-Q-TCN vs. FC-BorutaRF-TCN	P_{MAPE} (%)	6.1304	11.5312	14.6730
	P_{MAE} (%)	5.2906	11.2993	13.1106
	P_{RMSE} (%)	5.7475	7.7294	24.1311
FC-BorutaRF-Q-TCN vs. FC-Q-TCN	P_{MAPE} (%)	8.3800	13.2768	14.6732
	P_{MAE} (%)	5.3407	11.2830	13.7306
	P_{RMSE} (%)	10.0828	6.1967	23.7493
FC-BorutaRF-Q-TCN vs. TCN	P_{MAPE} (%)	12.1120	14.9892	16.1736
	P_{MAE} (%)	7.5177	14.3874	15.4049
	P_{RMSE} (%)	11.1289	10.6041	24.8420

Table 6. The promoting percentages of the Q-learning by heuristic algorithms.

Methods	Indexes	Data 1	Data 2	Data 3
FC-BorutaRF-Q-TCN vs. FC-BorutaRF-PSO-TCN	P_{MAPE} (%)	3.6737	8.7353	6.4459
	P_{MAE} (%)	1.6502	5.3744	8.7473
	P_{RMSE} (%)	4.3171	5.4914	3.9613
FC-BorutaRF-Q-TCN vs. FC-BorutaRF-GA-TCN	P_{MAPE} (%)	3.4723	8.7800	7.1132
	P_{MAE} (%)	2.3891	4.8602	9.8235
	P_{RMSE} (%)	0.7975	4.0930	7.0127

4.4. Contrast Experiment with Existing Algorithms

To prove that the FC-BorutarRF-Q-TCN model proposed in this paper is an advanced GDP forecasting model with excellent research prospects, it is compared with the four existing models. The four existing models include the classical time series model (ARIMA), the traditional machine learning model (SVM), and the two most advanced models (Yan's model [62] and Dong's model [63]). Figures 8–10 show the MAE, MAPE, and RMSE values of the proposed model and those of four existing models. Figures 11–13 show the prediction results of all the comparison models. Based on Figures 8–13, the following conclusions can be drawn:

(1) Compared with the classical ARIMA and SVM algorithms, all the mixed models can achieve more satisfactory prediction results. The experimental results fully prove the practicability and effectiveness of the hybrid model in GDP forecasting. The feasible reason is that the hybrid model effectively optimizes the input features of the GDP forecasting model from the perspective of feature analysis and data mining, which effectively improves the analysis and modeling capabilities of all predictors.
(2) The FC-BorutarRF-Q-TCN model proposed in this paper can achieve the best prediction accuracy in all cases. This fully proves the stability and advance of the FC-BorutarRF-Q-TCN model. First of all, the model outperforms the feature crossing algorithm to mine potential feature information from the original data. Then, the BorutarRF algorithm and the Q-learning algorithm screen the features obtained by the FC algorithm from two different angles. Finally, the screened features are used as the input of TCN to build the GDP prediction model and obtain the final prediction results. Overall, the model further improves the prediction performance from multiple perspectives. Therefore, the FC-BorutarRF-Q-TCN model can achieve excellent research value in the field of GDP prediction.

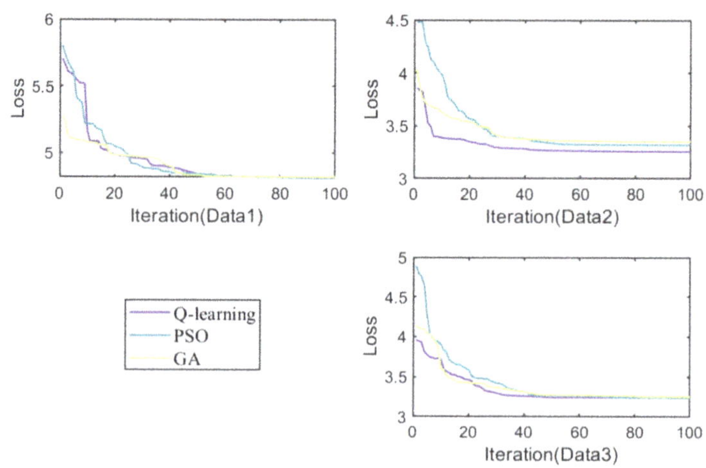

Figure 7. Loss of different feature selection algorithms during iteration.

Table 7. Feature selection results.

Series	Type of Feature	Number of Retained Features
	Historical GDP data	4
	Historical economic indexes	2
Data 1	Employment and population features	1
	Education	1
	Industrial structure features	3
	Historical GDP data	3
	Historical economic indexes	2
Data 2	Employment and population features	0
	Education	1
	Industrial structure features	3
	Historical GDP data	4
	Historical economic indexes	2
Data 3	Employment and population features	1
	Education	0
	Industrial structure features	4

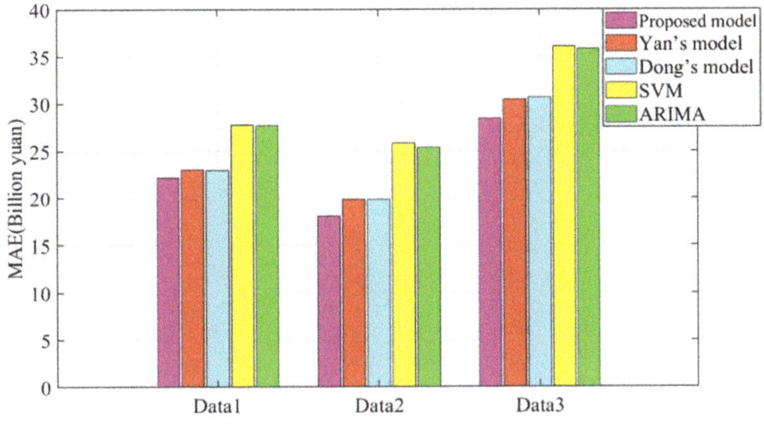

Figure 8. MAE values of the proposed model and existing models.

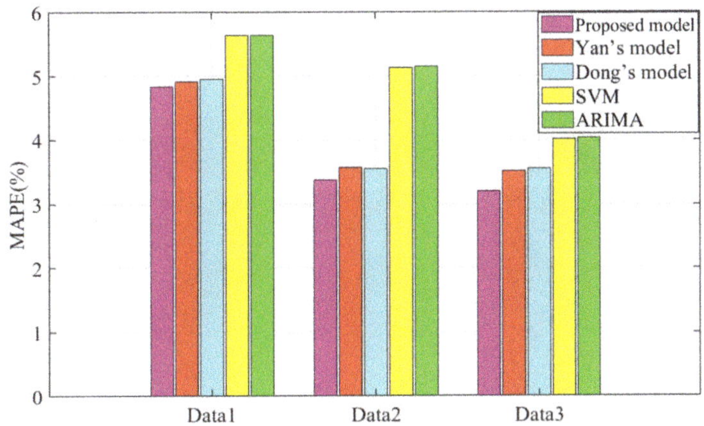

Figure 9. MAPE values of the proposed model and existing models.

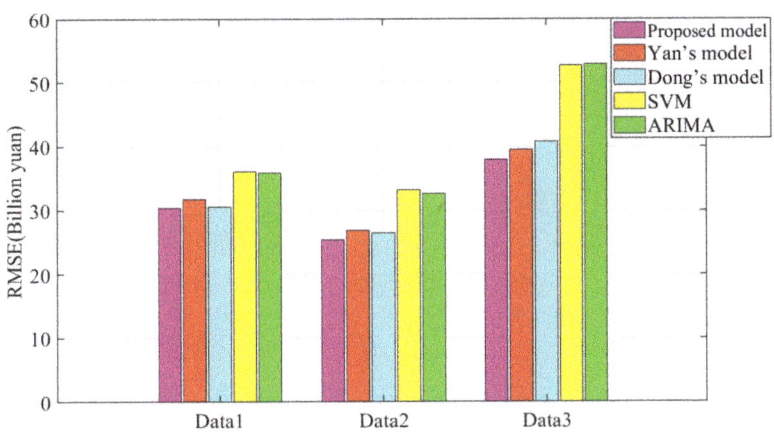

Figure 10. RMSE values of the proposed model and existing models.

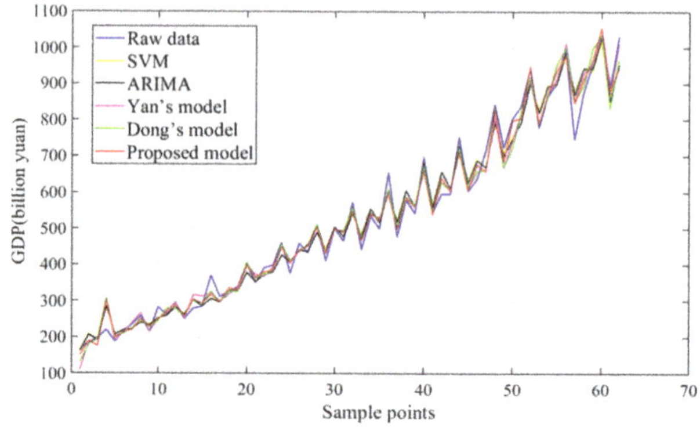

Figure 11. Predicted results of all algorithms (Data 1).

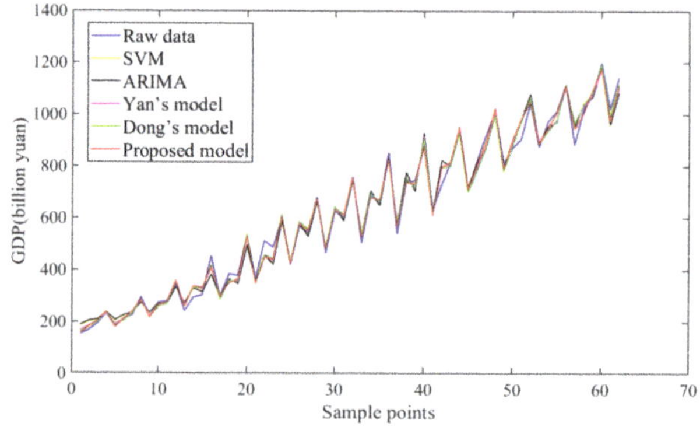

Figure 12. Predicted results of all algorithms (Data 2).

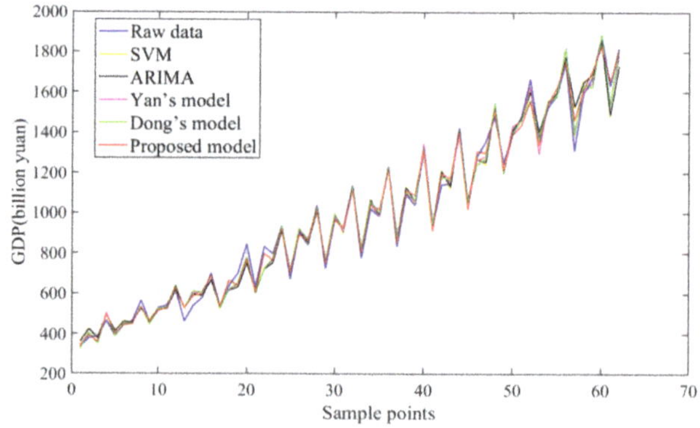

Figure 13. Predicted results of all algorithms (Data 3).

4.5. Discussion

Based on the above analysis of all experimental results, the following discussion and analysis are carried out in this section:

(1) The FC-BorutarRF-Q-TCN model proposed in this paper can achieve the best prediction accuracy in all cases. In addition, the stability and effectiveness of the model are fully proved by the results of ten-fold cross-validation and ten repeated experiments.

(2) Table 3 fully proves the predictive performance of TCN. Compared with other neural networks, TCN effectively combines the parallel computation capability of CNN and the recursive modeling capability of RNN. Therefore, TCN has achieved excellent results in the field of GDP forecasting modeling.
(3) Based on Tables 4–7 and Figure 7, it can be found that the proposed three-stage feature selection framework can effectively improve the prediction performance of TCN. In addition, based on the result of feature selection, it can be seen that historical GDP data and Industrial structure features are relatively crucial indexes affecting GDP prediction. Education and population have relatively little impact on GDP prediction. Therefore, the experimental results have certain help for the future government to formulate policies and promote economic development.
(4) Figures 8–13 fully show the application prospect of the proposed model in GDP prediction. As can be seen from Figures 8–10, the errors of the proposed FC-BorutarRF-Q-TCN model are significantly lower than that of other existing models. In addition, the MAPE values of the proposed model are less than 5% in all cases. In addition, based on Figures 11–13, it can be found that the GDP prediction result of the proposed FC-BorutarRF-Q-TCN model is extremely close to the real GDP data, which can prove the strong practicality of the model in this field.
(5) GDP prediction technology can provide technical support for regional economic development and policymaking. However, GDP prediction technology also has some limitations. GDP is a favorable indicator of physical production, and to some extent ignores the value of open-source products, services, free products, and other related industries. Therefore, GDP and other modern industries should be taken into consideration to comprehensively evaluate the regional economic level and formulate further development policies [64].

5. Conclusions and Future Work

As a comprehensive signal of the future economic situation, GDP forecasting technology provides technical support for national macro-economic regulation. In the paper, a new multi-data-driven GDP prediction model based on three-stage feature selection and a TCN network is proposed. The main contributions of this paper are summarized from the following perspectives:

(1) Different from the traditional single-variable GDP time series forecasting framework, this paper proposes a multi-data-driven GDP forecasting model. The model comprehensively considers the influence of other features on GDP and further optimizes the prediction performance of the model.
(2) Different from the traditional shallow neural network and recursive neural network, the TCN neural network adopted in the paper fully combines the training advantages of CNN and the timing sequence modeling ability of RNN. Therefore, the TCN algorithm could achieve a more excellent GDP prediction effect and is the most important predictor.
(3) A new three-stage feature selection framework is proposed to optimize the prediction performance of TCN. On the one hand, the framework uses the FC algorithm to further mine the potential features of the original data and expand the deep information of the data. On the other hand, Q-learning and BorutarRF algorithms screen features from different angles and ensure the quality of TCN inputs. The three-stage feature Selection framework improves TCN performance by more than 10%.
(4) The feature selection results show that historical GDP data and Industrial structure features are relatively crucial feature data for GDP prediction modeling. The feature selection results have important reference values for the government to adjust economic policy.
(5) In order to prove the advance and practicability of the proposed FC-BorutarRF-Q-TCN prediction framework, fourteen models used by other researchers were replicated and compared with the model proposed in this paper. The experimental results show

that the FC-BorutarRF-Q-TCN model is a GDP forecasting framework with excellent research prospects. The MAPE values are all less than 5%.

The proposed multi-factor data-driven GDP prediction framework provides a meaningful reference for regional economic development strategy. In the future, the model proposed can be further improved from the following perspectives to enhance its practical value:

(1) The GDP prediction framework is mainly obtained through multi-factor data-driven training. Therefore, when the amount of data increases and updates, the model also needs to be updated and trained constantly to ensure timeliness.
(2) The model can accurately predict the change in GDP in the next quarter. Based on the forecast results, the government makes relevant economic policies to realize the adjustment and development of the regional economic level. In the future, it is a very important step to formulate a reasonable sustainable development strategy based on the GDP prediction results.
(3) GDP prediction technology provides effective technical guidance for the sustainable development of the regional economy. Therefore, based on the research results, it can effectively drive the upgrading of regional industries and promote green development and sustainability in the future.
(4) GDP can effectively reflect the situation of physical production and regional economic development. However, GDP does not fully analyze related industries such as services, open-source products, and products provided by society for free. Therefore, comprehensive consideration of GDP and other industries is very indispensable for the sustainable development of the regional economy and prosperity level.

Author Contributions: Conceptualization, G.Y. and C.Y.; methodology, C.Y.; software, G.Y. and C.Y.; validation, C.Y., G.Y. and Q.L.; formal analysis, Q.L.; investigation, Q.L.; resources, Q.L.; data curation, Q.L. and C.Y.; writing—original draft preparation, Q.L.; writing—review and editing, Q.L., G.Y. and C.Y.; visualization, Q.L.; supervision, G.Y.; project administration, G.Y.; funding acquisition, Q.L. All authors have read and agreed to the published version of the manuscript.

Funding: This research received no external funding.

Institutional Review Board Statement: Not applicable.

Informed Consent Statement: Not applicable.

Data Availability Statement: Not applicable.

Acknowledgments: This study is fully supported by the Natural Science Foundation of Hunan Province (Grant No. 2021JJ60003).

Conflicts of Interest: The authors declare no conflict of interest.

References

1. Pavía-Miralles, J.M.; Cabrer-Borrás, B. On estimating contemporaneous quarterly regional GDP. *J. Forecast.* **2007**, *26*, 155–170. [CrossRef]
2. Jurun, E.; Pivac, S. Comparative regional GDP analysis: Case study of Croatia. *Cent. Eur. J. Op. Res.* **2011**, *19*, 319–335. [CrossRef]
3. Landefeld, J.S.; Seskin, E.P.; Fraumeni, B.M. Taking the pulse of the economy: Measuring GDP. *J. Econ. Perspect.* **2008**, *22*, 193–216. [CrossRef]
4. Chen, X.; Liu, C.; Yu, X. Urbanization, Economic Development, and Ecological Environment: Evidence from Provincial Panel Data in China. *Sustainability* **2022**, *14*, 1124. [CrossRef]
5. König, M.; Winkler, A. COVID-19: Lockdowns, fatality rates and GDP growth. *Intereconomics* **2021**, *56*, 32–39. [CrossRef]
6. Abbass, K.; Begum, H.; Alam, A.; Awang, A.H.; Abdelsalam, M.K.; Egdair, I.M.M.; Wahid, R. Fresh Insight through a Keynesian Theory Approach to Investigate the Economic Impact of the COVID-19 Pandemic in Pakistan. *Sustainability* **2022**, *14*, 1054. [CrossRef]
7. Kokolakakis, T.; Lera-Lopez, F.; Ramchandani, G. Measuring the Economic Impact of COVID-19 on the UK's Leisure and Sport during the 2020 Lockdown. *Sustainability* **2021**, *13*, 13865. [CrossRef]
8. Liu, X.; Liu, Y.; Yan, Y. China macroeconomic report 2020: China's macroeconomy is on the rebound under the impact of COVID-19. *Econ. Political Stud.* **2020**, *8*, 395–435. [CrossRef]

9. Yang, W.; Wang, X.; Zhang, K.; Ke, Z. COVID-19, Urbanization Pattern and Economic Recovery: An Analysis of Hubei, China. *Int. J. Environ. Res. Public Health* **2020**, *17*, 9577. [CrossRef]
10. Zheng, G.; Zhu, S. Research on the Effectiveness of China's Macro Control Policy on Output and Technological Progress under Economic Policy Uncertainty. *Sustainability* **2021**, *13*, 6844. [CrossRef]
11. Goolsbee, A.; Syverson, C. Fear, lockdown, and diversion: Comparing drivers of pandemic economic decline 2020. *J. Public Econ.* **2021**, *193*, 104311. [CrossRef] [PubMed]
12. Huynh, T.L. The COVID-19 risk perception: A survey on socioeconomics and media attention. *Econ. Bull.* **2020**, *40*, 758–764.
13. Gan, Z.; Li, C.; Zhou, J.; Tang, G. Temporal convolutional networks interval prediction model for wind speed forecasting. *Electr. Power Syst. Res.* **2021**, *191*, 106865. [CrossRef]
14. Sun, Q.; Yang, Z.; Chen, X.; Yu, C. Optical Performance monitoring using Q-learning optimized least square support vector machine in optical network. In Proceedings of the 2021 IEEE 6th International Conference on Signal and Image Processing (ICSIP), Nanjing, China, 22–24 October 2021; pp. 954–958.
15. Suganthi, L.; Samuel, A.A. Energy models for demand forecasting—A review. *Renew. Sustain. Energy Rev.* **2012**, *16*, 1223–1240. [CrossRef]
16. Kalantaripor, M.; Najafi Alamdarlo, H. Spatial Effects of Energy Consumption and Green GDP in Regional Agreements. *Sustainability* **2021**, *13*, 10078. [CrossRef]
17. Gómez, M.; Rodríguez, J.C. Energy consumption and financial development in NAFTA countries, 1971–2015. *Appl. Sci.* **2019**, *9*, 302. [CrossRef]
18. Bjørnland, H.C.; Ravazzolo, F.; Thorsrud, L.A. Forecasting GDP with global components: This time is different. *Int. J. Forecast.* **2017**, *33*, 153–173. [CrossRef]
19. Sezer, O.B.; Gudelek, M.U.; Ozbayoglu, A.M. Financial time series forecasting with deep learning: A systematic literature review: 2005–2019. *Appl. Soft Comput.* **2020**, *90*, 106181. [CrossRef]
20. Sadik-Zada, E.R.; Loewenstein, W. Drivers of CO_2-Emissions in Fossil Fuel Abundant Settings: (Pooled) Mean Group and Nonparametric Panel Analyses. *Energies* **2020**, *13*, 3956. [CrossRef]
21. Dai, S.; Niu, D.; Li, Y. Forecasting of energy consumption in China based on ensemble empirical mode decomposition and least squares support vector machine optimized by improved shuffled frog leaping algorithm. *Appl. Sci.* **2018**, *8*, 678. [CrossRef]
22. Abonazel, M.R.; Abd-Elftah, A.I. Forecasting Egyptian GDP using ARIMA models. *Rep. Econ. Financ.* **2019**, *5*, 35–47. [CrossRef]
23. Wu, C.; Chen, P. Application of support vector machines in debt to GDP ratio forecasting. In Proceedings of the 2006 International Conference on Machine Learning and Cybernetics, Dalian, China, 13–16 August 2006; pp. 3412–3415.
24. Ghanem, S.; Fandi, G.; Krepl, V.; Husein, T.; Rzek, O.; Muller, Z.; Kyncl, J.; Tlustý, J.; Smutka, L. The Impact of COVID-19 on Electricity Prices in Italy, the Czech Republic, and China. *Appl. Sci.* **2021**, *11*, 8793. [CrossRef]
25. Sadik-Zada, E.R.; Niklas, B. Business Cycles and Alcohol Consumption: Evidence from a Nonlinear Panel ARDL Approach. *J. Wine Econ.* **2021**, *16*, 429–438. [CrossRef]
26. Niklas, B.; Sadik-Zada, E.R. Income Inequality and Status Symbols: The Case of Fine Wine Imports. *J. Wine Econ.* **2019**, *14*, 365–373. [CrossRef]
27. Yusof, Y.; Kamaruddin, S.S.; Husni, H.; Ku-Mahamud, K.R.; Mustaffa, Z. Forecasting model based on LSSVM and ABC for natural resource commodity. *Int. J. Comput. Theory Eng.* **2013**, *5*, 906. [CrossRef]
28. Long, G. GDP prediction by support vector machine trained with genetic algorithm. In Proceedings of the 2010 2nd International Conference on Signal Processing Systems, Dalian, China, 5–7 July 2010; pp. V3-1–V3-3.
29. Guleryuz, D. Determination of industrial energy demand in Turkey using MLR, ANFIS and PSO-ANFIS. *J. Artif. Intell. Syst.* **2021**, *3*, 16–34.
30. Tuo, S.; Chen, T.; He, H.; Feng, Z.; Zhu, Y.; Liu, F.; Li, C. A Regional Industrial Economic Forecasting Model Based on a Deep Convolutional Neural Network and Big Data. *Sustainability* **2021**, *13*, 12789. [CrossRef]
31. Sa'adah, S.; Wibowo, M.S. Prediction of Gross Domestic Product (GDP) in Indonesia Using Deep Learning Algorithm. In Proceedings of the 2020 3rd International Seminar on Research of Information Technology and Intelligent Systems (ISRITI), Yogyakarta, Indonesia, 10–11 December 2020; pp. 32–36.
32. Liu, B.; Fu, C.; Bielefield, A.; Liu, Y.Q. Forecasting of Chinese primary energy consumption in 2021 with GRU artificial neural network. *Energies* **2017**, *10*, 1453. [CrossRef]
33. Wang, H.; Zhao, Y.; Tan, S. Short-term load forecasting of power system based on time convolutional network. In Proceedings of the 2019 8th International Symposium on Next Generation Electronics (ISNE), Zhengzhou, China, 9–10 October 2019; pp. 1–3.
34. Sun, Y.; Hong, Y.; Wang, S. Out-of-sample forecasts of China's economic growth and inflation using rolling weighted least squares. *J. Manag. Sci. Eng.* **2019**, *4*, 1–11. [CrossRef]
35. Yoon, J. Forecasting of real GDP growth using machine learning models: Gradient boosting and random forest approach. *Comput. Econ.* **2021**, *57*, 247–265. [CrossRef]
36. Bai, X.; Zhang, F.; Hou, J.; Xia, F.; Tolba, A.; Elashkar, E. Implicit multi-feature learning for dynamic time series prediction of the impact of institutions. *IEEE Access* **2017**, *5*, 16372–16382. [CrossRef]

37. Ortega-Bastida, J.; Gallego, A.-J.; Rico-Juan, J.R.; Albarrán, P. Regional gross domestic product prediction using twitter deep learning representations. In Proceedings of the IADIS International Conference Applied Computing, Bangkok, Thailand, 4–5 February 2020; pp. 89–96.
38. Nahil, A.; Lyhyaoui, A. Short-term stock price forecasting using kernel principal component analysis and support vector machines: The case of Casablanca stock exchange. *Procedia Comput. Sci.* **2018**, *127*, 161–169. [CrossRef]
39. Wang, J.; Li, Y. Multi-step ahead wind speed prediction based on optimal feature extraction, long short term memory neural network and error correction strategy. *Appl. Energy* **2018**, *230*, 429–443. [CrossRef]
40. Kosana, V.; Teeparthi, K.; Madasthu, S.; Kumar, S. A novel reinforced online model selection using Q-learning technique for wind speed prediction. *Sustain. Energy Technol. Assess.* **2022**, *49*, 101780. [CrossRef]
41. Xu, R.; Li, M.; Yang, Z.; Yang, L.; Qiao, K.; Shang, Z. Dynamic feature selection algorithm based on Q-learning mechanism. *Appl. Intell.* **2021**, *51*, 7233–7244. [CrossRef]
42. Maeda-Gutiérrez, V.; Galván-Tejada, C.E.; Cruz, M.; Valladares-Salgado, A.; Galván-Tejada, J.I.; Gamboa-Rosales, H.; García-Hernández, A.; Luna-García, H.; Gonzalez-Curiel, I.; Martínez-Acuña, M. Distal Symmetric Polyneuropathy Identification in Type 2 Diabetes Subjects: A Random Forest Approach. *Healthcare* **2021**, *9*, 138. [CrossRef] [PubMed]
43. Ortega-Bastida, J.; Gallego, A.J.; Rico-Juan, J.R.; Albarrán, P. A multimodal approach for regional GDP prediction using social media activity and historical information. *Appl. Soft Comput.* **2021**, *111*, 107693. [CrossRef]
44. Chiang, Y.H.; Tao, L.; Wong, F.K.W. Causal relationship between construction activities, employment and GDP: The case of Hong Kong. *Habitat Int.* **2015**, *46*, 1–12. [CrossRef]
45. Ifa, A.; Guetat, I. Does public expenditure on education promote Tunisian and Moroccan GDP per capita? ARDL approach. *J. Financ. Data Sci.* **2018**, *4*, 234–246. [CrossRef]
46. Wiemer, C.; Tian, X. The measurement of small-scale industry for China's GDP accounts. *China Econ. Rev.* **2001**, *12*, 317–322. [CrossRef]
47. Liang, J.; Hou, L.; Luan, Z.; Huang, W. Feature Selection with Conditional Mutual Information Considering Feature Interaction. *Symmetry* **2019**, *11*, 858. [CrossRef]
48. Aghaeepour, N.; Finak, G.; Hoos, H.; Mosmann, T.R.; Brinkman, R.; Gottardo, R.; Scheuermann, R.H.; Flow, C.A.P.C.; Consortium, D. Critical assessment of automated flow cytometry data analysis techniques. *Nat. Methods* **2013**, *10*, 228–238. [CrossRef] [PubMed]
49. Gholami, H.; Mohammadifar, A.; Golzari, S.; Kaskaoutis, D.G.; Collins, A.L. Using the Boruta algorithm and deep learning models for mapping land susceptibility to atmospheric dust emissions in Iran. *Aeolian Res.* **2021**, *50*, 100682. [CrossRef]
50. Ahmed, A.A.M.; Deo, R.C.; Feng, Q.; Ghahramani, A.; Raj, N.; Yin, Z.; Yang, L. Deep learning hybrid model with Boruta-Random forest optimiser algorithm for streamflow forecasting with climate mode indices, rainfall, and periodicity. *J. Hydrol.* **2021**, *599*, 126350. [CrossRef]
51. Park, J.; Kim, T.; Seong, S.; Koo, S. Control automation in the heat-up mode of a nuclear power plant using reinforcement learning. *Prog. Nucl. Energy* **2022**, *145*, 104107. [CrossRef]
52. Qi, C.; Zhu, Y.; Song, C.; Cao, J.; Xiao, F.; Zhang, X.; Xu, Z.; Song, S. Self-supervised reinforcement learning-based energy management for a hybrid electric vehicle. *J. Power Sources* **2021**, *514*, 230584. [CrossRef]
53. Subramanian, A.; Chitlangia, S.; Baths, V. Reinforcement learning and its connections with neuroscience and psychology. *Neural Netw.* **2022**, *145*, 271–287. [CrossRef] [PubMed]
54. Huynh, T.N.; Do, D.T.T.; Lee, J. Q-Learning-based parameter control in differential evolution for structural optimization. *Appl. Soft Comput.* **2021**, *107*, 107464. [CrossRef]
55. Schreiber, J.A.; Müller, S.L.; Westphälinger, S.E.; Schepmann, D.; Strutz-Seebohm, N.; Seebohm, G.; Wünsch, B. Systematic variation of the benzoylhydrazine moiety of the GluN2A selective NMDA receptor antagonist TCN-201. *Eur. J. Med. Chem.* **2018**, *158*, 259–269. [CrossRef]
56. Howell, B.A.; Puglisi, L.; Clark, K.; Albizu-Garcia, C.; Ashkin, E.; Booth, T.; Brinkley-Rubinstein, L.; Fiellin, D.A.; Fox, A.D.; Maurer, K.F.; et al. The Transitions Clinic Network: Post Incarceration Addiction Treatment, Healthcare, and Social Support (TCN-PATHS): A hybrid type-1 effectiveness trial of enhanced primary care to improve opioid use disorder treatment outcomes following release from jail. *J. Subst. Abus. Treat.* **2021**, *128*, 108315. [CrossRef]
57. Samal, K.K.R.; Panda, A.K.; Babu, K.S.; Das, S.K. Multi-output TCN autoencoder for long-term pollution forecasting for multiple sites. *Urban Clim.* **2021**, *39*, 100943. [CrossRef]
58. Ma, Q.; Wang, H.; Luo, P.; Peng, Y.; Li, Q. Ultra-short-term Railway traction load prediction based on DWT-TCN-PSO_SVR combined model. *Int. J. Electr. Power Energy Syst.* **2022**, *135*, 107595. [CrossRef]
59. Wu, X.; Zhang, Z.; Chang, H.; Huang, Q. A data-driven gross domestic product forecasting model based on multi-indicator assessment. *IEEE Access* **2021**, *9*, 99495–99503. [CrossRef]
60. Yan, G.; Yu, C.; Bai, Y. Wind Turbine Bearing Temperature Forecasting Using a New Data-Driven Ensemble Approach. *Machines* **2021**, *9*, 248. [CrossRef]
61. Liu, X.; Qin, M.; He, Y.; Mi, X.; Yu, C. A new multi-data-driven spatiotemporal PM2.5 forecasting model based on an ensemble graph reinforcement learning convolutional network. *Atmos. Pollut. Res.* **2021**, *12*, 101197. [CrossRef]

62. Yan, G.; Yu, C.; Bai, Y. A New Hybrid Ensemble Deep Learning Model for Train Axle Temperature Short Term Forecasting. *Machines* **2021**, *9*, 312. [CrossRef]
63. Dong, S.; Yu, C.; Yan, G.; Zhu, J.; Hu, H. A Novel Ensemble Reinforcement Learning Gated Recursive Network for Traffic Speed Forecasting. In Proceedings of the 2021 Workshop on Algorithm and Big Data, Fuzhou, China, 12–14 March 2021; pp. 55–60.
64. Jean-Paul, F.; Martine, D. *Beyond GDP Measuring What Counts for Economic and Social Performance: Measuring What Counts for Economic and Social Performance*; OECD Publishing: Paris, France, 2018.

sustainability

Article

Toward Sustainability: Dynamics of Total Carbon Dioxide Emissions, Aggregate Income, Non-Renewable Energy, and Renewable Power

Wan-Jiun Chen

Department of Economics, Chinese Culture University, No. 55, Hwa-Kang Rd., Taipei 11114, Taiwan; cwj@ulive.pccu.edu.tw; Tel.: +886-2861-0511

Abstract: The purpose of energy sustainability policy is to support both economic growth and environmental quality. With climate change accelerating, economies must reduce carbon emissions. Low-carbon economics can balance the oft-contradictory policy aims of income growth and carbon reduction. Carbon pricing and renewable substitutes can pave the way. This analysis probes the dynamics of the adjustments toward the ideals of low-carbon economics through Granger causality testing of total carbon emissions, income, nonrenewable energy consumption, and renewable power. Cointegration regressions and a panel data vector error correction model are used to demonstrate the aforementioned variables' long-term balance and short-term adjustment, respectively. Two panels of countries, namely 18 European Union and 32 Organization of Economic Co-operation and Development countries, are investigated with 1990–2021 data. Determinants for the success of low-carbon development and the implications of border regulations and taxation of carbon footprint are also discussed. Economic competitiveness, as well as increases in commodity prices, would initially emerge as interferences and then induce carbon reduction and accelerate the adoption and development of green technology.

Keywords: innovations; renewable power; rebound effect; low-carbon economy; carbon pricing policy; border carbon tax

Citation: Chen, W.-J. Toward Sustainability: Dynamics of Total Carbon Dioxide Emissions, Aggregate Income, Non-Renewable Energy, and Renewable Power. *Sustainability* **2022**, *14*, 2712. https://doi.org/10.3390/su14052712

Academic Editors: Elkhan Richard Sadik-Zada and Andrea Gatto

Received: 19 January 2022
Accepted: 22 February 2022
Published: 25 February 2022

1. Introduction

Since the industrial revolution, increasing amounts of carbon-based fuels have been used, with the combustion of fossil fuels first powering factories and later providing energy for electric power plants, transportation, and other uses. Carbon dioxide emissions naturally accompany the burning of these fuels. The carbon dioxide concentrates in the atmosphere and traps heat, causing climate change. This has become problematic for worldwide economic growth, as income and carbon emissions are positively correlated. Under the global warming emergency, strategies are required to mitigate climate change.

Nobel Laureate William Nordhaus [1] argued that carbon pricing and green technology progress are among the most effective approaches to reducing carbon emissions; a low-carbon economy is the goal for contemporary economic development. The contemporary economic reality is characterized by fast growth in renewable energy and renewable power, and carbon pricing policy variations locally within nations and expanding globally through border carbon price adjustment. Politics is focused on pursuing a low-carbon economy through energy transition [2]. Total carbon emissions contribute directly to the concentrations of atmospheric carbon dioxide and to the effects of climate change. Total carbon dioxide emissions is one of the key elements in climate change mitigation.

Sustainability of energy supports income and environmental quality. Energy securities and resilience are main governmental policies. Per capita income is widely used as a proxy to reflect the average standard of living in an economy. Researchers have affirmed the environmental Kuznets curve hypothesis in plotting per capita carbon emissions against per capita income [3,4]. The slope of the environmental Kuznets curve shares the same definition of the carbon intensity, namely the ratio of carbon emissions to income. The decreasing slope, as well as declining carbon intensity, has exhibited decreasing amounts of carbon emissions associated with one unit of income since the beginning of industrialization [3,4]. That the technology progress since then is a driving force to this declination verifies the insights of Nordhaus' argument that green technology progress is among one of the most effective approaches to reducing carbon emissions [1].

However, per capita carbon emissions cannot directly reflect the contribution of emissions to atmospheric concentrations. Actual observations indicate that the total amount of carbon emissions is rapidly increasing, and that induced effects of climate change from the total amount of carbon emissions are extremely severe. To reflect the pursuit of sustainability along the ongoing energy transition, the present study investigated the dynamics of total carbon dioxide emissions and its nexus with aggregate income, nonrenewable energy consumption, and renewable power.

The benign approach of carbon pricing policy, green technology development, and low-carbon economic development might mitigate climate change. To elucidate the mechanisms of this approach, the nexus and dynamics of key aggregate variables are investigated in this study. The variables included in the investigation are total carbon emissions, aggregate income, nonrenewable energy consumption, and renewable power. Their long-run relationships and short-run adjustment mechanisms are investigated by estimating cointegration equations and with a panel vector error correction model (PVECM).

Two carbon pricing measures are typically applied to reduce carbon emissions: carbon taxation, and emissions trading schemes (ETSs). A carbon tax is a tax on carbon emissions. With a tax, emitting carbon is no longer free, and it incentivizes firms to reduce emissions. The second policy is an ETS. With this policy, a country sets a maximum cap for carbon emissions, issues permits that match the cap, and restricts firms to trading emissions on a market. The permits are tradable, and the price of permits is set to reduce emissions to the cap level. In economic theory, these pricing systems are equivalent, although in their empirical implementations, a carbon tax is regarded as a stringent punishment of firms and emitters, whereas an ETS trading system allows emission flexibility.

After the climate issue drew attention and became recognized, the first international mitigation treaty, the Kyoto protocol, was adopted in 1997. Since then, many countries have adopted pricing measures, either carbon taxes or ETSs, in addition to developing environmental regulations through command-and-control policy. The commitment to and implementation of carbon taxes and ETSs can be observed at various spatial scales, such as cities, states, countries, and regions, throughout the world.

Asen [5] published a summary report on the European Union (EU) carbon taxation and the EU ETS. The information herein is based on Asen's report. Europe has the world's highest carbon tax coverage and its largest trading system [5]. Most European countries that levy a carbon tax are also part of the EU ETS. Finland was the first country worldwide to introduce a carbon tax, in 1990. Most European countries have followed suit and implemented carbon taxes, ranging from US$0.08 (Poland) to US$137 (Sweden) (USD 1 = EUR 0.84913 on 1 April 2021). All member states of the EU are part of the EU ETS. With the exception of Switzerland, Ukraine, and the United Kingdom (UK), European countries that levy a carbon tax are also part of the EU ETS (Switzerland has its own emissions trading system, which has been tied to the EU ETS since January 2020. Following Brexit, the UK implemented its own UK ETS as of January 2021).

Additionally, the EU Taxation and Customs Union announced the proposal of the world's first border carbon tax in 2021 [6]. An announced European Green Deal goal was to become "climate neutral" by 2050. International trade closely links the global economy. Hundreds of cities and private companies have already pledged to reach "net zero"—removing carbon emissions, thus following the EU's long-term strategy of achieving carbon neutrality by 2050. The decrease of carbon emissions outside EU countries is, hopefully, to be accelerated under the incentives of a border carbon tax [7].

In addition to carbon pricing policies, a strategy that Nordhaus [1] proposed is the development of green technology. Novel low-carbon, renewable-energy technologies, such as hydrological, wind, and solar power, are currently being implemented. They have been introduced on a relatively small scale but are flourishing in most developed and some developing countries. These technologies are part of a long-term plan for low-carbon energy development. Renewable energy is bounded by a series of challenges to address low generation capacity, unreliability, inefficiency, enormous capital requirements, large spaces for installation, expensive storage costs, lack of commercial viability, and pollution generation. Research into and applications of renewable-energy technology will take time.

Renewable energy can be used to ensure local energy abundance. However, the location of renewable energy generation is highly geographically restricted. Fortunately, renewable energy in the form of power can be connected with a power grid and transported without geographical restriction. Many countries aim to develop a power sector based largely on the renewable sources provided by European Commission [8], and the number of patents involving green energy is increasing. Carbon pricing policy and the promotion of renewable power as incentives and substitutes are applied to pave the way toward low-carbon economic patterns. Two questions present themselves here: whether renewable power is the ideal potential substitute for nonrenewable energy, and whether the availability of substitutes is the key determinant of the success of an extensive and intensive carbon pricing policy.

The objectives of this investigation are to demonstrate the long-run balance relationship and short-term adjustment patterns for the dynamic connections among (1) total carbon dioxide emissions, (2) aggregate income, (3) nonrenewable energy as represented by the sum of coal, gas, and oil consumption, and (4) renewable power. Policy implications for carbon and border taxes are addressed. The article aimed to investigate two groups of countries to obtain common empirical evidence. A panel of 18 countries in the EU (EU-18) and a panel of 32 countries in the Organization for Economic Co-operation and Development (OECD-32) were investigated over 1990–2021. The countries of EU-18 are Austria, Belgium, Czech Republic, Demark, Finland, France, Germany, Greece, Hungary, Ireland, Italy, Luxembourg, Netherlands, Poland, Portugal, Spain, Sweden, and the United Kingdom. The United Kingdom ended its membership on 31 January 2020. The countries of OECD-32 are Australia, Austria, Belgium, Canada, Chile, Czech Republic, Denmark, Estonia, Finland, France, Germany, Greece, Hungary, Iceland, Ireland, Israel, Italy, Japan, Luxembourg, Mexico, Netherlands, New Zealand, Norway, Poland, Portugal, South Korea, Spain, Sweden, Switzerland, Turkey, the US, and the United Kingdom. Various different new technologies have been initiated and adopted in different countries. Renewable power is of high heterogeneity between countries. Renewable energy has gradually become an effective source of electricity but remaining a small proportion. The EU countries are the most devoted countries to mitigating climate change. The dynamics of the selected variables in EU countries are of importance for climate mitigation policy. The wealthy OECD economies were also investigated for their higher propensity to develop advanced green technology for renewable power. Some countries overlap in the two groups. The abundance and availability of renewable power data in these two groups are relative high compared to other groups of less developed countries. Since common reliable data of renewable power in the data sources are highly available after 1990, the study year starts from 1990.

The goal of this study is to investigate linkages between total carbon emissions, income, nonrenewable energy consumption (coal, gas, and oil consumption), and renewable power in EU-18 and OECD-32 countries through cointegration regressions and Granger causality based on a PVECM. Studies employing a PVECM [9–12] have focused more on the relationship between renewable energies and income and have typically included the income nexus and ignored the carbon emissions nexus. The goal of carbon pricing and low-carbon development is to mitigate climate change without hindering income. Hence, in the present study, the nexus with carbon dioxide, as well as that with aggregate income, are investigated.

Most renewable energies provide energy for local use. However, the variable of renewable power is not adequately addressed in the literature [9–12]. In the context of severe climate change effects, adjustments in carbon emissions should be investigated; thus, renewable power is included in this study as a variable representing progress in energy efficiency.

The study also addresses questions regarding global carbon pricing policy, as well as the dynamics of adopting and developing energy-efficient technology, taking renewable power as an example. The contemporaneous dynamic responses of the variables in reacting to changes should be investigated, as should how a short-run adjustment moves back toward the long-run cointegration equilibrium and whether deviations disturb balanced long-run patterns.

This study also addresses the implication of rebound effects regarding energy efficiency increases and the carbon leakages of an economy through international trading. Technological progress should improve energy efficiency. However, empirical research, such as Greening et al. [13], has reported an increase in energy consumption associated with improvements in coal-firing efficiency. With rebound effects, a reduction in consumption cannot be expected when relevant technology progresses. Improved energy efficiency cannot reduce energy consumption if the rebound effect occurs. Questions thus surround the reasonability of expecting climate change mitigation with a reliance on technological progress.

Finally, carbon leakage frequently occurs in developed countries, especially in the aspect of consumption-based emissions [14]. These countries import and consume commodities produced (with carbon emitted) elsewhere, especially in developing or industrializing countries [14]. How carbon border adjustment affects international competition and carbon leakage merits investigation.

2. Literature Review

The energy driving the engine of growth is a vital component in economic development. Issues of energy security [15–18] and the resilience to energy vulnerability [19–21] are usually targets of concentrated policy efforts of a country, being undertaken to ensure economic sustainability. Both non-renewable and renewable energy sources are subjects of prudent policy and intensive research to strengthen the path of energy security to support the economy. However, energy services are essential for sustainable development as well. Three pillars are the foci of energy policy for sustainability: the economy, society, and the environment. Energy generation and use are strongly linked to all elements of sustainable development, including the sustainability of the economy, living standards, and the environment [22–26].

In fact, energy is one of the most important issues for human development. To review human history, human development closely relies on energy utilization and availability [24]. Over the energy transition history [27], the early use of biofuels and animal power that improved lives transformed to the present world with use of fossil fuels, electricity, and cleaner fuels. That large amounts of fossil fuel combustion produce carbon dioxide emissions has drawn much attention and concerns about the effects on global climate. International agreements, domestic policy, and local actions have been committed to reducing these emissions. Potential solutions to current environmental problems have been identified, along with the development of renewable energy technologies. Innovations promote energy transitions to the exploitation and promotion of renewable energy resources [28]. Kaygusuz [22,24,26] argued that innovations in renewable energy use may alleviate the growing concerns over energy security and climate change. These innovations are regarded as solutions to the challenges of global warming in the argument of Dincer [23]. Climate policies anticipate a transition to renewable energy, potentially achieving solutions to environmental problems that we face today. Green transitions to renewable sources for the purpose of reducing total greenhouse gas emissions requires long-term potential actions for the path of sustainable development [23]. In this regard, renewable energy and sustainable development are intimately connected, and the innovations of renewable energy resources appear to be some of the efficient and effective solutions [23].

Green innovations transform energy use in the energy transition [29]. Not only can renewable energy provide a clean, flexible power source at the micro-to-medium scales [24], but innovative technology progress has allowed electricity generated from renewable sources, and it has huge potential to create a multitude of meaningful uses. Rather than renewable energy, the present paper investigates the dynamic effects of electricity generated through the renewable sources. It is believed that the conclusions and recommendations drawn in the present study are useful for energy scientists and engineers and policy makers.

The energy transition patterns from wood to fossil fuels, and then to renewable energy are widely addressed in the literature [30,31]. Facing global climate change and scarce petroleum supplies, the world must switch to sustainable energy systems for the purpose of energy security [15]. Under threat of serious climate change crises over the world, a rapid renewable energy transition to renewable energy and renewable power provides an opportunity and is most often met with calls for innovation. Leach [31] suggested that human beings need a sustainable energy transition to avoid, first, a major crisis in energy supplies and, secondly, climate change with catastrophic consequences. The paths should be taken to ensure success in meeting these two goals of energy transition [31]. Hence, important studies are on the nexus and the dynamics of energy uses, along the energy transition, with income and carbon dioxide emissions. These issues are the aims that the present research is devoted to.

Sustainability issues of energy transition to support income and to mitigate climate change pose extraordinary challenges and opportunities for societies. With climate change intensifying, scientists in the United Nations Environment Programme (UNEP) [32] warned in the 2021 report that humanity is running out of time to limit global warming. With spatial and temporal externalities, the problems related to climate change have a global dimension, confront present and future generations, and are very difficult to solve [33]. The solution may require fundamental changes in consumption practices, lifestyles, technologies, infrastructures, business models, and policies. Energy transition from fossil fuels to renewables and to renewable power is in the focus of the sustainable transition. The present analysis makes contribution in the energy transition path towards sustainability.

Many energy transition studies focus on technical innovations [34], and many social economic studies on energy transition are on issues of income support and energy vulnerability, security, poverty, and justice [35–40]. The adaptation and initiations of renewable energy are often addressed in the aforementioned literature. Recent renewable energy developed in the form of renewable power that relies on much advanced feed-in grids is an example of the current technology innovation.

The motivation for this research is the currently serious crisis of climate change, renewable power as energy innovation technology, and ongoing policy formulation of carbon tax border adjustment for international commodity trade, as it is proposed as an incentive to mitigate climate change globally. There is a large body of research on the nexus of income, energy, and carbon dioxide emissions. The dynamics and nexus of economic growth, energy, renewable energy, carbon dioxide emissions, and carbon intensity have been examined in a variety of contexts. Carbon dioxide emissions have been studied in the scope of country panel groups, individual countries, and economic sectors, with a huge variety of empirical analysis techniques. Narayan and Popp [41] studied the nexus of income and energy consumption; Bilan et al. [42], Dogru et al. [43], and Balcilar et al. [44] studied the nexus of economic growth, renewable energy consumption, and carbon dioxide emissions. Dogru et al. [43] studied the OECD; Balcilar et al. [44] studied the G7 countries; Torvanger [45] studied the manufacturing sector of the OECD; Dogru et al. [43] studied the tourism sectors. These studies were on different economic scales, utilized a huge variation of different empirical analysis techniques, and reported a great variation of findings, accordingly.

Moreover, Cary [46] studied how tariffs impact carbon intensities, carbon dioxide emissions, and the environment under the US–China economic confrontation. Cary [46] indicated that imposing tariffs on other nations does not reduce local carbon dioxide emissions and would increase overall trade inefficiency. Since overall carbon dioxide emissions do not decline over time, concerns over environmental quality and health should be raised [46]. However, a study on carbon intensity can demonstrate the changes in the proportional rate of carbon emissions to income, suitable for analysis on an income basis. The present study is on the mechanisms of climate change mitigation in order to demonstrate the dynamics to environmental sustainability. Total emissions of carbon dioxide, as well as atmospheric carbon dioxide concentrations, contribute directly to climate change. The studies of Apergis and Payne [9,10] and Armeanu et al. [11] have focused more on the dynamics between renewable energies and income, and they found that renewable energy has induced income increases. A complete list tabulated in the study of Armeanu et al. [11] illustrates the previous related studies on renewable energy consumption and economic growth. However, these three analysis ignore the carbon emissions nexus.

The threat of climate change requires an energy transition from fossil fuels towards renewable energies and higher efficiency. The European energy transition includes a high-level of renewable energy installations. European energy transition policy is used as one tool to encompass both goals of climate change mitigation and economic stimulation, and as well as reducing trade deficits and increasing employment. Creutzig et al. [47] studied the dynamics of renewable energy, income, and carbon dioxide emissions for three categories of relevant countries: the European countries, the European candidate countries, and the potential European candidate countries. Creutzig et al. [47] found that (1) for the European countries, renewable energy has an impact on the GDP, and that (2) for the candidate and potential candidate countries for EU membership, developing affordable and effective instruments and mechanisms to intensify energy transition to renewables is necessary to decrease the impacts of climate change induced particularly from decreasing carbon dioxide emissions without any reduction in economic growth.

According to Wooldridge [48], stationary or non-stationary series can be used to study their cointegration tests, long-run relationships, and error correction mechanisms. Following the principles of panel data analysis [48], the present study investigated the dynamics of total carbon dioxide emissions, aggregate income, non-renewable energy consumption, and renewable power. The variable of renewable power was used to represent the advanced innovation of renewable energy at high energy efficiency. More specifically, through the perspective of energy sustainability, the primary purpose of this research was to estimate the long-run effects and short-run linkages among variables. The long-run balance of how aggregate income and total dioxide emissions are affected by the traditional nonrenewable energy consumption and the advanced innovations of renewable power was probed by adopting cointegration regressions. A Cobb–Douglas functional form was investigated; the data were investigated in their monotonic logarithm transformation [48]. The data are available at open data sources, namely the Penn World Table (PWT10) [49], and British Petroleum Global [50]. Then, the first step of the analysis was to take the logarithm of the original data to test the non-stationarity of the time series and then to test for cointegration. Each series of individual variables and their first difference were tested with panel unit root tests, specifically, the Levin, Lin, and Chu (LLC); Im, Pesaran, and Shin (IPS); Augmented Dickey–Fuller (ADF); and Phillips–Perron (PP) tests. After passing unit root tests, a group test was performed on the datasets of the four variables for their cointegration.

Econometric scholars have developed panel testing methodologies to determine the existence of cointegration. Kao [51] proposed a test for homogeneous cointegration through pooled regression permitting for individual fixed effects. Pedroni's method [52] tests for heterogeneous cointegration. However, the power of these tests remains unclear, as Gutierrez compared these two tests by using the Monte Carlo simulation method and reported inconclusive advantages over the test power [53]. A panel cointegration estimation technique is applied to estimate the single equation relationship. In the literature, single panel cointegration regression is typically estimated using ordinary least squares (OLS) or fully modified ordinary least squares (FMOLS), as outlined by Pedroni [54], for a heterogeneous cointegration panel, or dynamic ordinary least squares (DOLS), as outlined by Kao and Chiang [55] and Mark and Sul [56], for a homogeneous cointegration panel. If countries are assumed to exhibit heterogeneous cointegration, accordingly the FMOLS estimation is feasible for panel cointegration estimation, and a DOLS, developed especially for homogenous cointegration, is not suitable. The method of FMOLS was adopted in a series of studies [57–61], as was the method of DOLS [62–65]. The data in this investigation were assumed to have heterogeneity, and the model of the long-run relationship was estimated by using the FMOLS technique.

The mechanism of the short-run adjustment was probed by the method of PVECM, which is a capable and reliable econometric technique to estimate the dynamic linkages among variables and which has widely been applied in empirical data studies [66–85]. Granger causality can be adopted in the PVECM method by incorporating a vector of autoregressive model with lagged variables and the residuals of long term balance regressions. Granger causality based on PVECM was also applied to the aforementioned analyses on the dynamics of variables by Apergis and Payne [9,10], Armeanu et al. [11], and Creutzig et al. [47].

The present study estimated the long-run relationships by FMOL as proposed by Pederoni [52,54], with high heterogeneity in the variables across countries. Additionally, the linking effects among variables were probed via the Granger causality test, based on PVECM, by the procedures described in the literature [9–11].

In addition to parametric analysis of PVECM applied in the present study and a series of research in the literature [86–88], research based on nonparametric techniques are used in the analysis of climate change as well [89–97]. A combination of the two techniques has also been applied in the literature [98,99]. In a parametric analysis, the statistical information about the distribution of the population is known and represented based on a fixed set of parameters. On the contrary, in a nonparametric analysis, the statistical information about the distribution of a population is unknown, the parameters are not fixed, and it is necessary to test the hypothesis for the population.

Nonparametric analysis is an alternative technique with its own individual applicability, rather than being superior to others in all circumstances. The often-used time varying nonparametric methods are powerful to detect structure breaks, since they allow for flexibility in parameters over time. Nonparametric method can be used to identify structure breaks [91–94].

Chen et al. [98] used both parametric and nonparametric methods to capture the joint dynamics in their study, and Chen et al. [98] evidenced that parameters seemed to better capture the changes of the impacts than the nonparametric techniques in that study.

3. Methodology

After retrieving data from reliable sources, the first step was to test for the stationarity for the panel data series of the variables for the country groups under this investigation. If the non-stationarity and cointegration of the data were affirmed, their cointegration was estimated to represent their long-run equilibrium relationships, and the PVECM was applied with Granger causality for short-run patterns to adjust toward the long-run equilibrium relationships.

3.1. Data

This investigation demonstrated the long-run balance relationship and short-term adjustment patterns for the dynamic connections among (1) total carbon dioxide emissions, (2) aggregate income, (3) nonrenewable energy as represented by the sum of coal, gas, and oil consumption, and (4) renewable power.

The 1990–2021 raw data were retrieved from reliable open-access databases. Real income data of the study countries were retrieved from Penn World Table, PWT10 [49], represented by the output-side real GDP at current PPP (in mil. 2017 US$). Data for total carbon dioxide emissions and energy variables were retrieved from the Statistical Review of World Energy, British Petroleum Global [50]. The corresponding variable definitions were as follows:

1. $CO2_{MTCO2}$: Total carbon dioxide emissions in metric tons.
2. $RGDPO_{PWT10}$: National aggregate income represented by output-side real GDP compiled in PWT10.
3. $NREN_{EJ}$: Nonrenewable energy consumption (in ej, 10^{18} joules) as the sum of coal consumption, gas consumption, and oil consumption.
4. $REN_{POWER_{EJ}}$: Power generated from renewable energies in ej.

3.2. Long-Run Equilibrium Relationship

Two specifications were used to represent long-run cointegration relations among the variables. The long-run relationships between the variables were estimated in terms of carbon emissions and income as econometric specifications in Equations (1) and (2), respectively. The coefficient demonstrates the elasticity for a 1% change in the dependent variable to the percentage change of the independent variable, carbon emissions.

Cointegration (equilibrium) relations for carbon dioxide emissions are as follows:

$$\begin{aligned} LOG(CO2_{MTCO2})_{it} = \alpha_i + \beta_1 LOG(RGDPO_{PWT10})_{it} + \beta_2 LOG(POP)_{it} \\ +\beta_3 LOG(COALCONS_{EJ})_{it} + \beta_4 LOG(GASCONS_{EJ})_{it} \\ +\beta_5 LOG(OILCONS_{EJ})_{it} + \beta_6 LOG\left(REN_{POWER_{EJ}}\right)_{it} \\ +RESID_{CO2it} \end{aligned} \tag{1}$$

Cointegration (equilibrium) relations for income are as follows:

$$\begin{aligned} LOG(RGDPO_{PWT10})_{it} = \alpha_i + \beta_1 LOG(CO2_{MTCO2})_{it} + \beta_2 LOG(POP)_{it} \\ +\beta_3 LOG(COALCONS_{EJ})_{it} + \beta_4 LOG(GASCONS_{EJ})_{it} \\ +\beta_5 LOG(OILCONS_{EJ})_{it} + \beta_6 LOG\left(REN_{POWER_{EJ}}\right)_{it} \\ +RESID_{GDPit} \end{aligned} \tag{2}$$

where t = 1990, 1991, ..., 2019; the parameter α_i permits for country-specific fixed effects; subscript i denotes the ith country in the panel; and $RESID$ denotes the estimated residuals, which depict deviations from the long-run relationship.

3.3. Short-Run Dynamics

The short-run dynamics in the two panels of countries were investigated using a Granger causality test, based on PVECM. The deviations were captured by the residuals of the long-run cointegration regression. Then, this PVECM model was used (1) to demonstrate Granger causality between variables, and (2) as an error correction mechanism illustrating the long-run dynamics.

3.3.1. Vector Autoregression and Granger Causality

Granger causality tests based on vector autoregression (VAR) were adopted as the specifications for the present study. In the VAR method, each variable has an equation modeling its evolution over time. This study included one-period lagged (past) values of the first difference regarding the variable itself, the other variables, and an error term in the regression. VAR models do not require as much information regarding driving forces as structural models with simultaneous equations do. The only information required in a VAR model is a list of variables that can be hypothesized to affect each other over time. Granger "causality" does not measure cause–effect relationships but is a statistical correlation between the current value of one variable and the past values of others (i.e., Granger causality does not directly imply changes in one variable causing changes in another). Granger causality indicates the changes in one variable in response to one-period lagged changes in itself and other variables.

Granger causality relationships were verified between the following variables: (1) total carbon dioxide emissions, (2) aggregate income, (3) nonrenewable energy consumption (energy use measured by the sum of coal, gas, and oil consumption), and (4) green energy technology represented by the use of renewable power. The Granger causality model is a time series model of first differences that includes current values on the left-hand side and the past values of their first differences on the right-hand side. The Granger causality relationships between variables are indicated by the magnitudes and significance of the estimated coefficients.

3.3.2. Error Correction Mechanism

The error correction mechanism is an adjustment response to deviations from the long-term equilibrium [48]. The adjustment is indicated by how the contemporary changes of one variable responds to past deviations from the long-term equilibrium.

The error correction model is a time series model in first differences that also contains an error correction term, which works to bring the (1) series back into long-run equilibrium. This study assumed that the adjustment would occur as a response to deviations from the long-run equilibrium levels of carbon emissions and income. The aforementioned cointegration regression Equations (1) and (2) represent long-run balances. The forecasted residuals of these two regressions represent the deviations of actual carbon emissions and actual income from their balance equilibrium levels. Past values (one-period backward) of both forecasted residuals, as the error correction terms (ECTs), were simultaneously introduced into the Granger VAR model, which became a PVECM.

A two-step procedure was adopted. The first step was to estimate and forecast residuals of Equations (1) and (2) by using FMOLS for the EU and OECD countries. Subsequently, one-period lagged values of the residuals were included simultaneously as two ECTs in the PVECM. The estimated coefficients of ECTs demonstrate the adjustment backward (with a negative sign) or outward (with a positive sign) to the long-run equilibrium levels of carbon emissions and income.

3.3.3. PVECM Specification and Short-Run Dynamics

Briefly, this PVECM can demonstrate Granger causality by including the first differences of the included variables in a VAR system, as well as long-term adjustment patterns by including ECTs. The described long-run models are estimated by FMOLS for both panels. One-period lags of the deviation from equilibrium, represented by the error terms of Equations (1) and (2) and denoted as $RESID_{CO2it-1}$ and $RESID_{GDPit-1}$, are introduced into the PVECM as the ECTs. The coefficients of the ECTs reveal the dynamic adjustment patterns toward equilibrium.

The PVECM is set by including a one-period lagged VAR system with two ECTs. The PVECM equations are as follows:

For j = 1 (used to represent how current changes of total carbon dioxide emissions are affected by lagged changes (lagged first difference) of all selected variables and lagged residuals of the long-run system equilibrium),

$$\begin{gathered}\triangle LOG(CO2_{MTCO2})_{it} = \alpha_{1j} + \gamma_{11i} \triangle LOG(CO2_{MTCO2})_{it-1} \\ +\gamma_{12i} \triangle LOG(RGDPO_{PWT10})_{it-1} + \gamma_{13i} \triangle LOG(NREN_{EJ})_{it-1} \\ +\gamma_{14i} \triangle LOG(REN_{POWEREJ})_{it-1} + \varphi_{11i} RESID_{CO2it-1} \\ +\varphi_{12i} RESID_{GDPit-1} + e_{1it}\end{gathered} \quad (3)$$

For j = 2 (used to represent how current changes of national aggregate income are affected by lagged changes (lagged first difference) of all selected variables and lagged residuals of the long-run system equilibrium),

$$\begin{aligned}\triangle LOG(RGDPO_{PWT10})_{it} &= \alpha_{2j} + \gamma_{21i} \triangle LOG(CO2_{MTCO2})_{it-1} \\ &+\gamma_{22i} \triangle LOG(RGDPO_{PWT10})_{it-1} + \gamma_{23i} \triangle LOG(NREN_{EJ})_{it-1} \\ &+\gamma_{24i} \triangle LOG(REN_{POWEREJ})_{it-1} + \varphi_{21i} RESID_{CO2it-1} \\ &+\varphi_{22i} RESID_{GDPit-1} + e_{2it}\end{aligned} \tag{4}$$

For j = 3 (used to represent how current changes of nonrenewable energy consumption by lagged changes (lagged first difference) of all chosen variables and lagged residuals of the long-run system equilibrium),

$$\begin{aligned}\triangle LOG(NREN_{EJ})_{it} &= \alpha_{3j} + \gamma_{31i} \triangle LOG(CO2_{MTCO2})_{it-1} \\ &+\gamma_{32i} \triangle LOG(RGDPO_{PWT10})_{it-1} + \gamma_{33i} \triangle LOG(NREN_{EJ})_{it-1} \\ &+\gamma_{34i} \triangle LOG\left(REN_{POWER_{EJ}}\right)_{it-1} + \varphi_{31i} RESID_{CO2it-1} \\ &+\varphi_{32i} RESID_{GDPit-1} + e_{3it}\end{aligned} \tag{5}$$

For j = 4 (used to represent how current changes in renewable power are affected by lagged changes (lagged first difference) of all selected variables and lagged residuals of the long-run system equilibrium),

$$\begin{aligned}\triangle LOG\left(REN_{POWER_{EJ}}\right)_{it} &= \alpha_{4j} + \gamma_{41i} \triangle LOG(CO2_{MTCO2})_{it-1} \\ &+\gamma_{42i} \triangle LOG(RGDPO_{PWT10})_{it-1} + \gamma_{43i} \triangle LOG(NREN_{EJ})_{it-1} \\ &+\gamma_{44i} \triangle LOG\left(REN_{POWER_{EJ}}\right)_{it-1} + \varphi_{41i} RESID_{CO2it-1} \\ &+\varphi_{42i} RESID_{GDPit-1} + e_{4it}\end{aligned} \tag{6}$$

where the notation $\triangle$ denotes the first difference of the variable; the subscript j denotes the jth equations that represent the mechanism of the jth variable; and $j = 1, 2, \ldots,$ and 4.

4. Analysis

The analysis was based on aggregate country-level annual data of two panels of countries, EU-18 and OECD-32, from 1990 to 2019. The data were investigated in their monotonic logarithm transformation, as the original long-run relationship was specified as a Cobb–Douglas functional form. The data were tested for their individual panel stationarity and group cointegration. After affirmation of their non-stationarity and cointegration, their cointegration relationships were estimated to represent their long-run equilibrium relationships, and the PVECM was applied with Granger causality for short-run patterns to adjust the long-run equilibrium relationships.

4.1. Data Descriptive Statistics

Data definitions are as shown in Table 1 and descriptive statistics of the raw data are presented in the following Table 2.

Table 1. Definitions of variables.

Variable	$CO2_{MTCO2}$	$RGDPO_{PWT10}$	$NREN_{EJ}$	$RENPOWER_{EJ}$
(unit)	**(Metric ton)**	**(at Current PPPs in mil. 2017 US$)**	**(ej, 10^{18} joules)**	**(ej, 10^{18} joules)**
Definitions	Total carbon dioxide emissions	National aggregate income represented by output-side real GDP	Nonrenewable energy consumption as the sum of coal consumption, gas consumption, and oil consumption	Power generated from renewable energies
Data source	Statistical Review of World Energy, British Petroleum Global	Penn World Table, PWT10	Statistical Review of World Energy, British Petroleum Global	Statistical Review of World Energy, British Petroleum Global

Table 2. Descriptive statistics of variables as raw data (1990–2021).

Variable	$CO2_{MTCO2}$	$RGDPO_{PWT10}$	$NREN_{EJ}$	$RENPOWER_{EJ}$
(unit)	**(Metric ton)**	**(at Current PPPs in mil. 2017 US$)**	**(ej, 10^{18} joules)**	**(ej, 10^{18} joules)**
EU-18				
Mean	204.88	856,644.10	3.01	0.12
Median	105.62	351,442.50	1.37	0.04
Maximum	1007.60	4,275,312.00	13.33	2.00
Minimum	8.03	14,869.14	0.12	0.00
SD	213.91	968,394.60	3.12	0.24
OECD-32				
Mean	391.77	1,354,274.00	5.78	0.14
Median	105.62	412,131.50	1.37	0.03
Maximum	5884.15	20,595,844.00	84.88	4.37
Minimum	2.12	8453.93	0.03	0.00
SD	936.15	2,760,794.00	13.75	0.37

4.2. Testing for Non-Stationarity and Cointegration

The first step of the analysis was to take the logarithm of the original data to test the non-stationarity of the time series and to test for cointegration.

Each series of individual variables and their first difference were tested with panel unit root tests, including LLC, IPS, ADF, and PP tests, according to Wooldridge [48]. The results are listed in Tables 3 and 4. The results indicated that the individual variables were stationary in their first difference. The results for the level data were mixed, but most were nonstationary.

Both Kao's [51] and Pedroni's [52] tests were applied in this study to investigate cointegration in EU-18 (Table 3) and OECD-32 (Table 4). On the basis of the results of Kao's test, the variables were cointegrated, as the null of no cointegration was rejected.

A long-term equilibrium was observed for the aforementioned variables, as the hypothesis of the presence of a long-run relationship was confirmed. Then, the next step was to estimate this relationship with regressions.

Table 3. Panel unit root and cointegration tests for EU-18 (1990–2019).

Unit Root Test	LLC		IPS		ADF		PP	
VARIABLE								
$LOG(CO2_{MTCO2})$	−0.9871		0.4787		46.9686		46.6998	
$\triangle LOG(CO2_{MTCO2})$	−18.1172	***	−19.3402	***	340.8520	***	344.9060	***
$LOG(RGDPO_{PWT10})$	−4.1259	***	1.5725		30.3000		36.2469	
$\triangle LOG(RGDPO_{PWT10})$	−14.6981	***	−15.3259	***	264.6180	***	266.9580	***
$LOG(NREN_{EJ})$	−1.8735	*	−0.7238		59.9732	**	57.7107	*
$\triangle LOG(NREN_{EJ})$	−18.6400	***	−19.0193	***	333.9820	***	341.9630	***
$LOG\left(REN_{POWER_{EJ}}\right)$	−1.5272	*	2.5881		46.5231		81.4785	***
$\triangle LOG\left(REN_{POWER_{EJ}}\right)$	−12.9593	***	−13.3866	***	226.8940	***	236.2510	***
Pedroni Residual Cointegration Test (H_0: no cointegration)								
	within-dimension				between-dimension			
	Statistic		Weighted Statistic				Statistic	
Panel v-Statistic	0.45		0.17		Group rho-Statistic		1.47	
Panel rho-Statistic	1.01		0.15		Group PP-Statistic		−1.83	*
Panel PP-Statistic	−0.92		−2.81	**	Group ADF-Statistic		−0.69	
Panel ADF-Statistic	−0.44		−2.68	**				
Kao's Residual Cointegration Test (H_0: no cointegration)								
ADF t-Statistic	−7.79	***						
Residual variance	0.00							
HAC variance	0.00							

Source: Author's computation. Notes: Data are log transformed. △ denotes the first difference. * $p < 0.1$, ** $p < 0.01$, *** $p < 0.001$. LLC, IPS, ADF, PP respectively denote Levin, Lin, and Chu t* stat.; Im, Pesaran, and Shin W-stat.; ADF Fisher chi-square; and PP Fisher chi-square. LLC assumes common unit root process. IPS, ADF, and PP assume individual unit root process. The cointegration tests have a null hypothesis of no cointegration.

Table 4. Panel unit root and cointegration tests for OECD-32 (1990–2019).

Unit Root Test	LLC		IPS		ADF		PP	
VARIABLE								
$LOG(CO2_{MTCO2})$	1.5822		7.5111		53.3169	***	88.7943	***
$\triangle LOG(CO2_{MTCO2})$	−16.1028	***	−16.7259	***	378.8460	***	391.5040	***
$LOG(RGDPO_{PWT10})$	−4.7802	***	2.2249		64.4777	***	91.1717	*
$\triangle LOG(RGDPO_{PWT10})$	−20.2455	***	−19.7584	***	453.1790	***	458.7350	***
$LOG(NREN_{EJ})$	−7.4591	***	−3.4584	***	138.1830	***	187.2370	***
$\triangle LOG(NREN_{EJ})$	−24.8799	***	−25.3165	***	587.7110	***	596.7860	***
$LOG\left(REN_{POWER_{EJ}}\right)$	1.5822		7.5111		53.3169		88.7943	***
$\triangle LOG\left(REN_{POWER_{EJ}}\right)$	−16.1028	***	−6.7259	***	378.8460	***	391.5040	***

Table 4. *Cont.*

Pedroni Residual Cointegration Test (H_0: no cointegration)							
	within-dimension				between-dimension		
	Statistic		Weighted Statistic			Statistic	
Panel v-Statistic	1.98	*	1.14		Group rho-Statistic	1.20	
Panel rho-Statistic	0.53		−0.27		Group PP-Statistic	−3.52	***
Panel PP-Statistic	−1.68	*	−3.44	***	Group ADF-Statistic	−1.23	
Panel ADF-Statistic	−0.58		−2.42	**			
Kao's Residual Cointegration Test (H_0: no cointegration)							
ADF t-Statistic	−7.79	***					
Residual variance	0.00						
HAC variance	0.00						

Source: Author's computation. Notes: Data are log transformed. $\triangle$denotes the first difference. * $p < 0.1$, ** $p < 0.01$, *** $p < 0.001$. LLC, IPS, ADF, PP respectively denote Levin, Lin, and Chu t* stat.; Im, Pesaran, and Shin W-stat.; ADF Fisher chi-square; and PP Fisher chi-square. LLC assumes common unit root process. IPS, ADF, and PP assume individual unit root process. The cointegration tests have a null hypothesis of no cointegration.

4.3. Estimating the Long-Run Relationships

After the hypothesis of the presence of a long-run relationship among variables was affirmed, the estimated results of Equations (1) and (2) revealed the long-run equilibrium patterns for carbon emissions and income. A panel cointegration estimation technique was applied to estimate the single equation relationship. In the literature, single panel cointegration regression is typically estimated using OLS, FMOLS, or DOLS. The model of the long-run relationship in present study was estimated by using the FMOLS technique, as we assumed heterogeneity among countries. The estimated results of Equations (1) and (2) are presented in Table 5.

Table 5. Results of long-run cointegration relationship by FMOLS for EU-18 and OECD-32.

		Equation (1)		Equation (2)	
		Dependent variable : $LOG(CO2_{MTCO2})$		Dependent variable : $LOG(RGDPO_{PWT10})$	
	Independent variables	coefficient		coefficient	
EU-18					
	$LOG(CO2_{MTCO2})$	-	-	−0.2027	*
	$LOG(RGDPO_{PWT10})$	−0.0244	*	-	-
	$LOG(NREN_{EJ})$	1.0366	***	1.4247	**
	$LOG\left(REN_{POWER_{EJ}}\right)$	−0.0113	***	0.1425	***
	R-squared	0.99		0.99	
	Adjusted R-squared	0.99		0.99	
OECD-32					
	$LOG(CO2_{MTCO2})$	-		0.1612	
	$LOG(RGDPO_{PWT10})$	0.0040		-	
	$LOG(NREN_{EJ})$	1.0263	***	0.2685	
	$LOG\left(REN_{POWER_{EJ}}\right)$	−0.0138	***	0.1599	***
	R-squared	0.99		0.99	
	Adjusted R-squared	0.99		0.99	

Source: Author's computation. * $p < 0.1$, ** $p < 0.01$, and *** $p < 0.001$.

Carbon emission patterns emerged in the estimates of Equation (1). The findings for the EU-18 group were as follows. (1) Weak statistical significance and a small margin were present in the relation between income change and carbon emissions change (significant at 10% level). A 1% income increase would reduce aggregate country-level total carbon emissions by 0.0244%. (2) Carbon dioxide emissions in the EU were found to be closely connected with nonrenewable energy use. The magnitudes were large, and the coefficient was strongly statistically significant. The empirical estimate indicated that a 1.0366% increase in carbon emissions corresponds to a 1% increase in nonrenewable energy use in EU countries. Hence, if nonrenewable energy consumption were reduced in EU countries, carbon emissions would be reduced accordingly. Tremendous reductions in carbon emissions would be achieved if EU countries could promote and maintain decreasing trends in nonrenewable energy use. In this respect, countries face both profound challenges and numerous opportunities. (3) A slightly negative but strongly significant margin was reflected in the relationship between renewable power and carbon emissions. A 1% increase in renewable power use in the EU would reduce country-level total carbon emissions by 0.0113%. The elasticity magnitude was small, but the statistical significance was quite strong. With a sharply increasing trend in the use of renewable power, this evidence of the relationships between renewable power and carbon emissions in the EU is welcome news.

The empirical results for the OECD-32 group were similar to those for the EU-18 group.

(1) Carbon emissions did not significantly respond to income changes in the OECD, and as mentioned, a weak, small response was observed in the EU-18 group. The results confirmed that carbon emissions are not substantially affected by income change.
(2) In OECD countries, the nexus between carbon emissions and nonrenewable energy consumption had strong significance, with tremendous positive magnitude in elasticity. The empirical results indicated that a 1.0263% increase in carbon emissions corresponds in the same direction to a 1% increase in nonrenewable energy consumption in the OECD.
(3) Encouraging evidence was observed in renewable power and its long-run relationship with carbon emissions. A 1% increase in renewable power would decrease country-level total carbon emissions by 0.0138%. Thus, renewable power can play a positive role in reducing carbon emissions in both the EU and the OECD. Future prospects could rely on its accelerated development and wide adoption.

Three key findings were revealed in the evidence from the estimates in Equation (2).

(1) The evidence for a long-run relation between income and carbon emissions was weak in the EU-18 group and nonsignificant in the OECD-32 group.
(2) Nonrenewable energy consumption plays a major role in determining income for the EU, although this role is absent in the OECD. During 1990–2019, EU-18 income was significantly supported by nonrenewable energy consumption. By contrast, OECD-32 income was not statistically significantly supported by nonrenewable energy consumption. A 1% decrease in nonrenewable energy consumption would decrease income by an average of 1.4247% in the EU-18 group, but the effect was not as strong in the OECD group. This might suggest that the EU-18 has approached its carrying capacity in its carbon reduction and marginal abatement cost increases with the commitment to carbon pricing measures such as carbon taxation and its ETS, and that the OECD countries as a whole have made less effort in carbon reductions.
(3) Renewable power demonstrated a prominent contribution to income in both panels of countries. A 1% increase in renewable power use would increase incomes by 0.1425% and 0.1599% in the EU-18 and OECD-32 groups, respectively. The negative relations between renewable power and carbon emissions in estimating Equation (1) for the EU-18 and OECD-32 groups highlighted the essential role of renewable power in climate change mitigation. In EU and OECD countries, reliance on renewable power is growing. As mentioned, renewable energy, and therefore renewable power, still accounts for a small percentage of overall energy usage, but the future is full of opportunity.

4.4. PVECM and Short-Run Dynamics

The PVECM was estimated by allowing heterogeneous intercepts and no trend in each equation. The results revealed Granger causality between variables and adjustment to the balance equilibrium.

4.4.1. Findings in Granger Causality Test

The estimated coefficients of the lagged variables shown in Table 6 demonstrated short-run Granger causality. The value of the coefficient revealed the extent of the variable's impact in the contemporary period by 1% changes of itself ("self-causality") and other variables from the past. The Granger nexuses between variables are depicted in Figures 1 and 2 for the EU-18 and OECD-32 groups, respectively. The findings are summarized in the following text.

Table 6. Granger causality (based on PVECM) for EU-18 and OECD-32 (1990–2019).

Equation Number j	**(j = 1)**	**(j = 2)**	**(j = 3)**	**(j = 4)**
Dependent variable	$\triangle LOG(CO2_{MTCO2})$	$\triangle LOG(RGDPO_{PWT10})$	$\triangle LOG(NREN_{EJ})$	$\triangle LOG\left(REN_{POWER_{EJ}}\right)$
EU-18				
Short-run Granger causality				
$\triangle LOG(CO2_{MTCO2})$	−0.2145	0.0515	−0.2493	1.5469 *
$\triangle LOG(RGDPO_{PWT10})$	0.0338	0.1682 ***	0.0282	−0.2442
$\triangle LOG(NREN_{EJ})$	0.1652	0.0634	0.2102	−1.4551 *
$\triangle LOG\left(REN_{POWER_{EJ}}\right)$	0.0020	−0.0032	0.0027	0.1387 ***
Long-run Granger causality error correction term (ECT)				
$RESID_{CO2}$	−0.4486 ***	0.0409	−0.2235 *	−1.8821 ***
$RESID_{GDP}$	−0.0368 *	−0.0458 ***	−0.0352 *	0.2675 ***
OECD-32				
Short-run Granger causality				
$\triangle LOG(CO2_{MTCO2})$	−0.2026	0.1167	−0.2799 *	0.5332
$\triangle LOG(RGDPO_{PWT10})$	0.0527	0.1626 ***	0.0419	−0.2852
$\triangle LOG(NREN_{EJ})$	0.2111	−0.0592	0.3017 *	−0.4692
$\triangle LOG\left(REN_{POWER_{EJ}}\right)$	0.0096 *	0.0055	0.0079	0.1447 ***
Long-run Granger causality error correction term (ECT)				
$RESID_{CO2}$	−0.2866 ***	−0.1096 *	−0.1245 *	−1.1108 ***
$RESID_{GDP}$	−0.0150	−0.0420 ***	−0.0149	0.3718 ***

Source: Author's computation. △ denotes first difference. The *t* statistics are in parentheses. * $p < 0.1$, ** $p < 0.01$, and *** $p < 0.001$.

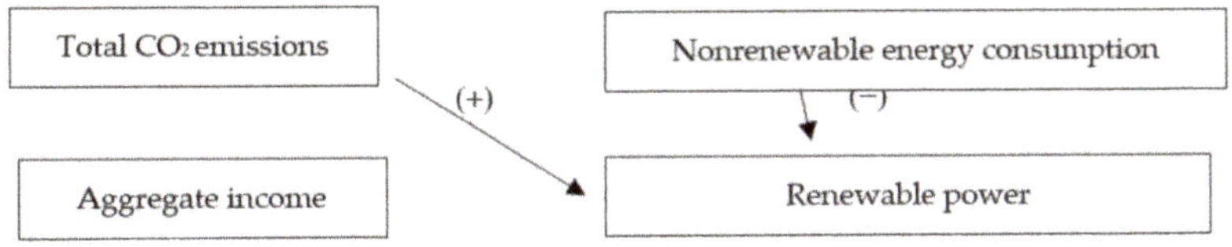

Figure 1. Granger nexus between selected variables in the EU-18.

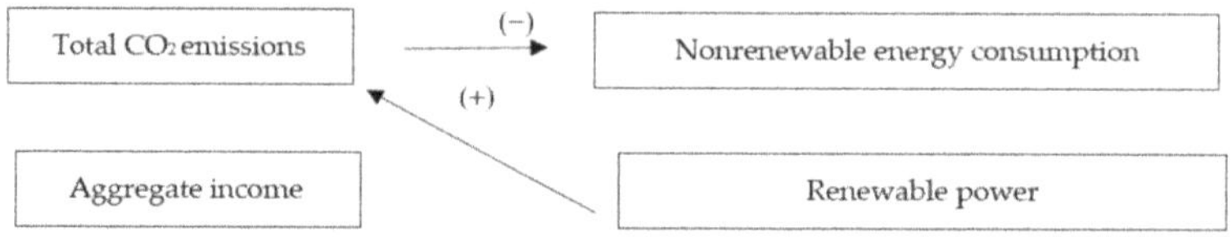

Figure 2. Granger nexus between selected variables in the OECD-32.

(1) None of the four variables of Granger caused the EU's carbon emissions. However, renewable power weakly Granger-caused the OECD's carbon emissions.
(2) Regarding self-causality, income and renewable power has significant self-Granger causality in the EU and OECD. Past income and renewable power each has a positive significant influence on their current value, and the coefficient magnitudes were high. Contemporary income responses of 0.1682% in the EU-18 and 0.1626% in the OECD with 1% self-changes were noted in the lagged period; current renewable power responses of 0.1387% in the EU-18 and 0.1447% in the OECD with 1% self-changes were observed in the lagged period. Both variables exhibited strong self-Granger causality in its evolution over time. Income growth would generate further income growth, and green technology progress would promote further green technology progress. Accumulation effects would result in strong increasing trends.
(3) Nonrenewable energy consumption in the OECD had weak self-Granger causality and the EU none. Additionally, self-causality was not present for total carbon emissions in either panel. Current changes in these two variables were not strongly influenced by their own past changes. Past changes in total carbon emissions would not affect current changes in total emissions. One can imagine carbon dioxide, emitted as trash and by-products in economic activities, having no feedback mechanism for past changes. However, the stringent carbon abatement policy in the EU has led to downward pressure on the region's nonrenewable energy consumption. Moreover, no downward or upward trends were observed in the EU's nonrenewable energy consumption and total carbon emissions.

The lack of self-Granger causality in the EU's nonrenewable energy consumption and total carbon emissions during the study period merits attention. After strict carbon abatement policies have eliminated most of the emissions that may be removed, policymakers must determine what areas the EU should continue to work on.

(4) The EU's carbon emissions significantly Granger-caused renewable power (with a positive sign) at the 10% significance level. A 1% change in lagged nonrenewable energy would Granger-cause 1.5469% changes in renewable power use. Furthermore, the EU's nonrenewable energy consumption significantly Granger-caused renewable power (with a negative sign) at a 10% significance level. A 1% decrease in lagged nonrenewable energy would Granger-cause a 1.4551% increase in the use of renewable energy. This finding affirmed that nonrenewable energy can be substituted by renewable power in the EU. However, this substitution effect was not significant in the OECD. Even in the EU, the substitution effect (between nonrenewable energy consumption and renewable power) had significance only at the 10% level. The evidence was statistically weak. Therefore, people have not strongly moved to use renewable power in response to carbon emissions, particularly in the OECD.

Thus, the Granger causality (from past changes of carbon emissions to contemporary renewable power) was not strong in the two panels of countries included in this study. Climate mitigation policies are expected to promote the development of renewable energy sources and thus curb the consumption of nonrenewable energy, help reduce carbon emissions, and ultimately counter climate change. However, these outcomes are not always realistic. In fact, renewable power would contribute to carbon emissions in the OECD with Granger causality at the 10% significance level.

(5) In the OECD, carbon emissions would inversely Granger-cause nonrenewable energy consumption. The response elasticity was found to be 0.2799% in the OECD. If a 1% increase in total carbon emissions occurred in the previous period, nonrenewable energy use would decrease by 0.27998% in the contemporary period. However, the effects in the EU were statistically insignificant.

Because none of the four variables Granger-caused the EU's reductions in carbon emissions, the EU may already be approaching its carrying capacity in carbon reduction through the reduction of nonrenewable energy use.

4.4.2. Findings from ECT

The direction and magnitude of the error correction toward long-run equilibrium can be indexed by the estimated coefficients of ECTs. This study included two ECTs, $RESID_{CO2it-1}$ and $RESID_{GDPit-1}$. The estimated coefficients of these two series of lagged residuals in PVECM captured how their deviations from the equilibria were adjusted. The value of the coefficient revealed how much of the adjustment in the contemporary period responded to 1% of the deviation of the actual values from their forecasted equilibrium level; that is, the corresponding adjustment in current changes toward the long-run equilibrium. A negative sign demonstrates an adjustment back toward the carbon cointegration equilibrium or toward the income cointegration equilibrium, or both. A positive sign demonstrates an outward movement that breaks the equilibrium. Several findings were evidence of short-run adjustments as deviations to the long-run equilibrium.

(1) According to the ECT estimates of carbon emissions in the EU-18, carbon emissions, nonrenewable energy consumption, and renewable power will adjust to the deviations of carbon dioxide from its long-run equilibrium. When 1% of the actual value of carbon emissions deviated from its balanced equilibrium in a period, then the next-period carbon emissions, nonrenewable energy, and renewable energy responded to this deviation by −0.4486%, −0.2235%, and −1.8821%, respectively, to return the emissions to equilibrium. The adjustment speed was quite high, especially that of renewable power. However, the response in next-period income was statistically insignificant.
(2) According to the ECT estimates of income in the EU-18, all variables (carbon emissions, income, nonrenewable energy, and renewable power) significantly responded to the income ECT, at speeds of −0.0368%, −0.0458%, and −0.0352%, and 0.2675%, respectively, in response to a 1% deviation. The response speed of renewable power was quite high in bringing the deviation outward from the equilibrium.
(3) According to the carbon dioxide ECT in the OECD-32, all variables significantly responded, at speeds of −0.2866%, −0.1096%, −0.1245%, and −1.1108%, respectively, in response to a 1% change in income ECT.
(4) In terms of the income ECT in the OECD-32, income and renewable power significantly responded at rates of −0.0420% and 0.3718%, respectively, in response to a 1% change of income ECT, whereas the responses of carbon emissions and nonrenewable energy were nonsignificant. The response speed of renewable power was quite high in bringing the deviation outward from the equilibrium.
(5) With positive significant ECTs of income, the changes of contemporary renewable power strongly pushed outward in response to the past deviation from long-run equilibrium in income. Renewable power could break long-run linkages of the income to other variables in this study. Income growth would trigger the rebound effects in renewable power. However, renewable power would bring a deviations in carbon emissions backward to its balance equilibrium, as indicated by negative ETCs of carbon emissions.

5. Discussion

(1) Long-run equilibrium relationships between variables

The results regarding long-run cointegration relationships indicate that the cointegrated linkage of carbon dioxide and income is not strong either in the EU or the OECD. Additionally, both renewable power and nonrenewable energy play central roles in supporting income in EU countries. Most importantly, renewable power has especially strong effects in alleviating carbon emissions and increasing income in both panels.

(2) Short-run Granger causality and equilibrium adjustment

After the long-term relationship is affirmed, the next question to be answered is how the variables adjust in response to changes. The PVECM provided evidence. The ECT results (Table 6) reveal that long-run adjustments statistically occur in the time series of all the variables and bring the series back to both equilibria; renewable power is the exception. Increases in renewable power use would be triggered by a positive deviation, as the value for actual renewable power is greater than that of the estimated equilibrium level.

Policy intended to decrease carbon emissions must be effectively devised because a reduction of emissions is not triggered by the variables included in the short-run PVECM used in this study. For EU carbon emissions, short-run causality from the included variables is not significant, and systematic adjustments are only noted in response to deviations in the long-run equilibrium of carbon emissions and income. For OECD carbon emissions, renewable power would mildly stimulate an increase of carbon emissions, and short-run adjustments would occur in response to deviation in carbon emissions; however, no significant adjustment is made in response to deviation in income equilibrium.

(3) Substitution effects between energies

Nonrenewable energy can be substituted by renewable power in the EU, while this substitution effect is not significant in the OECD.

(4) Role of renewable power

Renewable power in the EU and OECD demonstrates a strong, significant self-Granger causality in its evolution over time. Progress in green technology, indicated by renewable power in this study, would promote further progress over time. Accumulation effects would result in strong increasing trends.

Additionally, changes in contemporary renewable power strongly push deviations outward in response to past deviations from the long-run equilibrium of carbon emissions and income, as evidenced by the significant positive ECTs of emissions and income growth. Renewable power is capable of breaking long-run linkages among the selected variables. Renewable power has a lively role in its self-increase and in its response to the deviation from the long-run equilibrium. The outbreak of renewable power would indicate rebound effects with progress in green technology. Progress that improves energy efficiency would readily support income increases and further increase the demand for such power.

Potential carbon reduction arising from technological progress may be reduced by the rebound effect. Technological improvement and increases in its efficiency would increase the consumption of energy as discussed by Greening et al. [13]. This rebound effect (the Jevons paradox) was first proposed in 1865 in William Stanley Jevons's book *The Coal Question* [100], as he observed coal consumption in the United Kingdom soaring with coal-fired engine efficiency improvement [13].

The modern world is experiencing the problems associated with carbon-based energy. The problems have intensified as the world population's increasing needs are met. Humanity has long ignored the carrying capacity of the environment. Renewable energy is said to be limitless, but negative impacts might emerge when it is used on enlarged scales. Learning from our experiences, if it is time to develop a low-carbon economy, we should prepare for any negative impacts.

(5) Carbon pricing policy

Carbon pricing policies, such as carbon taxation, ETSs, and border carbon pricing adjustment, have been proposed to place a price on carbon emissions associated with commodities. Border carbon taxes, a measure of border carbon price adjustment, have been proposed in regional and global economic integration. The EU has proposed a border carbon tax that places a fee on global warming emissions embedded in goods produced outside the EU. The policy is aimed at incentivizing a reduction in carbon emissions beyond the EU.

According to the long-run connection patterns evidenced by Equations (1) and (2), an increase in renewable power use would suppress carbon emissions and support income increases. Renewable energy is limitless, and hopefully advances in technology will be triggered and inspired when the new demand emerges.

EU income exhibits short-run Granger causality one period ahead as a response to income changes. The same is true for the OECD. This is evidence that with an increasing trend, wealth in EU and OECD countries grew with the increase of prior wealth and not renewable power nor nonrenewable energy use. This is a noteworthy finding. These countries seem to insist on pursuing year-on-year income growth. However, income usually represents people's physical living standards.

The pollution haven hypothesis seems to hold, with emissions triggered in heavily industrialized developing countries. Because global production and consumption systems are closely integrated, the EU's border carbon pricing would be an effective remedy against pollution havens and carbon leakage by effectively imposing prices on emissions at each stage of supply and consumption chains. In addition to the explicated purposes of carbon reduction and climate change mitigation, the border carbon pricing policy can greatly increase the EU's international competitiveness. Stringent carbon reduction schemes might already have somewhat exhausted the carrying capacity of the EU's reduction without hindering its income. Implementing a border carbon pricing adjustment to drive dual advantages is reasonable. In this scenario, the price of imported commodities would increase. Whether the policy will lead to more application of local-made high-carbon products is another question.

The current results reveal that the adjustment of nonrenewable energy is barely significant in its connection to the prior-period changes of the included variables in the EU and OECD. Moreover, the long-run adjustment to deviations in income and carbon emissions (ECT) in the EU and OECD are not strongly significant. Indeed, nonrenewable energy is irresponsive to its own previous changes. "The dogs bark, but the caravan moves on." This appears to be unfortunate news for the mitigation of climate change. Persistent inertia is evidenced in nonrenewable energy consumption.

The expansion of carbon pricing, from local or regional carbon pricing to border adjustments to exert pricing effects on a global scale, will initially affect international competitiveness and result in changes in local consumer prices, even if the explicit goal is to mitigate climate change. Carbon pricing policy is also likely to have monetary effects. No matter how high the tax rates imposed, the effects on income may be greater than those on carbon reduction. Total carbon emissions is immovable. The current results indicate that the short-run Granger causality is such that (1) none of the four variables Granger-causes EU carbon emissions; and (2) renewable power even (weakly) Granger-causes OECD carbon emissions.

Questions remain regarding the types of policy most effective in combating climate change. An integration of current green technology into the designs of regulations would be effective in reducing total carbon emissions and mitigating climate change. However, recent policies of carbon pricing and the pursuit of net-zero emissions will have financial and trade impacts.

(6) Current carbon tax and border adjustment policy

Under current globalization and its interconnected international trade system, the proposed EU carbon border tax is intended to provide carbon reduction incentives to slash emissions far beyond Europe. Currently, the EU, South Korea, Singapore, Japan, and China have country-wide carbon pricing schemes, and the United States, Canada, and several other countries have local carbon pricing regulations. EU and OECD countries have enjoyed benefits from carbon leakage (emission outside and consumption inside). With the self-adjustment of renewable power continually increasing, as evidenced in the empirics of EU-18 and OECD-32 in this study, carbon pricing policy, including local, national, and border policy, would provide continuous and powerful reinforcement of the development of renewable power.

Border carbon taxation increases the price of imports and exports. International competitiveness is affected in the pursuit of explicit policy aims—climate mitigation. The study's finding of short-run adjustment to long-run equilibrium, occurring with respect to income and carbon emissions, indicates that governments, for international cooperation with global supply and demand chains, would react to increase competitiveness by seeking to reduce carbon emissions. Governments will, hopefully, outline the necessary steps for limiting global warming at the upcoming international climate meetings.

Taxing imported goods produced by high-carbon manufacturers could compel countries of production and the manufacturers to enact more aggressive climate rules. But is this fair to poor nations? This question remains unanswered. Furthermore, environmental capacity is also an important issue along with the expansion of renewable energy and renewable power. Rebound effects are affirmed to be embedded in the development of renewable power by their self-Granger causality. Our pursuit of income growth can trigger the rebound effects of renewable power. As the on-going green technology progress to meet needs of energy and pursuits of income increases, the related carrying capacity and environmental costs should also be addressed. The effects of tariffs were proposed by Carry [46], that tariffs would reduce trade deficits rather than promote sustainability, and tariffs would decrease economic efficiency and have no effects on reducing domestic total carbon dioxide emissions.

6. Conclusions

Nordhaus [1] posited that carbon pricing and green technology are the two most effective approaches for reducing carbon dioxide emissions to mitigate climate change. A decreasing in carbon intensity driven by the technology progress since industrial revolution in the study of Stefanski [3] is consistent with the insights of Nordhaus' argument that green technology progress is among one of the most effective approaches to reducing carbon emissions [1]. A low-carbon economy is the goal for contemporary economic development. Total carbon dioxide emissions is a direct key variable in determining climate change mitigation. Along the contemporary energy transition path, rapid technological progress promotes renewable energy, renewable power development, and energy efficiency boosting technology. Renewable power is promoted with high innovations, and of high heterogeneity between countries. Various different new technologies have been initiated and adopted in different countries. The current share of renewable energy, or renewable power, remains lower than that of nonrenewable energy, mostly fossil fuels. Renewable power is convenient when it is ready to feed into existing power grids. With progress in green technology and policies of carbon pricing triggering its advances, renewable power, as well as renewable energy, may have an opportunity to replace fossil fuels.

To probe the dynamics of key variables toward the low carbon development along the energy transition, this study investigates the long-run relationships and short-run linkages of total carbon emissions, aggregate output, nonrenewable energy consumption, and renewable power with a cointegration regression and Granger causality based on the PVECM model. The common dynamic empirics of the European Union countries and the Organization of Economic Co-operation and Development countries are addressed.

The Pedroni [52] heterogeneous panel cointegration tests reveal a long-run equilibrium relationship between the variables. Several findings are revealed regarding the long-term equilibrium and short-run adjustment linkage among variables.

Firstly, there is no strong long-term relation between aggregate income and total carbon dioxide emissions in both groups of countries, consistent with the evidences in the literature [101]. A negative but low significance relationship in the European Union countries is shown in Table 5, and it indicates a reversal in the directionality of the effects between income and total carbon dioxide emissions. The European Union countries, which are actively working on total carbon mitigations, appear to have approached a status where increases in income can be approached with decreasing total carbon dioxide emissions. For the group of the Organization of Economic Co-operation and Development countries, no linkage is presented at a low significance level (p-value < 0.1). This result states a week but bright vision for sustainable economic development of human beings to develop towards a low-carbon economy in the European Union countries.

According to this first finding in the long-run equilibrium, the loose long-term linkages between carbon dioxide emissions in its relationship to income, the viability of a carbon tax border adjustment policy might not be strong, especially to reduce carbon emissions by driven forces for the income effects of the policy. The feasible parts of the border regulation might generate from its mandatory effects of border regulation. Cary [46] argued that tariffs could be a viable political tool designed to reduce trade deficits rather than promote sustainability. Extending border taxation adjustment policy would, at first, have more effects on price changes and economic competition for the European Union countries, definitely.

The second findings evidenced the long-run relationships of the consumption of nonrenewable energy and carbon dioxide emissions. For European Union countries in 1990–2021, the consumption of non-renewable energy still remains its significant role in determining income, while this role has been sloughed off in the Organization of Economic Co-operation and Development countries. The European Union countries, which are actively working on CO_2 reduction, appear to have encountered a stringent reliance on non-renewable energy consumption to support income.

The third finding is regarding the connections of nonrenewable energy consumption and carbon dioxide emissions. Not surprisingly, nonrenewable energy consumption performs a strong intimate relation with total carbon dioxide emissions, since the combustion of fossil fuels directly emits carbon dioxides.

The fourth finding is related to the advanced innovation of renewable energy, represented by the renewable power. Renewable power seems important in both perspectives of supporting income and mitigating carbon dioxide emissions, in both groups of countries. This finding can answer a fundamental question: Will green technology progress allow us to see and to go for the dawn of mitigating climate change? Nowadays, renewable power, and also renewable energy, is in composite a very small part in people's energy use in most countries. However, the vision for the future will be full of opportunities. More investigations should be conducted on individual model countries that initiate and develop renewable power, as well as renewable energy.

Five findings are evidenced by the estimated results of Equations (3)–(6) as probing short-run adjustments with Granger causality tests based on a PVECM.

Firstly, strong self-causality of income and renewable power is evidenced along the short-run adjustment, in both country groups. An increase in income or renewable power would promote their own further increase. We can expect that spontaneous growth itself will create a quantum leap. In addition to the aforementioned strong self-causality of income and renewable power, some weak adjustments are evidenced at the 0.1 level. There are two folds for the European Union countries and another two for the Organization of Economic Co-operation and Development countries. In the European Union countries, (1) carbon dioxide emissions weakly Granger cause renewable power, and (2) consumption of nonrenewable energy have weak and negative effects on renewable power. In the Organization of Economic Co-operation and Development countries, (1) renewable power weakly Granger causes carbon dioxide emissions and (2) carbon dioxide emissions negatively, and renewable energy consumption itself positively Granger causes nonrenewable energy consumption. The European Union countries have actively adopted various methods to reduce carbon emissions and have achieved remarkable results, but it is also seen from their PVECM short-term adjustment estimates that they have been unable to adjust the nonrenewable energy consumption that causes carbon emissions through any variables.

In the Organization of Economic Co-operation and Development countries, nonrenewable energy consumption would still statistical weakly be Granger caused by total carbon dioxide emissions and nonrenewable energy consumption itself. Nonrenewable energy efficiency improvement is still possible, and renewable power is still a carbon dioxide generator in the Organization of Economic Co-operation and Development countries.

The long-run dynamics can be captured by the statistical significance of error correction terms in Equations (3)–(6), as seen by the estimates of the error correction term in the PVEVMs shown in Table 6. For the European Union countries, carbon dioxide emissions, nonrenewable energy consumption, and renewable power would response to a deviation of carbon dioxide equilibrium, while carbon dioxide emissions, income, nonrenewable energy consumption, and renewable power would response to a deviation of income equilibrium. For the Organization of Economic Co-operation and Development countries, the carbon dioxide emissions, income, nonrenewable energy consumption, and renewable power would response to a deviation of carbon dioxide emission equilibrium, while income and renewable power would respond to a deviation of income equilibrium. (1) The study on the European Union countries evidenced that a deviation of carbon dioxide emissions from its balanced equilibrium would move backward through adjustment of carbon dioxide emissions, nonrenewable energy consumption, and the use of renewable power. The adjustment speed of renewable power is especially high, and the adjustment of income is not significant. At the same time, a deviation from income equilibrium will be adjusted via all variables. However, the adjustment of renewable power pushes this deviation away from the balance. (2) The study of the Organization of Economic Co-operation and Development evidenced that deviations of carbon dioxide emissions from their equilibrium would be brought back through the adjustment of all variables, while a deviation from income equilibrium would only be adjusted back via income, and brought away by the adjustment of renewable power. However, the adjustment of carbon dioxide emissions and nonrenewable energy are insignificant.

The evidences of the aforementioned empirics indicated by income ECT in the European Union and the Organization of Economic Co-operation and Development affirm that only renewable power can break down the long-run equilibrium of income. The development of renewable power is adjusted for income supporting and expansion, rather than combating carbon dioxide emissions and rather than alleviating climate change. Green technology development supports income but has no effect on the reduction of domestic total carbon dioxide emissions. This is a sad story! The approach low carbon development through development of green technology leaves unfulfilled ideals and slogans. The question of what is the effective policy to break down the balance of carbon dioxide emissions for mitigating climate change remains barely answered by the development of renewable power to combat total carbon dioxide emissions. Policy recommendations work to bind and integrate current low carbon technology into income earnings, so as to develop prudent detailed government regulations. The study of carbon intensity and energy intensity already present declination trends due to technological innovations in the literature [2,4]. The internal reduction response effects of the enterprises and lifestyles of people would be another important issues to be probed.

Based upon the aforementioned dynamics of the selected variables, recent policies of carbon pricing and net-zero carbon emissions, which have intended to pursue suppressing the persistent increase of carbon dioxide emissions for combating climate change, have more role in terms of monetary effects rather than to increase the declination of total carbon dioxide emissions. As suggested by Cary [46], when it comes to extending carbon pricing to border adjustment from various local or regional carbon pricing to take effects globally, the effects would go to changes in international competitions and local consumption prices, even though the explicit goal is set to mitigate climate. Overall carbon dioxide emissions do not decline over time. The present study has evidenced a lack of connections to incentivize their reduction with the included aggregate variables. The concerns over climate changes should be raised.

The evidence in the present research is based on panel vector error correction model techniques and 1990–2021 data of 18 European Union countries and 32 Organization of Economic Co-operation and Development countries. The implementations of the study would represent the temporal and spatial scopes of the panel data applied. Renewable power is used to represent one advanced example of green technology progress. More important findings might be generated from extant studies by utilizing different empirical techniques on other economic scales by extending the research time frame. In order to find effective ways to mitigate climate change, more research is worthwhile on the dynamics of individual countries, on carbon intensity analysis, on the incentive effects of carbon pricing and carbon tax border adjustment, and on carbon reduction in enterprises and lifestyles. We would like to conclude this paper by outlining some further developments. The parametric estimation approach, which is used in the present study, can provide estimated value of a parameter to demonstrate the patterns for the linkages between variables. However, a time-varying nonparametric model would be suitable to detect the structural breaks as the changes in the parameter are captured [93]. The present study applies a parameter technique, and the parameters are fixed over time. The present paper would conclude further developments on the time varying techniques to detect the structure changes in the linkages of the variables.

Funding: This research did not receive any specific grant from funding agencies in the public, commercial, or not-for-profit sectors.

Institutional Review Board Statement: Not applicable.

Informed Consent Statement: Not applicable.

Data Availability Statement: The data were retrieved from Penn World Table and BP statistics.

Acknowledgments: The author wants to thank the anonymous reviewers and the editor that contributed to the quality of the paper with their insights.

Conflicts of Interest: The author declares no conflict of interest.

References

1. Nordhaus, W.D. *The Climate Casino*; Yale University Press: London, UK, 2013.
2. Rojey, A. *Energy and Climate: How to Achieve a Successful Energy Transition*; John Wiley & Sons: Hoboken, NJ, USA, 2009.
3. Stefanski, R. On the mechanics of the green solow model. *OxCarre Res. Pap.* **2013**, *47*. Available online: https://citeseerx.ist.psu.edu/viewdoc/download?doi=10.1.1.703.90&rep=rep1&type=pdf (accessed on 14 July 2021).
4. Chen, W.J. Is the green solow model valid for carbon dioxide emissions in the european union? *Environ. Resour. Econ.* **2017**, *67*, 23–45. [CrossRef]
5. Asen, E. *Carbon Taxes in Europe*; Tax Foundation: Washington, DC, USA, 2021.
6. EU Taxation and Customs Union. EU Taxpayers and Cross-Border Tax Issues. 2021. Available online: https://ec.europa.eu/taxation_customs/eu-taxpayers-and-cross-border-tax-issues_en (accessed on 14 July 2021).
7. Temple, J. How an EU Tax Could Slash Climate Emissions Far beyond Europe. 2020. Available online: https://www.technologyreview.com/2020/07/31/1005819/how-an-eu-tax-could-slash-emissions-far-beyond-its-borders/ (accessed on 7 March 2021).
8. European Commission. Energy and the Green Deal. 2021. Available online: https://ec.europa.eu/info/strategy/priorities-2019-2024/european-green-deal/energy-and-green-deal_en (accessed on 14 July 2021).
9. Apergis, N.; Payne, J.E. Renewable and non-renewable energy consumption-growth nexus: Evidence from a panel error correction model. *Energy Econ.* **2012**, *34*, 733–738. [CrossRef]
10. Apergis, N.; Payne, J.E. Renewable energy consumption and economic growth: Evidence from a panel of OECD countries. *Energy Policy* **2010**, *38*, 656–660. [CrossRef]

11. Armeanu, D.Ş.; Vintilă, G.; Gherghina, Ş.C. Does renewable energy drive sustainable economic growth? Multivariate panel data evidence for EU-28 countries. *Energies* **2017**, *10*, 381. [CrossRef]
12. Chen, C.W. Renewable Energy and Low Carbon Economic Development: A Panel Data Analysis. Master's Dissertation, Chinese Culture University, Taipei, Taiwan, 2017; p. 98.
13. Greening, L.A.; Greene, D.L.; Difiglio, C. Energy efficiency and consumption—The rebound effect—A survey. *Energy Policy* **2000**, *28*, 389–401. [CrossRef]
14. Peters, G.P. Policy Update: Managing carbon leakage. *Carbon Manag.* **2010**, *1*, 35–37. [CrossRef]
15. Kruyt, B.; Van Vuuren, D.P.; de Vries, H.J.; Groenenberg, H. Indicators for energy security. *Energy Policy* **2009**, *37*, 2166–2181. [CrossRef]
16. Cherp, A.; Jewell, J. The concept of energy security: Beyond the four As. *Energy Policy* **2014**, *75*, 415–421. [CrossRef]
17. Winzer, C. Conceptualizing energy security. *Energy Policy* **2012**, *46*, 36–48. [CrossRef]
18. Ang, B.W.; Choong, W.L.; Ng, T.S. Energy security: Definitions, dimensions and indexes. *Renew. Sustain. Energy Rev.* **2015**, *42*, 1077–1093. [CrossRef]
19. Andrews, C. National responses to energy vulnerability. *IEEE Technol. Soc. Mag.* **2006**, *25*, 16–25. [CrossRef]
20. Gnansounou, E. Assessing the energy vulnerability: Case of industrialised countries. *Energy Policy* **2008**, *36*, 3734–3744. [CrossRef]
21. Hall, S.M.; Hards, S.; Bulkeley, H. New approaches to energy: Equity, justice and vulnerability. Introduction to the special issue. *Local Environ.* **2013**, *18*, 413–421. [CrossRef]
22. Kaygusuz, K. Energy for sustainable development: A case of developing countries. *Renew. Sustain. Energy Rev.* **2012**, *16*, 1116–1126. [CrossRef]
23. Dincer, I. Renewable energy and sustainable development: A crucial review. *Renew. Sustain. Energy Rev.* **2000**, *4*, 157–175. [CrossRef]
24. Kaygusuz, K. Energy for Sustainable Development: Key Issues and Challenges. *Eng. Sources Part B Econ. Plan. Policy* **2007**, *2*, 73–83. [CrossRef]
25. Oyedepo, S.O. On energy for sustainable development in Nigeria. *Renew. Sustain. Energy Rev.* **2012**, *16*, 2583–2598. [CrossRef]
26. Mohamed, A.R.; Lee, K.T. Energy for sustainable development in Malaysia: Energy policy and alternative energy. *Energy Policy* **2006**, *34*, 2388–2397. [CrossRef]
27. Solomon, B.D.; Krishna, K. The coming sustainable energy transition: History, strategies, and outlook. *Energy Policy* **2011**, *39*, 7422–7431. [CrossRef]
28. Manfren, M.; Tagliabue, L.C.; Cecconi, F.R.; Ricci, M. Long-Term Techno-Economic Performance Monitoring to Promote Built Environment Decarbonisation and Digital Transformation—A Case Study. *Sustainability* **2022**, *14*, 644. [CrossRef]
29. Jiang, Z.; Lyu, P.; Ye, L.; Zhou, Y.W. Green innovation transformation, economic sustainability and energy consumption during China's new normal stage. *J. Clean. Prod.* **2020**, *273*, 123044. [CrossRef]
30. Gismondi, M. Historicizing transitions: The value of historical theory to energy transition research. *Energy Res. Soc. Sci.* **2018**, *38*, 193–198. [CrossRef]
31. Leach, G. The energy transition. *Energy Policy* **1992**, *20*, 116–123. [CrossRef]
32. United Nations Environment Programme (UNEP). The Emissions Gap Report 2021, United Nations Environment Program. Available online: https://www.unep.org/resources/emissions-gap-report-2021 (accessed on 14 July 2021).
33. Reid, W.V.; Chen, D.; Goldfarb, L.; Hackmann, H.; Lee, Y.T.; Mokhele, K.; Ostrom, E.; Raivio, K.; Rockström, J.; Schellnhuber, H.J. Earth system science for global sustainability: Grand challenges. *Science* **2010**, *330*, 916–917. [CrossRef]
34. Akin-Ponnle, A.E.; Pereira, F.S.; Madureira, R.C.; Carvalho, N.B. From macro to micro: Impact of smart turbine energy harvesters (STEH), on environmental sustainability and smart city automation. *Sustainability* **2022**, *14*, 1887. [CrossRef]
35. Davidson, D.J. Exnovating for a renewable energy transition. *Nat. Energy* **2019**, *4*, 254–256. [CrossRef]
36. Pérez, M.D.L.E.M.; Scholten, D.; Stegen, K.S. The multi-speed energy transition in Europe: Opportunities and challenges for EU energy security. *Energy Strat. Rev.* **2019**, *26*, 100415. [CrossRef]
37. Gatto, A.; Busato, F. Energy vulnerability around the world: The global energy vulnerability index (GEVI). *J. Clean. Prod.* **2020**, *253*, 118691. [CrossRef]
38. Gatto, A.; Drago, C. A taxonomy of energy resilience. *Energy Policy* **2020**, *136*, 111007. [CrossRef]
39. Gatto, A.; Drago, C. When renewable energy, empowerment, and entrepreneurship connect: Measuring energy policy effectiveness in 230 countries. *Energy Res. Soc. Sci.* **2021**, *78*, 101977. [CrossRef]
40. Gatto, A.; Loewenstein, W.; Sadik-Zada, E.R. An extensive data set on energy, economy, environmental pollution and institutional quality in the petroleum-reliant developing and transition economies. *Data Brief* **2021**, *35*, 106766. [CrossRef] [PubMed]
41. Narayan, P.K.; Popp, S. The energy consumption-real GDP nexus revisited: Empirical evidence from 93 countries. *Econ. Model.* **2012**, *29*, 303–308. [CrossRef]
42. Bilan, Y.; Streimikiene, D.; Vasylieva, T.; Lyulyov, O.; Pimonenko, T.; Pavlyk, A. Linking between renewable energy, CO2 emissions, and economic growth: Challenges for candidates and potential candidates for the EU membership. *Sustainability* **2019**, *11*, 1528. [CrossRef]
43. Dogru, T.; Bulut, U.; Kocak, E.; Isik, C.; Suess, C.; Sirakaya-Turk, E. The nexus between tourism, economic growth, renewable energy consumption, and carbon dioxide emissions: Contemporary evidence from OECD countries. *Environ. Sci. Pollut. Res.* **2020**, *27*, 40930–40948. [CrossRef] [PubMed]

44. Balcilar, M.; Ozdemir, Z.A.; Tunçsiper, B.; Ozdemir, H.; Shahbaz, M. On the nexus among carbon dioxide emissions, energy consumption and economic growth in G-7 countries: New insights from the historical decomposition approach. *Environ. Dev. Sustain.* **2020**, *22*, 8097–8134. [CrossRef]
45. Torvanger, A. Manufacturing sector carbon dioxide emissions in nine OECD countries, 1973–87: A Divisia index decomposition to changes in fuel mix, emission coefficients, industry structure, energy intensities and international structure. *Energy Econ.* **1991**, *13*, 168–186. [CrossRef]
46. Cary, M. Molecules of inefficiency: How tariffs impact carbon intensities, carbon dioxide emissions, and the environment. *Sci. Total Environ.* **2020**, *713*, 136531. [CrossRef]
47. Creutzig, F.; Goldschmidt, J.C.; Lehmann, P.; Schmid, E.; von Blücher, F.; Breyer, C.; Fernandez, B.; Jakob, M.; Knopf, B.; Lohrey, S.; et al. Catching two European birds with one renewable stone: Mitigating climate change and Eurozone crisis by an energy transition. *Renew. Sustain. Energy Rev.* **2014**, *38*, 1015–1028. [CrossRef]
48. Wooldridge, J.M. *Econometric Analysis of Cross Section and Panel Data*; MIT Press: London, UK, 2010.
49. Feenstra, R.C.; Inklaar, R.; Timmer, M.P. The next generation of the Penn World Table. *Am. Econ. Rev.* **2015**, *105*, 3150–3182. Available online: www.ggdc.net/pwt (accessed on 14 July 2021).
50. British Petroleum Global. Statistical Review of World Energy. 2021. Available online: https://www.bp.com/en/global/corporate/energy-economics/statistical-review-of-world-energy.html (accessed on 14 July 2021).
51. Kao, C. Spurious regression and residual-based tests for cointegration in panel data. *J. Econ.* **1999**, *90*, 1–44. [CrossRef]
52. Pedroni, P. Critical values for cointegration tests in heterogeneous panels with multiple regressors. *Oxford Bull. Econ. Stat.* **1999**, *61*, 653–670. [CrossRef]
53. Gutierrez, L. On the power of panel cointegration tests: A Monte Carlo comparison. *Econ. Lett.* **2003**, *80*, 105–111. [CrossRef]
54. Pedroni, P. Fully Modified OLS for Heterogeneous Cointegrated Panels. In *Nonstationary Panels, Panel Cointegration, and Dynamic Panels*; Emerald Group Publishing Limited: Bingley, UK, 2001. Available online: https://doi.org/10.1016/S0731-9053(00)15004-2 (accessed on 14 July 2021).
55. Kao, C.; Chiang, M.H. On the Estimation and Inference of a Cointegrated Regression in Panel Data. In *Nonstationary Panels, Panel Cointegration, and Dynamic Panels*; Emerald Group Publishing Limited: Bingley, UK, 2001. Available online: https://www.emerald.com/insight/content/doi/10.1016/S0731-9053(00)15007-8/full/pdf?title=on-the-estimation-and-inference-of-a-cointegrated-regression-in-panel-data (accessed on 14 July 2021).
56. Mark, N.C.; Sul, D. Nominal exchange rates and monetary fundamentals: Evidence from a small post-Bretton woods panel. *J. Int. Econ.* **2001**, *53*, 29–52. [CrossRef]
57. Pedroni, P. Purchasing power parity tests in cointegrated panels. *Rev. Econ. Stat.* **2001**, *83*, 727–731. [CrossRef]
58. Dinh, T.T.-H.; Vo, D.H.; Vo, A.T.; Nguyen, T.C. Foreign direct investment and economic growth in the short run and long run: Empirical evidence from developing countries. *J. Risk Financ. Manag.* **2019**, *12*, 176. [CrossRef]
59. Miśkiewicz, R. The impact of innovation and information technology on greenhouse gas emissions: A case of the visegrád countries. *J. Risk Financ. Manag.* **2021**, *14*, 59. [CrossRef]
60. Mitić, P.; Ivanović, O.M.; Zdravković, A. A cointegration analysis of real GDP and CO2 emissions in transitional countries. *Sustainability* **2017**, *9*, 568. [CrossRef]
61. Gómez, M.; Rodríguez, J.C. Energy consumption and financial development in NAFTA countries, 1971–2015. *Appl. Sci.* **2019**, *9*, 302. [CrossRef]
62. Stock, J.H.; Watson, M.W. A Simple Estimator of Cointegrating Vectors in Higher Order Integrated Systems. *Econ. Soc.* **1993**, *61*, 783. [CrossRef]
63. Hayakawa, K.; Kurozumi, E. The role of "leads" in the dynamic OLS estimation of cointegrating regression models. *Math. Comput. Simul.* **2008**, *79*, 555–560. [CrossRef]
64. Masih, R.; Masih, A.M. Stock-Watson dynamic OLS (DOLS) and error-correction modelling approaches to estimating long- and short-run elasticities in a demand function: New evidence and methodological implications from an application to the demand for coal in mainland China. *Energy Econ.* **1996**, *18*, 315–334. [CrossRef]
65. Al-Azzam, A.; Hawdon, D. *Estimating the Demand for Energy in Jordan: A Stock-Watson Dynamic OLS (DOLS) Approach (No. 97)*; Surrey Energy Economics Centre (SEEC), School of Economics, University of Surrey: Guildford, UK, 1999. Available online: https://kipdf.com/estimating-the-demand-for-energy-in-jordan-a-stock-watson-dynamic-ols-dols-appro_5b16594f7f8b9a1f098b45b6.html (accessed on 14 July 2021).
66. Apergis, N.; Payne, J.E. The renewable energy consumption–growth nexus in Central America. *Appl. Energy* **2011**, *88*, 343–347. [CrossRef]
67. Menegaki, A. Growth and renewable energy in Europe: A random effect model with evidence for neutrality hypothesis. *Energy Econ.* **2011**, *33*, 257–263. [CrossRef]
68. Inglesi-Lotz, R. The impact of renewable energy consumption to economic growth: A panel data application. *Energy Econ.* **2016**, *53*, 58–60. [CrossRef]
69. Kahia, M.; Ben Aïssa, M.S.; Lanouar, C. Renewable and non-renewable energy use—Economic growth nexus: The case of MENA Net Oil Importing Countries. *Renew. Sustain. Energy Rev.* **2017**, *71*, 127–140. [CrossRef]
70. Sadorsky, P. Renewable energy consumption and income in emerging economies. *Energy Policy* **2009**, *37*, 4021–4028. [CrossRef]

71. Shahbaz, M.; Rasool, G.; Ahmed, K.; Mahalik, M.K. Considering the effect of biomass energy consumption on economic growth: Fresh evidence from BRICS region. *Renew. Sustain. Energy Rev.* **2016**, *60*, 1442–1450. [CrossRef]
72. Lise, W.; Van Montfort, K. Energy consumption and GDP in Turkey: Is there a co-integration relationship? *Eng. Econ.* **2007**, *29*, 1166–1178. [CrossRef]
73. Li, Q.; Cherian, J.; Shabbir, M.S.; Sial, M.S.; Li, J.; Mester, I.; Badulescu, A. Exploring the relationship between renewable energy sources and economic growth. The Case of SAARC Countries. *Energies* **2021**, *14*, 520. [CrossRef]
74. Gherghina, Ş.C.; Onofrei, M.; Vintilă, G.; Armeanu, D.Ş. Empirical evidence from EU-28 countries on resilient transport infrastructure systems and sustainable economic growth. *Sustainability* **2018**, *10*, 2900. [CrossRef]
75. Gherghina, Ş.C.; Simionescu, L.N.; Hudea, O.S. Exploring foreign direct investment—Economic growth nexus—Empirical evidence from central and eastern European countries. *Sustainability* **2019**, *11*, 5421. [CrossRef]
76. Armeanu, D.Ş.; Gherghina, Ş.C.; Pasmangiu, G. Exploring the causal nexus between energy consumption, environmental pollution and economic growth: Empirical evidence from central and Eastern Europe. *Energies* **2019**, *12*, 3704. [CrossRef]
77. Hu, Y.; Guo, D.; Wang, M.; Zhang, X.; Wang, S. The relationship between energy consumption and economic growth: Evidence from China's industrial sectors. *Energies* **2015**, *8*, 9392–9406. [CrossRef]
78. Emam, M.A.; Leibrecht, M.; Chen, T. Fish exports and the growth of the agricultural sector: The case of south and southeast asian countries. *Sustainability* **2021**, *13*, 11177. [CrossRef]
79. Latief, R.; Kong, Y.; Javeed, S.A.; Sattar, U. Carbon emissions in the SAARC countries with causal effects of FDI, economic growth and other economic factors: Evidence from dynamic simultaneous equation models. *Int. J. Environ. Res. Public. Health* **2021**, *18*, 4605. [CrossRef]
80. Busu, M. Analyzing the Impact of the Renewable Energy Sources on Economic Growth at the EU Level Using an ARDL Model. *Mathematics* **2020**, *8*, 1367. [CrossRef]
81. He, Y.; Wu, R.; Choi, Y.-J. International Logistics and Cross-Border E-Commerce Trade: Who Matters Whom? *Sustainability* **2021**, *13*, 1745. [CrossRef]
82. Sung, B.; Choi, M.S.; Song, W.-Y. Exploring the Effects of Government Policies on Economic Performance: Evidence Using Panel Data for Korean Renewable Energy Technology Firms. *Sustainability* **2019**, *11*, 2253. [CrossRef]
83. Thathsarani, U.; Wei, J.; Samaraweera, G. Financial Inclusion's Role in Economic Growth and Human Capital in South Asia: An Econometric Approach. *Sustainability* **2021**, *13*, 4303. [CrossRef]
84. Socoliuc, M.; Cosmulese, C.-G.; Ciubotariu, M.-S.; Mihaila, S.; Arion, I.-D.; Grosu, V. Sustainability Reporting as a Mixture of CSR and Sustainable Development. A Model for Micro-Enterprises within the Romanian Forestry Sector. *Sustainability* **2020**, *12*, 603. [CrossRef]
85. Lv, Z.; Chu, A.M.Y.; McAleer, M.; Wong, W.-K. Modelling Economic Growth, Carbon Emissions, and Fossil Fuel Consumption in China: Cointegration and Multivariate Causality. *Int. J. Environ. Res. Public Health* **2019**, *16*, 4176. [CrossRef] [PubMed]
86. Natanian, J.; Aleksandrowicz, O.; Auer, T. A parametric approach to optimizing urban form, energy balance and environmental quality: The case of Mediterranean districts. *Appl. Eng.* **2019**, *254*, 113637. [CrossRef]
87. Amado, M.; Poggi, F. Solar urban planning: A parametric approach. *Energy Procedia* **2014**, *48*, 1539–1548. [CrossRef]
88. Wang, Q.; Zhou, P.; Shen, N.; Wang, S. Measuring carbon dioxide emission performance in Chinese provinces: A parametric approach. *Renew. Sustain. Energy Rev.* **2013**, *21*, 324–330. [CrossRef]
89. Iqbal, H.H.; Muhammad, H.; Tvaronavičienė, M.; Kittisak, J. The causal connection of natural resources and globalization with energy consumption in top Asian countries: Evidence from a nonparametric causality-in-quantile approach. *Energies* **2020**, *13*, 1–18.
90. Luo, N.; Mao, D.; Wen, B.; Liu, X. Climate change affected vegetation dynamics in the Northern Xinjiang of China: Evaluation by SPEI and NDVI. *Land* **2020**, *9*, 90. [CrossRef]
91. Hurvich, C.M.; Simonoff, J.S.; Tsai, C.-L. Smoothing parameter selection in nonparametric regression using an improved Akaike information criterion. *J. R. Stat. Soc. Ser. B* **1998**, *60*, 271–293. [CrossRef]
92. Li, D.; Chen, J.; Gao, J. Non-parametric time-varying coefficient panel data models with fixed effects. *Econom. J.* **2011**, *14*, 387–408. [CrossRef]
93. Liu, D.; Li, R.; Wang, Z. Testing for structural breaks in panel varying coefficient models: With an application to OECD health expenditure. *Empir. Econ.* **2011**, *40*, 95–118. [CrossRef]
94. Lee, J.; Robinson, P.M. Panel nonparametric regression with fixed effects. *J. Econ.* **2015**, *188*, 346–362. [CrossRef]
95. Lozano, S.; Gutierrez, E. Non-parametric frontier approach to modelling the relationships among population, GDP, energy consumption and CO2 emissions. *Ecol. Econ.* **2008**, *66*, 687–699. [CrossRef]
96. Sadik-Zada, E.R.; Loewenstein, W. Drivers of CO_2-emissions in fossil fuel abundant settings: (Pooled) Mean group and nonparametric panel analyses. *Energies* **2020**, *13*, 3956. [CrossRef]
97. Chen, B.; Hong, Y. Testing for smooth structural changes in time series models via nonparametric regression. *Econometrica* **2012**, *80*, 1157–1183.
98. Chen, H.; Liu, L.; Wang, Y.; Zhu, Y. Oil price shocks and U.S. dollar exchange rates. *Energy* **2016**, *112*, 1036–1048. [CrossRef]
99. Miao, Z.; Chen, X. Combining parametric and non-parametric approach, variable & source -specific productivity changes and rebound effect of energy & environment. *Technol. Forecast. Soc. Chang.* **2022**, *175*, 121368.

100. Jevons, W.S. *The Coal Question; An Inquiry Concerning the Progress of the Nation and the Probable Exhaustion of our Coalmines*; MacMillan and Co.: London, UK; Cambridge, UK, 1866. Available online: https://books.google.com.tw/books/about/The_Coal_Question.html?id=gAAKAAAAIAAJ&printsec=frontcover&source=kp_read_button&hl=en&redir_esc=y#v=onepage&q&f=false (accessed on 14 July 2021).
101. Chen, W.-J.; Wang, C.-H. A general cross-country panel analysis for the effects of capitals and energy, on economic growth and carbon dioxide emissions. *Sustainability* **2020**, *12*, 5916. [CrossRef]

sustainability

MDPI

Article

Does the Sino–US Trade Friction Promote Firm Innovation? The Role of the Export Grab Effect

Zhen Xu [1], Xin Zhong [1,*] and Zhongyan Zhang [2]

[1] Business School, Beijing Normal University, Beijing 100875, China; zhenxu@mail.bnu.edu.cn
[2] Department of Economics and Management, Taiyuan Institute of Technology, Taiyuan City 030008, China; 201831410011@mail.bnu.edu.cn
* Correspondence: 03022@bnu.edu.cn

Abstract: We examine and explain changes of firm innovation following the Sino–US trade frictions. Specifically, we use the unique microdata of Chinese technology and innovation firms from 2016 to 2019 to identify the response of firm innovation to the shock of the Sino–US trade friction and investigate the changes in firm innovation before and after the outbreak of the Sino–US trade friction. We found that, on average, export firms significantly increased innovation after the Sino–US trade friction. In particular, firms in the advanced manufacturing industry perform significantly better than firms in other technical fields. To explain this phenomenon, we compare firms' export performance in different technical fields and conduct a detailed mechanism test. It is found that the Sino–US trade friction has led to the export grab phenomenon in some technical fields, and the export grab effect is the reason for the significant increase in the innovation of export firms. The direct effect of export grab promotes the innovation of sustainable export firms, and the indirect effect of export grab leads to an increase in innovation by reducing the threshold of the export market. We verified the direct and indirect effects of export grab on firm innovation following the Sino–US trade frictions.

Keywords: Sino–US trade friction; export grab effect; firm innovation

Citation: Xu, Z.; Zhong, X.; Zhang, Z. Does the Sino–US Trade Friction Promote Firm Innovation? The Role of the Export Grab Effect. *Sustainability* **2022**, *14*, 2709. https://doi.org/10.3390/su14052709

Academic Editors: Elkhan Richard Sadik-Zada and Andrea Gatto

Received: 29 January 2022
Accepted: 23 February 2022
Published: 25 February 2022

1. Introduction

Innovation is a sustainable driving force for economic development [1], but the current international economic situation poses great challenges to innovative firms. Since the beginning of 2018, a series of adjustments of import tariffs by China and the United States have exposed the global economy to the great risks of trade policy uncertainty [2]. According to recent advances in trade policy uncertainty literature, the elimination of trade policy uncertainty benefits innovation [3–5]. Previous empirical studies have verified this using Chinese data at both the firm level and industry level [6,7]. Obviously, the theory reveals that there is an inverse linear relationship between trade policy uncertainty and innovation, which means that the rise of trade policy uncertainty will impair innovation. Therefore, does this indicate that the Sino–US trade frictions will damage the innovation of Chinese firms?

We find that trade policy uncertainty literature does not explain changes in Chinese firms' innovation following the Sino–US trade frictions, at least in the comparatively short term. By using the unique microdata of Chinese technology and innovation firms from 2016 to 2019, we find that export firms have significantly increased innovation after the outbreak of the Sino–US trade friction, which is evidently not in line with the internal logic of trade policy uncertainty literature. Further research also found that although the high-tech field is the focus of the Sino–US trade friction, the innovation performance of export firms in the advanced manufacturing industry shows an unusually prominent growth after the Sino–US trade friction. However, it seems difficult for the existing literature to make a reasonable explanation for such a phenomenon. The literature on competitive pressure in export markets may possibly explain this: the Sino–US trade friction intensifies the competitive pressure of export firms facing external markets [8], and escaping the competition effect may make such firms increase research and development (R&D) investment and innovation activities [9–11] to obtain greater market share and profits in the future [12].

However, the dataset used in this paper fails to reveal the competitive pressure in the export market faced by firms; in contrast, it shows that export firms in some technical fields have exceptionally prominent export performance after the Sino–US trade friction. For example, in the advanced manufacturing industry, which has outstanding innovation performance after the trade friction, the annual average export volume of its firms has increased by approximately twice as much as that before the trade friction, and the export share of total income has also increased significantly. Obviously, these indicators are not sufficient to help us obtain a sense of the pressure in the export market when firms are facing adverse shocks. However, by comparing the differences in exports of firms in different technical fields before and after the Sino–US trade friction, we find that those technical fields that have significantly increased innovation after the Sino–US trade friction are often accompanied by comparatively more prominent export business growth, while exports in the technical fields that do not have increased innovation after the Sino–US trade friction are generally not well performing and even have negative growth. Therefore, is this a coincidence?

In fact, after the outbreak of the Sino–US trade friction, Chinese firms have an obvious phenomenon of export grab. According to the annual statistics of China's National Bureau of Statistics, China's exports increased by 7.05% in 2018. Additionally, at the press conference in December 2018, a spokesman for China's Ministry of Commerce said that China's imports and exports achieved rapid growth in 2018. Empirical studies in the literature also reveal the existence of the export grab effect of the Sino–US trade frictions. For example, Xie and Wang [13] found that the Sino–US trade friction can promote the short-term export of products by using monthly data at the product level. Zhang et al. [14] verified the existence of the export grab effect of the Sino–US trade friction by using product-level custom data. Cavallo et al. [15] compared the commodity data of importers and retailers in the United States and found that after the Sino–US trade friction, importers increased their imports from China in the short term and expanded their inventory as much as possible to cope with the risk of further tariff increases in the future.

Export grab is a manifestation of the reallocation effect of the Sino–US trade friction on export firms. By probing into the export behaviour of firms in different technical fields, we find that firms in certain technical fields show an obvious export-grab phenomenon after the outbreak of the Sino–US trade friction, while others do not, and they even suffer a great export shrinkage. Theoretically, the export grab effect comes from the demand shock in the international commodity market. Foreign buyers overdraw their future demand to avoid the risk of future market uncertainty, resulting in a sharp rise in the demand for Chinese products in the short term, which may have two effects. The first is the direct effect. For firms that continue to export, due to the effect of economies of scale, the increase in output caused by demand expansion in the short term will reduce their marginal cost, and then the increased productivity will enable capable firms to carry out innovation activities [16,17]. The second is called the indirect effect. The shock of foreign demand leads to an insufficient supply of domestic sustainable exporters. As a result, export grab reduces the threshold for domestic firms to enter the export market, which will affect firms' entry and exit, export decision making and innovation [16]. The empirical study of this paper verifies that the direct effect and indirect effect of export grab are the mechanism of the increase in innovation of export firms after the Sino–US trade friction and explains the phenomenon that export firms increase innovation after the Sino–US trade friction.

The possible contributions of this paper are as follows: First, we use unique microdata of Chinese technology and innovation firms to provide new evidence that the Sino–US trade friction stimulates firms' innovation. Although previous studies have discussed trade friction and innovation [18,19], they have not yet formed a unified conclusion. In contrast to these studies, we use more detailed microdata of technology and innovative firms and take the outbreak of the Sino–US trade friction in 2018 as an exogenous shock. It is found that, on average, export firms increase innovation after the Sino–US trade friction. Second, we explain a new mechanism of trade friction affecting firm innovation, which is also the most important innovation of this paper. Wei and Liu [18] believe that trade friction has a compelling effect, which leads to an increase in firm innovation. We creatively find that the export grab effect is the reason for the increase in export firm innovation after the Sino–US trade friction. Third, this paper enriches the research on the relationship between exports and firm innovation. Studies in the field of trade have shown that exports could benefit firm innovation and technological upgrading [16,17,20,21]. We verify the significance of exports to firm innovation again.

The structure of the rest of the text is arranged as follows after the introduction part: the second part reviews the relevant literature; the third part introduces the empirical design and data; the fourth part reports the empirical benchmark results, the effectiveness test and robustness test of identification; the fifth part shows the results of the heterogeneity test and tests the export grab mechanism; the sixth part summarizes the entire text.

2. Literature Review

Since the purpose of this article is to test the response of firm innovation to the Sino–US trade friction, and explain the increase in firm innovation following the Sino–US trade friction from the perspective of export grab, we arrange our literature review as follows: We first review the relevant theories of trade friction and literatures evaluating the effects of trade friction in Section 2.1. We then summarize the main theoretical literature on the relationship between export and innovation in Section 2.2.

2.1. Related Economic Effects of the Sino–US Trade Frictions

The Sino–US trade friction represents the return of the United States to trade protectionism [22]. Discussions on trade protectionism in international economics have a long tradition and form a series of rich theories, but trade protectionism theories from different periods have different policy objectives due to the different economic contexts. According to the development context of trade protectionism theory, early mercantilism believed that primitive wealth could be accumulated by restricting imports and expanding exports, while List's infant industrial protection theory believed that trade policy did not need to restrict the import of all industries, and restrictions on agricultural products should be relaxed, while policies should be protected for promising industries [23]. However, when trade protection theory developed in the 1980s, technology transfer and technology restriction became an important part of the theory [24,25]. New trade protection theory and strategic trade protection theory believe that developed countries need to intervene in trade behaviour and restrict the trade of high-tech products by using policies such as technical barriers and intellectual property protection to reduce technology transfer from developed countries to developing countries and safeguard their international competitive advantage and interests [26,27].

However, free trade theory holds that large-scale trade frictions will hurt both sides. Quantitative studies based on this Sino–US trade friction have also verified this point. For example, in Chinese literature, the research of Li et al. shows that the increase in US tariffs will damage the welfare level of both China and the United States. In addition, the employment, import and export of the two countries will be damaged to a certain extent [28]. Lv et al. compared the economic effects of the strategic adjustment on both China and the United States. The study found that the Sino–US trade friction will reduce the scale of the Sino–US trade on a large scale, especially imports of the United States from China, which will be greatly damaged. However, compared to the United States, China's welfare loss will be more serious in trade frictions [29]. Cui et al. also confirmed the above view. Furthermore, they believe that China's retaliatory tariff will aggravate China's welfare loss in trade frictions [30]. In the United States, Amiti et al. studied the price mechanism of the impact of tariff increases on welfare in the United States and found that tariffs in the United States increase the pricing level and mark-ups of the American manufacturing industry and reduce the variety of products consumed by consumers, and that retaliatory tariffs from other countries will make American consumers face serious welfare losses [31]. Similarly, the research of Fajgelbaum et al. shows that the impact of tariff increases in the Sino–US trade frictions will fully penetrate the United States through the import price effect and will be absorbed by American consumers and import firms. In general, the United States will face serious welfare losses [22]. Cavallo et al. once again verified the price effect of tariffs on U.S. welfare with U.S. product level data. From the analysis of Chinese and American economists, it can be found that trade frictions will damage the overall welfare of China and the United States [15].

In addition to the macrolevel analysis of the Sino–US trade friction, the microlevel investigation in the literature also reveals that there is an obvious export grab effect following the Sino–US trade friction. For example, Xie and Wang verified that trade frictions have the effect of export grabbing by using monthly data at the product level, and that the Sino–US trade frictions can promote the short-term export of products [13]. Zhang et al. verified the existence of the export grab effect of trade friction by using data on customs products before and after the Sino–US trade friction [14]. Moreover, data from the United States also prove that trade frictions have an export grab effect in at least some specific industries. Cavallo et al. found that after the Sino–US trade friction, importers have increased their imports from China in the short term and expanded their inventory as much as possible to cope with the risk of further tariff increases in the future. Moreover, compared with Chinese firms exporting to the United States, American firms exporting to China have obvious price reduction behaviour in the face of China's retaliatory tariffs [15].

Therefore, what is the impact of trade friction on firm innovation? No clear and unified conclusion has been reached for this problem in the literature. For example, Wei and Liu used the data of China's listed companies from 2003 to 2016 to measure the trade environment uncertainty faced by firms by the number of trade frictions in the industry and studied the relationship between trade environment uncertainty and firm innovation [18]. The conclusion shows that trade environment uncertainty has a positive effect on firm innovation. The greater the uncertainty of the trade environment is, the higher the innovation of firms in the industry. In contrast, the research of Wan et al. shows that the Sino–US trade frictions will reduce the international R&D level of Chinese firms [19]. Therefore, the effect of trade friction on innovation is not generalized and should be analysed in detail.

2.2. Export Grab Effect and Firm Innovation

However, the export grab induced by the Sino–US trade friction may potentially lead to firm innovation. Although the relationship between export grab and firm innovation has not been addressed in the literature, there have been many discussions on the relationship between exports and innovation in the field of trade in recent years. First, exports have a learning effect. Export firms can improve technology and promote innovation through export learning. Many studies have verified the export learning effect [21,32,33]. On the one hand, the export learning effect affects firm innovation through economies of scale. Exports help in expanding firms' production, and then the role of economies of scale can be further brought into play, which will reduce the marginal cost of firms [20], and promote firm performance [32], and is reflected as productivity gains [34]. Therefore, with increasing performance, firms have an incentive to upgrade their technology, generating an export promotion technological upgrading effect [17]. On the other hand, the export learning effect can be improved through communication with foreign customers. This effect is particularly obvious for export firms in developing countries. Export firms learn from customers or competitors through export behaviour so that their productivity or technical level can be improved [10,21,35–38]. Second, exports have a selective effect. Engaging in innovation is conducive to increasing the productivity level of firms, allowing them to self-select in export markets [34,39]. Indeed, some firms that prefer to enter export markets will undertake ex ante R&D to achieve access to export markets [40], while sustainable exporters will also engage in innovative activities due to competitive pressures in international markets [8] to avoid being eliminated or losing market share in competitive markets [12]. Moreover, the shock of export demand will have a reallocation effect on export firms [11]. On the one hand, the demand expansion in foreign markets will increase the innovation rent of firms and stimulate the innovation investment behaviour of firms; on the other hand, the expansion of foreign demand reduces the threshold of the export market, and new export firms improve the competitive atmosphere of the export market, which also encourages firms' innovation behaviour. However, the shock of foreign demand tends to affect larger firms at the forefront of technology and may play a smaller role in promoting the innovation of new firms.

In fact, the phenomenon of export grabbing following the Sino–US trade friction is a reflection of the impact of foreign demand shocks. To hedge against the risk of further increases in future trade costs, foreign buyers overdrew future demand, which manifested itself in a sharp short-term rise in Chinese export volumes. Based on the above literature review, this paper theoretically argues that this will have two effects on exporters' innovation [11,16]. The first is the direct effect. The expanded export demand reduces the marginal cost of firms, promotes firm productivity and expands firm performance. Firms have the ability and motivation to carry out innovation activities and upgrade technology to obtain more market share and innovation rent in the future. The second is the indirect effect. The shock of foreign demand will change the distribution of firms and have a reallocation effect on domestic export firms, allowing new exporters to enter the export market and generate an export learning effect. Exports will be reallocated to larger firms, and innovation will converge with exporters at the technological frontier, thus leading to an overall increase in the level of innovation among exporters. This paper uses unique data on national technology and innovation firms from 2016 to 2019 to study the impact of the Sino–US trade frictions on firm innovation and verifies the direct and indirect effect mechanisms of firm innovation caused by export grabs following the Sino–US trade frictions.

3. Identification Strategy and Data

3.1. Identification Strategy

In this paper, the treatment group and the control group are divided according to whether the firm is an exporter or not. The control group is non-exporters, and the treatment group is exporters; that is, firms with exports in any period of the sample are defined as export firms in all periods to identify the differential responses of firms in different groups to the shock of the Sino–US trade friction, comparing the changes in firm innovation before and after the Sino–US trade friction. The specific identification equation is as follows:

$$ln(Innovation)_{fit} = \alpha + \beta_1 * EXP_{fi} * Post18_t + X_{ft} * \beta_2 + X_{it} * \beta_3 + \nu_f + Year_t + \varepsilon_{fit} \quad (1)$$

where subscript f means firm, i indicates the industry to which the enterprise belongs, and t indicates the year. The explained variable $ln(Innovation)_{fit}$ indicates the innovation level of firms. The explanatory variable $EXP_{fi} * Post18_t$ is the interaction of a firm's export status EXP_{fi} and the 2018 Sino–US trade friction shock variables $Post18_t$. EXP_{fi} takes 1 as exporters and 0 as non-exporters. $Post18_t$ takes 1 indicates years after the outbreak of the 2018 Sino–US trade friction and takes 0 otherwise. The coefficient of primary interest to us is β_1, which measures the innovation performance difference between export firms and non-export firms after being impacted by the Sino–US trade friction. X_{ft} and X_{it} represent the firm-level and industry-level control variables, respectively, which are used to control the firm characteristics and industry characteristics over time; ν_f is a firm-level fixed effect, which is used to control firm heterogeneity that does not change over time; $Year_t$ is the year fixed effect, which is used to control the interference of time factors on identification; and ε_{fit} is the error term. In addition, we also cluster the standard error according to the four-digit code of the Chinese industry.

It should be noted that, first, we divide the treatment group and the control group because we consider the selective effect of the export market. The Sino–US trade friction may affect the choice of firms to enter and exit the export market. The way we divide the treatment and control groups can fix the impact of the effect of firms entering and exiting the export market based on the results to a certain extent. Second, although export firms are intuitively more likely to be impacted by the Sino–US trade frictions, we cannot deny that non-export firms are not affected during the process of the Sino–US trade frictions. In identifying the effect of Sino–US trade friction on the innovation of exporting firms, using non-exporting firms as a control group requires an examination of whether exporting and non-exporting firms are comparable; therefore, we will verify in the identification validity below that there is a parallel trend between the two after controlling for as many potential influences as possible. Third, this paper chooses the beginning of 2018 as the shock time node of the Sino–US trade friction because the first import tariff increase of the United States was in January 2018; however, the "301" survey of the United States in 2017 may have had an impact on Chinese innovative firms. Therefore, setting 2018 as the shock node may only underestimate the effect of the Sino–US trade friction on firm innovation. In other words, if setting 2018 as the impact node can identify the effect of the Sino–US trade friction on firm innovation, we have reasons to believe that the real effect may be greater.

3.2. Data and Indicators

The empirical research mainly uses our unique mircodata of Chinese technology and innovation firms. This dataset is jointly investigated and recorded by China's National Bureau of Statistics and China's Ministry of Science and Technology. The respondents are technological and innovative firms in most cities of mainland China. The variables involved cover four categories of information: basic information of firms, economic overview of firms, employees of firms and intellectual property rights of firms. The data cover a wide range of industries, including many subindustries under agriculture, manufacturing and service industries. We select subsamples of tradable sectors of the manufacturing industry in the dataset. The sample year is 2016–2019, just two years before and after the Sino–US trade friction in 2018, which can be used to identify the changes in firm innovation before and after the Sino–US trade friction.

We use the logarithm of the number of intellectual property rights granted to a firm in the current year plus one as a measure of firm innovation $ln(Innovation)_{fit}$. The number of intellectual property rights granted in the current year can more accurately measure the innovation of a firm, which is the real output of the firms' innovation activities. In the robustness test, we also construct the innovation index based on the number of firms' intellectual property applications as the explanatory variable. Moreover, we try to control the characteristics of firms as much as possible; hence, we mainly include registered capital ($lnCapital_{fit}$), number of employees ($lnEmployment_{fit}$), R&D investment ($lnRD_{fit}$), angel investment ($lnAngel_{fit}$), and firm age ($lnAge_{fit}$). In addition, this paper also adds the type of firm taxpayer ($Type_{fit}$), technical field ($Field_{fit}$), whether it is a graduated firm ($Graduate_{fit}$), and whether it is a high-tech firm ($Hightech_{fit}$) in the regression.

Considering that import competition, export demand and domestic competition may affect firm innovation, we also control for the above potential effects at the industry level. First, we use China's import volume by industry ($lnImport_{it}$) and import growth rate (Imp_rate_{it}) to control import competition. Second, we use China's export volume by industry ($lnExport_{it}$) and export growth rate (Exp_rate_{it}) to control the impact of changes in export demand on firm innovation. Third, this paper uses the industry-level average profits of scaled Chinese industrial firms ($lnInd\_Competition_{it}$) as a proxy variable to measure domestic industry competition, and it can control the effect of domestic market competition.

Finally, we combine the above data to obtain an unbalanced panel dataset with 157,475 observation samples with a time span of 2016–2019. The descriptive statistics of the relevant indicators are shown in Table 1.

Table 1. Descriptive statistics of the main variables.

Variable	Mean	Standard Error	Minimum	Maximum
$ln(Innovation)_{fit}$	0.280	0.626	0	7.067
$EXP_{fi} * Post18_t$	0.029	0.168	0	1
$lnCapital_{fit}$	7.661	1.552	−4.605	23.49
$lnEmployment_{fit}$	2.658	0.971	0	10.65
$lnRD_{fit}$	2.951	3.036	0	14.86
$lnAngel_{fit}$	0.433	1.743	0	14.05
$lnAge_{fit}$	1.324	0.746	0	4.745
$Type_{fit}$	1.350	0.477	1	2
$Field_{fit}$	3.577	2.763	1	12
$Graduate_{fit}$	1.755	0.430	1	2
$Hightech_{fit}$	0.130	0.337	0	1
$lnExport_{it}$	11.45	1.062	7.191	12.89
Exp_rate_{it}	0.053	0.080	−0.197	0.335
$lnImport_{it}$	11.05	1.095	7.455	13.06
Imp_rate_{it}	0.052	0.129	−0.222	3.013
$lnInd\_Competition_{it}$	−1.854	0.565	−3.049	2.166

4. Empirical Results

4.1. Benchmark Regression Results

The benchmark regression will test the heterogeneous responses of the innovation behaviour of export firms and non-export firms to the Sino–US trade friction, comparing the changes in the innovation differences between export firms and non-export firms before and after the Sino–US trade friction, and then analysing whether export firms have strengthened innovation following the Sino–US trade friction.

The results are reported in Table 2. Column (1) only contains explanatory variables in the regression, with individual fixed effects and time fixed effects, and the estimated explanatory variable coefficient is 0.115, which is significant at the significance level of 1%, meaning that after the Sino–US trade friction, export firms have increased innovation compared with non-export firms. On this basis, Column (2) adds control variables at the firm level. Although the estimated coefficient of $EXP_{fi} * Post18_t$ is lower than Column (1), it is still significantly positive, indicating that after the occurrence of the Sino–US trade friction, export firms have better innovation performance than non-export firms.

Columns (3)–(5) further control the potential impact of foreign demand, import competition and domestic competition at the industry level on the basis of Column (2). Specifically, Column (3) further adds the logarithm of the industry's export volume and the industry's export growth rate to control the impact of the foreign demand shock. The estimated coefficient of $EXP_{fi} * Post18_t$ is still significantly positive, and the coefficients of industry-level foreign demand shock variables are not significant, which shows that export demand does not affect firm innovation. Similarly, Column (4) adds the logarithm of industry import volume and industry import growth rate to control the impact of import industry competition. The coefficient of $EXP_{fi} * Post18_t$ is still significantly positive, and the coefficients of import competition variables are not significant, indicating that industry-level import competition has not impacted firm innovation under the Sino–US trade frictions. Column (5) further controls the potential effect of domestic competition. The estimated coefficient of $EXP_{fi} * Post18_t$ is still significantly positive, and the coefficient of $lnInd\_Competition_{it}$ is still not significant, indicating that domestic industry competition has not had a significant impact on firm innovation. Column (6) contains all control variables. The estimated coefficient of $EXP_{fi} * Post18_t$ is still significantly positive, which is in accordance with the previous estimation results. Therefore, after controlling all potential factors affecting firm innovation as much as possible, we find that export firms strengthen innovation after the Sino–US trade friction more than non-export firms.

Table 2. Benchmark regression results.

	(1)	(2)	(3)	(4)	(5)	(6)
$EXP_{fi} * Post18_t$	0.115 ***	0.082 ***	0.086 ***	0.086 ***	0.082 ***	0.086 ***
	(0.022)	(0.023)	(0.024)	(0.024)	(0.023)	(0.024)
$lnCapital_{fit}$		0.004	0.004	0.004	0.004	0.004
		(0.004)	(0.004)	(0.004)	(0.004)	(0.004)
$lnEmployment_{fit}$		0.082 ***	0.082 ***	0.082 ***	0.082 ***	0.082 ***
		(0.006)	(0.006)	(0.006)	(0.006)	(0.006)
$lnRD_{fit}$		0.024 ***	0.024 ***	0.024 ***	0.024 ***	0.024 ***
		(0.001)	(0.001)	(0.001)	(0.001)	(0.001)
$lnAngel_{fit}$		0.012 ***	0.011 ***	0.011 ***	0.012 ***	0.011 ***
		(0.003)	(0.003)	(0.003)	(0.003)	(0.003)
$lnAge_{fit}$		0.099 ***	0.101 ***	0.101 ***	0.099 ***	0.101 ***
		(0.013)	(0.013)	(0.013)	(0.013)	(0.013)
$Type_{fit}$		−0.035 ***	−0.036 ***	−0.036 ***	−0.035 ***	−0.036 ***
		(0.010)	(0.010)	(0.010)	(0.010)	(0.010)
$Field_{fit}$		0.003	0.004	0.004	0.003	0.004
		(0.003)	(0.003)	(0.003)	(0.003)	(0.003)
$Graduate_{fit}$		0.036 ***	0.036 ***	0.035 ***	0.036 ***	0.036 ***
		(0.010)	(0.010)	(0.010)	(0.010)	(0.010)
$Hightech_{fit}$		0.162 ***	0.161 ***	0.161 ***	0.163 ***	0.161 ***
		(0.015)	(0.015)	(0.015)	(0.015)	(0.015)
$lnExport_{it}$			−0.002			−0.007
			(0.007)			(0.008)
Exp_rate_{it}			0.007			−0.005
			(0.054)			(0.055)
$lnImport_{it}$				0.006		0.009
				(0.006)		(0.008)
Imp_rate_{it}				0.023		0.020
				(0.025)		(0.024)
$lnInd\_Competition_{it}$					0.015	0.010
					(0.011)	(0.011)
$\_cons$	0.295***	−0.215 ***	−0.190 **	−0.282 ***	−0.188 ***	−0.208 **
	(0.001)	(0.043)	(0.094)	(0.080)	(0.051)	(0.098)
Individual FE	yes	yes	yes	yes	yes	yes
Year FE	yes	yes	yes	yes	yes	yes
N	157,475	134,869	131,953	131,953	134,869	131,953
R^2	0.620	0.641	0.640	0.640	0.641	0.641

Note: the dependent variable, firm innovation, is indexed by log (number of firm intellectual property granted + 1). The independent variable is $EXP_{fi} * Post18_t$. Column (1) contains only the independent variable. Column (2) controls all individual control variables. Column (3) further adds the export volume and export growth rate at the industry level to control the impact of industry-level foreign demand shocks. Column (4) further adds industry-level import volume and import growth rate to control import competition shock. Column (5) adds domestic industry competition to control for the impact of the domestic market competition effect. Column (6) contains all individual and industry control variables. All regressions control both individual and year fixed effects (FE). Standard errors are clustered at the four-digit code of the Chinese industry. ***, ** indicate significance levels of 1% and 5%, respectively. N stands for the number of observations and R^2 indicates the fitness of the model.

In short, the benchmark regression results show that after the outbreak of the Sino–US trade friction, export firms have increased innovation. Compared with non-exporters, exporters have evident innovation growth after being shocked by trade frictions. Based on the regression results in Column (6), on average, after the Sino–US trade friction, export firms have increased by 0.086% compared with non-export firms in the innovation measured by intellectual property rights granted. The rest of this section will verify whether export firms and non-export firms are comparable before trade friction and verify the robustness of the benchmark regression results.

4.2. Validation of Identification

The following methods will be used to test the effectiveness of the above identification. The results are reported in Table 3.

Table 3. Effectiveness test of identification.

	(1)	(2)	(3)	(4)
$EXP_{fi} * Post18_t$			0.090 ***	0.069 ***
			(0.031)	(0.025)
$EXP_{ft} * Post17_t$	−0.040			
	(0.032)			
$EXP_{ft} * YR2017_t$		0.014		
		(0.023)		
$EXP_{ft} * YR2018_t$		0.099 ***		
		(0.026)		
$EXP_{ft} * YR2019_t$		0.107 ***		
		(0.030)		
Control variables	yes	yes	yes	yes
Individual FE	yes	yes	yes	yes
Year FE	yes	yes	yes	no
City * Year FE	no	no	no	yes
N	35,242	131,953	56,460	131,872
R2	0.732	0.641	0.714	0.652

Note: The placebo test in Column (1) uses the samples of two years before the shock (2016–2017) to test whether the treatment group and the control group are comparable before trade friction. Column (2) replaces the explanatory variable with the intersections of the export state variable and year dummy variables to test the change in innovation difference between export firms and non-export firms over time. Column (3) regression was carried out using only the samples of the two years before and after the shock node (2017–2018). Column (4) further adds the city cross-year fixed effect (FE) to exclude the potential impact of time-varying city characteristics. Column (1)-(3) control both individual and year fixed effects (FE). Standard errors are clustered at the four-digit code of the Chinese industry. *** indicates significance levels of 1%. N stands for the number of observations and R^2 indicates the fitness of the model.

4.2.1. Placebo Test

Column (1) reports the results of a placebo test, which was conducted by using a subsample of the two years before the occurrence of the Sino–US trade friction in 2018. The purpose is to test whether the innovation difference between export firms and non-export firms also changes significantly over time before the shock. In the specification, we first take the beginning of 2017 as the node to generate a time indicator, $Post17_t$. Then, we interact it with export status variables to generate the placebo variable, $EXP_{ft} * Post17_t$. We replace the independent variable $EXP_{fi} * Post18_t$ in the benchmark regression by $EXP_{ft} * Post17_t$, combined with using samples from 2016 and 2017 in the regression. The result in Column (1) shows that the estimated coefficient of $EXP_{ft} * Post17_t$ is not statistically significant, indicating that after controlling all individual and industry control variables, the innovation difference between export firms and non-export firms has not changed significantly in approximately the year 2017; that is, the innovation of export firms and non-export firms is comparable before the Sino–US trade friction.

4.2.2. More Flexible Estimates

Column (2) adopts a more flexible identification method, replacing the independent variable in the benchmark estimation by a series of interactions of export status indicator and year dummy. The results show that the coefficient of $EXP_{ft} * YR2017_t$ is not significant, which shows that before the Sino–US trade friction in 2018, the innovation of export firms and non-export firms has a parallel trend, while the coefficients of both $EXP_{ft} * YR2018_t$ and $EXP_{ft} * YR2019_t$ are significantly positive, indicating that after the Sino–US trade friction in 2018, the innovation difference between export firms and non-export firms began to appear. As time passes, export firms tend to have more innovation than non-export firms.

4.2.3. Two-Phase Panel Estimation

Column (3) regresses the model using samples from the years before and after the Sino–US trade frictions (2017 and 2018). If the result is similar to the benchmark regression, we can then verify that the above benchmark regression results are effective to some extent. The results show that the coefficient of $EXP_{fi} * Post18_t$ is still significantly positive. Combined with the result of the placebo test, it can be inferred that around the beginning of 2018, the innovation difference between export firms and non-export firms began to increase significantly, and export firms strengthened innovation more than non-export firms.

4.2.4. Controlling the Fixed Effect of City Cross Year

Potential time-varying city features may affect the identification. To eliminate the influence of this factor and verify the effectiveness of the benchmark regression results, Column (4) further replaces the year fixed effect with the city cross-time fixed effect. The results show that the coefficient of $EXP_{fi} * Post18_t$ is still significantly positive, which shows that the potential city characteristics over time do not interfere with the basic empirical results of this paper.

4.3. Robustness Check

The results of robustness tests are reported in Table 4. Column (1) excludes the firms that do not have any innovation behaviour in the sample period. Column (2) selects a sub-sample of high-tech firms only. Column (3) controls the impact of possible tax incentives on firm innovation. Column (4) constructs the dependent variable with the number of intellectual property right applications and re-estimates the model.

Table 4. Robustness test.

	(1)	(2)	(3)	(4)
		$ln(Innovation)_{fit}$		$ln(Apply)_{fit}$
$EXP_{fi} * Post18_t$	0.084 ***	0.086 ***	0.075 ***	0.086 ***
	(0.026)	(0.025)	(0.023)	(0.031)
Tax_Policy_{it}			−0.001 *	
			(0.000)	
Control variables	yes	yes	yes	yes
Individual FE	yes	yes	yes	yes
Year FE	yes	yes	yes	yes
N	108,881	119,067	109,337	131,953
R^2	0.623	0.640	0.650	0.674

Note: Column (1) excludes the firms without innovative behaviour and only retains the firms with innovative behaviour. Column (2) only includes firms in high-tech industries, and Column (3) controls for the potential impact of tax incentives on firm innovation. Column (4) replaces the explanatory variable with log (number of intellectual property applications + 1) in the regression. All regressions control both individual and year fixed effects (FE). Standard errors are clustered at the four-digit code of the Chinese industry. ***, * indicate significance levels of 1% and 10%, respectively. N stands for the number of observations and R^2 indicates the fitness of the model.

4.3.1. Excluding Firms without Innovation Activity

Considering that there are some noninnovative firms in non-export firms, the average innovation level of non-export firms is lowered, which may interfere with the identification. To eliminate this impact, Column (1) excludes the firms that have no innovation activities in the data within four years, that is, the firms that have no R&D investment, no intellectual property application granted within four years, and we re-estimate the model. The result shows that the coefficient of $EXP_{fi} * Post18_t$ is still significantly positive. Under this sub-sample, the basic result of this paper is robust.

4.3.2. Re-Estimating with Sub-Sample of High-Tech Firms

High-tech firms are one of the focuses of the Sino–US trade frictions. Firms in industries with high scientific and technological content are more likely to be influenced by adverse shocks in foreign markets, which may reduce their innovation activities after trade frictions. At the same time, firms in other industries may continue to innovate without obvious impact, resulting in an increase in the innovation of the whole sample on average. Therefore, there may be opposite conclusions on the regression results of different sub-samples. Considering those potential effects, we exclude observations in comparatively low skill-intensive sectors, including textile, beverage, wood, etc., according to the Chinese industry classification. Column (2) regresses the model with the sample of industries with relatively high technology. The result shows that the coefficient of $EXP_{fi} * Post18_t$ is still significantly positive, indicating that the result of this paper is robust under the high-tech sub-sample.

4.3.3. Controlling the Impact of Tax Incentives

Preferential tax policies for firms are an important tool for the government to subsidize firms. Specifically, the preferential tax policies for firms within the sample years may affect the estimated results; that is, the innovation may come from the support of tax policies. To control the impact of preferential tax policies, this paper uses the annual actual tax payment of firms over firm income as the proxy variable of preferential tax policies. The specific indicators of preferential tax policies for firms are as follows:

$$Tax\_Policy_{it} = \frac{tax_{it}}{income_{it}}. \tag{2}$$

Tax_Policy_{it} measures the degree of tax preference. tax_{it} is the actual tax paid by the firm in the current year, and $income_{it}$ is the total income of the firm. The larger Tax_Policy_{it} is, the smaller the firm will enjoy preferential tax policies, and vice versa. Column (3) reports the regression results. The coefficient of Tax_Policy_{it} is -0.001, which is significant at the significance level of 10%, indicating that preferential tax policies have a positive impact on firm innovation. On average, firms with strong tax preferential policies support tend to have higher innovation output. However, according to the estimated coefficient, the effect of tax policy on firm innovation is comparatively small. At the same time, after controlling for preferential tax policies, the coefficient of $EXP_{fi} * Post18_t$ is still significantly positive, which shows that the results of this paper are robust.

4.3.4. Using the Number of Intellectual Property Applications as the Independent Variable

The number of intellectual property applications is also one of the indicators used to measure firm innovation. Column (4) uses the logarithm of the number of intellectual property applications plus 1 as the explanatory variable to re-estimate the model. The results show that the coefficient of $EXP_{fi} * Post18_t$ is still significantly positive, which shows that after the Sino–US trade friction, export firms have more intellectual property applications than non-export firms. This result also verifies once again that the results of this paper are robust.

5. Discussion on Heterogeneity and Mechanism

Through the previous analysis, we find that after the shock of the Sino–US trade friction, export firms have significantly increased innovation. In this section, we divide our sample by different technical fields to test the different performance of firm innovation after the Sino–US trade friction. More importantly, we will explain the reasons for the increased innovation of export firms after the Sino–US trade friction and the heterogeneity of the impact of the Sino–US trade friction on firm innovation in different technical fields.

5.1. Technical Field Heterogeneity

We divide the sample by different technical fields, as indicated by the dataset, to identify the heterogeneous innovation performance of export firms in different technical fields after the impact of the Sino–US trade friction. Specifically, the survey data used in this paper record firms' technical fields that belong to the following 12 categories, including: advanced manufacturing industry, modern transportation industry, biomedicine and medical devices, new materials, electronic information, aerospace, new energy and energy conservation, environmental protection, geospatial space and ocean, nuclear application technology, modern agriculture and others. This enables us to regress with sub-samples of each technical field separately. On the one hand, it helps to deeply explore the impact of the Sino–US trade friction on firm innovation in different technical fields. On the other hand, it further verifies the robustness of the empirical results from the response of firms in different fields to the Sino–US trade frictions. See Table 5 for specific test results.

Table 5. Technical field heterogeneity.

	(1)	(2)	(3)	(4)	(5)	(6)
	Advanced Manufacturing	**Modern Transportation**	**Biomedicine and Medical Devices**	**New Material**	**Electronic Information**	**Aerospace**
$EXP_{fi} * Post18_t$	0.097 ***	0.616 **	0.102 *	0.129 *	−0.013	−0.024
	(0.032)	(0.242)	(0.061)	(0.072)	(0.052)	(0.119)
Control variable	yes	yes	yes	yes	yes	yes
Individual FE	yes	yes	yes	yes	yes	yes
Year FE	yes	yes	yes	yes	yes	yes
N	50,705	998	19,829	14,079	24,699	1705
R^2	0.652	0.713	0.630	0.635	0.652	0.608
	(7)	**(8)**	**(9)**	**(10)**	**(11)**	**(12)**
	New Energy and Energy Conservation	**Environmental Protection**	**Geospatial Space and Ocean**	**Nuclear Application Technology**	**Modern Agriculture**	**Other**
$EXP_{fi} * Post18_t$	0.014	−0.101	0.049	0.000	−0.088	0.106
	(0.142)	(0.137)	(0.092)	(.)	(0.235)	(0.074)
Control variables	yes	yes	yes	yes	yes	yes
Individual FE	yes	yes	yes	yes	yes	yes
Time FE	yes	yes	yes	yes	yes	yes
N	4538	4278	224	105	3431	1755
R^2	0.673	0.633	0.838	0.608	0.649	0.617

Note: The above table examines the impact of the Sino–US trade friction on firm innovation in different technical fields divided according to the survey data. We arrange the order of different technical fields in the table according to the significance of the regression coefficients. All regressions control both individual and time fixed effects (FE). Standard errors are clustered at the four-digit code of the Chinese industry. ***, **, * indicate significance levels of 1%, 5% and 10%, respectively. N stands for the number of observations and R^2 indicates the fitness of the model.

The heterogeneity test results in Table 5 mainly reflect the following characteristics: First, the innovation performance of the advanced manufacturing industry is outstanding after the Sino–US trade frictions. Column (1) shows that the regression coefficient of $EXP_{fi} * Post18_t$ is significantly positive, which indicates that export firms in the advanced manufacturing industry have significantly increased innovation than non-export firms after the shock of the Sino–US trade friction. Comparing the results in other technical fields, it can be found that the innovation of export firms in the advanced manufacturing industry has increased most significantly after the impact of the Sino–US trade friction, which means that firms' innovation in the advanced manufacturing industry has a relatively greater response to the Sino–US trade friction. Second, the results estimated by using samples of the modern transportation industry, biomedicine and medical devices, and new material technology also show that after the Sino–US trade friction, export firms have increased innovation compared to non-export firms, but the significance of the regression coefficients is relatively low overall. Third, the results of the sub-samples of the other eight technical fields show that there is no significant change in the innovation difference between export firms and non-export firms before and after the Sino–US trade friction.

Overall, the heterogeneity test results show that the innovation performance of export firms in the advanced manufacturing industry is more prominent after being shocked by the Sino–US trade frictions. In addition, the observed value of advanced manufacturing firms entering the regression is 50,705, almost one-third of the national sample. Combined with the regression results of sub-samples in other technical fields, it can be inferred that the innovative effect of the Sino–US trade frictions on China's advanced manufacturing export firms is more evident.

5.2. Export Grab Effect

We hold that after the Sino–US trade friction, the export grab behaviour of firms has led to a significant increase in the innovation output of export firms after the shock. To verify the mechanism of the export grab effect affecting firm innovation, we first calculate the change in the average export situation of export firms in various technical fields from 2016 to 2019 according to the data used, compare the behaviour differences of export firms in different technical fields after the Sino–US trade friction, and verify that the "export grab" of firms following the Sino–US trade friction does exist from the microdata, especially for those firms in technical fields with prominent innovation output. Then, we construct relevant indicators to test the underlying mechanism by which export grabbing leads to firm innovation.

Table 6 shows the changes in export performance in various technical fields in the whole sample. After comparing the average export volume, export growth and the income share of exports, we find that first, on average, those firms in the technical fields that have significantly increased innovation have exceptionally prominent export performance after the Sino–US trade friction. Taking the advanced manufacturing industry as an example, before 2018, the average export volume remained at approximately 4600, and this index doubled in 2018 and soared to 13,708.90 in 2019. After the trade friction, it increased by 149% compared with that before the trade friction. At the same time, in the above heterogeneity test, the average export volume of firms in technical fields with a significant increase in innovation showed outstanding growth. After the trade friction, the average export volume of firms in the modern transportation industry, biomedicine and medical devices and new materials increased by 103%, 72% and 112%, respectively, while the average export growth of firms in other technical fields was comparatively slow, and even negative growth. Second, after the Sino–US trade friction, there has been a significant increase in innovative technology, and firms in the field have more export business growth, which is reflected in the obvious shift of firms' business focus to the export market. The last column of Table 6 calculates the changes in firms' average income share of exports in each technical field from 2017 to 2018. After comparing the indicators in different technical fields, we find that the technical fields with obvious innovation increases after the Sino–US trade friction have a relatively higher change in the income share of exports, which means that after the Sino–US trade friction, on average, firms in these technology fields are more inclined to increase export business. For example, in the most prominent advanced manufacturing industry, the average income share of exports in 2018 increased by 11.01% compared with that in 2017. Third, from the overall sample, after the Sino–US trade friction, the average export volume of firms increased by only 5%, and the proportion of export business income decreased by 5.89%. Obviously, those firms with negative trade growth after trade friction lowered the average export growth rate and the income share of exports, but this does not prevent us from seeing the existence of firms' export-grab behaviour through the heterogeneity of each technical field.

Table 6. Changes in exports of various technical fields before and after the Sino–US trade friction.

	Average Export Volume of Firms (Unit: Thousand RMB)				Export Growth Before and After Friction	Growth Rate	Changes of Export-Income Ratio (2017–2018)
	2016	2017	2018	2019			
1.Advanced manufacturing	4614.58	4550.54	9120.90	13,708.90	13,664.68	149%	11.01%
2.Modern transportation industry	2569.67	595.88	2389.68	4045.35	3269.48	103%	4.09%
3.Biomedicine and medical devices	7059.88	7082.30	10,664.42	13,699.95	10,222.19	72%	6.16%
4.New material	3055.57	3957.21	6722.93	8155.40	7865.55	112%	4.90%
5.Electronic information	62,834.50	41,128.35	13,575.33	42,553.66	−47,833.86	−46%	−30.01%
6.Aerospace	8552.50	7771.81	7425.47	4570.59	−4328.25	−27%	−1.45%
7.New energy and energy conservation	156.50	6289.41	4770.05	4034.61	2358.75	37%	−5.16%
8.Environmental protection	4246.35	2397.26	4796.42	3405.61	1558.42	23%	4.19%
All samples	13,892.59	13,010.03	9747.74	18,548.90	1394.02	5%	−5.89%

Note: this table only shows the export changes of eight technical fields, as the other four technical fields are not listed in the table due to small sample size or obvious exceptionality.

The analysis of the results in Table 6 means that export grab behaviour does exist in some specific technical fields, and those fields happen to have outstanding innovation performance. On the one hand, the export grab behaviour of firms is reflected in the substantial growth of the average export volume of firms; on the other hand, it is reflected in the increase in the proportion of export business. These indicators show that export firms have increased their export business in the short term, and the increase is large. Although we cannot intuitively observe the export grab behaviour of firms, the real data still enable us to speculate that export grab behaviour does exist in some specific technical fields.

The export grab may be the reason for the increase in firm innovation in some of the technical field, especially for advanced manufacturing industry, since the innovation promoting effect of export has been widely discussed in the extant trade literature. The sudden expansion in exports may benefit firm innovation through several channels as we have stated in the literature review, including export learning effect [21,32,35], cost reducing channel [17,20], market competition effect [9,11], reallocation effect [11,16], etc. In the following section, we try to verify the promoting effect of export grab on firm innovation by doing a detailed empirical test, which may explain the increase in firm innovation following Sino–US trade friction as well as the prominent performance of firms in advanced manufacturing industry.

5.3. Mechanism of the Export Grab Effect on Firm Innovation

In fact, the phenomenon of export grab is a manifestation of the shock of foreign demand. The previous analysis also shows that the export grab phenomenon is the overdraft of foreign demand for the future, and its purpose is to avoid the risk of rising trade costs. Therefore, after the break of the Sino–US trade friction, foreign procurement has increased rapidly, which is reflected in the export grab phenomenon of the Chinese firms. There are two effects on the innovation of export firms. The first is the direct effect. The direct effect comes from the diminishing marginal cost caused by the production expansion of export firms after the impact of foreign demand. Therefore, firms will choose to update their technology after obtaining a productivity advantage [16,17]. Atkeson and Burstein believe that the ratio of a firm's export quantity to its output can represent the impact of the direct effect [16]. Because our data only provide information on firms' export volume and income, we use firms' export income share $export\_rate_{fit}$ to indicate the direct effect, which can be used to reflect the changes in the proportion of the firms' export business in the total business. The second is the indirect effect. The indirect effect comes from the reallocation of domestic export firms by the impact of foreign demand. This effect will reduce the export threshold and allocate exports to capable firms [11,20], which is manifested in the concentration of export market share to large firms, so the innovation of large firms is more obvious. We use the logarithm of the export volume of firms $lnexport_{fit}$ to measure indirect effects.

5.3.1. Direct Effect

The direct effect of export grab on firm innovation comes from sustainable exporters. After the impact, due to the sharp increase in foreign demand, firms need to expand production and export relatively more products abroad in the short term, which is reflected in the increase in the proportion of the firm's export business. Therefore, this will reduce the marginal cost of production, and export firms will gain the advantage of productivity. In this case, firms will be likely to carry out innovation activities. To test the direct effect, we eliminate the samples of firms entering and exiting the export market after the Sino–US trade friction to fix the impact of the threshold of the export market on the results. In Column (1) of Table 7, the benchmark model is re-estimated with new samples, and the results show that the estimated coefficient of $EXP_{fi} * Post18_t$ is significantly positive, but its estimated coefficient of 0.044 is not only less than the level of 0.086 in the benchmark regression but also only 10% significant, which shows that after controlling the entry and exit threshold of the export market, sustainable export firms still increase innovation after the Sino–US trade friction, but the contribution of sustainable export firms to the overall innovation of export firms after the Sino–US trade friction may be relatively small. That is, the direct effect of export grab may be small. To further test the direct effect of export grab, we further add $export\_rate_{fit}$ to the regression, and the results are reported in Column (2) of Table 7. The estimated coefficient of $EXP_{fi} * Post18_t$ is lower than it is in Column (1) and is not significant. At the same time, the coefficient of $export\_rate_{fit}$ is significantly positive, which indicates that the significance of the estimated coefficient of $EXP_{fi} * Post18_t$ has been absorbed by the direct effect of export grab. This also means that the increase in sustainable export firms' innovation after the Sino–US trade friction can be explained by the direct effect of export grab. Therefore, combined with the results of Columns (1) and (2), it can be inferred that the direct effect of export grab is the mechanism of increasing innovation of export firms after the Sino–US trade friction.

To further test the existence of the direct effect of export grab, we retain only the firms that continue to export in the sample and use the method of long difference and two-stage least squares (2SLS) to verify the relationship between export grab and firm innovation again. The specific identification methods are as follows:

$$D.ln(Innovation)_{fi,2019\text{–}2017} = \gamma_0 + \gamma_1 D.export\_rate_{fi,2019\text{–}2017} + X_{fi,2017} * \gamma_2 + \theta_i + \varepsilon_{fi} \quad (3)$$

Among them, the explained variable, $D.ln(Innovation)_{fi,2019\text{–}2017}$, is the difference between firm innovation in 2019 and 2017, and the explanatory variable, $D.export\_rate_{fi,2019\text{–}2017}$, is the difference between the income share of exports of firms in 2019 and 2017. $X_{fi,2017}$ is a series of control variables at the firm and industry levels in 2017. θ_i is the fixed effect of the industry, and ε_{fi} is the residual term. The specific regression adopts the 2SLS method to calculate the income share of exports in 2017, $export\_rate_{fi,2017}$, as the instrumental variable for $export\_rate_{fi,2017}$.

The OLS regression results are shown in Column (3) of Table 7. The coefficient of $D.export\_rate_{fi,2019\text{–}2017}$ is significantly positive, which shows that there is an obvious positive relationship between the direct effect of export grab and firm innovation. The more firms increase the proportion of export business after the Sino–US trade friction, the more inclined they are to have more innovation. Column (4) further uses the 2SLS method to estimate the model, and the results show that the coefficient of $D.export\_rate_{fi,2019\text{–}2017}$ is significantly positive but slightly lower than that in Column (3). Therefore, combined with the estimation results of other columns in Table 7, we can basically judge the existence of the direct effect of export grab. Export grabbing after the Sino–US trade frictions leads to an increase in the innovation of sustainable export firms. However, from the estimated results of this sample, the direct effect of export grab is relatively small.

Table 7. Mechanism test of the export grab effect promoting firm innovation (direct effect).

	(1)	(2)	(3)	(4)
	$In(Innovation)_{fit}$		$D.In(Innovation)_{fi,2019\text{–}2017}$	
$EXP_{fi} * Post18_t$	0.044 *	0.035		
	(0.026)	(0.026)		
$export\_rate_{fit}$		0.320 ***		
		(0.098)		
$D.export\_rate_{fi,2019\text{–}2017}$			0.531 ***	0.705 **
			(0.190)	(0.287)
Control variables	yes	yes	yes	yes
Individual FE	yes	yes		
Industry FE			yes	yes
Year FE	yes	yes	yes	yes
N	12,7396	10,4905	672	672
R^2	0.638	0.649	0.062	0.060

Note: the direct effect of export grab comes from sustainable export firms. Therefore, Column (1) excludes the samples of newly entering and exiting export firms after the Sino–US trade friction to compare the changes in innovation of sustainable export firms before and after the friction. Column (2) further controls the variables of the direct effect mechanism. Therefore, it is found that the increase in innovation of export firms has been absorbed by the direct effect. Columns (3) and (4) only retain the samples of sustainable export firms and use the long difference method to estimate the direct effect of export grab, in which Column (3) is the OLS estimation result and Column (4) is the 2SLS estimation result. Column (1) and (2) control both individual and year fixed effects (FE). Column (3) and (4) control both industry and year fixed effects (FE). Standard errors are clustered at the four-digit code of the Chinese industry. ***, **, * indicate significance levels of 1%, 5% and 10%, respectively. N stands for the number of observations and R^2 indicates the fitness of the model.

5.3.2. Indirect Effects

The indirect effect of export grab on the innovation of export firms comes from the impact of foreign demand, which changes the distribution of export firms. The positive demand shock reduces the threshold for firms to enter the export market. On the one hand, it will lead to more new export firms entering the export market, and the export learning effect increases firms' innovation. On the other hand, large-scale export firms will be allocated more export shares, and firms will be more capable of innovation. Both of these aspects are reflected in the increase in firm export volume. We use the intersection of the logarithm of the export volume and $Post18_t$ to represent the indirect effect mechanism and replace it with the explanatory variable of the benchmark regression to test the indirect effect of export grab. Its estimation coefficient represents the impact of the change in the export market threshold on firm innovation after the Sino–US trade friction. It should be noted that the above has verified the existence of the direct effect mechanism, and the direct effect may be reflected in the growth of enterprise export volume. Therefore, to avoid the interference of the direct effect of export grab on the identification of indirect effects, we further control the direct effect, specifically, adding the income share of exports in the regression to control the direct effects of sustainable export firms.

Table 8 shows the test results of the indirect effect mechanism of export grab to promote firm innovation. Column (1) is the full sample regression result. The estimation coefficient of $EXP_{fi} * Post18_t$ is significantly positive, which means that after the Sino–US trade friction, firms with an expanded export scale have better innovation output. That is, on average, there is an indirect effect on firm innovation. At the same time, the estimated coefficient of the direct effect variables is also significantly positive, which once again verifies the existence of the direct effect of export grab, but the direct effect fails to absorb the indirect effect. To deeply verify the existence of the indirect effect mechanism and test the heterogeneity effect of the mechanism, Columns (2)–(5) of Table 8 also report the regression results of samples in different technical fields and select four representative technical fields, namely, advanced manufacturing industry, modern transportation industry, electronic information and new energy and energy conservation. This is because the results in Table 6 show that there is an obvious phenomenon of export grab in advanced manufacturing and modern transportation industries, while there is an obvious trade contraction in the fields of electronic information, new energy and energy conservation. Comparing the regression results of these four technical fields can test the difference in innovation between the technical fields with and without export grab, and further verify the effect of export grab on enterprise innovation. Column (2) gives the regression results of the advanced manufacturing sub-samples. The estimated coefficient of $lnexport_{fit} * Post18_t$ is significantly positive, which shows that the indirect effect of export grab has an obvious promoting effect on the innovation of export firms. Column (3) reports the sub-sample regression results of the modern transportation industry. Therefore, under the sub-sample of the modern transportation industry, the indirect effect of export grab has a positive effect on the innovation of export firms. Columns (4) and (5) are the regression results of electronic information and new energy and energy conservation sub-samples, respectively. The estimated coefficients of $lnexport_{fit} * Post18_t$ are not significant, which shows that there is no indirect effect of export grab in these two technical fields.

Table 8. Mechanism test of the export grab effect promoting firm innovation (indirect effect).

	(1)	(2)	(3)	(4)	(5)
	Total Sample	**Advanced Manufacturing**	**Modern Transportation Industry**	**Electronic Information**	**New Energy and Energy Conservation**
$lnexport_{fit} * Post18_t$	0.013 ***	0.014 ***	0.091 **	0.006	0.013
	(0.003)	(0.005)	(0.035)	(0.007)	(0.022)
$export\_rate_{fit}$	0.258 **	0.043	−0.762	0.443 **	0.902
	(0.100)	(0.155)	(0.895)	(0.210)	(0.714)
Control variables	yes	yes	yes	yes	yes
Individual FE	yes	yes	yes	yes	yes
Year FE	yes	yes	yes	yes	yes
N	109,337	43,425	790	20,569	3750
R2	0.651	0.661	0.726	0.658	0.687

Note: $lnexport_{fit}$ is the logarithm of the firm's export volume, which is multiplied by $Post18_t$ to form an explanatory variable. Column (1) is the full sample regression result, and Columns (2)–(5) are the sub sample regression results in the fields of advanced manufacturing, modern transportation, electronic information, new energy and energy conservation. All regressions controlled for direct effects. All regressions control both individual and year fixed effects (FE). Standard errors are clustered at the four-digit code of the Chinese industry. ***, ** indicate significance levels of 1% and 5%, respectively. N stands for the number of observations and R^2 indicates the fitness of the model.

Overall, the indirect effect of export grab explains the increase in innovation of export firms after the Sino–US trade friction. Due to the shock of foreign demand, the threshold for export firms to enter foreign markets has been reduced, which is manifested in the allocation of export volume to large firms, and new export firms have entered the market, which has promoted the innovation of export firms on average. By horizontally comparing the regression results of the columns in Table 8, it can be found that the indirect effect of export grab exists in the technical field where there is an obvious phenomenon of export grab, and the innovation of export firms in the technical fields where there is no export grab has not increased significantly after the Sino–US trade friction, which also verifies the role of export grab in promoting firm innovation on one side.

6. Concluding Remarks

This paper studies the short-term impact of the Sino–US trade frictions on China's firm innovation. By using the unique microdata of Chinese technology and innovation firms from 2016 to 2019, we find that the innovation of technological and innovative export firms has, on average, increased significantly after being impacted by the Sino–US trade friction, which contradicts the extant theoretical literature. For example, in the seminal work of Handley and Limão, the disappearance of trade policy uncertainty caused by the removal of tariff will promote technical upgrading [5], so the Sino–US trade friction will not be conducive to the growth of firm innovation, while aggravating trade policy uncertainty. Through the in-depth investigation of firm operational data, we also exclude the escaping competition effect induced by shocks in export market that might compel firms to innovate [11]. Therefore, external market export pressure, as it has been discussed in the literature [8,12], is not the reason for the increase in firm innovation in our case.

Through further study of sub-samples in different technical fields, we find that the advanced manufacturing industry has outstanding performance after the Sino–US trade frictions. The analysis results of the sub-samples of 12 technical fields in the sample show that the innovation of export firms in the advanced manufacturing industry is the most obvious in various technical fields, and the export firms in the modern transportation industry, biomedicine and medical devices, and new materials have also increased to varying degrees after the Sino–US trade friction. To explore the reasons for the innovative growth of export firms after the Sino–US trade friction, we conduct an in-depth investigation of the operations of firms and find that the technical fields that have increased innovation after the trade friction often have significant export business growth. Therefore, we speculate that export grab is the reason for the innovation growth of export firms after the Sino–US trade friction and test for it. Theoretically, export grab is the demand impact of foreign markets. Export grab has direct and indirect effects on the innovation of export firms. The direct effect is the productivity advantage brought by the decline of marginal cost after the impact, and the indirect effect is the reallocation effect brought by the impact on the adjustment of the threshold for firms to enter the export market. Both effects may lead to an increase in firm innovation. We construct indicators to verify the direct effect and indirect mechanism of export grab induced by the Sino–US trade friction on firm innovation. In line with the present research [11,16], we once again verified the great significance of export to firm innovation.

The limitation of this article is that we cannot test the long-term effect of the Sino–US trade friction on firm innovation due to the short-period data we used. The phenomenon of export grab is a relatively short response of exporters to the outbreak of the Sino–US trade friction. The economic relationship between China and the US has not been restored, which continues to impact on firm performance of both countries. Therefore, as the effect of export grab gradually fades, the long-term impact of the Sino–US trade friction on firm innovation may be different from what we have seen today.

In addition to the above long-term effect, there are some issues related to this paper that deserve attention. First, to study the new mechanism of the Sino–US trade friction on firm innovation. Although we have found that the export grab is a short-term channel through which the Sino–US trade friction affects firm innovation, we believe that there are more potential channels need to be found. Second, to study the effect of the Sino–US trade friction on the economic activities in other countries. Since the beginning of 2018, the Sino–US trade friction has been reallocating the international commodity market. Future research can focus on the impact of the Sino–US trade friction on economic activities in countries including but not limited to China and the US.

Author Contributions: Conceptualization, Z.X. and X.Z.; methodology Z.X.; software, Z.X.; formal analysis, Z.X., X.Z. and Z.Z.; resources, Z.Z.; data curation, Z.X. and Z.Z.; writing—original draft preparation, Z.X.; writing—review and editing, Z.X. and X.Z.; supervision, X.Z. All authors have read and agreed to the published version of the manuscript.

Funding: This research received no external funding.

Institutional Review Board Statement: Not applicable.

Informed Consent Statement: Not applicable.

Data Availability Statement: The data presented in this study are available on request from the corresponding author. The data are not publicly available due to privacy.

Acknowledgments: We are grateful to the helpful comments from Mi Dai, Guangming Ma, Xiufei Ding, Ke Ma, Hui Wang and Jingchen Jin.

Conflicts of Interest: The authors declare no conflict of interest.

References

1. Schumpeter, J.A. *The Theory of Economic Development*; Harvard University Press: Cambridge, MA, USA, 1932.
2. Handley, K.; Limão, N. Trade Under, T.R.U.M. P Policies. In *Economics and Policy in the Age of Trump*; Bown, C.P., Ed.; CEPR Press: Colchester, UK, 2017.
3. Handley, K. Exporting Under Trade Policy Uncertainty: Theory and Evidence. *J. Int. Econ.* **2014**, *94*, 50–66. [CrossRef]
4. Handley, K.; Limão, N. Policy Uncertainty, Trade and Welfare: Theory and Evidence for China and the U.S. *Am. Econ. Rev.* **2017**, *107*, 2731–2783. [CrossRef]
5. Handley, K.; Limão, N. Trade and Investment Under Policy Uncertainty: Theory and Firm Evidence. *Am. Econ. Rev.* **2015**, *7*, 189–222. [CrossRef]
6. Tong, J.D.; Li, S.Q. Effect of Trade Policy Uncertainty on Export Enterprises' Product Innovation. *Int. Trade Issues* **2016**, *6*, 25–32.
7. Liu, Q.; Ma, H. Trade Policy Uncertainty and Innovation: Firm Level Evidence from China' WTO Accession. *J. Int. Econ.* **2020**, *127*, 103387. [CrossRef]
8. Galdon-Sanchez, J.E.; Schmitz, J., Jr. Competitive Pressure and Labor Productivity: World Iron-Ore Markets in the 1980's. *Am. Econ. Rev.* **2002**, *92*, 1222–1235. [CrossRef]
9. Aghion, P.; Bloom, N.; Blundell, R.; Griffth, R.; Howitt, P. Competition and Innovation: An Inverted-U-Relationships. *Q. J. Econ.* **2005**, *120*, 701–728.
10. Aghion, P.; Blundell, R.; Griffith, R.; Howitt, P.; Prantly, S. The Effects of Entry on Incumbent Innovation and Productivity. *Rev. Econ. Stat.* **2009**, *91*, 20–32. [CrossRef]
11. Aghion, P.; Bergeaud, A.; Lequien, M.; Melitz, M. *The Heterogeneous Impact of Market Size on Innovation: Evidence from French Firm-Level Exports*; Working Paper; Harvard University: Cambridge, MA, USA, 2019.
12. Raith, M. Competition, Risk, and Managerial Incentives. *Am. Econ. Rev.* **2003**, *93*, 1425–1436. [CrossRef]
13. Xie, J.G.; Wang, X. The Trade Consequences of the Sino-US Trade Conflict: An Empirical Research Based on Segmented Product Data. *Theory Pract. Financ. Econ.* **2021**, *42*, 86–93.
14. Zhang, W.T.; Xu, R.H.; Wang, Q. The Impact of Sino-US Trade Friction on China' Exports from Uncertainty. *Theory Pract. Financ. Econ.* **2020**, *3*, 111–118.
15. Cavallo, A.; Gopinath, G.; Neiman, B.; Tang, J. *Tariff Passthrough at the Border and at the Store: Evidence Form Us Trade Policy*; NBER Working Papers No 26396; National Bureau of Economic Research, Inc.: Cambridge, MA, USA, 2019. [CrossRef]
16. Atkeson, A.; Burstein, A. Innovation, Firm Dynamics, and International Trade. *J. Political Econ.* **2010**, *118*, 433–484. [CrossRef]
17. Bustos, P. Trade Liberalization, Exports, and Technology Upgrading: Evidence on the Impact of MERCOSUR on Argentinian Firms. *Am. Econ. Rev.* **2011**, *101*, 304–340. [CrossRef]
18. Wei, M.H.; Liu, X.M. Trade-related Environmental Uncertainty and Firm's Innovation: Evidence from Chinese Listed Firms. *Nankai Manag. Rev.* **2021**, *24*, 16–25.
19. Wan YX Wei, Z.X.; Wang, Y.S. Does China-US Trade Dispute Affect Chinese Firms' R&D Internationalization. *Stud. Sci. Sci.* 2021; *submitted for publication*.
20. Lileeva, A.; Trefler, D. Improved Access to Foreign Markets Raises Plant-Level Productivity for Some Plants. *Q. J. Econ.* **2010**, *125*, 1051–1099. [CrossRef]
21. De-Loecker, J. Do Exports Generate Higher Productivity? Evidence from Slovenia. *J. Int. Econ.* **2007**, *73*, 69–98. [CrossRef]
22. Fajgelbaum, P.D.; Goldberg, P.; Kennedy, P.; Khandelwal, A. The Return to Protectionism. *Q. J. Econ.* **2020**, *135*, 1–54. [CrossRef]
23. List, F. *The National System of Political Economy*; OLL: London, UK, 1841.
24. Krugman, P. *Rethinking International Trade*; MIT Press: Cambridge, MA, USA, 1990.
25. Brezis, E.S.; Krugman, P.; Tsiddon, D. Leapfogging in International Competition: A Theory of Cycles in National Technological Leadership. *Am. Econ. Rev.* **1993**, *83*, 1211–1219.
26. Krugman, P. A Model of Innovation, Technology Transfer, and the World Distribution of Income. *J. Political Econ.* **1979**, *87*, 253–256. [CrossRef]
27. Dollar, D. Technological Innovation, Capital Mobility and the Product Cycle in North-South Trade. *Am. Econ. Rev.* **1986**, *76*, 177–190.
28. Li, C.D.; He, C.T.; Lin, C.W. Evaluating the Effects of China's Countermeasures to China-U.S. *Trade Frictions. China Ind. Econ.* **2018**, *10*, 137–155.
29. Lv, Y.; Lou, C.R.; Du, Y.X.; Tu, X.Q. An Analysis of the Impact of Sino-US Trade Friction Based on the Tariff Lists. *J. Financ. Econ.* **2019**, *45*, 59–72.
30. Cui, L.B.; Zhu, L.; Song, M.L.; Zheng, H.T. The Economic Evaluation of China-US Trade Frictions. *J. Financ. Econ.* **2018**, *44*, 4–17.
31. Amiti, M.; Redding, S.J.; Weinstein, D.E. The Impact of the 2018 Tariffs on Prices and Welfare. *J. Econ. Perspect.* **2019**, *33*, 187–210. [CrossRef]

32. Van Biesebroeck, J. Exporting Raises Productivity in Sub-Saharan African Manufacturing Firms. *J. Int. Econ.* **2005**, *67*, 373–391. [CrossRef]
33. Zhang, J.; Li, Y.; Liu, Z.B. Does Export Promote the Productivity of Chinese Firms? Empirical Evidence from Chinese Local Manufacturing Firms: 1999–2003. *Manag. World* **2009**, *12*, 50–64.
34. Melitz, M.J. The Impact of Trade on Intra-Industry Reallocations and Aggregate Industry Productivity. *Econometrica* **2003**, *71*, 1695–1725. [CrossRef]
35. De Loecker, J. Detecting Learning by Exporting. *Am. Econ. J. Microecon.* **2013**, *5*, 1–21. [CrossRef]
36. Hanley, A.; Perez, J.M. Are Newly Exporting Firms More Innovative? Findings from Matched Spanish Innovator. *Econ. Lett.* **2012**, *116*, 217–220. [CrossRef]
37. Li, B.; Yue, Y.S.; Chen, T. Export and Innovation: Empirical Evidence from Patent Data at Firm Level. *World Econ.* **2016**, *39*, 72–94.
38. Shi, Q.; Li, P.; Zong, Q.Q. Learning by Exporting: From the Perspective of Enterprise R & D Strategy Interaction. *World Econ.* **2017**, *40*, 72–97.
39. Aw, B.Y.; Roberts, M.J. Productivity and Turnover in the Export Market: Micro-Level Evidence from the Republic of Korea and Taiwan (China). *World Bank Econ. Rev.* **2000**, *14*, 65–90. [CrossRef]
40. Dai, M.; Yu, M. Firm R&D, Absorptive Capacity and Learning by Exporting: Firm-level Evidence from China. *World Econ.* **2013**, *36*, 1131–1145.

sustainability

MDPI

Article

Exploring the Effect of Family Life and Neighbourhood on the Willingness of Household Waste Sorting

Liyuan Zhao [1] and Hongsheng Chen [2,*]

[1] School of Architecture and Design, China University of Mining and Technology, Xuzhou 221116, China; Liyuan.Zhao@cumt.edu.cn
[2] School of Architecture and Urban Planning, Shenzhen University, Shenzhen 518060, China
* Correspondence: chenhongsh@szu.edu.cn

Abstract: For developing countries, garbage classification has become an important measure to handle the environmental pollution crisis. This empirical study examined urban and rural families' willingness to sort and deposit garbage at fixed recycling points. We found that urban residents demonstrated a significantly higher willingness to sort and deposit garbage at designated points compared to rural residents. The average number of family meals per month, average monthly household food expenditure, household cleanliness, and household crowding (spaciousness) are significantly related to a family's willingness to sort garbage. In terms of neighbourhood factors, families living in neighbourhoods with property management are more likely to have a higher willingness to sort garbage. The degree of air pollution in the neighbourhood also has an impact on the family's willingness to sort garbage. This study proposes that neighbourhood factors have a non-negligible influence on a household's willingness to sort and put garbage in designated locations. Especially in urban neighbourhoods, the willingness of residents living in commodity housing neighbourhoods to sort and place garbage at designated locations is significantly higher than that of residents living in other neighbourhoods. To improve the implementation effect of the waste sorting policy, we suggest that the configuration of neighbourhood garbage recycling services and facilities should be improved so that people can sort garbage more conveniently.

Keywords: garbage sorting; urban garbage disposal; rural garbage disposal; willingness; family life; neighbourhood

Citation: Zhao, L.; Chen, H. Exploring the Effect of Family Life and Neighbourhood on the Willingness of Household Waste Sorting. *Sustainability* **2021**, *13*, 13653. https://doi.org/10.3390/su132413653

Academic Editors: E. Richard Zada and Andrea Gatto

Received: 13 November 2021
Accepted: 9 December 2021
Published: 10 December 2021

1. Introduction

With rapid industrialisation and urbanisation, and the resultant improvements in people's incomes and living standards, household garbage generation has also increased, and the garbage composition process has increased in complexity [1]. The increase in the volume of household garbage has brought about a series of problems related to environmental pollution, public health crises, land resource occupation, and resource wastage, which have become global challenges [2]. Since the mid to late 20th century, in response to environmental problems caused by domestic waste, many developed countries have begun to pay attention to and explore social policies for waste management. For example, in Japan, which has high population density but scarce land, incineration is the most important method of waste disposal [3]. Japanese waste management measures also include pollution prevention, recycling, and reusing. In the 1990s, Finland set a national goal of garbage source classification and recycling. They also found that higher social participation and classification effectiveness could reduce garbage collection costs [4]. Reducing, reusing, and recycling, which are commonly referred to as the 3Rs, are regarded as effective waste management measures [5]. For instance, since the 1990s, Sweden took many measures to promote waste recycling activities. All Swedish citizens are required to sort their generated waste and participate in local recycling programs [6,7]. Germany formulated "The New German Closed Cycle Management Activity" to transform waste management into resource management and trigger the innovation of waste treatment technologies to improve the recycling capacity [8]. Garbage sorting has become one of the important requirements for sustainable development. However, the popularity of garbage sorting in developing countries is not high [9–11]. A large amount of waste generation had a great negative impact on developing countries' living environment. For example, in India, due to the lack of sufficient financial funds to build a garbage disposal system, a large amount of garbage accumulated at the outskirts of cities [12]. Vietnam also started a waste classification demonstration project in 2000, but it produced unsatisfactory results [13]. Furthermore, green circular economy has become the development trend of the global economy [14,15]. On the path to sustainable development, a green lifestyle is as important as green production and consumption. Among them, sorting domestic waste is one of the most important green lifestyles. Garbage classification is the basis for the establishment of a circular economy, which can also reduce environmental pollution. Waste sorting is not only a change of lifestyle, but also an exploration of establishing a circular economy system. Achieving green development requires extensive civic engagement [16]. To make waste sorting a widely accepted way of life requires the participation of all people. Therefore, research on the willingness of waste classification is helpful to better understand the micro-foundation of circular economy and sustainable society.

As the largest developing country in the world, China experienced rapid economic growth and industrialisation and, consequently, greatly increased its waste generation over the past decades [17]. Thus, in China, urban as well as rural areas now face great environmental pressures. Due to the many difficulties in implementing waste classification in China, landfilling and incineration are still the main methods of national waste disposal [18]. Landfills have the risk of infiltration of landfill leachate, which leads to pollution of the soil and groundwater around the landfill, which endangers the health of residents, and it is difficult to control pollution. Garbage incineration may also lead to the risk of heavy metal pollution in the soil and air pollution. Therefore, the implementation of garbage classification and recycling of garbage are important measures to protect the environment and reduce pollution. However, the promotion of waste sorting and waste reuse in China is still relatively low. In 2000, some pilot Chinese cities, including Beijing, Shanghai, Nanjing, Hangzhou, and Guangzhou, attempted to promote waste sorting in order to reduce waste generation and recycle waste sources [19]. In 2019, Shanghai and Beijing formulated a new series of stricter domestic waste sorting management policies, which were promoted at the neighbourhood level. In this context, many cities in China have begun to formulate policies for the classification of household waste, requiring residents to sort and put domestic waste in the designated place. However, because people's long-formed living habits are difficult to change in a short time, many cities have suspended their garbage classification requirements after implementing garbage classification policies for a period.

The transformation of garbage classification from "consensus" to "life habits" requires that most people in the society are willing to carry out garbage classification. However, what factors affect the residents' the willingness to sort garbage? The existing research has mainly analysed the individual's garbage sorting behaviour from the perspective of the individual's socioeconomic status and cultural customs [20,21]. However, since the disposal of domestic garbage is often the behaviour of the family, family lifestyle may affect garbage classification. One of the main theoretical innovations of this research is to analyse the willingness of family classification from the perspective of the family rather than the individual. Garbage sorting is not only an individual's behaviour, but also a family's behaviour. In addition, the behaviour of garbage classification may also be related to the external environment. For example, different types of neighbourhoods provide different garbage collection services, which may cause residents in different neighbourhoods to have different attitudes towards garbage classification. However, there is still a lack of research on the family's willingness to sort waste. To fill this gap, we will analyse the influence of family lifestyle and neighbourhood environment on family waste sorting willingness.

Furthermore, this research will help elucidate why the promotion of waste sorting is slow and will also provide guidance for improving the government's intervention activities to raise the level of urban and rural domestic waste classification in China.

2. Methods and Materials

The data used in this study were derived from the 2016 China Labor-force Dynamics Survey (CLDS), which was conducted by the Center for Social Science Survey at Sun Yat-sen University. The collected sample used a multistage cluster and stratified probability sampling strategy. The CLDS included information at the community, family, and individual levels. This database was China's first interdisciplinary, nationwide follow-up survey on the theme of labour. The survey covered many research topics such as family lifestyles and the neighbourhood environment. It accurately reflects the current basic conditions of social development in China. It was highly credible to use this database to analyse the household's willingness to sort garbage. In this study, we used data from the family level. After removing samples with missing values, 12,126 valid samples were included in this study. This study mainly analysed the influence of individual social and economic conditions on a family's willingness to classify garbage and their willingness to deposit garbage at designated places. The dependent variables included family willingness of waste sorting and family willingness to place waste at designated locations. We focused on the impact of household daily life and community living environment on household waste sorting willingness. The family's living habits may have an impact on the willingness to sort garbage. For example, if a family has many members living together, they may produce a lot of domestic waste, which may make it more difficult to classify waste, which will negatively affect the family's willingness to classify waste. Therefore, we selected independent variables, such as the number of family members living together, average number of family meals per month, average monthly household electricity consumption, average monthly household food expenditure, household cleanliness, and household crowding, which can reflect the daily life of the family. The degree of household cleanliness and crowding are evaluated on a 10-point scale, where 1 means very bad or very crowded, and 10 means very clean or very spacious.

In China, different types of neighbourhoods can obtain different public services. There is a huge gap between rural neighbourhoods and urban neighbourhoods in access to public services. Garbage collection is one of the public services that varies significantly by area. Urban neighbourhoods have specialised garbage disposal services, while many rural neighbourhoods do not have regular garbage disposal services. Therefore, we compare the willingness of residents living in different types of neighbourhoods to sort and place waste at designated locations. We selected neighbourhood types, neighbourhood property management, the level of air pollution, the level of water pollution, the level of noise pollution, the level of soil pollution as independent variables, which can reflect the neighbourhood living environment. The types of neighbourhoods are divided into rural neighbourhoods and urban neighbourhoods. Urban neighbourhoods can be further divided into commercial housing neighbourhoods, old neighbourhoods, unit neighbourhoods, and security housing neighbourhoods, shantytown neighbourhoods and others. Because the respondents' willingness to sort waste is related to their perception of environmental quality, the data on the pollution degree of the living environment in this study come from the respondents' subjective judgments, which are subjective indicators. The level of pollution (air pollution, water pollution, noise pollution, and soil pollution) is evaluated on a 4-point scale, where 1 means very serious pollution and 4 means not serious pollution. As the dependent variable was a dichotomous variable, this study used logistic regression models to analyse the influencing factors on respondents' willingness to sort waste and willingness to place waste at designated locations.

3. Results

3.1. Descriptive Statistics

Table 1 shows the descriptive statistics of household waste sorting willingness. Among the respondents, those who were willing to classify their household garbage accounted for 83.04%, while those who are not willing to classify their household garbage accounted for only 16.96%. Most respondents were willing to deposit their household waste at the designated collection point, and only 5.41% of respondents were unwilling to deposit their household waste at the designated collection point. Regarding family lifestyle, the average number of family members living together and family meals per month were 4.14 and 19.42, respectively. The average monthly household electricity consumption and food expenditure were 137 KWH and 14,420 yuan, respectively. In terms of family living environment, the mean values of household cleanliness and household crowding were 6.26 and 6.27, respectively. Regarding the neighbourhood living environment, 21.35% of participating families lived in neighbourhoods with property management, while 78.65% lived in neighbourhoods without property management. While 59.84% of surveyed households were in rural neighbourhoods, 40.16% were in urban neighbourhoods. The average evaluations of the environmental quality of the air, water, noise, and soil of the neighbourhoods in which the interviewed households live were 3.13, 3.21, 3.20, and 3.39, respectively. Among the four types of pollution, air pollution is the most serious, while soil pollution is the least, probably because people perceive air quality most directly. Table 2 summarises the cross-tabulation results for willingness regarding household waste sorting between rural and urban residents. The proportion of rural respondents who were willing to conduct household waste sorting was significantly lower than that of urban residents (78.39% and 89.98%, respectively). The proportion of rural residents who were willing to deposit their household waste at designated collection points was also significantly lower than that of urban residents (92.05% and 98.38%, respectively). Figure 1 shows the distribution of household waste sorting willingness in seven types of urban neighbourhoods. Households living in commodity housing neighbourhoods have the highest willingness to sort garbage, reaching 93.89%, followed by living in security housing neighbourhoods (92.74%), unit neighbourhoods (90.34%), urban shantytown neighbourhoods (90.14%), residential neighbourhood changed from a rural neighbourhood (88.09%), and old residential neighbourhoods (86.13%).

Table 1. Descriptive statistics for the variables (N = 12,126).

	Mean Value/Proportion
Neighbourhood types (%)	
Rural neighbourhood	59.84
Urban neighbourhood	40.16
Household waste sorting willingness (%)	
Willing	83.04
Unwilling	16.96
Willingness to deposit household waste at a fixed collection point (%)	
Willing	94.59
Unwilling	5.41
Number of family members living together (1–18)	4.14 (SD = 2.10)
Average number of family meals per month (0–90)	19.42 (SD = 27.86)
Average monthly household electricity consumption (100 KWH)	1.37 (SD = 1.29)
Average monthly household food expenditure (1000 yuan)	14.42 (SD = 23.83)
Household cleanliness (1–10)	6.26 (SD = 1.76)

Table 1. *Cont.*

	Mean Value/Proportion
Household crowding (1–10)	6.27 (SD = 1.77)
Neighbourhood property management (%)	
Have	21.35
Do not have	78.65
The level of air pollution (1–4)	3.13 (SD = 0.85)
The level of water pollution (1–4)	3.21 (SD = 0.79)
The level of noise pollution (1–4)	3.20 (SD = 0.85)
The level of soil pollution (1–4)	3.39 (SD = 0.69)

N is an abbreviation for Number. SD is an abbreviation for standard deviation.

Table 2. Cross-tabulation for household waste sorting willingness between rural and urban residents (N = 12,126).

	Rural Residents	Urban Residents	Chi-Square	*p* Value
Household waste sorting willingness (%)			277.97	0.000
Willing	78.39 (N = 1568)	89.98 (N = 4382)		
Unwilling	21.61 (N = 5688)	488 (N = 10.02)		
Willingness to deposit household waste at a fixed collection point (%)			228.17	0.000
Willing	92.05 (N = 6679)	98.38 (N = 4791)		
Unwilling	7.95 (N = 577)	1.62 (N = 79)		

N is an abbreviation for Number.

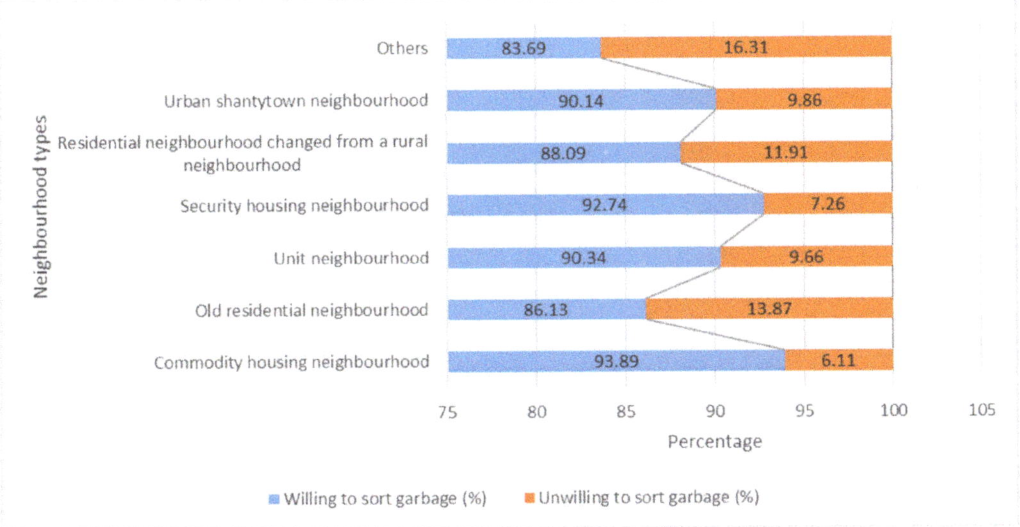

Figure 1. Distribution of households' willingness to sort garbage in different types of urban neighbourhoods.

3.2. Regression Results for Respondents' Waste Sorting Willingness

Table 3 shows the regression results for respondents' waste sorting willingness. Model 1 presents the results for all samples. Compared to residents living in rural neighbourhoods, residents living in urban neighbourhoods are more willing to sort household waste. The following variables of family lifestyle significantly and positively affected respondents' waste sorting willingness: average number of family meals per month, average monthly household food expenditure, household cleanliness, and household crowding. The more family meals and food expenditure, the more domestic waste may be generated, but the family's willingness to sort waste will also be higher. The cleanliness of the household can also directly reflect the willingness of household waste sorting. In terms of neighbourhood living environment, compared with families living in neighbourhoods with property management, families living in neighbourhoods without property management have relatively lower willingness to sort garbage. In addition, respondents who think that air pollution is not serious are more likely to have a lower willingness to sort garbage. Model 2 presents the regression results for rural respondents. The more the average monthly household food expenditure, the more likely the family will be willing to sort garbage. The degree of household cleanliness and household crowding can also significantly affect the household's willingness to sort garbage. Families living in neighbourhoods without property management have relatively lower willingness to sort garbage than those living in neighbourhoods with property management. Respondents who think that air pollution is not serious are more likely to have a lower willingness to sort garbage. Model 3 presents the analysis results for urban respondents. The number of family meals per month, family food expenditure, and house spaciousness have a positive impact on the household's willingness to sort garbage. Property management in the neighbourhoods will also increase residents' willingness to sort garbage. Model 4 of Table 4 shows the results of the garbage sorting willingness of households living in different types of urban neighbourhoods. Compared with households living in commodity housing neighbourhoods, households living in old residential neighbourhoods, unit neighbourhoods, and a residential neighbourhood changed from a rural neighbourhood are more likely to have a lower willingness to sort waste. This shows that the family willingness to sort garbage is significantly affected by the neighbourhood environment.

Table 3. Regression results for respondents' waste sorting willingness.

	Model 1		Model 2		Model 3	
	OR	95% CI	OR	95% CI	OR	95% CI
Urban neighbourhoods (ref: rural neighbourhoods)	1.814 ***	[1.577, 2.086]				
Number of family members living together	0.983	[0.961, 1.006]	0.988	[0.963, 1.014]	0.961	[0.913, 1.010]
Average number of family meals per month	1.002 ***	[1.001, 1.004]	1.002	[0.999, 1.004]	1.005 ***	[1.001, 1.008]
Average monthly household electricity consumption	0.996	[0.956, 1.039]	1.003	[0.952, 1.057]	0.976	[0.911, 1.046]
Average monthly household food expenditure	1.005 **	[1.001, 1.009]	1.005 *	[0.999, 1.010]	1.006 *	[0.999, 1.012]
Household cleanliness	1.118 ***	[1.077, 1.162]	1.145 ***	[1.096, 1.197]	1.041	[0.967, 1.122]
Household crowding	1.082 ***	[1.042, 1.123]	1.060 ***	[1.014, 1.108]	1.141 ***	[1.061, 1.227]
Neighbourhood property management (ref: have)	0.779 ***	[0.652, 0.930]	0.610 *	[0.369, 1.007]	0.806 **	[0.662, 0.983]
The level of air pollution	0.891 ***	[0.819, 0.968]	0.838 ***	[0.758, 0.927]	1.013	[0.873, 1.175]
The level of water pollution	1.024	[0.945, 1.109]	1.007	[0.918, 1.104]	1.127	[0.957, 1.327]

Table 3. *Cont.*

	Model 1		Model 2		Model 3	
	OR	95% CI	OR	95% CI	OR	95% CI
The level of noise pollution	0.948	[0.876, 1.026]	0.922	[0.834, 1.020]	0.970	[0.852, 1.105]
The level of soil pollution	1.063	[0.968, 1.168]	1.134 **	[1.015, 1.268]	0.926	[0.777, 1.103]
Number of samples	12126		7256		4870	
Log likelihood	−5260.662		−3699.207		−1550.743	
χ^2	517.623		175.740		69.214	

OR, odds ratio; CI confidence interval. * $p < 0.10$, ** $p < 0.05$, *** $p < 0.01$. χ^2 represents the chi-square test value.

Table 4. Regression results for urban respondents' willingness to sort waste and willingness to place waste at designated locations.

	Model 4: Willingness to Sort Waste		Model 5: Willingness to Place Waste at Designated Locations	
	OR	95% CI	OR	95% CI
Classification of urban neighbourhood (ref: commodity housing)				
Old residential neighbourhood	0.492 ***	[0.371, 0.652]	0.217 ***	[0.099, 0.478]
Unit neighbourhood	0.688 **	[0.511, 0.926]	0.253 ***	[0.114, 0.560]
Security housing	0.942	[0.462, 1.920]	0.745	[0.093, 5.965]
Residential neighbourhood changed from a rural neighbourhood	0.593 ***	[0.417, 0.842]	0.338 **	[0.132, 0.870]
Urban shanty town	0.671	[0.367, 1.226]	0.232 **	[0.060, 0.891]
Other	0.402 ***	[0.271, 0.597]	0.422	[0.124, 1.437]
Number of family members living together	0.974	[0.925, 1.026]	0.970	[0.859, 1.095]
Average number of family meals per month	1.004 **	[1.001, 1.008]	1.003	[0.995, 1.012]
Average monthly household electricity consumption	0.975	[0.910, 1.045]	0.925	[0.793, 1.079]
Average monthly household food expenditure	1.005	[0.999, 1.012]	0.999	[0.991, 1.006]
Household cleanliness	1.021	[0.947, 1.100]	1.242 **	[1.029, 1.498]
Household crowding	1.132 ***	[1.052, 1.218]	1.003	[0.834, 1.207]
Neighbourhood property management (ref: have)	0.968	[0.786, 1.191]	1.624 **	[1.004, 2.627]
The level of air pollution	1.010	[0.870, 1.172]	0.803	[0.557, 1.156]
The level of water pollution	1.104	[0.937, 1.301]	1.610 **	[1.120, 2.315]
The level of noise pollution	0.978	[0.858, 1.114]	0.871	[0.631, 1.202]
The level of soil pollution	0.931	[0.782, 1.109]	1.184	[0.799, 1.756]
Number of samples	4870		4870	
Log likelihood	−1534.096		−379.190	
χ^2	102.509		49.513	

OR, odds ratio; CI, confidence interval. ** $p < 0.05$, *** $p < 0.01$. χ^2 represents the chi-square test value.

3.3. Regression Results for Willingness Regarding Deposition of Household Waste to a Fixed Collection Point

Table 5 shows the regression results for respondents' willingness to deposit household waste at a fixed collection point. Model 6 presents the results for all the samples. Compared to rural respondents, urban respondents are more willing to deposit household garbage at a designated garbage collection point. Respondents who have more family meals each month are more likely to deposit their household waste at designated collection points. Households with higher electricity consumption are also more likely to have a higher willingness to place garbage at a fixed point. The household cleanliness and spaciousness (crowding) can significantly reflect the family's willingness to deposit waste at designated collection points. Respondents who think that air pollution is not serious are more likely to have a lower willingness to deposit waste at designated collection points, but those who think that soil pollution is not serious are more likely to have a higher willingness to deposit waste at designated collection points. This may be mainly because air pollution is more easily perceived by individuals, which can directly affect people's behaviour. Model 7 presents the results for the rural respondents. The results of Model 7 are like those of Model 6. Model 8 presents the analysis results for urban respondents. The household cleanliness can significantly reflect the urban family's willingness to deposit waste at designated collection points. Different from rural respondents, urban respondents who think that water pollution is not serious are more likely to have a higher willingness to deposit waste at designated collection points. To a certain extent, this shows that the better the environmental quality, the more likely people will be to protect the environment and avoid littering in urban areas. Model 5 of Table 4 shows that households living in commodity housing neighbourhoods are more likely to have a higher willingness to deposit waste at designated collection points than those living in other types of neighbourhoods (old residential neighbourhood, unit neighbourhood, residential neighbourhood changed from a rural neighbourhood, and urban shantytown).

Table 5. The regression results for respondents' willingness to place waste at designated locations.

	Model 6		Model 7		Model 8	
	OR	95% CI	OR	95% CI	OR	95% CI
Urban neighbourhoods (ref: rural neighbourhoods)	3.553 ***	[2.632, 4.795]				
Number of family members living together	0.995	[0.959, 1.032]	0.996	[0.958, 1.035]	0.964	[0.856, 1.085]
Average number of family meals per month	1.005 ***	[1.001, 1.008]	1.005 **	[1.001, 1.008]	1.004	[0.996, 1.013]
Average monthly household electricity consumption	1.086 *	[0.998, 1.182]	1.122 **	[1.016, 1.239]	0.941	[0.805, 1.099]
Average monthly household food expenditure	1.004	[0.997, 1.012]	1.007	[0.997, 1.017]	1.000	[0.991, 1.008]
Household cleanliness	1.183 ***	[1.111, 1.258]	1.166 ***	[1.091, 1.246]	1.275 **	[1.059, 1.536]
Household crowding	1.055 *	[0.992, 1.122]	1.058 *	[0.990, 1.131]	1.029	[0.858, 1.234]
Neighbourhood property management (ref: have)	0.933	[0.637, 1.368]	0.436	[0.160, 1.190]	1.262	[0.791, 2.014]
The level of air pollution	0.679 ***	[0.583, 0.791]	0.645 ***	[0.544, 0.766]	0.841	[0.587, 1.206]
The level of water pollution	1.037	[0.909, 1.183]	0.994	[0.862, 1.148]	1.615 ***	[1.130, 2.310]
The level of noise pollution	0.924	[0.801, 1.066]	0.930	[0.790, 1.095]	0.866	[0.628, 1.192]
The level of soil pollution	1.214 **	[1.036, 1.423]	1.221 **	[1.026, 1.453]	1.180	[0.797, 1.747]
Number of samples	12126		7256		4870	
Log likelihood	−2336.612		−1933.485		−389.600	
χ^2	429.664		161.509		28.693	

OR, odds ratio; CI, confidence interval. * $p < 0.10$, ** $p < 0.05$, *** $p < 0.01$. χ^2 represents the chi-square test value.

4. Discussion

4.1. Family Lifestyle and Neighbourhood Affect the Family's Willingness to Sort Waste

This study shows that urban residents' willingness to classify garbage is significantly higher than that of residents living in rural neighbourhoods. The emergence of this phenomenon is related to the family lifestyle and the living environment of urban and rural neighbourhoods. This study proposes that garbage classification implementation should pay attention to the willingness of households to sort waste, because compared with the difficulty of personal waste sorting, household waste sorting is more difficult. For example, kitchen waste is generated in family life, which is difficult to handle in garbage classification. Unlike other studies that focus on the influence of individual socioeconomic status on the willingness to sort waste [22,23], we put forward that the influence of family lifestyle and neighbourhood living environment on the household waste sorting willingness cannot be underestimated. For instance, the results of all samples show that the more family meals per month, the higher the household's willingness to sort garbage, but this result is only significant in the urban samples, and not in the rural samples. This is mainly due to the differences in the garbage collection facilities and public service provision between urban and rural areas. Compared with rural neighbourhoods, public service facilities in urban neighbourhoods are better [18,24]. For example, there will be dedicated personnel to collect and manage garbage in the neighbourhoods. Furthermore, in urban areas, systematic garbage collection, transhipment, and disposal infrastructure have advanced over decades, and the domestic garbage can be quickly recycled after it has been classified. However, the government's investment in infrastructure has been insufficient in rural areas [25–27]. The facilities and management systems for rural households' garbage remain underdeveloped, and rural residents cannot access facilities or services easily. All these conditions could result in greater difficulties regarding garbage classification implementation in rural areas.

In addition, residents' willingness to sort waste is different between urban and rural areas. There are also significant differences in the willingness of residents in different types of urban neighbourhoods to sort waste. In this research, we further divide urban neighbourhoods into six types of neighbourhoods, including commodity housing neighbourhoods, old residential neighbourhoods, unit neighbourhoods, security housing neighbourhoods, residential neighbourhood changed from a rural neighbourhood, and urban shantytowns. Among them, the commodity housing neighbourhood is a new type of market-oriented housing that emerged in China in the 1990s [28]. Commodity housing neighbourhoods generally have better property management. However, many scholars call such neighbourhoods "gated neighbourhoods" because they are often managed in a closed manner (or in an exclusive manner) [29]. Most of the old residential neighbourhood have been standing for a long time. The built environment of this kind of neighbourhood is quite serious. Unit (Danwei) neighbourhood is a kind of residential neighbourhood originating from China's planned economy period [30]. The housing in this neighbourhood is provided by the unit to the unit workers. However, this model of housing supply is no longer in place. There are complex housing property rights in the unit neighbourhood, and the property management of the neighbourhood is chaotic. The security housing neighbourhood is built for low-income urban residents. Although this kind of neighbourhood is provided to low-income families, most of these neighbourhoods are managed in accordance with the commodity housing neighbourhoods, but the level of property management and the quality of public service facilities are relatively low. The residential neighbourhoods that changed from rural neighbourhoods generally appear in the urban fringe areas. Due to China's dual urban–rural land use system, the supply mode of public service facilities in urban and rural areas is different. With the spread of urban areas, rural land was converted into urban land after being expropriated by the local government. As a result, some rural neighbourhoods are restructured into urban neighbourhoods. Relatively speaking, the quality of public service facilities in these neighbourhoods is worse than those in commodity housing neighbourhoods. Through comparative analysis, we found that the willingness of residents living in commodity housing neighbourhoods to sort waste is significantly stronger than that of residents living in other neighbourhoods. The public service facilities of the commodity housing neighbourhoods are complete, and garbage collection points will be set up more scientifically and reasonably in the residential design. The residents' willingness to sort garbage is not only affected by family living habits, but also related to the living environment of the neighbourhood in which they live. The neighbourhood differentiation and neighbourhood effect of the willingness to sort waste in China can be understood from two aspects. On the one hand, there are huge differences in property management in different types of neighbourhoods, which directly affects residents' garbage sorting behaviour. On the other hand, this is also one of the manifestations of the social spatial differentiation of Chinese cities. People of different socioeconomic status and lifestyles live in different types of neighbourhoods, leading to distinct neighbourhood differentiation in garbage sorting behaviour.

4.2. Policy Suggestion and Research Limitations

Waste sorting is one of the important ways for developing countries to achieve sustainable development. However, the formation of personal waste sorting habits is a slow process that not only needs to change the family's living habits, but also needs to provide more convenient public service facilities for waste sorting. First, garbage classification must become the consensus of the society, and make garbage classification a daily habit of everyone. Secondly, it is necessary to solve the problem of unequal garbage collection in urban and rural neighbourhoods. Local governments should increase investments in rural waste management systems and build the infrastructure necessary for sorting, transferring, and disposing waste in rural areas. Third, it is necessary to address the shortcomings of different types of urban neighbourhoods in garbage recycling and garbage disposal management. Some neighbourhoods have arranged staff to help residents sort garbage, but some neighbourhoods do not have such guidance. Different neighbourhood public services will cause residents in different neighbourhoods to have different degrees of willingness for garbage classification, which will lead to unsatisfactory implementation of the garbage classification policy. Developing different waste sorting assistance strategies for different types of neighbourhoods can better improve the implementation effect of waste sorting policies. There are two main research limitations in this study. First, the results of this study are limited by the database used. Because some neighbourhoods have not actually begun to implement mandatory garbage classification requirements, the residents' willingness to sort garbage may change after the mandatory garbage classification requirements are implemented. Second, the residents' willingness to sort garbage is affected by many factors. Although this study proposes that the external living environment has an important influence on the residents' willingness to sort garbage, the residents' living habits and socioeconomic status may also affect their garbage classification willingness.

5. Conclusions

Garbage classification is one of the important ways to achieve sustainable development in developing countries, but it the process of garbage sorting becoming part of people's way of life is a gradual one. The results of the study indicate that the family lifestyle and neighbourhood environment have a significant correlation with a family's willingness to sort garbage and to put garbage in designated locations. We conclude that in China, most people agree with the necessity of implementing garbage classification. However, the slow promotion of garbage classification may be mainly due to the relatively low quality of community garbage collection and management services, and the failure to associate community garbage collection with family life. In the process of implementing the waste sorting policy, not only must the residents' living habits be changed, but also more neighbourhood public service facilities must be provided for waste sorting, so that people can more conveniently sort waste.

Author Contributions: Conceptualisation, H.C.; Data processing and methodology, H.C. and L.Z.; Supervision, H.C.; Writing—original draft, H.C. and L.Z.; Writing—review and editing, L.Z. and H.C. All authors have read and agreed to the published version of the manuscript.

Funding: This research was funded by the National Natural Science Foundation of China (grant number 51908114) and the doctoral program of first-level discipline of Architecture, School of Architecture and Urban Planning, Shenzhen University.

Institutional Review Board Statement: Not applicable.

Informed Consent Statement: Not applicable.

Data Availability Statement: The data used in this study were derived from the 2016 China Labor-force Dy-namics Survey (CLDS), which was conducted by the Center for Social Science Survey at Sun Yat-sen University. http://css.sysu.edu.cn (accessed on 1 December 2021).

Conflicts of Interest: The authors declare no conflict of interest.

References

1. Grazhdani, D. Assessing the variables affecting on the rate of solid waste generation and recycling: An empirical analysis in prespa park. *Waste Manag.* **2016**, *48*, 3–13. [CrossRef]
2. Iyamu, H.; Anda, M.; Ho, G. A review of municipal solid waste management in the bric and high-income countries: A thematic framework for low-income countries. *Habitat Int.* **2020**, *95*, 102097. [CrossRef]
3. Okuda, I.; Thomson, V.E. Regionalization of municipal solid waste management in Japan: Balancing the proximity principle with economic efficiency. *Environ. Manag.* **2007**, *40*, 12–19. [CrossRef]
4. Tanskanen, J.-H.; Kaila, J. Comparison of methods used in the collection of source-separated household waste. *Waste Manag. Res.* **2001**, *19*, 486–497. [CrossRef]
5. Bai, R.; Sutanto, M. The practice and challenges of solid waste management in singapore. *Waste Manag.* **2002**, *22*, 557–567. [CrossRef]
6. Bernstad, A.; la Cour Jansen, J.; Aspegren, H. Life cycle assessment of a household solid waste source separation programme: A swedish case study. *Waste Manag. Res.* **2011**, *29*, 1027–1042. [CrossRef] [PubMed]
7. Dahlén, L.; Vukicevic, S.; Meijer, J.-E.; Lagerkvist, A. Comparison of different collection systems for sorted household waste in sweden. *Waste Manag.* **2007**, *27*, 1298–1305. [CrossRef]
8. Nelles, M.; Gruenes, J.; Morscheck, G. Waste management in germany–development to a sustainable circular economy. *Procedia Environ. Sci.* **2016**, *35*, 14. [CrossRef]
9. Zen, I.S.; Noor, Z.Z.; Yusuf, R.O. The profiles of household solid waste recyclers and non-recyclers in kuala lumpur, malaysia. *Habitat Int.* **2014**, *42*, 83–89. [CrossRef]
10. Charuvichaipong, C.; Sajor, E. Promoting waste separation for recycling and local governance in thailand. *Habitat Int.* **2006**, *30*, 579–594. [CrossRef]
11. Marshall, R.E.; Farahbakhsh, K. Systems approaches to integrated solid waste management in developing countries. *Waste Manag.* **2013**, *33*, 988–1003. [CrossRef] [PubMed]
12. Zurbrügg, C.; Drescher, S.; Patel, A.; Sharatchandra, H. Decentralised composting of urban waste–an overview of community and private initiatives in indian cities. *Waste Manag.* **2004**, *24*, 655–662. [CrossRef]
13. Nguyen, T.T.P.; Zhu, D.; Le, N.P. Factors influencing waste separation intention of residential households in a developing country: Evidence from hanoi, vietnam. *Habitat Int.* **2015**, *48*, 169–176. [CrossRef]
14. D'Amato, D.; Droste, N.; Allen, B.; Kettunen, M.; Lähtinen, K.; Korhonen, J.; Leskinen, P.; Matthies, B.D.; Toppinen, A. Green, circular, bio economy: A comparative analysis of sustainability avenues. *J. Clean. Prod.* **2017**, *168*, 716–734. [CrossRef]
15. Sadik-Zada, E.R. Natural resources, technological progress, and economic modernization. *Rev. Dev. Econ.* **2021**, *25*, 381–404. [CrossRef]
16. Radtke, J. A closer look inside collaborative action: Civic engagement and participation in community energy initiatives. *People Place Policy* **2014**, *8*, 235–248. [CrossRef]
17. Zhuang, Y.; Wu, S.-W.; Wang, Y.-L.; Wu, W.-X.; Chen, Y.-X. Source separation of household waste: A case study in china. *Waste Manag.* **2008**, *28*, 2022–2030. [CrossRef]
18. Zhang, D.Q.; Tan, S.K.; Gersberg, R.M. Municipal solid waste management in china: Status, problems and challenges. *J. Environ. Manag.* **2010**, *91*, 1623–1633. [CrossRef]
19. Tai, J.; Zhang, W.; Che, Y.; Feng, D. Municipal solid waste source-separated collection in china: A comparative analysis. *Waste Manag.* **2011**, *31*, 1673–1682. [CrossRef]
20. Padilla, A.J.; Trujillo, J.C. Waste disposal and households' heterogeneity. Identifying factors shaping attitudes towards source-separated recycling in bogotá, colombia. *Waste Manag.* **2018**, *74*, 16–33. [CrossRef]
21. Massarutto, A.; Marangon, F.; Troiano, S.; Favot, M. Moral duty, warm glow or self-interest? A choice experiment study on motivations for domestic garbage sorting in italy. *J. Clean. Prod.* **2019**, *208*, 916–923. [CrossRef]

22. Nutbeam, D.; Aaro, L.; Wold, B. The lifestyle concept and health education with young people: Results from a who international survey. *J. Inst. Health Educ.* **1991**, *29*, 98–103. [CrossRef]
23. Vereecken, C.A.; Maes, L.; De Bacquer, D. The influence of parental occupation and the pupils' educational level on lifestyle behaviors among adolescents in belgium. *J. Adolesc. Health* **2004**, *34*, 330–338. [CrossRef]
24. Chung, S.-S.; Poon, C.-S. A comparison of waste-reduction practices and new environmental paradigm of rural and urban chinese citizens. *J. Environ. Manag.* **2001**, *62*, 3–19. [CrossRef] [PubMed]
25. Zeng, C.; Niu, D.; Zhao, Y. A comprehensive overview of rural solid waste management in china. *Front. Environ. Sci. Eng.* **2015**, *9*, 949–961. [CrossRef]
26. Shenggen, F.; Zhang, X. Infrastructure and regional economic development in rural china. *China Econ. Rev.* **2004**, *15*, 203–214.
27. Yu, X. Is environment 'a city thing'in china? Rural–urban differences in environmental attitudes. *J. Environ. Psychol.* **2014**, *38*, 39–48. [CrossRef]
28. Wu, F. Residential relocation under market-oriented redevelopment: The process and outcomes in urban china. *Geoforum* **2004**, *35*, 453–470. [CrossRef]
29. Li, Z.; Wu, F. Tenure-based residential segregation in post-reform chinese cities: A case study of shanghai. *Trans. Inst. Br. Geogr.* **2008**, *33*, 404–419. [CrossRef]
30. Bjorklund, E. The danwei: Socio-spatial characteristics of work units in china's urban society. *Econ. Geogr.* **1986**, *62*, 19–29. [CrossRef]

sustainability MDPI

Article

Political Economy of Green Hydrogen Rollout: A Global Perspective

Elkhan Richard Sadik-Zada [1,2,3]

[1] Institute of Development Research and Development Policy, Faculty of Management and Economics, Ruhr-University, 44789 Bochum, Germany; sadikebd@ruhr-uni-bochum.de
[2] Centre for Studies on European Economy (AIM), Azerbaijan State University of Economics (UNEC), Baku 1007, Azerbaijan
[3] Center for Environmental Management, Resources and Energy (CURE), Faculty of Management and Economics, Ruhr-University, 44801 Bochum, Germany

Abstract: The present paper dwells on the role of green hydrogen in the transition towards climate-neutral economies and reviews the central challenges for its emancipation as an economically viable source of energy. The study shows that countries with a substantial share of renewables in the energy mix, advanced natural gas pipeline infrastructure, and an advanced level of technological and economic development have a comparative advantage for the wider utilization of hydrogen in their national energy systems. The central conclusion of this review paper is that a green hydrogen rollout in the developed and oil-exporting developing and emerging countries is not a risk for the rest of the world in terms of the increasing technological disparities and conservation of underdevelopment and concomitant socio-economic problems of the Global South. The targets anchored in Paris Agreement, but even more in the EU Green Deal and the European Hydrogen Strategy will necessitate a substantial rollout of RESs in developing countries, and especially in the countries of the African Union because of the prioritization of the African continent within the energy cooperation frameworks of the EU Green Deal and the EU Hydrogen Strategy. Hence, the green hydrogen rollout will bridge the energy transition between Europe and Africa on the one hand, and climate and development targets on the other.

Keywords: green hydrogen; green hydrogen value chain; developing countries; renewable energy sources; hydrogen strategy

Citation: Sadik-Zada, E.R. Political Economy of Green Hydrogen Rollout: A Global Perspective. *Sustainability* **2021**, *13*, 13464. https://doi.org/10.3390/su132313464

Academic Editor: Doug Arent

Received: 4 September 2021
Accepted: 1 December 2021
Published: 6 December 2021

1. Introduction

Hydrogen (H) is the simplest, lightest, and most widespread chemical element on earth. Hydrogen is not a source, but rather the carrier of energy. It does not exist separately and is found in compounds with other elements [1]. It can be obtained as molecular dihydrogen (H_2) from water, biomass, and hydrocarbons [2,3]. To obtain pure hydrogen, it must be separated from its compounds. Hydrogen production is predicated on fossil fuels and electrolysis. Currently, hydrogen is mostly utilized in the chemical sector for the production of ammonia and refining the hydrocracking and desulphurization of fossil fuels [4].

Dependent on the underlying production technology of the colorless hydrogen, hydrogen is divided into eight categories, which are labeled by different colors. These are black, brown, grey, turquoise, blue, yellow, pink, and green hydrogen types. The black, brown, turquoise and blue hydrogen are based on fossil fuels. Yellow and pink hydrogen are produced using electrolysis. The used electricity is, however, generated in nuclear power plants. Green hydrogen is entirely electrolysis-based, whereby the required electricity originates from solar or wind energy [4,5].

All the eight shades of hydrogen, but especially the green one offers a realistic opportunity for the reduction of the greenhouse gas footprints of the hard-to-abate and/or hard-to-electrify sectors, such as the steel industry, chemicals, long-haul transport, shipping, and aviation [6,7]. In contrast to oil, natural gas, or coal-based hydrogen, only the uptake of the green hydrogen, which is generated by electrolysis triggered by renewable energy sources can lead to the real climate neutrality of the energy sector itself and the above-mentioned hard-to-electrify sectors.

Electrolysis is an established technology to produce hydrogen through water and electricity [6]. Hydrogen is easy to store and this feature makes hydrogen complementary to solar and wind energy in the energy mix. Hence, hydrogen could be a potential solution for the volatility and intermittence of wind and solar energy that necessitate a substantial portion of fossil fuels, especially natural gas, as back-up energy in the energy mixes of individual countries [8–10]. Green hydrogen can be easily stored for longer periods and stabilize wholesale electricity prices. In addition, the storage of surplus renewable energy could prevent zero or below-zero electricity prices and by doing so increasingly protect renewable energy producers from losses [7].

The notion of a hydrogen economy suggested in 1970 by the U.S. electrochemist John Bockris seemed to be a futurist vision for more than four decades [11,12]. In his recent article on the perspectives of hydrogen, Mike Scott concludes that "Now it looks like the future has arrived" and backs his opinion by the large-scale advances in terms of investments in the development of hydrogen infrastructure [13]. Currently, there are more than 170 operational hydrogen projects in 162 countries [2]. Global demand for pure hydrogen has increased from 20 million metric tons (Mt) in 1975 to 70 Mt in 2019 [5].

On 8 December 2020, seven leading green hydrogen developers, Saudi clean energy group ACWA Power, Australian energy project developer CWP Renewables, Chinese wind turbine producer Envision, Italian gas group Snam, Norwegian fertilizer producer Yara, and the energy giants Iberdrola and Ørsted launched the Green Hydrogen Catapult Initiative, which envisages a 50-fold scale-up of the production of green hydrogen before 2026 [14]. In absolute numbers, the surge of output implies the deployment of 25 gigawatts of green hydrogen and the reduction of the production costs of hydrogen below the threshold of 2 USD per kilogram [6]. In 2020 the global green hydrogen market had a turnover of 0.3 billion USD. Following the projections of the IEA, in the coming years, the global green hydrogen market is going to grow at exponential rates and reach the threshold of 10 billion USD in 2028. In 2050 hydrogen could cover 24% of global energy demand [15].

A trillion-dollar European Green Deal Package, adopted in 2019, lead to the sky-rocketing interest of investors in green hydrogen [16]. In the subsequent EU Hydrogen Strategy released in 2020, green hydrogen is deemed to be the key driver of the EU's carbon neutrality target by 2050 [17]. The hydrogen strategy of the EU is the most ambitious green hydrogen strategy worldwide, envisaging 40 GW of green hydrogen production by 2030 (Figure 1). This strategy is in line with the national hydrogen strategies of the individual EU member states, Austria, France, Germany, Italy, the Netherlands, Poland, Portugal, and Spain. In addition to the EU member states, Australia, Canada, Chile, Morocco, Norway, and Spain also have recently adopted their national hydrogen strategies [18,19]. The U.S. and Japan also adopted recently their national comprehensive renewable hydrogen plans and a number of countries are on the verge of doing this [20].

The contemporary hydrogen production is mostly fossil-fuel-based [21]. This is grey, carbon-intensive hydrogen. According to International Energy Agency (IEA), only 0.1% of global hydrogen production can be attributed to water electrolysis. The inexorable growth of global demand for renewable hydrogen leads to the reduction of electrolyzer equipment [22]. In the past five years, the costs of electrolyzers have declined by more than 50%. According to Hydrogen Council and IEA, by 2030, green hydrogen production costs are going to decrease significantly. Dependent on the employed scenario the cost reductions will range between 30% and 70% [23]. This, in combination with the steadily decreasing production costs of renewables and public support for clean energy sources, contribute to an increasing interest in electrolytic hydrogen [24]. Analogously, hydrogen-based fuel cells are also becoming cheaper and more durable, and secure. Fuel cell technologies enable low carbon transition in the transport sector and can potentially contribute to energy supply security in remote areas [22].

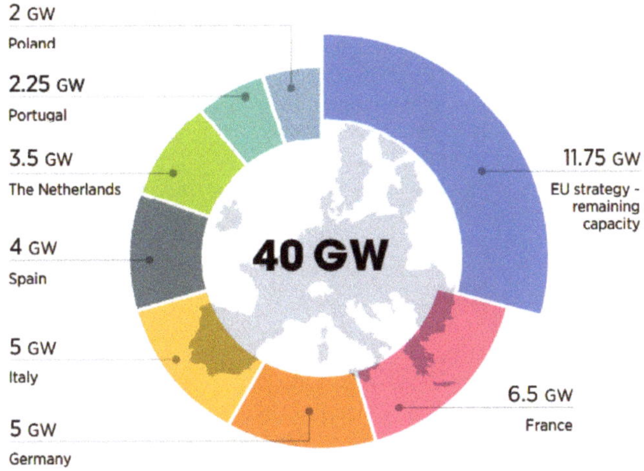

Figure 1. Electrolyzer capacity targets in European hydrogen strategies, 2030. Source: IRENA (2021), p. 26.

The central issue within the respective national hydrogen strategies is the roadmap for the rollout of green hydrogen production capacities [25]. The reason for the pronounced importance of the green hydrogen rollout emanates from the fact that green hydrogen is the only zero-carbon hydrogen. Carbon capture and storage have a potential of 85–95% [6]. Since the market for hydrogen consists of large parts of plants, which produce conventional hydrogen as a by-product, it is necessary to initiate large-scale additions of green hydrogen capacities if a hydrogen economy shall provide fuel for a significant part of the national economies [26]. The major problem related to the expansion of green hydrogen is that under free-market conditions green hydrogen is currently not competitive. The current production costs of green hydrogen are relatively high across the entire value chain, i.e., from electrolysis to fuel cells, and include the lack of infrastructure for transport and storage [21]. The costs of electrolysis facilities, especially, have to decrease by 40% in the short- and by 80% in the long term [6].

To assure the competitiveness of green hydrogen in the long-term the provision of public support schemes is indispensable. Within the above-mentioned hydrogen strategies, promotion of the explicit electrolysis capacities and cost targets occupy the central position [6]. These schemes have been anchored in the national and regional hydrogen strategies that the paper dwells on in Section 3. Due to the grave repercussions of the energy transition triggered by the wider use of green hydrogen, in this paper, we envisage the assessment of the effects of increasing hydrogen in the energy mixes of the Global North on economic development and energy resilience of the Global South. This is the first study, which explicitly addresses the effects of the uptake of green hydrogen on the problems of development in the Global South.

The article proceeds as follows: Section 2 reviews the global state of play of the national hydrogen strategies and contemporary policies towards the rollout of hydrogen in the national or regional energy mixes. Section 3 discusses the repercussions of hydrogen uptake in the advanced economies on the economies of less developed countries. Section 4 analyzes the national and regional hydrogen strategies. Section 5 delves into the repercussions of these strategic roadmaps on the economies of the developing countries. Section 5 concludes.

2. Green Hydrogen Value Chain

Green hydrogen is produced through water electrolysis, whereby electrolysis is fueled by renewable electricity, mostly solar and wind. There are four established electrolyzer technologies. These are alkaline, proton exchange membrane (PEM), anion exchange membrane (AEM), and solid oxide electrolyzer cells (SOEC). The overwhelming majority of the installed electrolysis technology is alkaline and PEM [27]. SOEC and AEM are still at the research and development phase and are less efficient than alkaline or PEM.

Green hydrogen cannot take off without comprehensive, reasonably balanced and at the same time sophisticated support across the energy generation value chain. Green hydrogen is two to three times more expensive than blue hydrogen with carbon capture and storage, which is generated from fossil fuels. The central drivers of the costs of on-site green hydrogen production are the costs of renewable electricity, which is required to power the electrolyzer. Additionally, harnessing hydrogen for end users is also more expensive than using fossil fuels. For instance, hydrogen- or fuel-cell-driven vehicles are one-and-a-half to two times more expensive than conventional fossil-fuel-driven vehicles [7,8].

The second barrier for the uptake of the green hydrogen is the lack of pipeline infrastructure. The gross length of global hydrogen pipelines is just 5000 kilometers. For comparison, the length of natural gas pipelines is 3 million kilometers. Globally there are fewer than 500 hydrogen refueling stations and more than 200,000 gasoline and diesel refueling stations in the EU and in the US. The latest decision of the German supreme court allowed the use of the natural gas pipelines for hydrogen. This kind of conversion could contribute to the competitiveness of the hydrogen-based energy supply systems in Europe [7].

The third essential barrier on the way to green hydrogen rollout is that over the hydrogen value chain, a substantial share of energy is lost due to technical reasons. More than 30% of energy used for electrolysis lost. In addition, the conversion of hydrogen to ammonia leads to an up to 25% energy loss. The employment of hydrogen in fuel cells causes further energy losses of 40–50% [7,8].

There are three production models for renewables-based hydrogen. These are full on-site production, electricity from the power grid, and a hybrid solution (Figure 2). In contrast to full on-site hydrogen production, connecting the electrolyzers to the power grid would enable the counteracting of the volatility of the wind and solar energy supply and stabilize hydrogen production.

Hence, sustainably low prices of the electricity from renewables are the prerequisite for a successful hydrogen economy. The second precondition for the gradual development of the hydrogen economy are reductions in the cost of electrolysis [6]. The fulfillment of these conditions would require substantial public support. According to IRENA's report on public policies on green hydrogen, in order to achieve the threshold for market penetration nations have to employ an integrated energy sector development strategy epitomized in four pillars. These are: 1. Adopting national hydrogen strategies; 2. Formulation of policy priorities; 3. Establishment of a governance system and enabling policies; and 4. Creation of the system that identifies the origin of green hydrogen [8].

There are, nevertheless, no universal policies for the green hydrogen development. The concrete public support measures for the uptake of green hydrogen depend on the level of hydrogen sector development. To address this issue, IRENA proposed the "policy stage" concept in the context of the electrolysis and infrastructure deployment [9]. At the initial stage green hydrogen output is at the megawatt level; green hydrogen is used in niches, and is not competitive with grey hydrogen. There is no substantial pipeline infrastructure and hydrogen is transported by hydrogen tracks. The share of renewables in the energy mix is far below its potential. This phase is termed Stage 1 or the "technology readiness" phase.

At the second stage, economies of scale in hydrogen production come fully into their own. Electrolyzers are at the gigawatt level and the production of renewables is cheap and their share in the energy mix is significant. This leads to a production surplus during sunny and/or windy periods. Hydrogen is no more demanded only by niches. There is a large-scale demand for hydrogen in regions with hard to decarbonize sectors. This leads to the advancement of the hydrogen infrastructure and especially pipelines. At that stage, the countries with a dense gas pipeline infrastructure start to convert these pipelines for hydrogen. At the subsequent stage green hydrogen is fully comparable with grey hydrogen and the power sector is a zero or negative emission sector. Green hydrogen continues its expansion until the transition to a climate-neutral energy system is complete.

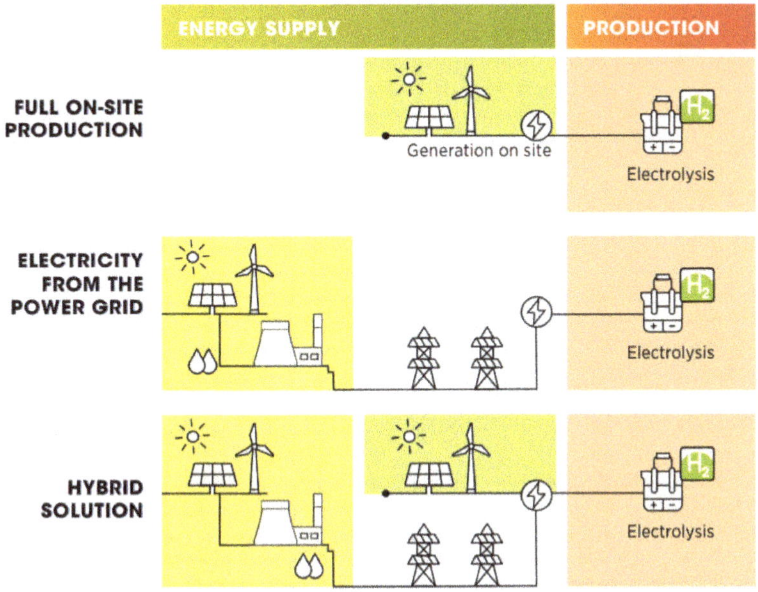

Figure 2. Production models. Source: IRENA (2021), p. 29.

Hence, the ideal typical transition to green hydrogen is a process, which consists of two major phases. These are the first phase of the renewables rollout and the second stage is the uptake of the green hydrogen economy.

As shall be shown later, application of the subsidization schemes that have been applied to the rollout of the renewable sources, such as solar and wind energy for green hydrogen, could lead to allocative inefficiencies epitomized in the double subsidization of hydrogen-based electricity production.

3. Central Challenges for the Rollout of the Green Hydrogen Economy

The central problem for the rollout of green hydrogen emanates from the relatively high production, conversion, transport, and storage costs of green hydrogen. Production costs depend on the price of the renewable electricity, costs of electrolysis and their capacities. Green hydrogen can achieve cost-competitiveness with blue hydrogen only if the operating hours per year exceed at least 3000 [27]. The conversion and transport costs of hydrogen are also important drivers of hydrogen; especially with increasing distance between the location of electrolysis and end demand site, the costs of transportation and conversion increase linearly (Figure 3).

The employment of trucks for the transport of hydrogen makes sense only for short distances because trucks can carry relatively small amounts of hydrogen. The carrying capacity of trucks could be substantially increased if instead of hydrogen gas they could transport liquified hydrogen. This, however, would require additional conversion costs, which also have to be considered in calculations.

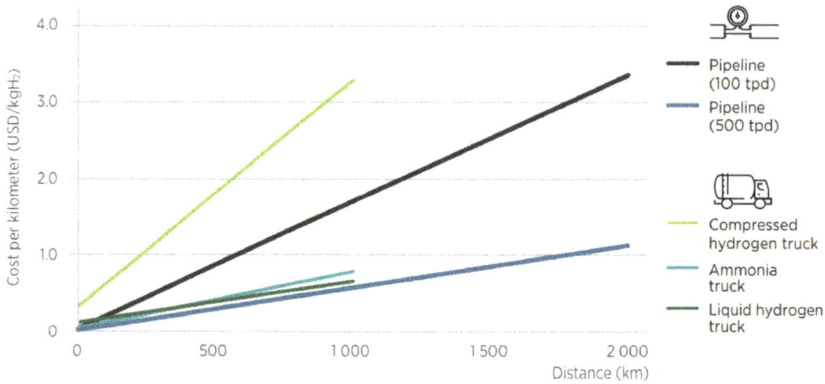

Figure 3. Costs for hydrogen transport as a function of distance by selected transport mode. Source: IRENA (2021) [27].

The cheapest transportation option for hydrogen is transportation via natural gas pipelines. Especially in the initial phases of infrastructure development and hydrogen technology proliferation long-distance hydrogen delivery plays a central role [28]. There are two possibilities for the utilization of the natural gas pipelines for the deliveries of hydrogen. These are hydrogen blending and the use of natural gas pipelines for pure hydrogen transport [29]. Construction of the pipeline infrastructure is very capital-intensive and would necessitate more investments than those for the construction of new gas pipelines [30]. Repurposing the existing natural gas pipeline infrastructure for the transportation of green hydrogen is deemed to be a relatively cost-effective transport option for large amounts of hydrogen. This would imply only 10–25% of the greenfield cost of new pipeline construction.

Cerniauskas et al. find that 80% of natural gas pipelines can be reassigned for hydrogen and reduce the hydrogen transportation costs by at least 60% [31].

There are three major fields of private and public investment in hydrogen. These are investments in hydrogen infrastructure, investments in research and development, and deployment [2].

4. Global Implications of the National Hydrogen Strategies

In 2017, Japan was the first country to propose a national hydrogen strategy and by doing so sparked interest in hydrogen in the Asian-Pacific region. In 2019, Australia and South Korea followed Japan's lead in dealing with hydrogen economy development and put forward their own national hydrogen strategies. The pioneer in Europe in terms of hydrogen strategy was Germany, which adopted its comprehensive hydrogen strategy as first European country to do so and pushed the EU hydrogen strategy during its presidency of the EU Council. In Latin America Chile has been promoting hydrogen rollout perhaps more than any other country in the world and in 2020 these policies came into their own within the framework of the first adopted national hydrogen strategy in Latin America. At present, there are already 13 countries and one regional integration block which have adopted national hydrogen strategies. Besides the already-mentioned countries and the EU, these are France, Netherlands, Norway, Portugal, Spain, Hungary, Canada, and the UK. All of them adopted their national hydrogen strategies in 2020 and 2021. This was to a large extent catalyzed by the United Nations Climate Change Conference in Glasgow (COP26) between 31 October and 12 November 2021 [32]. Twenty further countries are currently preparing their hydrogen strategies. The majority of the countries currently pushing a hydrogen strategy are countries with advanced or emerging economies. There are also more than 30 countries where political discussions are taking place, including official statements on hydrogen economy and initial hydrogen projects.

Figure 4 is an overview map of the degree of activity in terms of hydrogen strategy development. The map shows clearly that there is a clear gap in terms of hydrogen economy development between Africa, the Middle East, Central Asia, and South East Asia on the one hand and the rest of the world on the other.

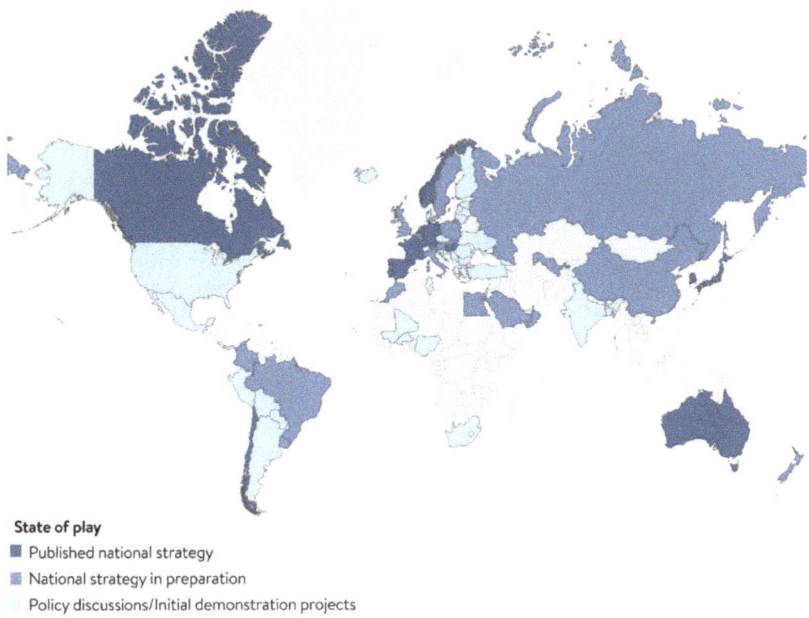

Figure 4. Country activities with regards to hydrogen development policies. Source: World Energy Council, 2021 [31].

According to a report of UNEP, which was prepared a decade and half ago, due to the financial leeway and qualified engineering staff gap, the developing world has been unable to participate in the research, development and deployment (RDD) of hydrogen and related technologies, and the transition to hydrogen economy will take place later in these developing countries [1]. The geography of the adopted and planned hydrogen strategies vindicates that this expectation was plausible. Without massive financial and technical support for the development of the prerequisites for the hydrogen economy, Sub-Saharan Africa, Central Asia, and South-East Europe will not be able to advance the hydrogen economies. This could lead to even greater discrepancies between the level of per capita income of developing and advanced nations and aggravate the problem of poverty and undernutrition in the respective regions. As a possible solution, the experts of UNEP recommend the large-scale support of international organizations and especially international development agencies for the rollout of hydrogen in developing countries.

5. Green Hydrogen and Industrialization of Developing Areas

Electric power is deemed to be the central trigger and the durable driver of the "Second Industrial Revolution" that took place the last quarter of the 19th and first decades of the 20th centuries [33]—hence, the rollout of the famous GOERLO-Plan, the first Soviet plan for national economic recovery and modernization, which in its essence was a roadmap for a large-scale electrification in post-revolutionary Russia, i.e., the Soviet Union. The source of the electrification in the advanced economies was the combustion of the fossil fuels and nuclear power. These sources are capable of delivering electricity uninterruptedly, which is required for industrial processes. The central shortcoming of these energy sources is their environmental risks and their negative impact on environmental quality and the health status of human beings. In this regard, renewables offer a superior alternative to conventional energy sources. The central shortcoming of the renewable energy sources

(RESs) is, however, their intermittency and their inability to provide sufficient electricity for industrial processes. This is why solar off-grid-based or solar-based electricity is mostly utilized for the consumptive electricity needs of households. Because of inadequate storage capacities, the surplus electricity from RESs in disproportionately sunny or windy phases cannot be made use of. Hence, the rollout of green hydrogen in developing areas could close this gap because there are a plenty of technical ways for hydrogen gas to store surplus energy. In addition, hydrogen is capable of generating energy for hard-to-electrify sectors, such as heavy industry, which is focal in terms of the rapid industrialization of the underdeveloped areas. The central merit of hydrogen is that in contrast to solar and wind energy projects, green hydrogen could really trigger industrial uptake in fossil-fuel-poor developing countries, which are more vulnerable to oil and gas price spikes than the rest of the world.

Nevertheless, the least developed areas are not capable of triggering the hydrogen rollout without massive investments in electrolyser capacity building and a surge in renewable energy generation. Furthermore, developing countries lack the advanced natural gas pipeline infrastructure and sufficient stock of human capital for the development of the hydrogen economy. In addition, developing countries, in contrast to the advanced ones, do not have sophisticated and seasoned hydrogen strategies. These are all comparative disadvantages of the Global South in terms of hydrogen. Their implications are the increasing technological disparity of the energy systems of the Global North and Global South and the aggravation of the economic underdevelopment of the developing areas.

This picture changes, nevertheless, if we consider that advanced countries, and especially Europe, cannot achieve the envisaged surge of green hydrogen in its energy mix (up to 24 percent by the year 2050) without imports of renewable electricity or hydrogen imports from overseas. Due to the solar and wind abundance in large parts of Africa and Asia, Europe is interested in the large-scale rollout of RESs in developing countries. This kind of bridging of the energy transition between advanced and developing countries is anchored in the EU Green Deal and the European Hydrogen Strategy, whereby energy sector cooperation with Africa and the members of the African Union has been given priority. By supporting RESs and green hydrogen exports from Africa to the EU developing African countries' comparative advantage in the generation of solar and wind electricity translates in the long-term to a comparative advantage in the production and export of green hydrogen. This is the central explanation for the recent projections of PWC and the Word Bank, which indicate that the African continent and large parts of the non-African developing countries will be the potential exporters of green hydrogen and produce it with a significant cost advantage. Hence, despite substantial comparative short-term disadvantages in promoting hydrogen, especially for the fossil-fuel importing developing countries, hydrogen could imply a substantial improvement of the trade balance and also an increase in energy security and resilience.

In addition, the green hydrogen rollout will contribute to the food security of the Global South. The experience of developing countries with hydrogen for the production of fertilizers confirms this expectation. Developing countries, such as Egypt, India, Turkey, Costa Rica, and Zimbabwe have made substantial advances in terms of their food sovereignty due to the production of green hydrogen [22]. In 1958 India installed 106.0 MW, Zimbabwe installed 1975 74.6 MW, and Egypt installed 1960 115.0 MW electrolyzer capacities. These and many other electrolysis-based hydrogen production projects were mostly supported by international development institutions, in the first line the World Bank, to support food security and the domestic production of fertilizers. The IGSAS fertilizer project in Turkey and the Fauji fertilizer project in Pakistan the 1980s were financed by the World Bank. As a natural-gas-abundant country, Malaysia has extensively employed alkaline electrolyzers to fuel domestic manufacturing. The polysilicon plant Sarawak in Indonesia exhibits the largest current 25 MW electrolyser worldwide [32].

6. Conclusions

The central conclusion of this review is that a green hydrogen rollout in the developed and oil-exporting developing and emerging economies is not a risk for the rest of the world in terms of the increasing technological disparities and conservation of underdevelopment and concomitant socioeconomic problems of the Global South. The targets, anchored in the Paris Agreement, but even more in the EU Green Deal and the European Hydrogen Strategy, will give positive impulses for the rollout of RESs in developing countries, and especially in the countries of the African Union because of the prioritization of the African continent within the energy-cooperation frameworks of the EU Green Deal and the EU Hydrogen Strategy. Hence, the green hydrogen rollout will bridge the energy transition between Europe and Africa on the one hand, and climate and development targets on the other.

In addition to this long-term, bird's eye perspective, based on the literature analysis, it can be concluded that the countries with a sizeable share of renewables in their energy mixes, whereby renewables are overwhelmingly sustainable without subsidies, have a cost advantage in green hydrogen production. In addition, countries with an advanced gas pipeline infrastructure will be able to make use of the existing natural gas pipelines for the unfolding hydrogen infrastructure. Hence, advanced countries with large solar and wind energy generation capacities and gas pipeline infrastructure are currently more capable of developing green hydrogen infrastructure with amenable costs.

According to the scholars of the Energy Sector Management Assistance Program (ESMAP) of the World Bank Group, developing areas with sizeable infrastructure for the production of natural gas and natural-gas-pipeline networks have a comparative advantage for the unfolding of green hydrogen. They mention explicitly Argentina, China, Europe, the Gulf Cooperation Council countries, Japan, Indonesia, Malaysia, North America, and Thailand as countries with favorable starting conditions for the unfolding of a hydrogen economy [22].

What are the implications of the prohibitive green hydrogen costs in a number of advanced, transition, and developing countries? Should these countries subsidize green hydrogen despite existence of cheaper alternatives? Such a strategy does not make sense economically and is not sustainable. Public policies in these countries must focus on the rollout of renewable energy sources and trigger a large-scale subsidization of hydrogen at the more advanced stages of renewable energy sector development.

Fossil-fuel-abundant developing and transition economies, on contrary, should follow a strategy which differs from that of the oil and gas importing countries. The development of the infrastructure for nonrenewable hydrogen and the advancement of solar and wind parks could lead to accelerated pathways of energy transition in this group of countries because of the complementarity of the infrastructure for green and nonrenewable hydrogen. The same holds also for countries that rely predominantly on nuclear energy in their energy mixes and are capable of producing pink hydrogen. The development of the infrastructure for pink hydrogen and rising shares of renewable energy sources could serve as a fruitful soil for surging green hydrogen in long-term energy mixes.

Funding: This research received no external funding.

Institutional Review Board Statement: Not applicable.

Informed Consent Statement: Not applicable.

Data Availability Statement: Data sharing is not applicable to this article.

Conflicts of Interest: The authors declare no conflict of interest.

References

1. UNEP. The Hydrogen Economy. A Non-Technical Review. 2006. Available online: https://www.unep.org/resources/report/hydrogen-economy-non-technical-review (accessed on 4 November 2021).
2. Ochu, E.; Woodall, C.M.; Braverman, S.; Smith, G.; Friedmann, J. Hydrogen Fact Sheet: Policy Support and Investments in Low-Carbon Hydrogen. Available online: https://www.energypolicy.columbia.edu/sites/default/files/pictures/HydrogenProduction_CGEP_FactSheet3_052521.pdf (accessed on 4 November 2021).
3. Kalamaras, C.M.; Efstathiou, A.M. Hydrogen Production Technologies. Current State and Future Developments. Hindawi Publishing Corporation Conference Papers No. 690627. 2013. Available online: https://downloads.hindawi.com/archive/2013/690627.pdf (accessed on 4 November 2021).
4. Bhagwat, S.R.; Olczak, M. *Green Hydrogen: Bridging the Energy Transition in Africa and Europe*; European University Institute: Florence, Italy, 2020. Available online: https://africa-eu-energy-partnership.org/wp-content/uploads/2021/04/AEEP_Green-Hydrogen_Bridging-the-Energy-Transition-in-Africa-and-Europe_Final_For-Publication_2.pdf (accessed on 4 November 2021).
5. Noussan, M.; Rimondi, P.P.; Scita, R.; Hafner, M. The Role of Green and Blue Hydrogen in the Energy Transition—A Technological and Geopolitical Perspective. *Sustainability* **2021**, *13*, 298. [CrossRef]
6. IRENA. *Green Hydrogen Cost Reduction: Scaling up Electrolysers to Meet the 1.5 °C Climate Goal*; International Renewable Energy Agency: Abu Dhabi, United Arab Emirates, 2020. Available online: https://irena.org/publications/2020/Dec/Green-hydrogen-cost-reduction (accessed on 4 November 2021).
7. IRENA. *Hydrogen from Renewable Power Technology Outlook for the Energy Transition*; International Renewable Energy Agency: Abu Dhabi, United Arab Emirates, 2018.
8. Sadik-Zada, E.R.; Gatto, A. Energy Security Pathways in South East Europe: Diversification of the Natural Gas Supplies, Energy Transition, and Energy Futures. In *From Economic to Energy Transition. Energy, Climate and the Environment*; Mišík, M., Oravcová, V., Eds.; Palgrave Macmillan: Cham, Switzerland, 2020. [CrossRef]
9. Eichman, J.; Harrison, K.; Peters, M. *Novel Electrolyzer Applications: Providing More Than Just Hydrogen*; National Renewable Energy Laboratory: Golden, CO, USA, 2014.
10. Baker McKenzie. Shaping Tomorrow's Global Hydrogen Market. Vie De-Risked Investments. 2020. Available online: https://www.bakermckenzie.com/-/media/files/insight/publications/2020/01/hydrogen_report.pdf?la=en (accessed on 4 November 2021).
11. Renssen, S. The Hydrogen Solution? *Nat. Clim. Chang.* **2020**, *10*, 799–801. [CrossRef]
12. Goldman Sachs. Green Hydrogen. The Next Transformational Driver of the Utilities Industry. Equity Research. 2020. Available online: https://www.goldmansachs.com/insights/pages/gs-research/green-hydrogen/report.pdf (accessed on 4 November 2021).
13. Scott, M. Green Hydrogen. The Fuel of the Future, Set for 50-Fold Expansion. Forbes. 2020. Available online: https://www.forbes.com/sites/mikescott/2020/12/14/green-hydrogen-the-fuel-of-the-future-set-for-50-fold-expansion/?sh=3392b87e6df3 (accessed on 4 November 2021).
14. Coleman, K. Green Hydrogen Catapult. Race to Zero. 2020. Available online: https://racetozero.unfccc.int/green-hydrogen-catapult/ (accessed on 4 November 2021).
15. Shan, R.; Signore, S.R.; Smith, B.T. *Analysis of Hydropower Pllnat Revenues for Independent System Operator New England (ISO-NE)*; No. ORNL/SPR-2019/1294; Oak Ridge National Lab (ORNL): Oak Ridge, TN, USA, 2019.
16. Deign, J. So What Exactly Is Green Hydrogen? Available online: https://www.greentechmedia.com/articles/read/green-hydrogen-explained (accessed on 4 November 2021).
17. EU. *A Hydrogen Strategy for Climate-Neutral Europe*; European Commission: Brussels, Belgium, 2020. Available online: www.hydrogen_strategy.pdf (accessed on 4 November 2021).
18. Bridle, R. Should Governments Subsidize Hydrogen? Global Subsidies Initiative. International Institute for Sustainable Development. 2021. Available online: https://www.iisd.org/gsi/subsidy-watch-blog/should-governments-subsidize-hydrogen (accessed on 4 November 2021).
19. IRENA. *Green Hydrogen: A Guide to Policy Making*; International Renewable Energy Agency: Abu Dhabi, United Arab Emirates, 2020. Available online: https://www.irena.org/-/media/Files/IRENA/Agency/Publication/2020/Nov/IRENA_Green_hydrogen_policy_2020.pdf (accessed on 4 November 2021).
20. Nagashima, M. Japan' Hydrogen Society Ambition: 2020 Status and Perspectives. IFRI. 2020. Available online: https://www.ifri.org/sites/default/files/atoms/files/nagashima_japan_hydrogen_2020.pdf (accessed on 4 November 2021).
21. IRENA. *Hydrogen: A Renewable Energy Perspective*; International Renewable Energy Agency: Abu Dhabi, United Arab Emirates, 2019.
22. ESMAP. *Green Hydrogen in Developing Countries*; International Bank of Reconstruction and Development/The World Bank: Washington, DC, USA, 2020.
23. Hydrogen Council. *Path to Hydrogen Competitiveness—A Cost Perspective*; Hydrogen Council Report: Brussels, Belgium, 2020.
24. IEA. *The Future of Hydrogen. Seizing Today's Opportunities*; Report Prepared by the IEA for the G20, Japan; International Energy Agency: Paris, France, 2019.
25. BWMI. *Die Nationale Wasserstoffstrategie*; German Federal Ministry for Economics Affairs and Energy (BMWI): Berlin, Germany, 2020.

26. Velazquez Abad, A.; Doods, P.E. Production of Hydrogen. In *Encyclopedia of Sustainable Technologies*; Abraham, M.A., Ed.; Elsevier: Amsterdam, The Netherlands, 2017; pp. 293–304.
27. IRENA. *Green Hydrogen Supply. A Guide to Policy Making*; International Renewable Energy Agency: Abu Dhabi, United Arab Emirates, 2021. Available online: https://irena.org/-/media/Files/IRENA/Agency/Publication/2021/May/IRENA_Green_Hydrogen_Supply_2021.pdf (accessed on 4 November 2021).
28. Cerniauskas, S.; Grube, T.; Praktiknjo, A.J.; Stolten, D.; Robinius, M. Future hydrogen markets for transportation and industry: The impact of CO_2 taxes. *Energies* **2019**, *12*, 4707. [CrossRef]
29. Cerniauskas, S.; Chavez Junko, A.J.; Grube, T.; Robinius, M.; Stolten, D. Options of natural gas pipeline reassignment for hydrogen: Cost assessment for a Germany case study. *Int. J. Hydrog. Energy* **2020**, *45*, 12095–12107. [CrossRef]
30. Guidehouse. European Hydrogen Backbone. Utrecht, The Netherlands. 2020. Available online: https://gasforclimate2050.eu/wp-content/uploads/2020/07/2020_European-Hydrogen-Backbone_Report.pdf (accessed on 4 November 2021).
31. World Energy Council. Hydrogen on the Horizon: Ready, Almost Set, Go? National Hydrogen Strategies. World Energy Council Working Paper. 2021. Available online: https://www.worldenergy.org/assets/downloads/Working_Paper_-_National_Hydrogen_Strategies_-_September_2021.pdf (accessed on 4 November 2021).
32. Buttler, A.; Spliethof, H. Current Status of Water Electrolysis for Energy Storage, Grid Balancing and Sector Coupling via Power-to-Gas and Power-to-Liquids: A Review. *Renew. Sustain. Energy Rev.* **2018**, *82*, 2440–2454. [CrossRef]
33. Rosenberg, N. The Role of Electricity in Industrial Revolution. *Energy J.* **1998**, *19*, 7–24. [CrossRef]

sustainability

MDPI

Article

Food: Not Only Safety, but Also Sustainability. The Emerging Trend of New Social Consumers

Gennaro Civero [1], Vincenzo Rusciano [1], Debora Scarpato [1,*] and Mariarosaria Simeone [2]

[1] Department of Economic and Legal Studies, University of Naples Parthenope, 80133 Naples, Italy; gennaro.civero@uniparthenope.it (G.C.); vincenzo.rusciano@uniparthenope.it (V.R.)

[2] Department of Law, Economics, Management and Quantitative Methods (DEMM), University of Sannio, 82100 Benevento, Italy; msimeone@unisannio.it

* Correspondence: debora.scarpato@uniparthenope.it

Abstract: Consumers' purchasing choices are no longer based only on economic factors but also on ethical reasons related to environmental sustainability and food safety. However, nutritional information on food labels is underused by consumers. Often the lack or incompleteness of information available on the market obstructs the complex transition towards sustainable consumption patterns. This empirical study analysed a sample of 359 consumers from an area in Southern Italy (city of Naples) to identify homogeneous consumer clusters with respect to the assessment of the level of consumer attention to sustainable environmental, social behaviours in daily life, and also to safety attributes. The most important sources of information influencing the consumers' choices, food safety knowledge, and future purchasing behaviour were analysed. The research sample was self-selected, and the questionnaire for the survey was administrated through a non-probability sample from a reasoned choice. The results indicate that the ideal solution is a five-cluster partition that confirms a good level of attention to intrinsic attributes, in particular food expiry, transparency of food information, food traceability, and seller confidence. In addition, the research could provide an opportunity to consider collaborative actions between policy makers and industries to increase consumer awareness of environmental attributes.

Keywords: consumers; cluster analysis; sustainable consumption; mass media; social media

Citation: Civero, G.; Rusciano, V.; Scarpato, D.; Simeone, M. Food: Not Only Safety, but Also Sustainability. The Emerging Trend of New Social Consumers. *Sustainability* **2021**, *13*, 12967. https://doi.org/10.3390/su132312967

Academic Editors: E. Richard Zada and Andrea Gatto

Received: 20 October 2021
Accepted: 21 November 2021
Published: 23 November 2021

Publisher's Note: MDPI stays neutral with regard to jurisdictional claims in published maps and institutional affiliations.

1. Introduction

The modern agri-food consumers are more inclined to make informed and conscious purchasing decisions regarding products that favour a level of environmental, social, and economic sustainability while not neglecting economic aspects [1,2]. Numerous studies highlight the lack of or incomplete information available on the market that creates [3] information asymmetry issues that obstruct the complex acquisition of sustainable consumption patterns [4]. According to recent studies [5–7], the nutritional value of food remains a key factor for consumers in their food purchases. Nevertheless, even when nutritional information on food labels is comprehensive, it is underused by consumers [8,9].

The agri-food sector attaches great importance to the quality of raw materials, with respect for and protection of the environment and human resource, involving the stakeholders [10–12]. Consumers' choices are also ethical choices related to the protection of the environment and working conditions, protecting the territory and resources to be handed over to future generations [13]. Consumers do not base their purchasing choices only on economic motivations in relation to quality/price [14–20] but, above all, on the benefits to the community for the development of a more environmentally friendly society. Important international reports, such as the International Assessment of Agricultural Knowledge, Science, and Technology (IAASTD); the Climate Change Manifesto; and the Future of Food Safety [21], highlight the link between food security and environmental sustainability.

This connection is explained in these studies by stating that agriculture has a fundamental footprint on all major environmental issues [22,23]: climate change, biodiversity, land degradation, water quality, etc. These issues present difficult development challenges that need to be resolved because we must reduce the climate change footprint of agriculture; we must not degrade our soil and water, which has negative effects on biodiversity. [24,25]. These major challenges can be addressed by integrating local and traditional knowledge with institutional knowledge.

The new needs expressed by consumers and acknowledged by the actors of the production chain and in synergy with national governments encourage the development of increasingly articulated certification systems [26,27]. In the context of the green economy, such products have higher costs to operate in an ecological and social way. A good digital green marketing strategy and a good use of social media as a communication channel are essential to communicate the value of the product and to achieve a higher level of civic engagement [28]. The aim is to encourage the purchase of eco-friendly products and therefore more ethical than those with only economic benefits.

This study aims to analyse the degree of awareness of a sample of 359 Italian consumers regarding the adoption of sustainable behaviour in their lifestyle and attention to food safety attributes. In addition, it analyses the consumer's understanding of the main sources of information and their influence on the respondent's food safety knowledge and future purchasing behaviour.

2. Theoretical Framework

Food has an essential role in society by influencing people's lifestyle, health, and habits [29] and is an essential part of socialization. Today, the consumer defined as "critical" or "responsible" can show his or her attention to food safety, environmental protection issues, justice, human rights, and everything related to the ethical content of commercial activities to the production world [30,31].

In addition, the critical consumer avoids products that are dangerous to the health or cause damage to the environment during manufacture, use, or disposal. In this context, there are many events that focus on the valorisation and importance of food. For example, Expo 2015 "Feeding the Planet, Energy for Life" connected food issues with sustainability [32–34]. This event analysed the role of food in addressing nutrition, food production, management, and distribution as well as global and regional food governance, promoting an international dialogue on nutrition and natural resources. Furthermore, some studies [35,36] have supported that gastronomy can help build social relationships with local communities by clarifying the different motivational factors that drive consumers to choose experiences that promote a healthy lifestyle. According to Beldad and Hegner (2018) [37], the propensity toward purchasing food safety products is higher among women than men, as women are more willing to pay more. According to Monier-Dilhan (2018) [38], consumers with higher levels of education are more likely to buy organic products for health reasons, product quality, and environmental protection [39]. In accordance with previous studies, [40] the level of education is rather more relevant [41] than the age variable in obtaining a good level of attention to food safety issues. In the field of communication, according to Panagiotopoulos et al., (2015) [42], before digital media, food companies had to use expensive marketing media to reach their audience with limited feedback and low possibility of target knowledge.

Since the 2000s, digital green marketing strategies related to the optimal and widespread use of social media to effectively promote product value have become widely established [43,44]. However, previous studies show that employees (especially if they are married) use social media much less and are more sceptical about this digital evolution [45].

Academic research is developing, and related concepts are being explored, such as the impact of such information sources on consumers' purchasing motivations [46]. The aim is to induce consumers' purchasing motivations towards more sustainable and environmentally friendly products. Based on these assumptions, agri-food systems are seeking new levels of sustainability to respond to new levels of wellbeing to monitor the health of the planet [47,48].

The study focuses on the following Research Questions (RQs):

RQ1: The level of consumer attention to intrinsic attributes of the product purchased and the reasons for purchasing food products.

RQ2: The implementation of sustainable environmental and social behaviours in the daily life of respondents.

RQ3: The most important sources of information that influence the consumer's choices.

3. Materials and Methods

This study investigates the respondents' level of awareness of sustainable behaviours in their daily lives, consumers' understanding of and attention to intrinsic attributes, and the influence of information sources on future purchasing behaviour. To collect information, a questionnaire was prepared for a population of 359 Italian consumers residing in Campania (a southern Italian region), surrounding the city of Naples. The interviews were conducted directly face-to-face during the period January–February 2021.

The research sample was self-selected; it was an exploratory study without inferential objectives. The questionnaire was administrated through a non-probability sample from a reasoned choice.

The final questionnaire was divided into four sections consisting of multiple-choice questions measured on a Likert scale from 1 to 5 points. The first part investigated the safety of food consumed while the second part the tendency to implement sustainability-oriented behaviours in the daily lives of respondents.

In addition, the greater willingness to purchase food products when food safety-oriented production behaviours are adopted was analysed. Lastly, which sources of information on food, between mass media and social media, were most appreciated by the interviewees was analysed, highlighting a possible influence of information sources on future purchasing behaviours. Using the data from the questionnaires, a first univariate exploratory analysis was elaborated (analysis of the frequency and use of synthetic indicators), then a multivariate analysis [49], and an analysis of the main components and cluster analysis were performed. Principal component analysis (PCA) was used as a way of reducing dimensionality [50]. It made it possible to restrict the number of quantitative variables into a smaller set of factors or principal components (PCs). After identifying the key components (with PCA analysis) that influence the awareness of respondents on intrinsic attributes, a cluster analysis was finalized [51] to segment the statistical units [52,53]. In this scheme, the K-means approach was chosen, in which objects were divided into separate sub-sets, and each object is part of one and the same cluster. With this analysis, an ideal partition of 5 clusters was identified, analysed, and discussed.

4. Results

Socio-demographic characteristics of the sample are summarized in Table 1. The consumers participating in this study were mostly single, employed women (57.66%), aged between 18 and 25, with a high school diploma and a total net annual income of 10,000–20,000 €. It can be assumed that the female population is still responsible for food purchase [54], according to the study of Beldad and Hegner (2018) [37].

The sample of respondents is "very" concerned (35.9%) about food safety standards, with more attention paid to aspects such as food expiry (33.1%) and transparency in food information (28.69%). Other aspects, such as food traceability (29.53%) and seller confidence (31.5%), were only "moderately important". From the sample interviewed, it was shown that among the most considered purchasing motivations were the place of origin (28.69%) and the presence of certifications (23.96%), while the method of production (28.97%) and the method of conservation (24.79%) were of "moderate importance". The questions in the second part of the questionnaire investigated the use of sustainable behaviour adopted daily by citizens, and 28.41% stated that the habit of buying eco-labelled products is "very important" in their daily lives. The purchase of fair-trade products and the use of eco-labels are still niche issues, according to Monier-Dilhan (2018) [38] and Van't Veld (2020) [39]. In the third survey section, respondents were asked how much more they would be willing to spend in the purchase of a product if the company adopted food safety behaviours. Looking at the responses, 48.75% said they would be willing to spend at least 10 % more than the average selling price for products with safety attributes; 30.92% said they would be willing to spend at least 20% more than the average price; 25.07% stated that they would have "moderate" willingness to spend 30% more than the average selling price.

Table 1. Socio-demographic variables of the sample.

Attributes	No. Consumers	Percentage of Sample
Males	152	42.34%
Females	207	57.66%
18–25	129	35.93%
26–35	75	56.82%
36–45	44	12.26%
46–55	64	17.83%
56–65	27	7.52%
>65	20	5.57%
Single/Bachelor	203	56.55%
Married	156	43.45%
Absence of school certificate	32	8.91%
Secondary school certificate	56	15.60%
Diploma	200	55.71%
Degree	65	18.11%
Postgraduate degree	6	1.67%
Student	112	31.20%
Employee	114	31.75%
Freelancer	48	13.37%
Unemployed	16	4.46%
Housewife	34	9.47%
Retired	15	4.18%
Other	20	5.57%
$X \leq$ €10,000	93	25.91%
€10,000 < $X \leq$ 20,000	136	37.88%
€20,000 < $X \leq$ 30,000	90	25.07%
€X > 30,000	40	11.14%

Source: Authors' elaboration of data from survey.

The final part of the survey questions investigated on the information sources on food safety evaluating their level of clarity and reliability. First, the information sources with most clarity and reliability were doctors (34.54%), Ministry of Health (30.08%), and health associations (28.69%), while the least reliable were the food blogger (32.59%) and social media (32.59%), whose clarity and reliability were evaluated "for nothing". Looking at traditional media, we can see that TV (34.26%), the Internet (31.48%), and magazines (33.43%) present a "moderate" score.

In this fourth section of the questionnaire, the influence of information sources on the future purchasing behaviour of the respondents was investigated. In addition, how the reputation of food companies influences the respondents' cultural enrichment on food safety matters was investigated. The results showed that 30.92% of the sample buy food safety products having been influenced by the source of information, while 33.15% are influenced to a "moderate" extent in their purchase by the reputation of the companies, and 30.64% of the sample are "very" inclined to acquire more information about environmental and social sustainability, according to Van't Veld (2020) [39].

4.1. Principal Component Analysis

The sensitization of the sample on the topics of food safety and the related propensity towards greater purchase is influenced by several variables. Therefore, the use of principal component analysis provides an overall clearer and more immediate interpretation [50]. In particular, the components identified can be interpreted as follows:

1. Purchasing motivations describe the consumer purchasing motivations on food products, such as place of origin or method of production.
2. Sustainable behaviour describes the implementation in daily life of the interviewees whose behaviour is strongly oriented toward sustainability, for example, the purchase of fair-trade products and the use of trademarks protecting workers.
3. Propensity to purchase describes consumer's willingness to pay for a food safety product.
4. Attention to safety attributes is a component that describes the principal attributes for which consumers pay more their attention, for example, food expiration and food traceability.
5. Media sources satisfaction describes the consumers comprehension of the informative source for a food safety product. In this component, we discuss traditional media, such as TV, the Internet, and magazines.
6. Awareness is a component that describes if the information source is influencing the consumer's food safety knowledge and future buying behaviour.
7. Institutional sources satisfaction describes the consumers comprehension on the informative source for a food safety product. In this component, we talk about doctors, Ministry of Health, and health associations.
8. Social media satisfaction describes the consumers comprehension of the informative source for a food safety product. In this component, we talk about social networks and food bloggers.

The choice of the number of components occurs by considering three joint criteria: values of community, the amount of cumulative variance explained by the factors (Table 2), and the Eigenvalues of the components. The eight principal components thus identified (Table 3), in fact explain 65.929% of the variance in the original variables.

Table 2. Total of the explained variance.

Total of the Explained Variance			
Component	**Initial Eigenvalues**		
	Total	**% Variance**	**% Variance Cumulative**
1	5.443	20.936	20.936
2	2.377	9.143	30.079
3	2.135	8.213	38.292
4	1.746	6.717	45.009
5	1.696	6.522	51.531
6	1.365	5.249	56.780
7	1.250	4.809	61.589
8	1.128	4.340	65.929
9	0.862	3.314	69.244
10	0.785	3.020	72.264

Table 2. *Cont.*

Total of the Explained Variance			
Component	**Initial Eigenvalues**		
	Total	**% Variance**	**% Variance Cumulative**
11	0.698	2.683	74.947
12	0.635	2.444	77.391
13	0.610	2.347	79.738
14	0.574	2.208	81.946
15	0.510	1.962	83.907
16	0.505	1.943	85.850
17	0.505	1.941	87.792
18	0.469	1.804	89.595
19	0.434	1.669	91.264
20	0.421	1.619	92.884
21	0.386	1.486	94.370
22	0.359	1.382	95.752
23	0.318	1.222	96.974
24	0.310	1.193	98.167
25	0.299	1.151	99.318
26	0.177	0.682	100.000

Source: Authors' elaboration of data from survey.

Table 3. The rotated component matrix.

The Rotated Component Matrix									
Variables	**Component**								
	1	**2**	**3**	**4**	**5**	**6**	**7**	**8**	**Communality**
Food expiry	0.132	−0.080	−0.013	**0.688**	0.141	0.066	0.138	−0.081	0.547
Transparency of food information	0.285	0.122	0.091	**0.734**	0.003	0.075	0.048	−0.083	0.658
Food traceability	0.257	0.328	0.065	**0.615**	−0.015	0.017	0.017	0.034	0.558
Seller confidence	−0.012	0.183	0.055	**0.658**	0.053	0.188	−0.082	0.151	0.537
Environmental trademarks	0.209	**0.737**	0.038	0.073	0.079	0.031	0.151	−0.035	0.624
Fair trade	0.229	**0.765**	0.054	0.040	0.108	0.153	0.043	0.030	0.680
Avoid unfair brands	0.060	**0.734**	0.177	0.103	0.091	−0.020	0.063	0.003	0.597
Trademarks protecting workers	0.192	**0.727**	−0.002	0.180	−0.042	0.087	−0.012	0.122	0.623
Spending + 10%	0.089	0.028	**0.787**	0.116	0.055	0.178	0.073	−0.077	0.688
Spending + 20%	0.009	0.098	**0.927**	0.033	0.051	0.089	0.016	−0.017	0.881
Spending + 30%	0.091	0.114	**0.805**	0.009	0.021	−0.026	0.094	0.050	0.682
TV source satisfaction	0.014	0.109	0.073	0.044	**0.793**	0.177	−0.020	0.110	0.692
Internet source satisfaction	−0.103	0.024	0.113	0.098	**0.730**	0.031	−0.049	0.284	0.650
Magazine source satisfaction	0.111	0.084	−0.043	0.036	**0.800**	−0.058	0.155	0.055	0.693
Doctors source satisfaction	0.042	−0.012	0.043	0.100	0.260	0.133	**0.629**	−0.159	0.520
Ministerial source clarity	0.035	0.090	0.097	0.017	−0.023	0.101	**0.857**	−0.055	0.768
Association source clarity	0.042	0.159	0.058	0.007	−0.104	−0.001	**0.777**	0.308	0.740
Food blogger source satisfaction	−0.004	0.112	−0.055	−0.011	0.124	0.047	0.096	**0.831**	0.733
Social media source satisfaction	0.085	−0.037	0.009	0.006	0.281	0.005	−0.075	**0.778**	0.698
After info–more purchases	0.122	0.093	0.038	0.040	0.130	**0.807**	0.030	−0.035	0.696
After info—more reputation	0.154	0.043	0.149	0.045	0.059	**0.772**	0.068	0.148	0.675
After info—more environmental and social knowledge	0.107	0.071	0.050	0.262	−0.054	**0.712**	0.143	−0.050	0.620

Table 3. *Cont.*

The Rotated Component Matrix									
Variables	Component								
	1	2	3	4	5	6	7	8	Communality
Place of origin	**0.728**	0.117	0.027	0.197	0.035	0.106	0.058	−0.027	0.599
Certifications	**0.786**	0.170	0.112	0.043	−0.032	0.125	0.027	−0.076	0.684
Method of conservation	**0.738**	0.168	0.023	0.210	0.005	0.105	0.071	0.035	0.636
Method of production	**0.729**	0.249	0.058	0.121	0.047	0.083	−0.031	0.193	0.660

Source: Authors' elaboration of data from survey.

The values of community explain the amounts of variance of each variable explained by extracted factors and such values, being always greater or equal to 50% of the initial variance, demonstrate the fact that the variables are well represented by the factors. The scree plot constructed (Figure 1) shows a slope change at the point where the components become negligible, demonstrating the fact that the first eight factors are explicative since they presented an Eigenvalue greater than 1.

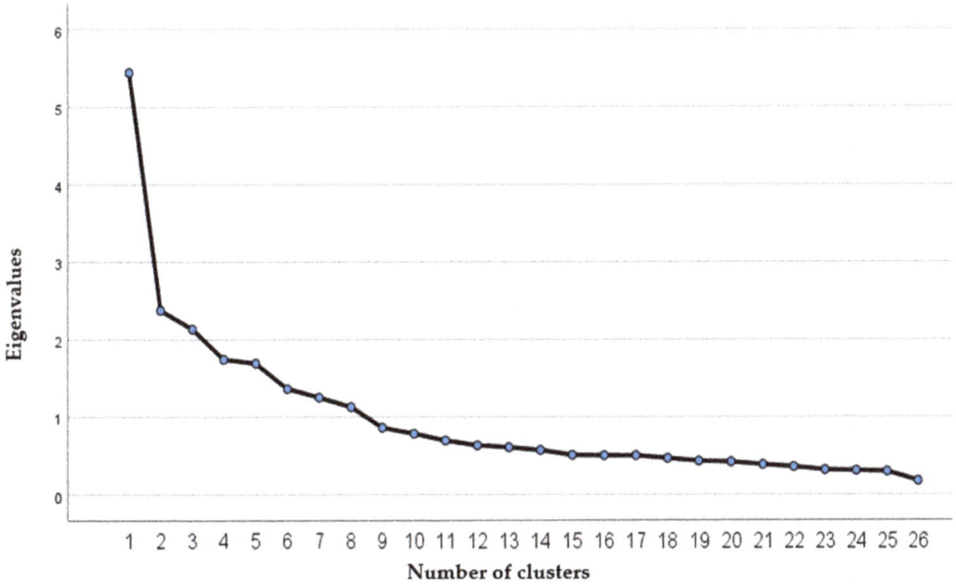

Figure 1. The scree plot. Source: Authors' elaboration of data from survey.

4.2. Cluster Analysis

In this empirical survey, the ideal solution is a five-cluster partition, and Table 4 presents the final cluster centres. Having identified the key components that influence the sensitization of the interviewees to the topics analysed, the aim of a cluster analysis was [55] to segment the statistical units [56].

Table 4. The final cluster centres.

Components	Final Cluster Centres				
	Cluster				
	1	2	3	4	5
1. Purchasing motivations	0.65606	−0.85465	−0.40100	0.31052	0.63766
2. Sustainable behaviour	0.31252	−0.29437	0.14175	−0.57718	0.41056
3. Propensity to purchase	0.81462	−0.03139	−0.31679	0.01042	−0.51417
4. Attention to food safety	0.20225	−0.36736	0.24880	−0.06051	0.18865
5. Media sources satisfaction	0.81769	0.07587	0.38269	−0.80027	−0.59981
6. Awareness	−0.10956	−0.36072	0.79324	−0.30242	0.26063
7. Institutional sources satisfaction	0.08879	0.41438	−1.18549	0.03991	0.13734
8. Social media satisfaction	−0.01540	−0.29603	0.52680	1.28354	−0.81811

Note: Values close to 1 or higher indicate that the specific component represents that group of clusters. Extremely negative values indicate that the specific component has little or no representation in that cluster group. Source: Authors' elaboration of data from survey.

The clusters can be described as follows:

- Cluster 1 can be defined in summary as the group of "food safety virtuous", represented by 20.05% of the sample (72 of 359 consumers) and characterized mainly by component 3 (value of 0.81462). In fact, their propensity to buy food safety products is much higher, and their motivation to buy them is considerable. They are sustainable in their daily life and present a good level of comprehension of information sources, such as the traditional media, while having a negative level of social media satisfaction (−0.01540).
- Cluster 2 can be defined in summary as the group of "consumer sensitivity" to institutional information. The propensity to purchase and the purchasing motivations are negative in respect to all the components evaluated, especially with reference to orientation towards sustainability. This cluster is more numerous, comprising 106 of 359 interviewees (29.52% of the sample) and is extremely far from component 3 (value of −0.03139), named propensity to purchase. The component 8 (social media satisfaction) presents a negative value (−0.29603), while component 7 (Institutional sources satisfaction) presents a positive value (0.41438).
- Cluster 3 can be defined as the group of "sceptical consumer", representing 14.76% of the sample (53 of 359 consumers) and is extremely far from component 7 (value of −1.18549). Therefore, this means that the cluster does not trust institutional sources for food product safety information. This clustering is the least numerous with an elevated awareness level (component 6 with value of 0.79324) and positive social media satisfaction (component 8 with value of 0.52680). However, there is no willingness to pay more for a food safety product because propensity to purchase (component 3) presents a negative value (−0.31679).
- Cluster 4 can be defined as the group of "social media consumer" since this consumer does not pay attention to sustainability and to food safety; they choose to maintain a motivation purchase (component 3 with a positive value) based on social media information rather than mass media. This cluster represents 57 consumers out of 359 (14.48%) and is characterized mainly by component 8, social media satisfaction (value of 1.28354), and it is extremely far from component 5 (media sources satisfaction with value of −0.80027).
- Cluster 5 can be defined as the group of "distrustful consumer" since they do not see the information source as an advantage and an opportunity to be followed. In fact, component 3, propensity to purchase, presents a negative value. This cluster comprises 76 consumers out of 359 (21.16 %) and is characterized mainly by component 1 (purchasing motivations with value of 0.63766), but the most negative components evaluated are number 8, the social media satisfaction (value of −0.81811) and component 5 (media sources satisfaction with value of −0.59981).

The ANOVA table shows which variables contributed the most in the identification of the clusters. Social media satisfaction, purchasing motivations, and media sources satisfaction are the three variables associated significantly with the clusters identified. These are followed by institutional sources satisfaction and propensity to purchase. However, attention to food safety and sustainable behavior are the least influential in the division of the groups (Table 5).

Tables 6 and 7 show the composition of the cluster groups according to socio-demographic variables. Cluster 1 composed of 72 statistic units (20.5% of the sample) and includes a concentration of unmarried females (age 18–25 years old, 32% of the consumers). Regarding instruction, most of the sample have a diploma and are represented by students (32% of the sample), while 43% have an income between 10,000 and 20,000 euros. This cluster presents the highest level of instruction in the survey because the percentage of degrees and postgraduates is more than 30%. This important information confirms the theory concerning a higher attention to food safety based on education level, according to the study by Monier-Dilhan (2018) [38], because cluster 1 is the most virtuous of all concerning food safety issues. In addition, it is confirmed that the education of women is fundamental in the sensitization and promotion of food safety issues, according to Beldad and Hegner (2018), [37] because this cluster presents a higher concentration of women graduates.

Table 5. ANOVA table.

ANOVA						
	Cluster		Error		F	Sig.
	Mean Square	Df	Mean Square	df		
1. Purchasing motivations	38.213	4	0.580	354	**65.940**	0.000
2. Sustainable behaviour	11.854	4	0.877	354	13.511	0.000
3. Propensity to purchase	18.325	4	0.804	354	22.786	0.000
4. Attention to food safety	5.857	4	0.945	354	6.197	0.000
5. Media sources satisfaction	29.289	4	0.680	354	**43.051**	0.000
6. Awareness	14.481	4	0.848	354	17.083	0.000
7. Institutional sources satisfaction	23.693	4	0.744	354	31.863	0.000
8. Social media satisfaction	40.137	4	0.558	354	**71.961**	0.000

Source: Authors' elaboration of data from survey.

In cluster 2, composed of 106 units (29.52% of the sample), the distribution with respect to age is the same as cluster 1, but the age class 18–25 is more numerous with 46 units (43.39% of the sample), with a good percentage of the students (40.56%) having a diploma. This cluster has a higher concentration of young people and men (that represent 46.22% of the sample with 49 statistic units), but at same time, it has the worst instruction level because only 11% have a degree or are postgraduates. In addition, it is also the cluster that presents the worst incomes because only 31 units out of 106 earn more than 20,000 euros. This confirms the theory (Vlontzos et al., 2018) [57] that young people (mainly the men) earn less than adults, do not show an interest in the food safety issues, and do not have an increasing propensity to purchase. Therefore, this cluster is the most negative regarding food safety attention, and it needs more sensibilisation on these issues.

In cluster 3 (N = 53), 56% of the consumers are unmarried women in the age group between 18 and 25 years; the prevalent level of education is the diploma, while their income is medium-high, with 44% of these consumers (24 statistic units) earning over 20,000 euros p.a.

Table 6. Composition of cluster groups I, II and III.

	Cl. I		Cl. II		Cl. III	
	72 Units (20.05%)		106 Units (29.52%)		53 Units (14.76%)	
	Absolute Value	Percentage Value	Absolute Value	Percentage Value	Absolute Value	Percentage Value
Male	27	37.5	49	46.22	23	43.39
Female	45	62.5	57	53.77	30	56.60
18–25	23	32	46	43.39	21	39.62
26–35	18	25	19	17.92	10	18.86
36–45	13	18	12	11.32	9	16.98
46–55	13	18	10	9.43	8	15.09
56–65	4	5.5	10	9.43	5	9.4
>65	1	1.4	9	8.49		
Single/Bachelor	44	61	66	62.26	31	58.49
Married	28	39	40	37.73	22	41.50
Without secondary school certificate	2	2.7	13	12.26	3	5.66
Secondary school certificate	11	15.3	10	9.43	10	18.86
Diploma	37	51	71	66.98	30	56.60
Degree	18	25.1	12	11.32	10	18.86
Postgraduate	4	5.5				
Student	23	32	43	40.56	15	28.30
Employed	22	30.5	31	29.24	17	32.07
Freelancer	8	11	12	11.32	8	15.09
Unemployed	3	4	1	0.94	6	11.32
Housewife	7	9.7	6	5.66	6	11.32
Retired	1	1,3	7	6.60	0	0
Other			6	5.66	1	1.8
$X \leq 10{,}000$	15	20.8	29	27.35	11	20.75
$10{,}000 < X \leq 20{,}000$	31	43	46	43.39	18	33.96
$20{,}000 < X \leq 30{,}000$	18	25	24	22.64	15	28.30
$X > 30{,}000$	8	11	7	6.6	9	16.98

Source: Authors' elaboration of data from survey.

In fact, this is the potentially interesting cluster in food safety issues because they present a good predisposition to sustainability behaviours and a positive attitude versus food safety attributes. Although their incomes are very high, there is no predisposition to increase their purchase of food safety products. This could be due to a low level of instruction; in fact, only 18% (10 statistic unit) have a degree or are postgraduates.

Cluster 4 is composed of 52 statistic units (14.48%), mainly of female students with a diploma. In this cluster, the percentage of married women is relevant, nearly 50%. This group is defined as the group of "social media consumer" since they do not think that attention to food safety is a goal to be pursued for sustainability; they choose to maintain a motivation purchase based on social media, according to Heinonen 2011 [58]. This point is particularly important to demonstrate that age is not a relevant variable for a good level of attention to food safety issues, according to previous studies [40,59], confirming that instruction level is rather more fundamental [41].

Finally, cluster 5 composed of 76 units (21.16% of the sample), including mostly married consumers, with a percentage of 53.94 (41 units). This group is characterized mainly by women in the age group of 18–25 years and who are prevalently employees. This cluster shows the dissatisfaction versus the source's information (mainly social media), and this could confirm the theory that employees (especially if married) spend less time on social media and so are more sceptical of this digital evolution [45]. Their motivation to purchase is quite solid, and their sustainable behaviour discreet, but the cluster does not show them to be believers in the food safety information sources (media and social media) to obtain the achievement of sustainability.

Table 7. Composition of cluster groups IV and V.

	Cl. IV 52 Units (14.48%)		Cl. V 76 Units (21.16%)	
	Absolute Value	**Percentage Value**	**Absolute Value**	**Percentage Value**
Male	23	44.23	30	39.47
Female	29	55.76	46	60.52
18–25	17	32.69	22	28.94
26–35	12	23.07	16	21.05
36–45	5	9.6	5	6.57
46–55	11	21.15	22	28.94
56–65	2	3.84	6	7.89
>65	5	9.61	5	6.57
Single/Bachelor	27	51.92	35	46.05
Married	25	48.07	41	53.94
Without secondary school certificate	7	13.46	7	9.2
Secondary school certificate	14	26.92	11	14.47
Diploma	19	36.53	43	56.57
Degree	12	23.07	13	17.10
Postgraduate			2	2.63
Student	15	28.84	16	21.05
Employed	12	23.07	32	42.10
Freelancer	10	19.23	10	13.15
Unemployed	4	7.69	2	2.63
Housewife	8	15.38	7	9.21
Retired	1	1.92	6	7.89
Other	2	3.84	3	3.94
$X \leq 10{,}000$	16	30.76	22	28.94
$10{,}000 < X \leq 20{,}000$	17	32.69	24	31.57
$20{,}000 < X \leq 30{,}000$	15	28.84	18	23.68
$X > 30{,}000$	4	7.69	12	15.78

Source: Authors' elaboration of data from survey.

4.3. Discussion

For three of the five cluster groups, attention to food safety has a positive value with respect to sustainable behaviour and purchasing motivations. In addition, consumers' level of education (university degree or postgraduate degree), when accompanied by high levels of earnings and employment status, increases the level of attention to the quality of food consumption and the link between food and health, according to Monier-Dilhan (2018) [38]. Regarding the information sources, the consumers show a preference for institutional sources, so four out of five clusters present a positive value for this attribute, while for social media, only two clusters out of five present a positive value. The fundamental role is attributed to educated women (more virtuous, in the cluster 1) who present a higher level of attention to food safety issues, preferring institutional sources to media sources, according to Beldad and Hegner (2018) [37], and with a good propensity to purchase food safety products. This survey reveals that the classic role of the consumer has been surpassed and that the consumer is now seen as an active producer of corporate value. In particular, the study by Heinonen (2011) [58] shows that consumer behavior is not exclusively influenced by traditional marketing communications. Therefore, companies can be expected to participate more in the activities of their target groups on social media to understand the impact of social media on their brand and their image. However, digital marketing in the circular economy and optimal use of social media still have a long way to go to educate, inform, and engage consumers in sustainability. Duit and Galaz and Walker et al., (2008) [60,61] affirm that to meet the challenges of current society, more sustainable approaches, such as new management and governance systems, are needed, especially for consumers who are not very active online and commonly show a low level of participation and contribution on social media, according to Schau (2009) [62]. However, other studies have drawn attention to the importance of social networking (Schau et al., 2009) [63]. The negative effect of lower income on food consumption decisions, evidenced in this empirical research, can be compensated for by food safety education. Especially for young people (males), it is necessary to promote ad-hoc information programmes within the formal educational processes since, according to the study by Vlontzos et al., (2018) [57], they are the ones most influenced by the media in their purchasing decisions.

In Italy, this issue has been approached by the Ministry of Health, which, to raise the level of food safety information among young people, proposes a link between food risk communication and social media [43,64]. The European Food Safety Authority (EFSA) risk communication guidelines mentions that social networks should inform and involve stakeholders in communicating easy and clear messages to a wide variety of consumers. It provides information to consumers so that they are more aware of the risks associated with a food product. The aim is to establish confidence in risk assessment through accurate and appropriate information. This allows consumers to choose from a variety of options that may meet their risk criteria.

These messages are also effectively spread through "online community discussions" that influence consumer behavioural change.

According to Heinonen (2011) [58], this can be achieved by promoting discussions on different everyday purchasing choices and encourage consumers to ask questions and share experiences. Relating the offer to consumers' daily lives in this way can also promote the food company and its image. Through consumer engagement and information sharing on social media, it is possible to obtain real-time product reviews, inviting consumers to share their opinions.

Information shared on social media plays a key role in improving consumer empowerment and knowledge in the food industry (Stefanidis et al., 2013) [65,66]. According to Dijkmans et al., (2015) [67], companies' engagement in the promotion of their products through the optimal use of social media increases and facilitates the development of a company's corporate reputation.

Consequently, the use of social networks should be a strategic part of food risk communication projects. The results of this empirical analysis can support the companies in the introduction of new marketing strategies to propagate emerging food safety trends among post-modern consumers.

5. Conclusions and Political Implications

The study confirms a good level of attention towards intrinsic attributes, especially for all the aspects concerning food expiry, transparent food information, food traceability, and seller confidence. Although the purchase of environmentally friendly products is increasing, there are still many consumers who do not use a high degree of environmental sustainability as a purchasing parameter.

The results of this study may stimulate the strengthening of information campaigns to increase food safety knowledge and improve awareness levels in consumer choice. Moreover, research may contribute to the understanding how social media can affect consumer awareness of environmental and social sustainability and of intrinsic attributes to define a business strategy. In addition, such studies could provide an opportunity to consider collaborative actions between institutions and industries to increase consumer awareness of environmental attributes. The work presents some limitations with reference to empirical research: the numerosity of the samples of consumers analysed (defined with random criteria) as well as the reference context (Naples, Italy). Possible future developments of the research could extend the analysis of consumers over the entire Italian territory in order to evaluate in more detail all the dynamics for the achievement of sustainable consumption models as well as designing an empirical survey that evaluates the sensitivity to the issue (intended as willingness to spend, role of online/offline communication channels, and so on), showing more differences (if existing) between males and females by analysing the level of awareness of sustainability.

Author Contributions: Conceptualization, G.C., V.R., D.S., and M.S.; methodology, G.C., V.R., D.S., and M.S.; software, G.C., V.R., D.S., and M.S.; validation, G.C., V.R., D.S., and M.S.; formal analysis, G.C., V.R., D.S., and M.S.; investigation, G.C., V.R., D.S., and M.S.; resources, G.C., V.R., D.S., and M.S.; data curation, G.C., V.R., D.S., and M.S.; writing—original draft preparation, G.C., V.R., D.S., and M.S.; writing—review and editing, G.C., V.R., D.S., and M.S.; visualization, G.C., V.R., D.S., and M.S.; supervision, G.C., V.R., D.S., and M.S.; project administration, G.C., V.R., D.S., and M.S. All authors have read and agreed to the published version of the manuscript.

Funding: This research received no external funding.

Institutional Review Board Statement: Not applicable.

Informed Consent Statement: Informed consent was obtained from all subjects involved in the study.

Data Availability Statement: The data that support the findings of this study are available on request from the corresponding author.

Conflicts of Interest: The authors declare no conflict of interest.

References

1. Pengue, W.A.; Gemmill Herren, B.; Balázs, B.; Ortega, E.; Acevedo, F.; Diaz, D.N.; Díaz de Astarloa, D.; Fernandez, R.; Garibaldi, L.A.; Giampietro, M.; et al. Eco-agri-food systems: Today's realities and tomorrow's challenges. In *TEEB for Agriculture & Food: Scientific and Economic Foundations*; UN Environment: Geneva, Switzerland, 2018; Chapter 3; pp. 57–109.
2. Zhang, W.; Thorn, J.P.R.; Gowdy, J.; Bassi, A.; Santamaria, M.; DeClerck, F.; Adegboyega, A.; Andersson, G.; Augustyn, A.M.; Bawden, R.; et al. *Systems Thinking: An Approach for Understanding 'Eco-Agri-Food Systems'*; The Economics of Ecosystems and Biodiversity: Geneva, Switzerland, 2018.
3. Van Langen, S.K.; Vassillo, C.; Ghisellini, P.; Restaino, D.; Passaro, R.; Ulgiati, S. Promoting circular economy transition: A study about perceptions and awareness by different stakeholders' groups. *J. Clean. Prod.* **2021**, *316*, 128166. [CrossRef]
4. Stiglitz, J.E. *La Globalizzazione che Funziona [Making Globalization Work]*; Einaudi: Turin, Italy, 2006.
5. Rana, J.; Paul, J. Consumer behavior and purchase intention for organic food: A review and research agenda. *J. Retail. Consum. Serv.* **2017**, *38*, 157–165. [CrossRef]
6. Chu, K.M. Mediating influences of attitude on internal and external factors influencing consumers' intention to purchase organic foods in China. *Sustainability* **2018**, *10*, 4690. [CrossRef]
7. Hati, S.R.H.; Zulianti, I.; Achyar, A.; Safira, A. Perceptions of nutritional value, sensory appeal, and price influencing customer intention to purchase frozen beef: Evidence from Indonesia. *Meat Sci.* **2021**, *172*, 108306. [CrossRef]
8. Miller, L.M.S.; Cassady, D.L. The effects of nutrition knowledge on food label use. A review of the literature. *Appetite* **2015**, *92*, 207–216. [CrossRef]
9. Simeone, M.; Scarpato, D.; Rotondo, G. Consumer attitudes to food labelling: Opportunities for firms and implications for policymakers. *Calitatea* **2015**, *16*, 312.
10. Freeman, I.; Hasnaoui, A. The meaning of corporate social responsibility: The vision of four nations. *J. Bus. Ethics* **2011**, *100*, 419–443. [CrossRef]

11. Cattivelli, V. *Not Just Food: The, New, Importance of Urban Gardens in the Modern Socio-Economic System: A Brief Analysis*; Ledizioni: Milan, Italy, 2015; pp. 53–74.
12. Scarpato, D.; Civero, G.; Rusciano, V.; Risitano, M. Sustainable strategies and corporate social responsibility in the Italian fisheries companies. *Corp. Soc. Responsib. Environ. Manag.* **2020**, *27*, 2983–2990. [CrossRef]
13. Civero, G.; Rusciano, V.; Scarpato, D. Orientation of Agri-Food Companies to CSR and Consumer Perception: A Survey on Two Italian Companies. *Recent Pat. Food Nutr. Agric.* **2018**, *9*, 134–141. [CrossRef] [PubMed]
14. Ferrell, O.C.; Fraedrich, J. *Business Ethics: Ethical Decision Making & Cases*; Nelson Education: Toronto, ON, Canada, 2015.
15. Fabris, G. *Il Nuovo Consumatore: Verso il Postmoderno*; Franco Angeli: Milan, Italy, 2003; Volume 39.
16. Unnevehr, L. (Ed.) *Food Safety in Food Security and Food Trade*; International Food Policy Research Institute: Washington, DC, USA, 2003; Volume 10.
17. Scarpato, D.; Simeone, M.; Rotondo, G. The challenge of Euro-Mediterranean integration for Campania agribusiness sustainability. *Agric. Econ.* **2019**, *6*, 539–549. [CrossRef]
18. Gatto, A. A pluralistic approach to economic and business sustainability: A critical meta-synthesis of foundations, metrics, and evidence of human and local development. *Corp. Soc. Responsib. Environ. Manag.* **2020**, *27*, 1525–1539. [CrossRef]
19. Gravagnuolo, A.; Micheletti, S.; Bosone, M. A Participatory Approach for "Circular" Adaptive Reuse of Cultural Heritage. Building a Heritage Community in Salerno, Italy. *Sustainability* **2021**, *13*, 4812. [CrossRef]
20. Seo, H.J.; Son, M.; Hong, A.J. Trends in Civic Engagement Disaster Safety Education Research: Systematic Literature Review and Keyword Network Analysis. *Sustainability* **2021**, *13*, 2505. [CrossRef]
21. Menozzi, D.; Mora, C.; Chryssochoidis, G.; Kehagia, O. Rintracciabilità, qualità e sicurezza alimentare nella percezione dei consumatori. *Econ. Agro-Aliment.* **2010**, *12*, 137–158. [CrossRef]
22. Cisco, G.; Gatto, A. Climate Justice in an Intergenerational Sustainability Framework: A Stochastic OLG Model. *Economies* **2021**, *9*, 47. [CrossRef]
23. Zhang, L. China, and the UN Food System Summit: Silenced Disputes and Ambivalence on Food Safety, Sovereignty, Justice, and Resilience. *Development* **2021**, 1–5. [CrossRef] [PubMed]
24. Hilbeck, A.; Tisselli, E. The emerging issue of "digitalization" of agriculture. In *Transformation of Our Food Systems*; Herren, H.R., Haerlin, B., The IAASTD+ 10 Advisory Group, Eds.; Zukunftsstiftung Landwirtschaft: Bochum, Germany, 2020; Chapter 59.
25. Turnhout, E.; Duncan, J.; Candel, J.; Maas, T.Y.; Roodhof, A.M.; DeClerck, F.; Watson, R.T. Do we need a new science-policy interface for food systems? *Science* **2021**, *373*, 1093–1095. [CrossRef]
26. Di Martino, P. Rintracciabilità obbligatoria e rintracciabilità volontaria nel settore alimentare. *Dirit. E Giurisprud. Agrar. E Dell'ambiente* **2005**, *3*, 141–149.
27. Charlier, C.; Valceschini, E. Coordination for traceability in the food chain. A critical appraisal of European regulation. *Eur. J. Law Econ.* **2008**, *25*, 1–15. [CrossRef]
28. Sadik-Zada, E.R.; Gatto, A. The puzzle of greenhouse gas footprints of oil abundance. *Socio-Econ. Plan. Sci.* **2021**, *75*, 100936. [CrossRef]
29. Fanelli, R.M. Seeking Gastronomic, Healthy, and Social Experiences in Tuscan Agritourism Facilities. *Soc. Sci.* **2020**, *9*, 2. [CrossRef]
30. Sadik-Zada, E.R.; Ferrari, M. Environmental policy stringency, technical progress and pollution haven hypothesis. *Sustainability* **2020**, *12*, 3880. [CrossRef]
31. Rusciano, V.; Civero, G.; Scarpato, D. Social and ecological high influential factors in community gardens innovation: An empirical survey in Italy. *Sustainability* **2020**, *12*, 4651. [CrossRef]
32. Civero, G.; Rusciano, V.; Scarpato, D. Consumer behaviour and corporate social responsibility: An empirical study of Expo 2015. *Br. Food J.* **2017**, *119*, 1826–1838. [CrossRef]
33. Gargano, G.; Licciardo, F.; Verrascina, M.; Zanetti, B. The Agroecological Approach as a Model for Multifunctional Agriculture and Farming towards the European Green Deal 2030—Some Evidence from the Italian Experience. *Sustainability* **2021**, *13*, 2215. [CrossRef]
34. Massari, S. Transforming research and innovation for sustainability: Transdisciplinary design for future pathways in agri-food sector. In *Transdisciplinary Case Studies on Design for Food and Sustainability*; Woodhead Publishing: Rome, Italy, 2021; pp. 315–326.
35. Civitello, L. *Cuisine and Culture: A History of Food and People*; John Wiley & Sons: Hoboken, NJ, USA, 2011.
36. Leong, Q.L.; Ab Karim, S.; Awang, K.W.; Bakar, A.Z.A. An integrated structural model of gastronomy tourists' behaviour. *Int. J. Cult. Tour. Hosp. Res.* **2017**, *11*, 573–592. [CrossRef]
37. Beldad, A.; Hegner, S. Determinants of fair-trade product purchase intention of Dutch consumers according to the extended theory of planned behaviour. *J. Consum. Policy* **2018**, *41*, 191–210. [CrossRef]
38. Monier-Dilhan, S. Food labels: Consumer's information or consumer's confusion. *OCL* **2018**, *25*, D202. [CrossRef]
39. Van't Veld, K. Eco-labels: Modeling the consumer side. *Annu. Rev. Resour. Econ.* **2020**, *12*, 187–207. [CrossRef]
40. Bourn, D.; Prescott, J. A comparison of the nutritional value, sensory qualities, and food safety of organically and conventionally produced foods. *Crit. Rev. Food Sci. Nutr.* **2002**, *42*, 1–34. [CrossRef]
41. Ruby, G.E.; Abidin, U.F.U.Z.; Lihan, S.; Jambari, N.N.; Radu, S. A cross sectional study on food safety knowledge among adult consumers. *Food Control* **2019**, *99*, 98–105. [CrossRef]

42. Panagiotopoulos, P.; Shan, L.C.; Barnett, J.; Regan, Á.; McConnon, Á. A framework of social media engagement: Case studies with food and consumer organisations in the UK and Ireland. *Int. J. Inf. Manag.* **2015**, *35*, 394–402. [CrossRef]
43. Simeone, M.; Scarpato, D. Sustainable consumption: How does social media affect food choices? *J. Clean. Prod.* **2020**, *277*, 124036. [CrossRef]
44. Su, B.C.; Wu, L.W.; Chang, Y.Y.C.; Hong, R.H. Influencers on social media as References: Understanding the Importance of Parasocial Relationships. *Sustainability* **2021**, *13*, 10919. [CrossRef]
45. Eastman, J.K.; Iyer, R.; Eastman, K.L.; Gordon-Wilson, S.; Modi, P. Reaching the price conscious consumer: The impact of personality, generational cohort and social media use. *J. Consum. Behav.* **2021**, *20*, 898–912. [CrossRef]
46. Šikić, F. Using Instagram as a Communication Channel in Green Marketing Digital Mix: A Case Study of bio&bio–Organic Food Chain in Croatia. In *The Sustainability Debate*; Emerald Publishing Limited: Bingley, UK, 2021.
47. Olsson, P.; Galaz, V.; Boonstra, W.J. Sustainability transformations: A resilience perspective. *Ecol. Soc.* **2014**, *19*, 1. [CrossRef]
48. Pereira, L.; Karpouzoglou, T.; Doshi, S.; Frantzeskaki, N. Organising a safe space for navigating social-ecological transformations to sustainability. *Int. J. Environ. Res. Public Health* **2015**, *12*, 6027–6044. [CrossRef]
49. Gilg, A.; Barr, S.; Ford, N. Green consumption, or sustainable lifestyles? Identifying the sustainable consumer. *Futures* **2005**, *37*, 481–504. [CrossRef]
50. Fanelli, R.M.; Romagnoli, L. Customer satisfaction with farmhouse facilities and its implications for the promotion of agritourism resources in Italian municipalities. *Sustainability* **2020**, *12*, 1749. [CrossRef]
51. De Lillo, A. *Analisi Multivariata per le Scienze Sociali*; Pearson Italia Spa: Turin, Italy, 2007.
52. Brunner, T.A.; Siegrist, M. A consumer-oriented segmentation study in the Swiss wine market. *Br. Food J.* **2011**, *113*, 353–373. [CrossRef]
53. Amato, M.; Fasanelli, R.; Riverso, R. Emotional profiling for segmenting consumers: The case of household food waste. *Age* **2019**, *18*, 79.
54. Peta, E.A. *Consumi Agro-Alimentari in Italia E Nuove Tecnologie*; Quadro Strategico Nazionale; Ministero dello Sviluppo Economico, Dipartimento per le Politiche di Sviluppo e di Coesione, Unità di Valutazione degli Investimenti Pubblici (UVAL): Rome, Italy, 2007.
55. Bollani, L.; Peira, G.; Pairotti, M.B.; Varese, E.; Nesi, E.; Pairotti, M.B.; Bonadonna, A. Labelling and sustainability in the green food economy: Perception among millennials with a good cultural background. *Rev. Stud. Sustain.* **2017**, *2*, 83–101. [CrossRef]
56. Di Vita, G.; Chinnici, G.; D'Amico, M. Clustering attitudes and behaviours of Italian wine consumers. *Calitatea* **2014**, *15* (Suppl. 1), 54.
57. Vlontzos, G.; Kyrgiakos, L.; Duquenne, M.N. What are the main drivers of young consumers purchasing traditional food products? European field research. *Foods* **2018**, *7*, 22. [CrossRef]
58. Heinonen, K. Consumer activity in social media: Managerial approaches to consumers' social media behavior. *J. Consum. Behav.* **2011**, *10*, 356–364. [CrossRef]
59. Todua, N.; Jashi, C. Influence of Social Marketing on the Behavior of Georgian Consumers Regarding Healthy Nutrition. *Bull. Georgian Natl. Acad. Sci.* **2018**, *12*, 183–190.
60. Duit, A.; Galaz, V. Governance and complexity—Emerging issues for governance theory. *Governance* **2008**, *21*, 311–335. [CrossRef]
61. Walker, B.H.; Barrett, S.; Galaz, V.; Polasky, S.; Folke, C.; Engström, G.; Ackerman, F.; Arrow, K.; Carpenter, S.R.; Chopra, K.; et al. Looming global-scale failures and missing institutions. *Science* **2009**, *325*, 1345–1346. [CrossRef]
62. Shao, G. Understanding the Appeal of User-Generated Media: A uses and Gratification Perspective. *Internet Res.* **2009**, *19*, 7–25. [CrossRef]
63. Schau, H.J.; Muñiz, A.M.; Arnould, E.J. How Brand Community Practices Create Value. *J. Mark.* **2009**, *73*, 30–51. [CrossRef]
64. Russo, C.; Simeone, M. The growing influence of social and digital media: Impact on consumer choice and market equilibrium. *Br. Food J.* **2017**, *119*, 1766–1780. [CrossRef]
65. Stefanidis, A.; Cotnoir, A.; Croitoru, A.; Crooks, A.; Rice, M.; Radzikowski, J. Demarcating new boundaries: Mapping virtual polycentric communities through social media content. *Cartogr. Geogr. Inf. Sci.* **2013**, *40*, 116–129. [CrossRef]
66. Gatto, A.; Drago, C. When renewable energy, empowerment, and entrepreneurship connect: Measuring energy policy effectiveness in 230 countries. *Energy Res. Soc. Sci.* **2021**, *78*, 101977. [CrossRef]
67. Dijkmans, C.; Kerkhof, P.; Beukeboom, C.J. A stage to engage: Social media use and corporate reputation. *Tour. Manag.* **2015**, *47*, 58–67. [CrossRef]

sustainability MDPI

Article

The Key Organizational Factors in Healthcare Waste Management Practices of Libyan Public Hospitals

Mohammed Khalifa Abdelsalam [1], Ibrahim Mohammed Massoud Egdair [2], Halima Begum [3,*], Diara Md. Jadi [1], Hussein-Elhakim Al Issa [4], Omar Saad Saleh Abrika [5] and A. S. A. Ferdous Alam [6]

1 Department of Banking and Risk Management, School of Economics, Finance and Banking, Universiti Utara Malaysia UUM, Sintok 06010, Kedah, Malaysia; m.khalifa.abdelsalam@uum.edu.my (M.K.A.); diara@uum.edu.my (D.M.J.)
2 Department of Business Administration, Faculty of Economic and Accounting-Murzq, Sebha University, Sebha 00218, Libya; Ibr.egdair@sebhau.edu.ly
3 School of Economics, Finance and Banking, Universiti Utara Malaysia UUM, Sintok 06010, Kedah, Malaysia
4 Business School, Queen Margaret University, Edinburgh EH21 6UU, UK; alissaht@gmail.com
5 Faculty of Pharmacy, Sebha University, Sebha 00218, Libya; oma.abrika@fpharma.sebhau.edu.ly
6 School of International Studies, Universiti Utara Malaysia UUM, Sintok 06010, Kedah, Malaysia; ferdous@uum.edu.my
* Correspondence: dr.halima.begum@uum.edu.my; Tel.: +6-04-9286566

Abstract: This study aims to investigate factors contributing to healthcare waste management practices among Libyan public hospitals. The organizational culture and structure are proposed to have their effect upon hospital organizational units in charge of healthcare waste production by a theoretical review to develop two main hypotheses. Hence, this study used the stratified random sampling technique to select respondents such as top management officials, heads of departments, and administrators who work in all the hospitals located in the south of Libya, from whom data was collected. The data for the study was gathered via a survey questionnaire from Libyan public hospitals in the country's southern region. A total of 210 questionnaires were distributed and 171 usable responses were received, yielding a 70% response rate. Though the findings of the study show some inconsistency, the two dimensions of the culture examined in this study are found to have a positive relationship and significant influence on the management practices of health waste. Besides, it shows the positive relationship between organizational structure and healthcare waste management practices (HWMP). However, the findings of this study suggested that nurses and cleaners' practices should critically consider structure dimensions such as formalization as well as moderating variables such as hospital location and type of services supplied on the interactions to improve the management of healthcare waste in Libya's public hospitals.

Keywords: organizational factors; healthcare waste management practices; public hospitals; Libya

Citation: Abdelsalam, M.K.; Egdair, I.M.M.; Begum, H.; Md. Jadi, D.; Al Issa, H.-E.; Abrika, O.S.S.; Alam, A.S.A.F. The Key Organizational Factors in Healthcare Waste Management Practices of Libyan Public Hospitals. *Sustainability* **2021**, *13*, 12785. https://doi.org/10.3390/su132212785

Academic Editors: E. Richard Zada and Andrea Gatto

Received: 14 October 2021
Accepted: 10 November 2021
Published: 19 November 2021

1. Introduction

A report by the World Health Organization [1] uncovered the critical shortcomings that need immediate attention in Libyan hospitals: scaling up hygiene standards and healthcare waste collection and disposal, training of selected staff, technical support for disposal of large amounts of expired drugs and strengthening and developing medical waste management including waste segregation, collection, treatment, and disposal.

The basic assumption known by various scholars is that many organizational factors affect healthcare waste management practice. For instance, organizational structure is found to be a clear factor influencing healthcare waste management [2]. Similarly, the author [3] has classified organizational culture (OC) as human interaction and organizational arrangements. Some other previous research suggests that several factors will influence the management of healthcare waste [4]. It is essential to consider the influence of the system components on each other to arrive at an optimal plan for the hazardous waste management system [5]. Our findings illustrate that OC does have a significant impact on the adoption of HWMP. Furthermore, the authors [6] asserted that centralization may reduce creative solutions and impede communication between departments and the sharing of ideas. This has a clear impact on healthcare facilities when, for example, a healthcare facility has accumulated expired medications that must be handled. On the other hand, a decentralized organizational structure can be more beneficial in allowing employees to bring in full participation for the building of spontaneous processes [7].

The influence of internal and external factors such as culture and human factors on healthcare waste management practice have been investigated by existing studies [8]; the factors that influence HWMP among the Libyan public hospitals have been given limited attention. Therefore, the present research is considered essential to many stakeholders in Libya who oversee the management of waste since the last update by the authors in their studies [9,10]. Where they recorded that all hospital surveyed in their study were having poor waste management in terms of regulation and concerning adequate ways to handle waste and even though waste disposal has not taken place due to lack of awareness by employees who tend to perform all related activities without proper safety, training, and direction. Therefore, the current research is timely in its quest to gain insights into the development of effective healthcare waste management practices by seeking to better understanding its antecedents and outcomes like organizational culture and environmental condition.

The authors [11] mentioned that the main issues and challenges that affect healthcare waste management are organizational structure and infrastructure in the National Health Service (NHS) in Cornwall, UK. We note that their study did not concentrate on environmental factors that were highlighted in previous studies. Additionally, in their study, the waste manager and an administration assistant are responsible for observing the logistic documents along with other Cornwall Healthcare Estates and Support Services functions. This will result in inefficient communication to guarantee dissemination from trust management to all employees, which will also result in the manager having to communicate with each worker from all the trusts. A study of existing authors [11] described the current waste management by Cornwall NHS from the perspective of organizational structure and barriers to recycling and reusable materials for the internal factors, whereas in this paper the conceptualization is forwarded to include more multidimensional factors and approach as presented in the later part of this study. Recently, the author [12] examined the effect of education on compliance and waste generation in European healthcare. His findings illustrate low compliance and education is the greatest policy affecting compliance with proper healthcare waste handling. On the other hand, the author [9] conducted a study in three different cities in Libya on the management of hospitals. Their findings show that the targeted hospitals transport their containers via uncovered trolleys. Containers are being dumped in total insanitary conditions and the final disposal practice of waste was discarded along with massive local waste in a bare state outside of city limits. The findings, in addition, revealed that 85% of the personnel surveyed (including managers, cleaning staff, and environmental workers) were not trained in hospital waste management. The situation in Libya is even worse. Data is not available on the prevalence of hepatitis B virus (HBV), hepatitis C virus (HCV), or human immunodeficiency virus (HIV) among handlers of healthcare waste [10].

The purpose of this paper is to examine the influence of organizational factors and culture and structure on healthcare waste management practice in Libyan public hospitals. It ought to be noted that the present research is dissimilar from the study conducted by previous authors [10] in two main parts: First, our sample covers a more extensive area, which includes five states in the southern region. Secondly, it focused only on internal factors (such as transport, onsite storage, segregation, and training).

The current study's research gap is evident in its examination of the influence of organizational factors, namely, culture and structure, on healthcare waste management practices in Libyan hospitals. The findings can help hospital administrators and Libyan policymakers by providing insights into the effects of cultural factors on waste management practices and, as a result, finding ways to improve existing waste systems.

Thus, this paper is structured as follows. First, researchers begin by reviewing the empirical and theoretical background of organizational structure and culture as factors influencing healthcare waste management practice among Libyan hospitals. Next, the research hypotheses are developed. This is followed by the methods used in the study by shedding some light on the sampling technique used and research design. Before finally discussing implications and future research direction, the authors discuss important findings.

2. Theoretical Framework and Hypotheses Development

The criterion as the dependent variable for this study is healthcare waste management practices whereas the predictor variables are the organizational culture and structural factors. A discussion was then provided in the following sections on all the variables included in this study.

2.1. Organizational Structure and Healthcare Waste Management Practices

According to the authors [13], organizational structure could show an enduring configuration of activities and tasks. Organizational structure as being described in literature refers to an organization's internal way of relationships, communication, and authority [14]. Organizational structure is also known as the formal allocation of work policies and administrative mechanisms for controlling and integrating work activities [15]. In short, the organizational factors of an organization refer to how activities such as task distribution, management, and supervision are headed for the achievement of the organization's aims and goals [15]. Even though the structural dimensions include the scope of centralization, formalization, and specialization, the most studied dimensions are centralization and formalization [16,17]. Based on the statements above, this study examined merely centralization and formalization as the structural dimensions of the organizations among waste management practices in Libyan hospitals.

2.2. Centralization

Centralization is defined as "the concentration of power or decision-making authority in an organization" [18] (p. 68). According to the authors [19], centralized organizations will enhance work alienation, which reduces friendship; as per other authors [20], it is an important factor in employees' willingness to assist each other [21]. Since workplace settings involve informal and personal interactions, the resultant friendships increase the support given and resources shared between individuals in the workplace. Scholars agree that centralization concentrates decision making and the evaluation of activities for improving best practices in healthcare waste management [11]. Centralization is beneficial in that it ensures standardization, clear documentation, responsibility for best practice, and minimizes the number of interested parties who are missing certain information or skills. It allows individuals to take advantage of the skills of central and specialized experts while maintaining stronger control over operations [22].

From the perspective of waste management, the body in charge practices centralization structures in which only the authority personnel oversee the decision making and full empowerment lies in the hands of top managers. Therefore, as a result, the benefit of centralization is to prevent employees or their bosses from being more adaptable and resourceful in initiating new courses of action when carrying out their job duties [23].

2.3. Formalization

Formalization describes the degree to which an organization uses rules and procedures to specify behavior and how, where, and by whom duties are to be performed and so reduces role ambiguity [24]. This definition has a negative insinuation because it shows that formalization can restrict when exacting formal guidelines take over an organization [25]. Another drawback to formalization is that it increases the likelihood that a strategic process will be motivated by reactive instead of proactive behavior [26]. On the other hand, formalization can increase teamwork and partnering among employees [27]. Moreover, formalization could shape interaction structure and scope and provide helpful insights for organizational management improvement [28]. Formalization measures the extent to which an organization uses rules and procedures to prescribe behavior [29]. However, the authors [30] suggested that formalization and organizational routines possess certain similarities in the sense that they both symbolize a manner of behavior, action, procedures, or interaction. However, they diverge in those routines that are considered a type of implicit knowledge while formalization is explicit and systematized [31]. In many cases, both formalization and organizational routines could go contrary to feasibility and hinder effectiveness by driving organizational inflexibility and static and mechanistic forms and patterns of activity. Organizational routines, according to some theorists, are rather more dynamic systems than static objects [32]. The adopted definition of formalization in this present research is that formalization is referred to the degree to which decisions and working relationships are directed by formal rules and standard policies and procedures in the management of waste healthcare facilities [28].

From the standpoint of safe healthcare handling, health waste management with proper structure and clear rules and procedures will allow management to first ease the circulation of properly handling the waste in which it is produced in different departments [33] and secondly reduce ambiguity [34]. Lastly, with formal procedures, employees tend to address contingencies more effectively because they can rely on set procedures well defined from experience that have been integrated into corporate memory [35]. So, in this context, formalization control and regulation practices to steady and distribute consistent programs will enable employees to follow them regularly and increase the quality of their performance. An instance of the strong association that can exist between formalization and waste management in published work is found in Total Quality Management (TQM) literature. Total Quality Management means the analysis and evaluation of all the activities improved within an organization [36], so that assessment may produce ideas and novel ways destined to be categorized in a sequence of formal records that result in the development of the quality in the chain of medical waste management. The authors [37] demonstrated that formalization is positively related to the quality of products or services offered by the organization.

2.4. Organizational Culture and Medical Healthcare Management Practice

Scholars accept that organizational culture has a major impact on the lasting effectiveness of organizations. The concept of culture has received sufficient attention only at the beginning of the 1980s from the relevant scholars. Not many areas have been fortunate enough to have the agreement of practitioners recognized as a critical factor shaping organizational performance. Organizational culture (OC) is an important concept in revealing direction and leadership for managers as they look for ways to boost the effectiveness of their organizations [38]. Studies have already reported a clear distinction between continents and countries based on definite key dimensions [39]. OC refers to shared assumptions and the glue that holds the organization together as a source of identity and distinctive competence [29].

Within the context of an organization, culture represents the behavior of human beings who are a part of an organization and the meaning that people connect or attach to their actions in the chain of medical waste management. The literature clearly states the relationship between an organization's culture and its management [40].

According to the authors [41], OC is the set of assumptions, values, attitudes, and beliefs shared by people in an organization. Along with the same line, culture could also be defined as the collective mind program which distinguishes members of one category or group of people from another [42]. According to the author, this kind of definition is not yet completed, but it covers what he intends to measure. However, culture, in his sense, consists of systems of values, and values are among the building blocks of culture. Furthermore, culture will be characterized by specific problems arising from the inexhaustible nature of its components. Therefore, in analyzing the cultural impact on the behavior of the members of any sub-culture, we select the dimensions that could most be applied to the perspective of the cultural behavior being studied. Four types of organizational cultures have been identified in the literature, namely individualism, power distance, uncertainty avoidance, and masculinity [42].

According to the dimensions, the first two dimensions (individualism vs. collectivism and power distance) are chosen because of their relevant values for studying the evaluation and management of waste by the leadership style [43]. Furthermore, they had been recorded and developed through empirical measurement, which was tested for their validity and reliability [41]. Conversely, some other scholars mentioned that OC does not impact organizational effectiveness directly; instead, it shapes member behavior in an unclear uncertain world. OC helps organizations absorb information by structuring the unknown and so contributes to the most crucial element of organizational decision making [44]. Sharing values and belief in an organization has a great influence on waste management [45].

The current research conceptualizes individualism vs. collectivism and power distance as per Hofstede [42] as the two main dimensions of the MWMP, so that management can be incorporated into an organizational memory. The entire process is conditioned by the organizational culture, as the authors [38] identified that the values and behavioral norms held by organizational members serve as a filter in the sense-making and meaning-construction processes.

2.5. Individualism vs. Collectivism

The cultural dimension of Individualism-Collectivism (IC) describes the association among members of societies as it pertains to collectivity and the individuality that exists in each group. Individualism describes how people seek their own interests as opposed to the group's and so have lower loyalty levels to organizations and higher dependability on themselves. Collectivism, on the other hand, refers to the group's rather than the state's dominance, with higher levels of collaboration and loyalty for the firm and lower competition [46]. Individualism/collectivism stands for factors which could be essential and important in an ideal organization, such as: challenge, training, physical conditions, and the use of skills, according to [47].

From the perspective of medical waste management, training and education programs, for instance, must be available for all hospital staff, as proper training will enhance the development of consciousness about health, safety, and environmental issues [10]. Furthermore, the authors [48] indicated that if the understanding of medical waste disposal methods is increased using skills and advanced technology, medical waste management will be greatly enhanced.

2.6. Power Distance

Power distance refers to the formal way or approach in which a society or organization handles inequality and subsequently the way people build their institutions and organizations. In addition to that, power distance is divided into parts. These are large distances and small distances. At a large distance, an organization or society tends to accept a hierarchical order where everybody has a space which does not need any justification. On the other hand, in small power distances, an organization or society tends to strive for power equalization and justification for the existence of those power inequalities [42]. An example of this dimension regarding MWMP is the classification into the administrative and technical aspects [49]. The administrative waste management of healthcare facilities is related to the components affecting the social system and members of the organization, such as the rules, roles, procedures, and structures concerning communication and exchange between the members. The technical part of medical waste management refers to the operating constituent affecting the technical system. Examples of these components are equipment as well as methods of operation utilized in their production process.

2.7. Healthcare Waste Management Practices

The author [50] proposed that by separating waste into reusable and non-reusable, harmful and non-lethal components, healthcare waste management can drastically reduce risks in healthcare facilities. They also suggested that the institutionalization of an active management system can go a long way; it allows for the elimination or minimization of undue waste manufacture, the evasion of risky substances wherever possible, preservation of the safety of workers, use of safe waste collection and transportation methods, and setting up a functioning waste treatment and removal system. Others, like similar studies [51], have suggested specific waste management steps that involve the proper handling, segregation, mutilation, disinfection, storage, transportation, and final disposal of medical waste. They argued that these are vital measures or steps that need to be undertaken for the sake of safe and scientific medical waste management in any institution. Other authors have advocated other methods of managing medical waste, including appropriate techniques for disposal [52,53], an internal management system, and training program for related personnel.

Following the World Health Organization [1], we define healthcare waste to include all the waste generated by healthcare institutions, laboratories, and research facilities. It also includes waste from minor or scattered sources (for example, waste generated during home healthcare) [54]. Figure 1 illustrates the conceptual framework of proposed relationships in the current study.

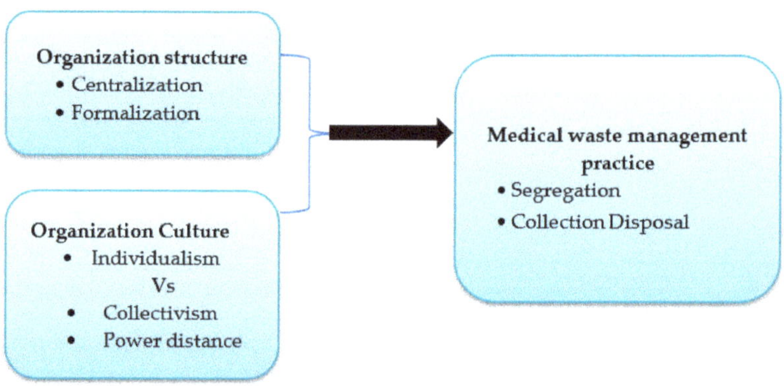

Figure 1. Theoretical framework.

Keeping this discussion in mind, the following two hypotheses (H1 and H2) are presented, one of which was the hypothesis for organizational structure construct (H1) followed by two sub-hypotheses H1a and H1b and the second hypothesis for organizational culture construct (H2) followed by two sub-hypotheses H2a and H2b:

Hypothesis 1a (H1a). *Organizational Structure (OS) is related to healthcare waste management practices (HWMP) in Libyan public hospitals. Centralization is related to HWMP.*

Hypothesis 1b (H1b). *Organizational Structure (OS) is related to healthcare waste management practices (HWMP) in Libyan public hospitals. Formalization is related to HWMP.*

Hypothesis 2a (H2a). *Organizational Culture (OC) is related to HWMP in Libyan public hospitals. Individualism vs. collectivism is related to HWMP.*

Hypothesis 2b (H2b). *Organizational Culture (OC) is related to HWMP in Libyan public hospitals. Power distance is related to HWMP.*

3. Materials and Method

3.1. Sampling

The questionnaire was distributed among the five southern states in Libya on January 20th and the survey has taken five months to complete. The population (respondents) for the current study included different levels of healthcare waste respondents (top management, head departments, administration, and doctors). The sample frame consists of names and addresses of hospitals obtained from the health website of Libya (www.health.gov.ly, accessed on 20 June 2015) where the data was published in 2010 and we were able to access the data on 20 June 2015. We physically sent out 210 questionnaires to all selected hospitals and received back 171, which yielded a more than 50% response rate. The internal consistency of items in the current study was above 0.65, which is the minimum acceptance value recommended by the author [55].

3.2. Measures and Variables

Organization culture was measured against 18 items, categorized under the two dimensions: Individualism vs. collectivism and power distance, which were adopted from Hofstede [42]. OC was operationalized as the shared assumptions and the glue that holds the organization together as a source of identity and distinctive competence about individuality, collectivism, and power distance [56,57]. The five-point Likert-like scale extending from "strongly disagree" to "strongly agree" was employed and had an internal consistency of 0.65.

The organizational structure (OS) was assessed using 14 items [58]. OS was operationalized as the degree to which an organization is centralized and formalized. The five-point Likert scale was employed while assessing the internal consistency, which was 0.874.

Healthcare waste management practices were measured using 12 items, four of which measured segregation, four items measured collection, and four items measured disposal, as adopted [59]. The standard five-point Likert-like scale extending from "strongly disagree" to "strongly agree" was employed, showing an internal consistency of 0.742.

4. Results and Discussion

An exploratory factor analysis (EFA) was performed to discover the underlying structure of the organizational structure and culture and healthcare waste management Practice measures. During the EFA, the 14 items of the OS construct were exposed to principal components analysis (PCA) using SPSS v20. To aid in the analysis of the two components of the construct, namely, centralization and formalization, Varimax rotation was carried out. The correlation matrix showed the items' coefficients were at 0.3 or above. The scree plot showed an obvious break after the second component and the existence of two factors with robust loadings. The researcher then opted to retain two components for further investigation [60]. This was strengthened by the parallel analysis showing the eigenvalue generated from EFA stopped exceeding the random criterion value created by parallel analysis at the third component (14 variables 171 respondents) (Tables 1–3). Factor loadings were all acceptable for OS, OC, and HWMP ranging from 0.468 to 0.784, 0.433 to 0.749, and 0.523 to 0.793, respectively (Tables 4–6). To ensure the reliability of the scales, internal consistency confirmation of the scales was performed by checking the Cronbach 's alpha coefficient. The cut-off points for measuring the reliability for the current research is coefficient alpha of above 0.65 as recommended by the author [55]. There was a total of two statistical measures to assess the factorability of the data conducted through Kaiser-Meyer-Olkin (KMO) to determine the measure of sampling adequacy|| value. The value reported was 0.887 for OC, 0.843 for OS, and 0.788 for HCWMP respectively, exceeding the recommended value of 0.6 [61–63]. Bartlett's test of sphericity [64,65] is significant at $p < 0.001$. Therefore, the sample size here is adequate for factor analysis for all variables.

Table 1. Factor analysis of organizational structure (Part-1).

Factor/Item	Loading	
	1	**2**
Factor 1: Formalization (α * = 0.723)		
b1b	0.702	
b2b	0.646	
b3b	0.701	
b4b	0.543	
b5b	0.606	
b6b	0.784	
b7b	0.612	
b8b	0.638	
Factor 2: Centralization (α = 0.825)		
b1a		0.585
b3a		0.472
b4a		0.482
b5a		0.468
b6a		0.750
Eigenvalues	4.81	1.33
Percentage	37.06	10.23
KMO	0.843	
Barlett's test of sphericity	683.97	
Sig.	0.000	

Note: * α = Cronbach's alpha coefficient.

Table 2. Factor analysis of organizational culture (Part-2).

Factor/Item	Loading	
	1	**2**
Factor 1: Power Distance (0.695)		
c1b	0.749	
c2b	0.684	
c3b	0.686	
Factor 2: Individualism and Collectivism (0.891)		
c1a		0.615
c2a		0.553
c3a		0.493
c4a		0.433
c5a		0.588
c6a		0.660
c7a		0.692
c8a		0.743
c9a		0.702
c10a		0.716
c11a		0.669
c12a		0.504
c13a		0.570
c14a		0.770
Eigenvalues	6.75	1.65
Percentage	39.71	9.72
KMO	0.887	
Barlett's test of sphericity	1195.47	
Sig.	0.000	

Table 3. Factor analysis of healthcare waste management practice.

Factor/Item	Loading		
	1	2	3
Factor 1			
e10	0.523		
e11	0.530		
e12	0.711		
e14	0.635		
Factor 2			
e5		0.757	
e6		0.793	
e7		0.725	
e8		0.604	
Factor 3			
e1			0.704
e2			0.672
e3			0.528
e4			0.701
Eigenvalues	1.837	1.696	1.069
Percentage	31.976	14.130	8.906
KMO	0.788		
Barlett's test of sphericity	533.202		
Sig.	0.000		

The charts in Appendix A and Table 4 show the demographic statistics of respondents' background at the hospitals surveyed. Out of 171 respondents who returned the completed questionnaires, 70.7% of the participants were from the District General Hospital, while 7.0% came from teaching hospitals and 5.7% were from specialist hospitals. They held various positions in the hospital. Most of them were heads of departments (41.1%) and doctors (12.5%). Males made up 55.2 percent of respondents, while females made up 44.8 percent. Most of the participants had finished their tertiary education and had more than 8 years of experience. It could also be found that most of the hospitals were old hospitals, and they were established for more than 20 years. According to the number of employees, most of the participants were from the hospital with 300 employees.

Table 5 presents the correlation between medical waste management practices and structure. It was found that there was a significant relationship between both variables (r = 0.609, $p < 0.01$). However, there was also an indication of the significant relationship between centralization and collection (r = 0.525, $p < 0.01$) and disposal (r = 0.193, $p < 0.05$). Another significant relationship can also be found between formalization collection (r = 0.486, $p < 0.01$) and disposal (r = 0.208, $p < 0.01$). Centralization and formalization were found to have no significant relationship with separation. The outcomes then provide the statistical proof to support H1 and H2.

Table 6 reveals that the OC-HWMP association was significant (r = 0.739, $p < 0.01$). Another strong association appears between individualism and collection (r = 0.506, $p < 0.01$) and disposal (r = 0.374, $p < 0.05$) and between power distance collection (r = 0.282, $p < 0.01$) and disposal (r = 0.436, $p < 0.01$). The outcomes then provide the statistical evidence to support H_A1 (8), H_A1 (9).

Table 7 shows finding from the regression analysis which examined the OS-HWMP association. It was found that organizational structure explained 37.1 percent of MWMP (R^2 = 37.1, F = 49.03, $p < 0.01$). Both dimensions significantly predicted the MWMP in public hospitals in Libya as follows: centralization (B = 0.313, t = 3.805, $p < 0.01$) and formalization (B = 0.355, t = 4.316, $p < 0.01$).

Table 4. Background of the respondents.

	Frequency	Percentage
Type of the Hospital		
Teaching Hospital	11	7.0
Specialist Hospital	9	5.7
District General Hospital	111	70.7
Others	26	16.6
Position		
Head of Hospital	3	1.8
Hospital Manager	8	4.8
Head of Hospital Department	69	41.1
Inflection Control Officer	11	6.5
Hospital Engineer	6	3.6
Chief Pharmacist	13	7.7
Radiation Officer	7	4.2
Senior Nursing Officer	11	6.5
Financial Controller	6	3.6
Waste Management Officer	3	1.8
Doctor	21	12.5
Others	10	6.0
Gender		
Male	74	55.2
Female	60	44.8
Education		
High School	5	3.6
High Diploma	49	35.8
University	83	60.6
Experience		
1–3 years	17	11.4
4–7 years	32	21.5
>8 years	100	67.1
Years of Established		
<10	24	14.0
10–20	15	8.8
21–30	88	51.5
31–40	43	25.1
>40	1	0.6
Number of Employees		
<100	9	5.3
100–200	13	7.6
201–300	18	10.5
301–400	45	26.3
401–500	54	31.6
>500	32	18.7

Table 5. The correlation between medical waste management practices and structure.

	MWNP	Separation	Collection	Disposal	Organizational Structure	Centralization	Formalization
HMP	1						
Separation	0.485 **	1					
Collection	0.643 **	0.272 **	1				
Disposal	0.699 **	0.319 **	0.180 *	1			
Organizational Structure	0.609 **	−0.001	0.549 **	0.220 **	1		
Centralization	0.548 **	0.016	0.525 **	0.193 *	0.879 **	1	
Formalization	0.563 **	−0.013	0.486 **	0.208 **	0.940 **	0.663 **	1

Note: * $p < 0.05$, ** $p < 0.01$.

Table 6. Relationships between waste management practices and organizational culture.

	WNP	Separation	Collection	Disposal	Organizational Culture	Individualism and Collectivism	Power Distance
HMP	1						
Separation	0.485 **	1					
Collection	0.643 **	0.272 **	1				
Disposal	0.699 **	0.319 **	0.180 *	1			
Organizational Culture	0.739 **	0.102	0.491 **	0.414 **	1		
Individualism and Collectivism	0.697 **	0.086	0.506 **	0.374 **	0.985 **	1	
Power Distance	0.672 **	0.130	0.282 **	0.436 **	0.751 **	0.623 **	1

Note: * $p < 0.05$, ** $p < 0.01$.

Table 7. Effect of organizational structure on medical waste management practices.

	B	t	Sig.
Centralization	0.313	3.805	0.000
Formalization	0.355	4.316	0.000
R^2	0.371		
F	49.03		
Sig.	0.000		

Likewise, Table 8 showed the results of regression analysis that examined the OC-HWMP association. It was found that the organizational culture explained 57.8 percent of MWMP ($R^2 = 0.578$, F = 113.655, $p < 0.01$). Both dimensions had also significantly predicted MWMP in public hospitals in Libya as follows: individualism and collectivism (B = 0.455, t = 7.508, $p < 0.01$) and power distance (B = 0.388, t = 6.018, $p < 0.01$).

Table 8. Effect of organizational culture on healthcare waste management practices.

	B	t	Sig.
Individualism and Collectivism	0.455	7.058	0.000
Power Distance	0.388	6.018	0.000
R^2	0.578		
F	113.655		
Sig.	0.000		

The present study was conducted with two objectives in mind. First, it examined the relationship between organizational structure and organizational culture and healthcare waste management practice. In this study, organizational structure was conceptualized as formalization and centralization. Organizational culture was conceptualized as power distance and collectivism vs. individualism. Nevertheless, the output of factor analyses shows that segregation and collection disposal were highly correlated and were subsequently collapsed as a dimension of safe management practice of healthcare waste. The strong relationship between variables was addressed in the literature. The authors [11] mention that the main challenges and issues, such as collection infrastructure, can affect waste management through the organizational structure of the National Health Service (NHS). Furthermore, the authors [66] argued that the differences in medical waste management practice in terms of generation rate may be due to living habits, standards, availability of treatment facilities, and ways to categorize wastes. A study [67] also reported that the medical waste generation rate depends on the size and the type of the medical institution and level of economic development. To have better practice of healthcare waste management, Libyan hospitals need to put into consideration employee training for old and new works, provide continuous education and have a management evaluation process for system and workers. Waste management practices were dimensional in this research: segregation, collection, and disposal. The importance of these processes is highly reported in literature. For example, to have a successful program with respect to HCWMP, a good plan must be available at the source of segregation, disinfection at the earliest opportunity, safe handling during transportation (within or outside the premises), and eco-friendly disposal [68]. To have an effective medical waste management strategy, it is essential to understand current hospital practices concerning the segregation of a variety of waste category streams [69]. In addition, the current study revealed that organizational culture has significant influence on HWMP in Libya. This might be due to the strength of relationship in which correlation coefficient between the two variables is (r = 739). The author [70] states that when studying the situation in any advanced or less advanced country, it is essential to consider the cultural beliefs, degree of awareness of health issues, and the practices and technology during healthcare waste management. The entire process is conditioned by organizational culture because the values and behavioral norms held by organizational members serve as a filter in the sense-making and meaning-construction processes [71].

5. Conclusions

The present research examined the relationships between organizational culture, structure, and healthcare waste management practice among Libyan public hospitals with the aim of assessing the relationships between organizational structure, culture, and healthcare waste management practice among Libyan public hospitals. The population for organizational study is southern Libyan public hospitals. Simple stratified random sampling was utilized for the hospitals selected. In addition, self-administrated structured questionnaires were physically distributed to 210 selected hospitals in the five states, followed by phone calls and reminders, seeking to get back positive feedback. A total of 171 respondents were returned. The data was also analyzed using a variety of methods, including correlation, regression, and descriptive analysis. The results of the reliability analysis showed that all the variables were reliable for this study by checking the Cronbach's alpha coefficient, which held values greater than 0.65 for all variables. The regression analysis revealed that all the variables (organizational culture and structure) significantly predicted healthcare waste management practices in the Libyan hospitals. The results of the correlation analysis showed that organizational culture and organizational structure have a strong impact on HCWMP. The result of the descriptive analysis for the mean score ranged from 1.6 to 4.7, and all the standard deviations were low, except for Q6 about segregation of waste and Q14 about penalizing following Standard Operation Procedures (SOP) on healthcare waste, which suggested utilizing the data. In general, the current practice of healthcare waste management within Libyan hospitals indicates what has been previously stated: there was an emphasis on some of the major and priority needs in Libya's primary healthcare structure and hospitals as per the authors [2]. These include expanding sanitation standards and healthcare waste collection, disposal, training of selected staff, technical support for dumping significant amounts of expired drugs, creating medical waste separation, treatment, and disposal. In another study consistent with this research, conducted by the authors [61,62], it was determined that in most of the healthcare facilities surveyed in Turkey, top management, managers, and senior nurses did not pay any attention to hospital waste, due to their insufficient knowledge and the significance of medical waste and their lack of interest. However, in examining the extent of the relationship between organizational culture, structure, and healthcare waste management practice among Libyan public hospitals, in some recent and past studies, the focus was on type of healthcare facilities, thereby identifying health waste management practices from a one-dimensional approach, such as organizational structure [11], organizational culture [72], and organizational size [73]. This one–dimensional concept may have brought about outcomes on the definition of healthcare waste management and research instruments that do not gather all the dimensions influencing the management of healthcare waste in the context of best practices. Therefore, this study empirically established the relationship between organizational structure, organizational culture, and healthcare waste management practice. Previous studies have shown that certain internal and external factors of organization do influence best practice of healthcare waste [11,66,67]. This study will not be exhaustive enough without examining certain organizational internal factors that have been found to influence waste management as a best practice in previous studies. Therefore, there is a need to assess the extent of waste management from a multidimensional approach; this study seeks to fill the research gap created by scarcity of literature on healthcare waste management practice factors in Libyan industry. This study therefore aims at examining the influence of organizational structure and culture factors on healthcare waste management practices in southern public hospitals in Libya. Based on the results and the findings, this study succeeded in positioning the direct relationship between organizational internal factors of organizations and healthcare waste in Libya as far as the practice management is concerned. The method used in assessing the extent of healthcare waste management practice in this study can be useful in ranking Libyan public hospitals according to their level of waste management practice.

6. The Practical Implications of the Study

Practical implications of the current study are evident in that the findings can help improve the practice of healthcare waste management among Libyan hospitals. All the interested parties in the field of safe management of healthcare facilities, including managers, medical staff, nurses, environmental officers, and waste management officers, need to seriously give attention to factors such as organizational structure (centralization and formalization). For instance, previous studies [63,74] have illustrated that centralization may reduce the original solutions and impede communication between departments and the more regular sharing of information and ideas because of the existence of time-consuming formal communication channels.

This may clearly be noticed when a healthcare facility has accumulated expired medications that must be handled. In a similar vein with formalization, healthcare waste management with proper structure and clear rules and procedures will allow the management to easily handle the waste properly where it is generated from different departments, and secondly, it will decrease ambiguity [34,75]. Lastly, with formal procedures, employees tend to handle situations more effectively because they comprise best practices learnt from experience and which are fused into organizational memory [35]. Furthermore, this research plays an essential role that is important to whoever oversees medical waste management, such as the Ministry of Health, the Ministry of Environment, healthcare facilities' managers, and the lower-level staff.

To summarize, in gathering all the relevant information for this research, there are limitations that arise from using single respondents to collect data and employing the survey method. Thus, our future directions in this field of research should consider multiple respondents for gathering the relevant data. Additionally, moderating variables can be examined to increase our understanding of the relationships studied, like hospital location and type of services offered, and the relationships between organizational factors and medical waste management practices.

Author Contributions: Conceptualization, M.K.A.; methodology, M.K.A., I.M.M.E.; software, M.K.A., I.M.M.E.; formal analysis, M.K.A.; investigation, H.B.; resources, writing—original draft preparation, M.K.A. and I.M.M.E.; writing—review and editing, H.B., O.S.S.A. and H.-E.A.I.; visualization, M.K.A., I.M.M.E., H.-E.A.I.; supervision, H.B. and A.S.A.F.A.; funding acquisition, H.B., I.M.M.E., O.S.S.A. and A.S.A.F.A., English Check & Proofreading, H.B.; added professionally, D.M.J. All authors have read and agreed to the published version of the manuscript.

Funding: This research received no external funding.

Data Availability Statement: Not Applicable.

Conflicts of Interest: The authors declare no conflict of interest.

Appendix A

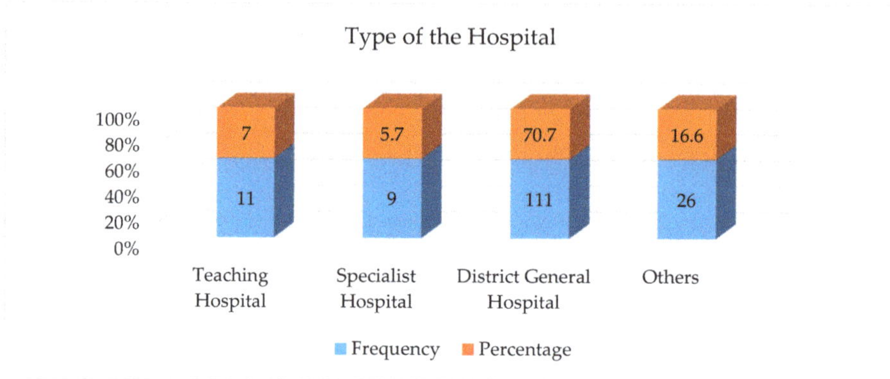

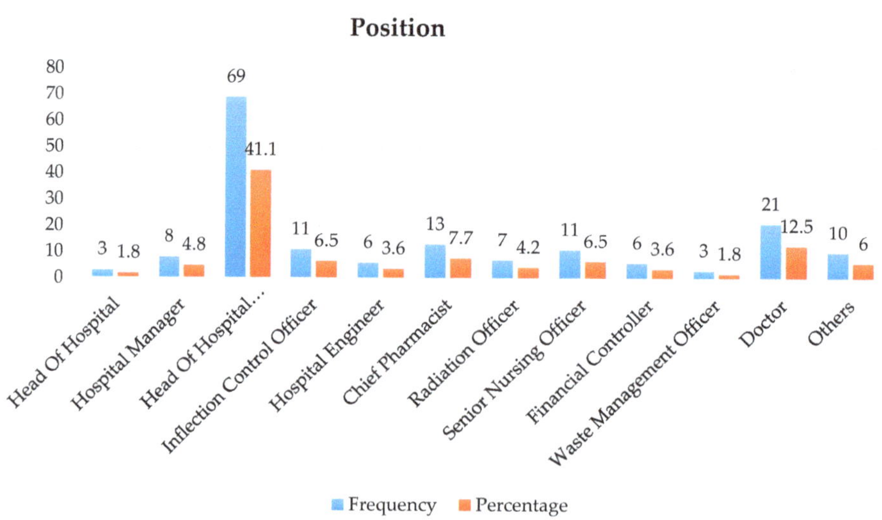

Figure A1. *Cont.*

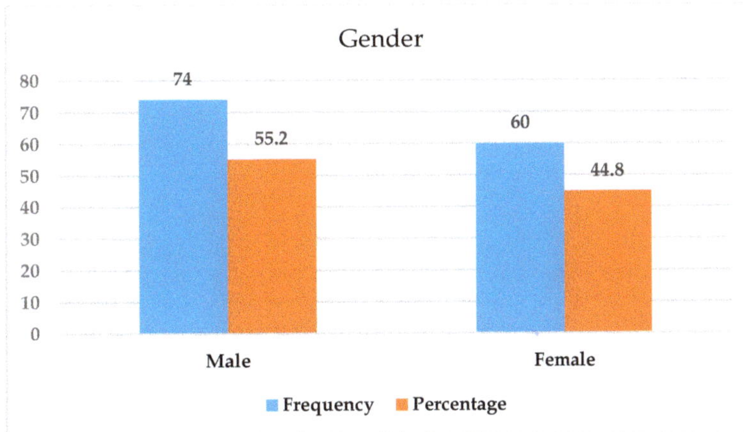

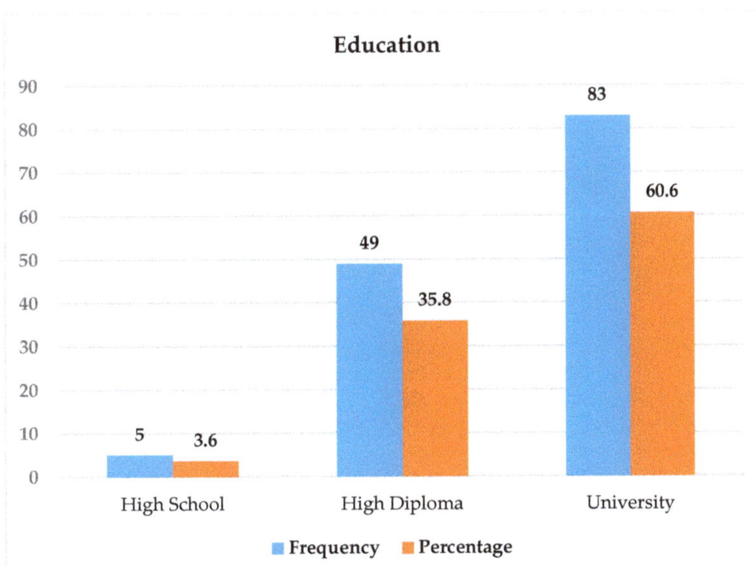

Figure A1. *Cont.*

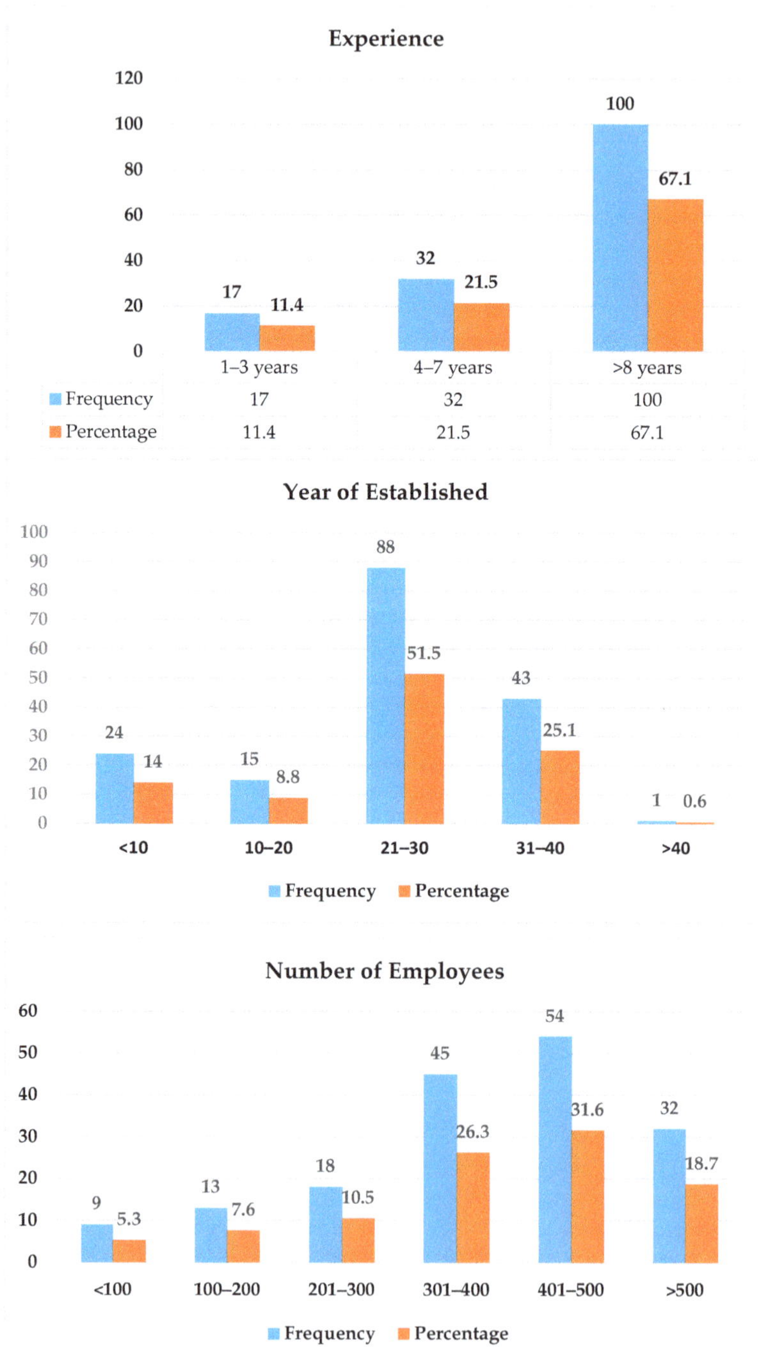

Figure A1. Background of the Respondents.

References

1. WHO. Health-Care Waste. 2018. Available online: https://www.who.int/news-room/fact-sheets/detail/health-care-waste (accessed on 23 June 2021).
2. Thakur, V.; Ramesh, A. Healthcare waste management research: A structured analysis and review (2005–2014). *Waste Manag. Res.* **2015**, *33*, 855–870. [CrossRef] [PubMed]
3. Bode, I. Processing Institutional Change in Public Service Provision the Case of the German Hospital Sector. *Public Organ. Rev.* **2012**, *13*, 323–339. [CrossRef]
4. Manyele, S.V.; Lyasenga, T.J. Factors affecting medical waste management in low- level health facilities in Tanzania. *Afr. J. Environ. Sci. Technol.* **2010**, *4*, 304–318.
5. Priyaranjan, H.S.; Mago, M.; Pandey, S.; Mishra, A. Awareness of Dental Health Care Waste Management among Dentists—A Cross Sectional Study. *J. Adv. Med. Dent. Sci. Res.* **2019**, *7*, 179–182.
6. Righi, S.; Oliviero, L.; Pedrini, M.; Buscaroli, A.; Della Casa, C. Life Cycle Assessment of management systems for sewage sludge and food waste: Centralized and decentralized approaches. *J. Clean. Prod.* **2013**, *44*, 8–17. [CrossRef]
7. Jokar, A.; Ghafori, D.; Malekian, N.; Namdar, H. Investigating the relationship between knowledge management processes and organizational culture. *Life Sci. J.* **2012**, *9*, 1701–1706.
8. Fobil, J.N.; Armah, N.A.; Hogarh, J.N.; Carboo, D. The influence of institutions and organizations on urban waste collection systems: An analysis of waste collection system in Accra, Ghana (1985–2000). *J. Environ. Manag.* **2008**, *86*, 262–271. [CrossRef] [PubMed]
9. Elmabrouk, F. Medical Waste Management in Libya Northeastern Region Hospitals as a Case Study. *Middle East* **2015**, *1*, 3–10. [CrossRef]
10. Sawalem, M.; Selic, E.; Herbell, J.D. Hospital waste management in Libya: A case study. *Waste Manag.* **2009**, *29*, 1370–1375. [CrossRef]
11. Tudor, T.L.; Noonan, C.L.; Jenkin, L.E.T. Healthcare waste management: A case study from the National Health Service in Cornwall, United Kingdom. *Waste Manag.* **2005**, *25*, 606–615. [CrossRef]
12. Botelho, A. The impact of regulatory compliance behavior on hazardous waste generation in European private healthcare facilities. *Waste Manag. Res.* **2013**, *31*, 996–1001. [CrossRef] [PubMed]
13. Skivington, J.E.; Daft, R.L. A Study of Organizational 'Framework' And 'Process' Modalities for The Implementation of Business-Level Strategic Decisions. *J. Manag. Stud.* **1991**, *28*, 45–68. [CrossRef]
14. Hage, J.; Aiken, M. Program change and organizational properties. A comparative analysis. *Am. J. Sociol.* **1967**, *72*, 503–519. [CrossRef] [PubMed]
15. Gougui, A. Organizational Structure, Human Resource Practices, Organizational Culture and Innovation Activity Among Small and Medium-Sized Enterprises in Algeria Ayoub Gougui Master of Science (Management) Universiti Utara Malaysia. Master's Thesis, 2017. Available online: https://etd.uum.edu.my/6370/1/s818420_01.pdf (accessed on 24 June 2021).
16. Adeleke, A.Q.; Taiwo, T.T.; Kanimoli, A. Organizational Structure Influence on Construction Waste Management Among Penang Malaysian Construction Industry: An Approach Via Partial Least Square Structural Equation Modeling. *J. Bus. Manag. Econ. Res.* **2020**, *4*, 56–73. [CrossRef]
17. Adiharja, H.S.; Hendarsjah, H. Analysis of The Relationship Between Personal Competencies and Individual Work Performance with Job Specialization, Formalization, and Centralization as Moderators. *Artic. Asian J. Technol. Manag.* **2020**, *13*, 243–256. [CrossRef]
18. Schminke, M.; Cropanzano, R.; Rupp, D.E. Organization structure and fairness perceptions: The moderating effects of organizational level. *Organ. Behav. Hum. Decis. Process.* **2002**, *89*, 881–905. [CrossRef]
19. Yen, C.H.; Teng, H.Y. The effect of centralization on organizational citizenship behavior and deviant workplace behavior in the hospitality industry. *Tour. Manag.* **2013**, *36*, 401–410. [CrossRef]
20. Sias, P.M.; Cahill, D.J. From coworkers to friends: The development of peer friendships in the workplace. *West. J. Commun.* **1998**, *62*, 273–299. [CrossRef]
21. Bowler, W.M.; Brass, D.J. Relational correlates of interpersonal citizenship behavior: A social network perspective. *J. Appl. Psychol.* **2006**, *91*, 70–82. [CrossRef]
22. Aminu, I. Mediating Role of Access to Finance and Moderating Role of Business Environment on The Relationship Between Strategic Orientation Attributes and Performance of Small and Medium Enterprises in Nigeria. 2015. Available online: http://etd.uum.edu.my/5308/2/s94308_abstract.pdf (accessed on 17 October 2016).
23. Katsikea, E.; Theodosiou, M.; Perdikis, N.; Kehagias, J. The effects of organizational structure and job characteristics on export sales managers' job satisfaction and organizational commitment. *J. World Bus.* **2011**, *46*, 221–233. [CrossRef]
24. Benzer, J.K.; Charns, M.P.; Hamdan, S.; Afable, M. The role of organizational structure in readiness for change: A conceptual integration. *Health Serv. Manag. Res.* **2017**, *30*, 34–46. [CrossRef]
25. Zhang, L.; Lee, M.K.O.; Zhang, Z.; Banerjee, P. Critical success factors of enterprise resource planning systems implementation success in China. In Proceedings of the 36th Annual Hawaii International Conference on System, Big Island, HI, USA, 6–9 January 2003; p. 10.
26. Fredrickson, J.W. The Strategic Decision Process and Organizational Structure. *Acad. Manag. Rev.* **1986**, *11*, 280–297. [CrossRef]

27. Chen, C.A.; Rainey, H.G. Personnel Formalization and the Enhancement of Teamwork: A public-private comparison. *Public Manag. Rev.* **2014**, *16*, 945–968. [CrossRef]
28. Pertusa-Ortega, E.M.; Zaragoza-Sáez, P.; Claver-Cortés, E. Can formalization, complexity, and centralization influence knowledge performance? *J. Bus. Res.* **2010**, *63*, 310–320. [CrossRef]
29. Fischer, R.; Ferreira, M.C.; Van Meurs, N.; Gok, K.; Jiang, D.Y.; Fontaine, J.R.; Harb, C.; Cieciuch, J.; Achoui, M.; Mendoza, M.S.D.; et al. Does organizational formalization facilitate voice and helping organizational citizenship behaviors? It depends on (national) uncertainty norms. *J. Int. Bus. Stud.* **2019**, *50*, 125–134. [CrossRef]
30. Feldman, M.S.; Pentland, B.T.; D'Adderio, L.; Lazaric, N. Beyond routines as things: Introduction to the special issue on routine dynamics. *Organ. Sci.* **2016**, *27*, 505–513. [CrossRef]
31. Reynaud, B. The void at the heart of rules: Routines in the context of rule-following. The case of the Paris Metro Workshop. *Ind. Corp. Chang.* **2005**, *14*, 847–871. [CrossRef]
32. Novak, L.; Brooks, J.; Gadd, C.; Anders, S.; Lorenzi, N. Mediating the intersections of organizational routines during the introduction of a health IT system. *Eur. J. Inf. Syst.* **2012**, *21*, 552–569. [CrossRef]
33. Giusti, L. A review of waste management practices and their impact on human health. *Waste Manag.* **2009**, *29*, 2227–2239. [CrossRef]
34. Pozo, E.C.; Morales, V.J.G.; Correa, J.A.A. Inter-departmental collaboration and new product development success: A study on the collaboration between marketing and R&D in Spanish high-technology firms. *Int. J. Technol. Manag.* **2006**, *35*, 52. [CrossRef]
35. Adler, P.S.; Borys, B. Two types of bureaucracy: Enabling and coercive. *Adm. Sci. Q.* **1996**, *41*, 61–89. [CrossRef]
36. Ruggie, J.G. Protect, respect, and remedy: The UN framework for business and human rights. In *International Human Rights Law: Six Decades after the UDHR and Beyond*; Ashgate Publishing Limited: Farnham, UK, 2013; pp. 519–538.
37. Bretschneider, S.; Parker, M. Organization formalization, sector and social media: Does increased standardization of policy broaden and deepen social media use in organizations? *Gov. Inf. Q.* **2016**, *33*, 614–628. [CrossRef]
38. Zheng, W.; Yang, B.; McLean, G.N. Linking organizational culture, structure, strategy, and organizational effectiveness: Mediating role of knowledge management. *J. Bus. Res.* **2010**, *63*, 763–771. [CrossRef]
39. Hofstede, G. Culture and organizations. *Int. Stud. Manag. Organ.* **1980**, *10*, 15–41. [CrossRef]
40. Fondas, N.; Denison, D.R. Corporate Culture and Organizational Effectiveness. *Acad. Manag. Rev.* **1991**, *16*, 203. [CrossRef]
41. Davies, H.T.O.; Mannion, R.; Jacobs, R.; Powell, A.E.; Marshall, M.N. Exploring the relationship between senior management team culture and hospital performance. *Med. Care Res. Rev.* **2007**, *64*, 46–65. [CrossRef] [PubMed]
42. Hofstede, G. Empirical models of cultural differences. In *Contemporary Issues in Cross-Cultural Psychology*; Swets & Zeitlinger Publishers: Lisse, The Netherlands, 1991; pp. 4–20.
43. Taras, V.; Kirkman, B.L.; Steel, P. Examining the Impact of Culture's Consequences: A Three-Decade, Multilevel, Meta-Analytic Review of Hofstede's Cultural Value Dimensions. *J. Appl. Psychol.* **2010**, *95*, 405–439. [CrossRef]
44. Nieswandt, M. Changing Organisational Culture: A Review of the Literature. In *Fast Cultural Change*; Palgrave Macmillan: London, UK, 2015; pp. 5–60.
45. Esmaeilian, B.; Wang, B.; Lewis, K.; Duarte, F.; Ratti, C.; Behdad, S. The future of waste management in smart and sustainable cities: A review and concept paper. *Waste Manag.* **2018**, *81*, 177–195. [CrossRef]
46. Massoud, I.M.; Abdelsalam, M.K. The Impact of Corporate Culture on Service Quality of Islamic Banks in Libya: The Mediating Role of Work Engagement. *Hamdard Islam* **2020**, *43*, 71–93.
47. Mora, C. Cultures and Organizations: Software of the Mind Intercultural Cooperation and Its Importance for Survival. *J. Media Res.* **2013**, *6*, 65.
48. Amos, D.; Musa, Z.N.; Au-Yong, C.P. Modelling the performance of waste management services in Ghana's public hospitals: A facilities management perspective. *Facilities* **2020**, *38*, 9–10. [CrossRef]
49. Nemathaga, F.; Maringa, S.; Chimuka, L. Hospital solid waste management practices in Limpopo Province, South Africa: A case study of two hospitals. *Waste Manag.* **2008**, *28*, 1236–1245. [CrossRef] [PubMed]
50. Abor, P.A. Managing healthcare waste in Ghana: A comparative study of public and private hospitals. *Int. J. Health Care Qual. Assur.* **2013**, *26*, 375–386. [CrossRef]
51. Awodele, O.; Adewoye, A.A.; Oparah, A.C. Assessment of medical waste management in seven hospitals in Lagos, Nigeria. *BMC Public Health* **2016**, *16*, 269. [CrossRef] [PubMed]
52. Khan, B.A.; Cheng, L.; Khan, A.A.; Ahmed, H. Healthcare waste management in Asian developing countries: A mini review. *Waste Manag. Res.* **2019**, *37*, 863–875. [CrossRef]
53. Mmereki, D.; Baldwin, A.; Li, B.; Liu, M. Healthcare waste management in Botswana: Storage, collection, treatment and disposal system. *J. Mater. Cycles Waste Manag.* **2017**, *19*, 351–365. [CrossRef]
54. Bendjoudi, Z.; Taleb, F.; Abdelmalek, F.; Addou, A. Healthcare waste management in Algeria and Mostaganem department. *Waste Manag.* **2009**, *29*, 1383–1387. [CrossRef]
55. Nunnally, J. *Psychometric Methods*; McGraw: New York, NY, USA, 1978.
56. Cortina, K.S.; Arel, S.; Smith-Darden, J.P. School Belonging in Different Cultures: The Effects of Individualism and Power Distance. *Front. Educ.* **2017**, *2*, 56. [CrossRef]
57. Infortunio, F.A. An exploration of the correlations between industrial fatal accident rates across nations and the cultural dimensions of power distance and individuality. *Trans. Am. Nucl. Soc.* **2013**, *109*, 2161.

58. Kohli, A.K.; Jaworski, B.J. Market Orientation: The Construct, Research Propositions, and Managerial Implications. *J. Mark.* **1990**, *54*, 1–18. [CrossRef]
59. Ngamlert, K.; Sinthuwattanawibool, C.; McCarthy, K.D.; Sohn, H.; Starks, A.; Kanjanamongkolsiri, P.; Anek-vorapong, R.; Tasaneeyapan, T.; Monkongdee, P.; Diem, L.; et al. Diagnostic performance and costs of capilia TB for mycobacterium tuberculosis complex identification from broth-based culture in Bangkok, Thailand. *Trop. Med. Int. Heal.* **2009**, *14*, 748–753. [CrossRef]
60. Bourke-Taylor, H.; Pallant, J.F.; Law, M.; Howie, L. Relationships between sleep disruptions, health, and care responsibilities among mothers of school-aged children with disabilities. *J. Paediatr. Child Health* **2013**, *49*, 775–782. [CrossRef] [PubMed]
61. Kaiser, H.F. A second-generation little jiffy. *Psychometrika* **1970**, *35*, 401–415. [CrossRef]
62. Alagöz, A.Z.; Kocasoy, G. Determination of the best appropriate management methods for the health-care wastes in İstanbul. *Waste Manag.* **2008**, *28*, 1227–1235. [CrossRef]
63. Moreira, A.M.M.; Günther, W.M.R. Assessment of medical waste management at a primary health-care center in São Paulo, Brazil. *Waste Manag.* **2013**, *33*, 162–167. [CrossRef] [PubMed]
64. Souitaris, V. Strategic Influences of Technological Innovation in Greece. *Br. J. Manag.* **2001**, *12*, 131–147. [CrossRef]
65. Barlett, R.G.; Bohr, V.C.; Helmendach, R.H.; Foster, G.L.; Miller, M.A. Evidence of an emotional factor in hypothermia produced by restraint. *Am. J. Physiol.* **1954**, *179*, 343–346. [CrossRef]
66. Mbarki, A.; Kabbachi, B.; Ezaidi, A.; Benssaou, M. Medical waste management: A case study of the Souss-Massa-Drâa region, Morocco. *J. Environ. Prot.* **2013**, *4*. [CrossRef]
67. Marinković, N.; Vitale, K.; Holcer, N.J.; Džakula, A.; Pavić, T. Management of hazardous medical waste in Croatia. *Waste Manag.* **2008**, *28*, 1049–1056. [CrossRef]
68. Verma, L.K.; Mani, S.; Sinha, N.; Rana, S. Biomedical waste management in nursing homes and smaller hospitals in Delhi. *Waste Manag.* **2008**, *28*, 2723–2734. [CrossRef]
69. Abdulla, F.; Abu Qdais, H.; Rabi, A. Site investigation on medical waste management practices in northern Jordan. *J. Waste Manag.* **2008**, *28*, 450–458. [CrossRef] [PubMed]
70. WHO. *Safe Management of Wastes from Health-Care Activities*; World Health Organization: Geneva, Switzerland, 1999.
71. De Long, D.W.; Fahey, L. Diagnosing cultural barriers to knowledge management. *Acad. Manag. Perspect.* **2000**, *14*, 113–127. [CrossRef]
72. North, C.D. Institutions, institutional change and economic performance. In *The Political Economy of Institutions and Decisions*; Alt, J., North, D., Eds.; Cambridge University Press: Cambridge, UK, 1999.
73. Manowan, V. Awareness and management of hospital waste in developing countries: A case study in Thailand. Ph.D. Thesis, The George Washington University, Washington, DC, USA, 2009.
74. Franka, E.; El-Zoka, A.H.; Hussein, A.H.; Elbakosh, M.M.; Arafa, A.K.; Ghenghesh, K.S. Hepatitis B virus and hepatitis C virus in medical waste handlers in Tripoli, Libya. *J. Hosp. Infect.* **2009**, *72*, 258–261. [CrossRef] [PubMed]
75. Cohendet, P.; Creplet, F.; Diani, M.; Dupouët, O.; Schenk, E. Matching communities and hierarchies within the firm. *J. Manag. Gov.* **2004**, *8*, 27–48. [CrossRef]

sustainability

Review

Fresh Insight through a Keynesian Theory Approach to Investigate the Economic Impact of the COVID-19 Pandemic in Pakistan

Kashif Abbass [1,2], Halima Begum [3,*], A. S. A. Ferdous Alam [4], Abd Hair Awang [5], Mohammed Khalifa Abdelsalam [6], Ibrahim Mohammed Massoud Egdair [7] and Ratnaria Wahid [4]

1 School of Economics and Management, Nanjing University of Science and Technology, Nanjing 210094, China; kashifabbass@njust.edu.cn
2 Riphah School of Business and Management, Riphah International University, Lahore 54000, Pakistan
3 Schools of Economics, Finance, and Banking, Universiti Utara Malaysia, Sintok 06010, Kedah, Malaysia
4 School of International Studies, Universiti Utara Malaysia, Sintok 06010, Kedah, Malaysia; ferdous@uum.edu.my (A.S.A.F.A.); ratnaria@uum.edu.my (R.W.)
5 Faculty of Social Sciences & Humanities, Universiti Kebangsaan Malaysia, Bangi 43600, Selangor, Malaysia; hair@ukm.edu.my
6 Department of Banking and Risk Management, School of Economics, Finance and Banking, Universiti Utara Malaysia, Sintok 06010, Kedah, Malaysia; khalifaedri@gmail.com
7 Department of Business Administration, Faculty of Economic and Accounting-Murzq, Sebha University, Sebha 00218, Libya; Ibr.egdair@sebhau.edu.ly
* Correspondence: dr.halima.begum@uum.edu.my; Tel.: +60-182-487-410

Abstract: Beyond the immediate impositions of dealing with COVID-19, this disease represents a severe and significant challenge confronting Pakistan's economy. The study's objective was to evaluate the coronavirus epidemic's effect on Pakistan's economy and measures devised to mitigate the damage done by this disease. The study research design used the elementary concept of Keynesian theory comprising of the mapping of systematic behavior of the COVID-19 pandemic. Issues were formally underpinned, described, and visualized through the Keynesian theory concept. The eruption of COVID-19 has jolted the national and international economy. Pakistan is included, causing millions of people to stay at home, lose their jobs, and suspend or end business operations. Unemployment in Pakistan has reached nearly 25 million people, driving many towards conditions of hunger and poverty as the major economic damage in several sectors is anticipated at around PKR 1.3 trillion. The hardest-affected sectors comprise industries such as tourism and travel, financial markets, entertainment, manufacturing, etc., having a devastating effect on gross domestic product (GDP). It is mainly daily-wage earners and people running small businesses that have been seriously exploited and subjected to a curfew-like situation. However, the Keynesian theory suggests that supportive macroeconomic policies must restore trust, demand recovery, and provide interest-free loans to overcome Pakistan's currently upcoming crisis.

Keywords: COVID-19; Pakistan economy; economic scale; Keynesian theory; mitigation; economic revival measures

Citation: Abbass, K.; Begum, H.; Alam, A.S.A.F.; Awang, A.H.; Abdelsalam, M.K.; Egdair, I.M.M.; Wahid, R. Fresh Insight through a Keynesian Theory Approach to Investigate the Economic Impact of the COVID-19 Pandemic in Pakistan. *Sustainability* **2022**, *14*, 1054. https://doi.org/10.3390/su14031054

Academic Editors: Elkhan Richard Sadik-Zada and Andrea Gatto

Received: 13 October 2021
Accepted: 10 December 2021
Published: 18 January 2022

1. Introduction

In recent decades, normal economic activities have become threatened and radicalized by expanding human enterprises that are risky and imprudent [1–4]. Consequently, the natural environment is at significant risk and impacts people's communities [5,6]. Historically, the dangers to human wellbeing originate in people's contact with the natural habitat and affect it in such a way as to increase the risk and spread of infections that are difficult to control [7]. Thus, on a global scale, the world has, at times, not resisted certain diseases and sicknesses, known and obscure, despite the tremendous medical advances made in eliminating lethal organisms in the 1970s [8]. For instance, the World Bank group in China (2020) [9] only recently described China as one of the world's cloudiest nations due to its vast population and despite now being very wealthy [10]. The authors [11–13] reported that China has been susceptible to spreading diseases like African pig fever, which is highly transmissible in the form of influenza, rabies, plague, and other zoonotic pathogens that are potentially pandemic [14–16]. The authors [17] stated that the most probable cause of the COVID-19 infection was the bats, which ended up being sold at a Wuhan market, where interactions between humans and live creatures occurred. This subsequent infection spread from China's Hubei area and affected around 211,000 individuals by 31 December 2019 [18], and it rapidly spread from human to human on a global scale. Previously known as the "2019 Novel Coronavirus," it became the infamous COVID-19 [19,20], though vaccines have been invented to treat victims.

Nonetheless, the world's economies have been now extensively damaged since March 2020 due to this pandemic, and the response needs to avoid a possible health catastrophe [21,22]. Several studies document evidence that the economic disadvantages or destruction associated with this epidemic are worldwide [23–28]. The epidemic's effects go beyond death and indisposition and have an enormous impact on economic systems with dangerous social consequences [26]. More than 100 countries are now affected, with a fatality rate of 5.5% reported in recent weeks due to COVID-19 [29]. Due to its highly infectious nature, governments worldwide have implemented lockdowns in various measures, which means locking down economic activities [30]. As a result, [31] found that commercial, industrial, and professional activities are operating at less than half their previous rates due to restrictions on people's movements imposed by federal, provincial, regional, state, and local governments, which have the most impact on aggregate demand (AD) and supplies.

The Worldometer [18] evaluated that COVID-19 spread very quickly, with 248,983,180 cases, 5,040,413 deaths, and 225,552,585 recovered to date. It started in Wuhan, China, on 17 November 2019, and three weeks later, cases were reported in Japan, South Korea, and the U.S., which declared their first instances of COVID-19. It took 42 days for the confirmed cases to reach 100 in the U.S., while Japan and South Korea took 31 and 29 days, respectively. Elsewhere, India had 34,320,142 confirmed cases as of 8 November 2021, while Italy had over 4,782,802 [18]. Regarding COVID-19 issues, the U.K., the fourth most-devastated European nation, has outpaced France. Spain has more than 5,019,255 affirmed cases and is currently the second most infected E.U. nation. The number of cases in Germany, Russia, Turkey, and Brazil is 4,662,011, 8,673,860, 8,121,226, and 21,835,785, respectively [18]. Meanwhile, [32,33] reported that the infection's spread throughout the populace in different nations occurs at different rates, much depending on the nature of social, economic, and political variables, such as people's responses, the density of the population, the size of family units, etc.

Figure 1 below illustrates that the COVID-19 pandemic entered Pakistan on 26 February 2020, when an undergraduate returned from Iran to Karachi. By 18 March, each of the four regions, (the two autonomous regions), and Islamabad's government region had reported coronavirus cases. Meanwhile, more than 1,277,160 patients were confirmed by 8 November 2021, in Pakistan. However, Punjab had the most victims, with 441,176 reported, while Khyber Pakhtunkhwa suffered 178,643 casualties as of 8 November 2021. Nonetheless, after the partial lockdown, according to the government of Pakistan's official portal, there are a total number of cases of 1,277,160 and 28,547 deaths. It is obvious the cases and death rates are increasing rapidly. As a result, in order to save the country's economy, new strategies and policies must be implemented.

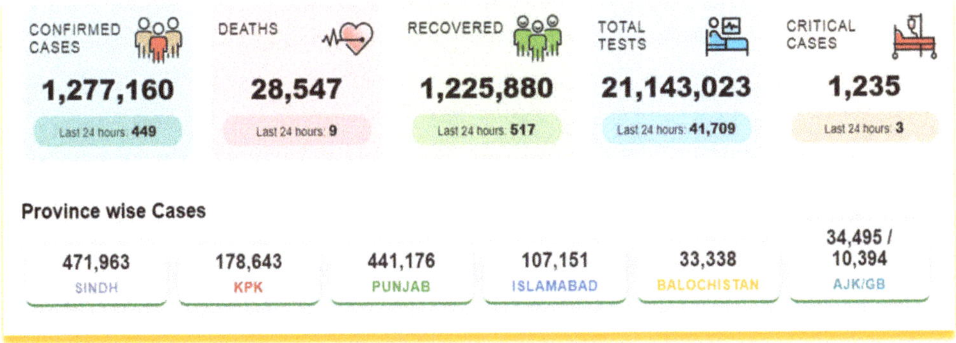

Figure 1. Cases of coronavirus in Pakistan. Source: [34].

It is important to note that contradictions have been observed among the governments, private-sector companies, and the public regarding the emergence of the pandemic above worldwide. However, it is pertinent that in many sections of Pakistan, particularly the provinces, a fierce debate has erupted concerning the impact of the coronavirus epidemic on the nation's economy [35,36]. Pakistan is experiencing challenges due to a lack of resources to address the current scenario. While the coronavirus (COVID-19) is being handled through systematic initiatives and policies, these are constrained by limited resources [37]. Besides, economic uncertainty could be lessened by following other countries' strategies, such as the authors [21] asserted in a context of uncertainty about the end of the pandemic and complemented by economic, fiscal, and monetary policies that mitigate the economic recession. The authors propose shifting strategies and initiatives from an indiscriminate suppression strategy to a targeted, effective, and intelligent mitigation strategy that minimizes the risk of human-life costs and socioeconomic costs.

The pandemic affects the world economy and creates a hazard to economic development among low- and middle-income countries like Pakistan [38,39]. Apart from the significant inconvenience of combating COVID-19, Pakistan, like all other countries, is struggling to find the funds to deal with this situation [40,41]. The authors opined that Pakistan's economy could overcome the vast challenges, while others author [42] suggested to come out with more advanced economic strategies and several theoretical and logical arguments to bolster aggregate demand.

The study outcome would suggest reliable opportunities for Pakistan and highlight the importance of international cooperation to promote cooperative and supportive macroeconomic policies and the benefits of joint macroeconomic actions. The critical aspect to be developed is "a best action" plan that helps Pakistan adopt the best strategy for post-COVID to promote better recovery development and related bilateral initiatives.

This study attempted to compute the impending Pakistani financial cost of COVID-19 under various probable situations. The objective was to offer direction to the experts and policymakers for internationally coordinated strategy retorts to tame the case. The study constructs the understanding obtained from gauging SARS's economics and the influenza epidemic's economics [43]. The research initially sheds light on Pakistan's different significant sectors most affected by COVID-19. Section 3 used Keynesian theory for theoretical support, stressing its strengths to measure the epidemic's international cost. In Section 4, the study development framework, like the type of policy challenges the Pakistani government faces and the kind of policy responses the Pakistani government took to mitigate the pandemic's effect, is described. Section 5 discusses the study's recommendations in several directions. Section 6 of the article determines the main findings of the study.

2. The Macroeconomic Effects of a Pandemic in Pakistan

Around the world, Pakistan's economy is struggling in a severe downturn. COVID-19's impacts on Pakistan's economy could result in a sharp drop in GDP growth [42], a skewed current and fiscal balance, disrupted supply chains, and increased unemployment across the board [37]. Several macroeconomic effects are explained in more detail below in Figure 2, considering these issues.

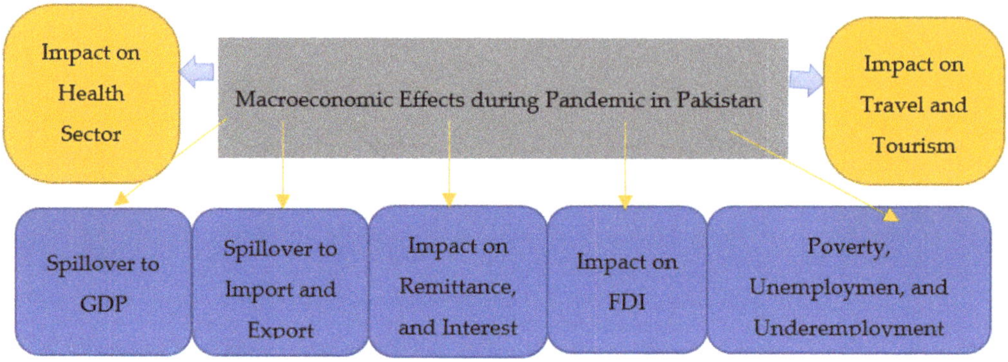

Figure 2. Several macroeconomic factors that impact Pakistan's economy. Source: author's development.

2.1. Spillover to Gross Domestic Product (GDP)

The worldwide pandemic has resulted in national output falling by 2% below the benchmark for the world. It is estimated that the manufacturing and technical sectors contribute 2.5% and 1.8%, respectively, to the national GDP. Surprisingly, local administrations were affected by the pandemic [10]. The global epidemic is anticipated to decrease China's GDP by 3.7% and Pakistan's GDP by 4.64% [38]. Subsequently, China's falling GDP has had "knock-on" effects on Pakistan's total aggregates because both countries exchange in trade [44]. The authors [45] reports that a fall in GDP leads to a subsequent decline in business (and capital), prompting lower manufacturing and productivity, lower import–export, and an across-the-board reduction in families' livelihoods, employment, and take-home pay. The total national output in Pakistan was worth USD 320 billion in 2019 [46], and the GDP was estimated to be 0.26% of the world [47]. In Figure 3, the bar chart estimates that GDP was expected to cause a 10% economic loss (estimated at PKR 1.1 tri.) during the last quarter (April–June 2020) of the 2020 financial year. As calculated by PIDE, if imports and exports fell by 20% during the epidemic, a 6.64% loss was observed in Pakistan's GDP [48,49].

2.2. Spillover to Import and Export of Pakistan

At present, Pakistan's imports have a 2:1 tariff restriction ratio [47], with oil accounting for 25% of the country's total import bill [50]. Recurring companies for individuals dependent on hardware imports, metals such as iron and steel, and other substances imported will almost certainly fail because no one wants to conduct business in the current environment [51]. Imports decreased by 16.9%, from USD 43,447 million in July–April 2018–2019 to USD 36,091 million in July–April 2019–2020 [52]. Pakistan is likely to see a decline in imports, which would have an impact on its GDP due to a lack of available materials.

Exports are also an essential aspect of a country's economy. World trade has dropped by 4.6%. Various economies that now have greater than mediocre global losses of business to the European region include China (9.8%), Hong Kong SAR, Singapore (8.5%), Cambodia (7.4%), the Russian Federation (7.3%), Lao PDR (7.3%), Thailand (6.8%), and the Philippines (6.4%). On the contrary, Europe, Canada, and the United States have been projected to decrease by 4.5% [10,53]. Pakistan's exports of textiles amount to 55% compared to fish, rice, wheat, sports, and mineral fuels [47]. Exports are undeniably harmed when a pandemic occurs [51], and a 2.4% decrease means a loss of USD 19,650 million from July to April 2019–2020, compared to the previous fiscal year [52].

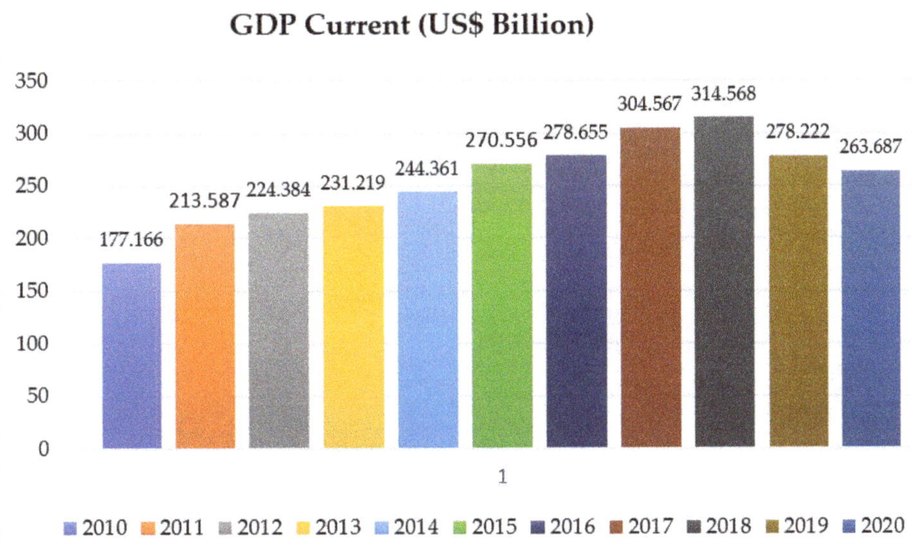

Figure 3. GDP of Pakistan. Source: [49].

2.3. The Effect on Remittances and Interest Rates

Remittances represented over 5% of GDP in 2019, and this applied to 66 nations, especially the developing economies [54–57]. As indicated by the State Bank of Pakistan [58], Pakistan's internal settlements were down 4.4% m-o-m (monthly) by February 2020, at USD 1824 million during the COVID-19 pandemic. During 2020, most remittances sent to Pakistan came from the UAE (USD 388.1 million), KSA (USD 422.0 million), and the USA (USD 333.5 million). Undeniably, these kinds of payments have been a significant boost to Pakistan's economy for many years, and, overall, USD 1 billion has been sent every month over the last two decades, cresting in the 2018 financial year when settlements were worth approximately USD 20 billion [59,60]. The World Bank group [61] noted that sending remittances may be mainly due to the widespread shutdowns caused by COVID-19 outbreaks and that these financial settlements to Pakistan will decline [48]. Moreover, the SBP's Monetary Policy Committee (MPC) amended the benchmark interest rate in March 2020 as financing costs rose by USD 75 billion [48]. It is observed that losses amounted to USD 150 billion during the pandemic, and, subsequently, the MPC implemented an aggregate rate of 11.0% to combat the problem [51]. Thus, Pakistan's economic development in the wake of this global epidemic necessitates more scrutiny.

2.4. Foreign Direct Investment (FDI) Impact

Since the peak of the 2007–2008 fiscal year, FDI inflows to Pakistan have been relatively low, decreasing from USD 4 billion in 2007 to USD 2.3 billion in 2018 [62]. Furthermore, FDI inflows increased by 68.3 percent to USD 1.34 billion in the first six months of the 2019–2020 fiscal year (July–December 2019), compared to USD 796.8 million the previous year, owing to the Chinese market's dominance [45]. Nonetheless, according to [59,63], financial flows were higher near the end of the year as the Pakistani government abandoned its year-long policy of allowing the rupee to weaken against the US dollar. The government's political vulnerability improved significantly because of the general election in July 2018.

However, the situation has changed due to the pandemic, as many governments have taken severe measures, and Pakistan is no exception [64,65]. Additionally, necessary financial arrangements have been made to ensure that the widespread health emergency can be paid for [66]. As a result, flexibility is critical during emergencies, as subsidiary and parent organizations require access to funds to operate [67,68]. Thus, FDI plays a significant role in supporting economies but might significantly threatened by the pandemic's effects on FDI streams due to the worldwide lockdown.

2.5. Poverty, Unemployment, and Underemployment

According to the "Employment Trends" report published by the Pakistan Bureau of Statistics (PBS) in 2018, the total labor force in Pakistan stands at 63.4 million, of which vulnerable employment was 26.41 million (41.6%). Vulnerable employment is measured as the proportion of own-account workers and contributing family workers in total employment (poor workers generally dependent on daily wages). These workers are likely to be the largest impacted individuals and could lose their employment due to the COVID-19 pandemic [69]. The COVID-19 epidemic, according to [48], led in millions of job losses across the country. Because of the slowdown in the economic activity, which has placed employment in a precarious situation, there was a significant increase in poverty and unemployment [70]. According to a PIDE report, approximately 12.3 million people are expected to face unemployment under a scenario of moderate restrictions by the government [48]. According to [71], this is a proximately 46.3% of the total vulnerable employment and 19.4% of the total employment in Pakistan. Wholesale retail trade is expected to witness the highest layoff of 4.55 million people. Thus, the poverty rate in Pakistan could increase from 23.4% to 44.2%.

2.6. The Effect on the Travel and Tourism Industry

The tourism industry in pre-COVID-19 times was a worldwide, huge profit-earning industry [72]. As of late 2019, people's movements and the travel industry have been stopped by as much as 10.4% of the worldwide GDP, and 319 million people worldwide who work in the travel industry have lost their jobs as of late 2019 [73]. Remarkably, Pakistan is a varied geographical area, ranging from the coastline to the high mountain slopes of the Himalayas. Pakistan's travel industry began in 1972, with the creation of a range of services for different parts of the country [74,75]. However, in 2012, 46.7 million travelers went to Pakistan, and Khyber Pakhtunkhwa was 19%. During 2004–2011, the travel and tourism industry gradually improved and reached its peak in 2019 [69]. However, the tourism industry is now largely redundant, and many people will be unemployed because COVID-19 has natural and long-lasting ramifications for social interaction and entertainment.

2.7. The Effect on the Health Sector

Pakistan's border provinces were the first to experience the COVID-19 outbreak [66,76]. On 26 February 2020, Pakistan's Ministry of Health reported the first case of COVID-19 in Karachi, followed by Sindh province and Islamabad [77]. Later, a total of 98,943 confirmed positive cases with 4960 critical mortalities and 2002 deaths were reported by 7 June 2020. To ensure patients' good healthcare as much as possible, Punjab and KPK allocated more beds—955 and 856, respectively. However, [78] noted that Pakistan's current situation is unsatisfactory, requiring further action [79]. Other than China, the United States, the United Kingdom, and Russia, Pakistan is a developing country that lacks the financial resources to combat the COVID-19 pandemic. Furthermore, it does not have anywhere near enough hospitals and quarantine facilities, which are urgently needed [79]. Transmission of the virus can only be controlled by updating and disbursing the required medical facilities. For both arrivals and departures, Pakistan needs more screening facilities [77].

3. Methodology

The research methods mainly included a review of previous studies, determining Pakistan's economic recession because of the natural disaster; determining economic key indicators, such as GDP, imports and exports of Pakistan, interest rates, and remittances; and determining strategies for reducing poverty and unemployment because of the disaster. The study used a Keynesian approach to examine the impact of COVID-19 on public health and the economic consequences of this coronavirus outbreak [80,81]. According to the consensus, global supply chains are disrupted, and more factories will have to shut down because of the virus. In addition to affecting the country's overall production, this disruption has a direct impact on the country's aggregate demand. How can Pakistan's government deal with this situation through monetary and fiscal policy now that it has been raised? Keynesian and real-business-cycle models can be used to answer this question. COVID-19's supply disruption was the focus of this model, which we hope will be short-lived. A demand-driven downturn could result from the virus' spread, opening the door to stagnation traps set up by pessimistic animal spirits. [81].

4. Keynesian Theory and the 2020 Great Economic Recession

As a result of the novel coronavirus (COVID-19) spreading over the world, the World Health Organization (WHO) declared a worldwide pandemic. On the one hand, its effects on public health are likely to devastate all types of economic systems. This disruption affects aggregate demand and the overall level of manufacturing production and related industries. Now the question is: how can Pakistan's government cope with this situation through monetary policy and fiscal policy? To answer this, [42] suggests that it may be possible to solve economic breakdown for limited resources by implementing a modern financial strategy. This study investigated the real consequences of COVID-19 on the Pakistani economy using the Keynesian AD-AS theory, which is based on theoretical and logical premises [80–82]. According to Keynesian theory, COVID-19 has created a scenario in Pakistan where the supply and demand for goods and services have temporarily halted. The spread of COVID-19 has caused a reduction in demand (for goods and services), bringing countries on the verge of economic recession [81,82].

4.1. Coronavirus's Impact on Aggregate Demand (A.D.)

The new Keynesian model (stripped-down version) has been used as an economic theory standard [80]. As per the Keynesian theory, aggregate demand specifies employment and output levels. On the other hand, the concept of positive, productive growth determines the A.D. as faster productivity growth enhances future incomes and encourages the creators of goods and services to devote additional resources in the present to investment projects [83]. Aggregate demand consists of government purchases, investment, consumption, and net exports. Anything that changes C, I, G.P., or N.X. will shift aggregate demand.

The Keynesian theory says a recession is caused by an aggregate demand shock. At the end of February, the Pakistan government announced a complete lockdown to stop the virus's spread. After the announcement, all industries, hotels, gyms, offices, and borders were closed and restricted them from staying at home. This situation, firstly, hurts the aggregate demand at any level of prices. This situation made Pakistanis more pessimistic about their future survival because nobody knew when the case would become familiar. This situation decreased their current preferred consumption; consistently, they increased their current preferred saving.

On the other hand, the coronavirus (COVID-19) outbreak resulted in a long-term decrease in productivity. The government of Pakistan shut down all industries for a specific period. The other reason was that if the government permitted them to work, industrialists did not want them to work because of low demand. They could not afford the cost of production at low prices. Manufacturing and production investment are positively connected to collective needs. Businesses invest more when they expect to get high returns on their investments, and the profit ratio is high when predictable demand is high. The decreases in investment further decrease production progress and, alternatively, collective need, and the cycle will continue. As shown in Figure 4, the coronavirus causes a temporary decline in production because of aggregate demand shocks.

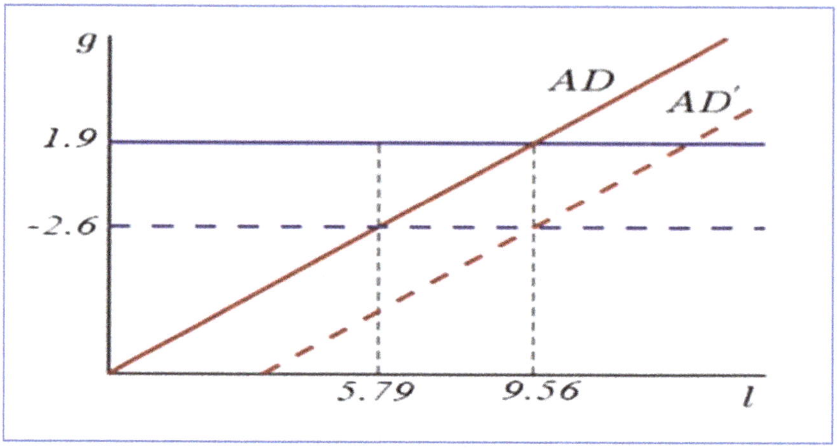

Figure 4. Impact of coronavirus on aggregate demand. Source: [83]. Note: unemployment on the horizontal axis; productivity growth rate on the vertical axis.

Already in the grip of a recession, the situation is deteriorating, particularly with the prospect of joblessness looming (up from 9.56 percent to 12.3 million to 18.53 million). These downsizings will likely reduce households' discretionary income while simultaneously causing a demand shock across the country. This risky condition diminishes not only local people's disposable income but also overall estimated remittances. A drop in private consumption expenditures of 4 to 8% is expected, affecting the country's economic picture and altering people's way of life as a result.

4.2. The Supply–Demand Doomsday Cycle

The beginning of COVID-19 badly impacted the domestic and international economy. This virus affected the demand side, and, at the same time, numerous companies were encountering various catastrophes with a specific ratio of income losses. Predominantly, businesses face multiple issues such as inventory-supply disturbances, termination of international trade orders, raw material scarcity, conveyance disturbances, and much more. Pakistan is a developing country, and small to medium industries play a vital role in boosting the trade sector. Furthermore, many small to medium businesses are fiscally vulnerable, with limited operations and resources. In comparison to their counterparts, i.e., huge corporations, their sensitivity to the current economic crisis is greater [84,85]. External financial situations, including epidemic diseases, earthquakes, floods, and other concerns, have unpleasantly impacted manufacturing activities and their existence. As we know, China is the world's largest export-oriented country, exporting raw materials and delicate, heavy machinery all over the world. According to the World Bank, almost 25% of world trade is affected by China's lockdown. Pakistan exchanged goods with China for approximately USD 11.8 billion in the 2019 fiscal year, but this was reduced to approximately USD 6 dollars in the current fiscal year during COVID-19 [86]. Pakistan's economic situation is very feeble after the four to five-month lockdown.

The Pakistan government announced a partial lockdown on May 31 to cope with this dire economic situation, but this strategy did not effectively overcome Pakistan's financial woes. Companies are functioning at less than half of their usual volumes because of the scarcity of inventory, a deficiency of labor, or limitations on people's travel forced by the provincial governments, taking a toll on cumulative supply. In the lockdown, people's demand also reduces, and a decrease in demand reduces private investment and productivity [87]. This pandemic has become a supply-side shock in Pakistan primarily because the lockdown situation has directly affected the production of the goods-and-services sector, even though some highly contact-intensive sectors cannot fulfil individuals' private desires. Figure 5 depicts the absence of the Pakistan government, damaging monetary and economic policy participation. Such persistent stock disorder triggers a demand-driven decrease, triggering a supply and demand doom cycle and opening the door to sluggishness traps aided by unenthusiastic animal spirits.

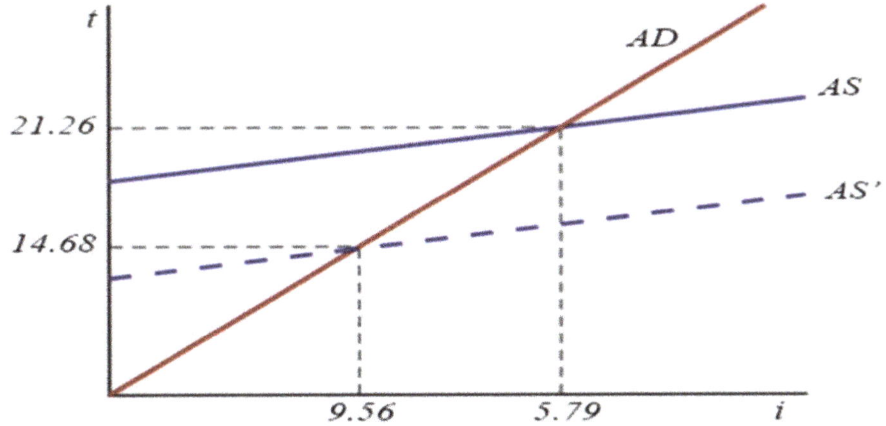

Figure 5. The supply–demand doom loop. Source: [83]. Note: unemployment on the horizontal axis; balance of goods trade on the vertical axis.

4.3. Animal Spirits and Stagnation Traps

In times of economic stress or uncertainty, the famous British economist John Maynard Keynes coined the term "animal spirits" to explain how people make financial decisions, such as purchasing and selling assets. In both the world and Pakistani economies, COVID-19 has left serious scars. The most prominent and most recent impact of the lockdown has been to create difficulties in current economic developments. Sindh was the first province to be placed under lockdown on 23 March 2020. In Pakistan, it is also a significant economic center. The country's most important industrial zone handled most international trade, accounting for more than 30% of total international trade. Due to fractional lockdown, just about 50 of Karachi's 2700 manufacturing units were functioning on the first working day [88]. The firms were facing many difficulties, like supply disruptions and the unavailability of workers. According to the recent survey conducted by the Pakistan Labor Force (2017–2018), the joblessness ratio in Pakistan was 5.8%, As for the ongoing current economic situation and lockdown, the joblessness ratio is likely to increase by 9.5% in the financial year 2020–2021 [89]. Small businesses, also known as self-employed individuals, include small shop holders, domestic companies, street sellers, and others, who severely depend on their daily business and are adversely impacted by the epidemic. According to the currently presented data, self-employed individuals total 35.7% (2017–2018) of the total Pakistan employment ratio [90].

Moreover, 59.6% (2019–2020) of these firms are defenseless. Furthermore, more than 87% of employment is in agriculture, a third–fourth of retail and wholesale business employees, more than 50% of restaurant employees, and others who work in business and real estate. The transportation and communication sectors employ more than two-fifths of the workforce [90]. Statistics on the sector-by-sector distribution of precarious employment in small industries are depicted in Figure 6 [91].

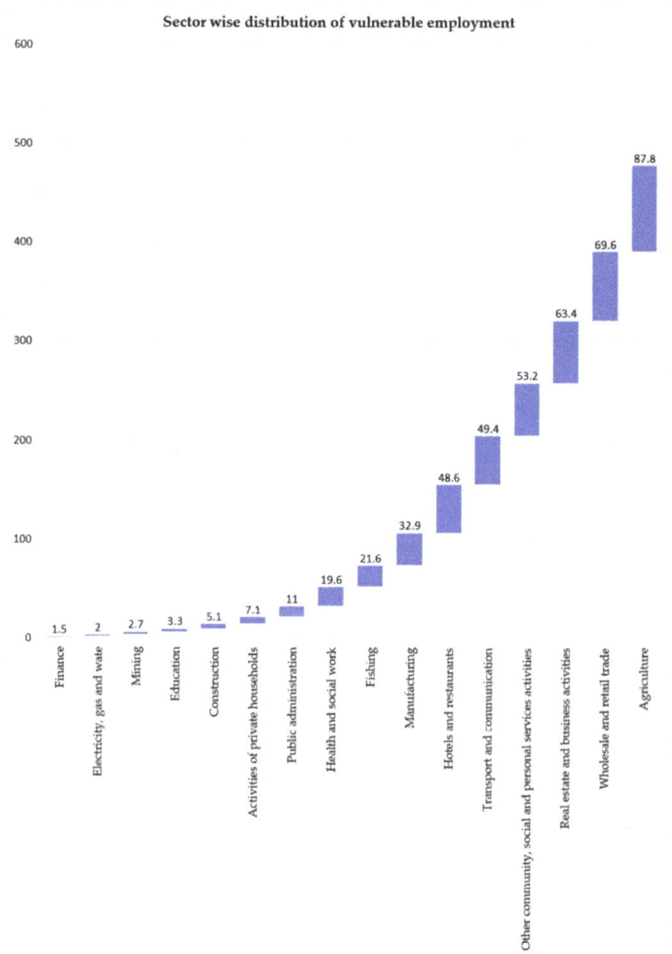

Figure 6. Distribution of vulnerable employment by sector. Source: [91].

On the other hand, many large- and medium-sized businesses are experiencing severe financial difficulties. The fabric and apparel industries, for example, have been severely harmed because of the unexpected lockdown. More than half of Pakistan's manufacturing sector's trade is in beverages, food, tobacco, and fabric; a drop in international demand for these industries will have a significant impact on Pakistan [61].

Similarly, the current crisis in other sectors is not diverse. Most prominently, in the blessed month of Ramadan, various industries such as fabrics, electronics, beauty, food, and shoes, along with others, were harshly impacted. These small- and mid-sized firms are likely to encounter liquidity crises and are severely hit by the current situation. However, those small businesses that are still operating their units face higher costs for buying masks to cover their faces, gloves, sanitizers, and other items to ensure their employees' health and shelter.

However, the above analysis indicates that Pakistan suffers disproportionately. So, the State Bank of Pakistan adopts a zero lower-bound-constraint monetary policy to mitigate the coronavirus's negative impact. The situation worsens because the money supply suddenly increases at a 7% interest rate, but the income level remains unchanged. In this scenario, Pakistan's economy, which also faces a huge debt burden, subsequently faces a depreciation of the rupee, and upsurges the hyperinflation rate by 14.56 percent [71]. Then, Pakistan's economy will stick in a liquescency trap, and both progress and hire rates will be dejected auxiliaries.

Consider what might happen if Pakistani industrialists and consumers lose faith in future efficiency growth. Due to the zero lower bound, the SBP is unable to alleviate the corresponding reduction in demand. As a result, jobs and financial activity decrease, and firms respond by reducing savings, negatively impacting production growth. As shown in Figure 7, the coronavirus is widespread and can bring expectation-driven inaction traps, specifically by flagging the economy's evolution rudiments.

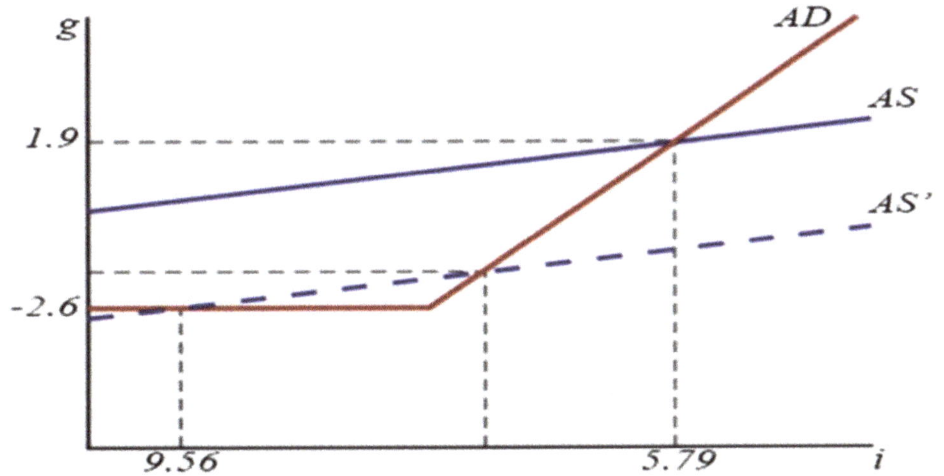

Figure 7. Animal spirits and stagnation traps. Source: [83]. Note: unemployment on the horizontal axis; productivity growth on the vertical axis.

An unaided conservative fiscal strategy may be insufficient or fail to converse in the loop when interest rates are near zero, as is debatably the case now. Dealing with such a situation will therefore require helicopter money or an expansionary monetary strategy, or both. When the economy is in a liquidity trap, helicopter money is a proposed unconventional monetary strategy that is occasionally recommended as an alternative to quantitative easing (QE) (when interest rates are near zero and the economy remains in recession). Although the term "helicopter money" was coined to describe central banks making direct payments to individuals, economists have used it to describe a variety of policy ideas, including the "permanent" monetization of budget deficits—with the added element of attempting to shock beliefs about future inflation or nominal GDP growth in order to change expectations.

4.4. Stabilization Policy under Keynesian Theory during Pakistan's Unsteady Economy

Following [85], the global economic uncertainty has triggered COVID-19, but the illustrative exercises and theories have future effects by aggregating business models. Hence, the method includes reviewing the previous studies of COVID-19 by following Keynes universal philosophy of occupation, interest, and coinage to determine the economic impact of the natural disaster on Pakistan, determining the vital economic indicators, such as GDP, imports, exports, interest rates, remittance, poverty, and unemployment and taking measures to reduce the economic impact by incorporating this set of factors into the development of concepts for dealing with and managing the epidemic. However, ref. [84]

makes evident that Keynes's theory significantly affects reassessing business models for overcoming economic crises.

4.4.1. Keynesian Policy Stabilization to Mitigate the Worse Effects of COVID-19

Now the query is: what strategy involvement can prevent the enchanting place's torpor trap? Because the policy rate is lowered due to the zero lower limit, a careful fiscal policy could only go so far. Still, under the burden of a huge debt, there is no room for Pakistan's state bank to reduce the interest rate to close to zero. It is the fifth time in four months that the State Bank of Pakistan has reduced the strategy rate (the per annum rate given in the Performance Swap Information Table) [86]. The State Bank of Pakistan (SBP) has now lowered the policy rate by a gigantic 625 points, from the comparatively extraordinary 13.25 pc to 7 pc, and this monetary action has shown its adverse effect. The price rise rate has hit a record of 14.56 pc, according to data published by Pakistan's official statistics department (PBS). According to Keynes (1958), only one financial strategy was ineffective in enhancing the economy—because decreasing interest rates governed it. In a depression, interest rates were previously adjacent to zero. Economic policy can be an influential support to transferring the economy because the outcome of an upsurge in outlays or a decrease in taxes would be multiplied by encouraging surplus demand for consumable goods by families.

4.4.2. Pakistan's Fiscal Policy in the Keynesian Legacy

The Keynesian model was established during the Great Depression as economic experts resisted elucidating the global financial failure and designing strategies to aid the economic recovery. This pandemic also created the situation of economic recession globally. The Pakistani government has made different fiscal changes to combat this financial crunch.

4.4.3. Increased Government Spending

Pakistan is amid a potential slump. Cash relocations and other economic measures may have an enormous optimistic effect on the international economy. Keeping in mind the current situation, the Pakistan government declared an aspiring financial inducement package in March intended to provide much-required aid to the helpless customers and firms within the realm. The USD 8 billion (PKR 1.2 trillion) aid package was designed to prevent Pakistan from falling into a self-perpetuating cycle of financial stress and reliance because of pre-emptive steps to curb the COVID-19 epidemic. Cash-transmission expenses under the Ehsaas Emergency Cash program, costing USD 894 million (PKR 144 billion), account for USD 894 million (PKR 144 billion) of the USD 8 billion economic incentive. Though this is a praiseworthy step, additional financial growth is still taking place [92,93].

The primary objective of this money injection is to unexpectedly increase household consumption, which immediately raises the demand for goods from the market and decreases the desired national saving at any level of real interest rate. As seen in Figure 8, the market curve moves from IS1 to IS2 to the right, but this capital infusion temporarily decreases the severity of the crisis. Other relief packages announced by the Pakistani government include PKR 572 billion for mass transportation, urban rail, and road transport systems; PKR 267 billion for solid waste management and storm water drainage systems; PKR 141 billion for sewage treatment plants; PKR 92 billion for water supply projects; and PKR 41 billion for disaster relief.

Like increasing government purchases, these fiscal stimuli eventually increase market demand and allow industrialists to invest. It has also decreased gas charges for exports concerned with zero-rated activities in July and August to help them recommend their construction activities after the COVID-19-associated lockdown [94]. So, this fiscal-policy change adjusts prices because companies encounter the developed demand at the static price level and shift the market demand curve upward. This expansionary change raises output and employment.

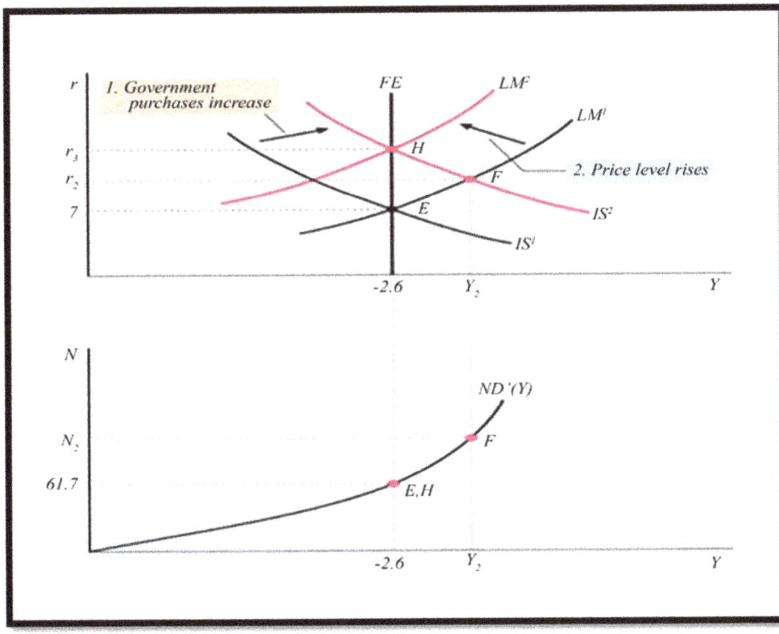

Figure 8. An increase in government purchases. Source: [83]. Note: interest rate and employment on the vertical axis; output on the horizontal axis.

5. Development of a Proposed Study Framework

The proposed research framework suggests focusing on the economic policies that have prolonged the medical crisis, as well as a grouping of financial, economic, and other systems and border cessations, self-isolation, and social connection constraints to resolve global and monetary market disputes. As a result, the study draws a framework in Figure 9 that is portrayed graphically with the discussions.

Policy Challenges

The problems for economic experts have been one of the uses of besieged policies that can address what are likely to be short-term anomalies in countries that can survive the impact of the epidemic itself without forming falsifications. Experts around the world have been left speechless by the quick shift in the condition of the global health emergency, which appears to be evolving into an international corporate trade and economic disaster with increasing consequences for the global economy. As the epidemic's fiscal impact grows, experts are focusing increasingly on remedies that, rather than longer-term considerations like debt accretion, may address immediate monetary consequences. Because of the low elasticity of fiscal and monetary support under conservative standards, many strategists have restricted their capacity to respond to the current crisis, given the corresponding recent decline in global economic development, particularly in output and trade, since the beginning of the flu epidemic. The epidemic has an influence on international policymaking. In the beginning, the fiscal impact of the pandemic was likely to be a temporary supply of products as factory production stopped because workers were self-quarantined by social contact to decrease the scope of the outbreak.

These rising financial impacts are likely to increase liquidity constraints and credit-market constraints in international monetary markets as businesses accumulate cash, with adverse effects on economic growth. Unlike the financial crisis of 2008–2009, lower consumer demand, labor market issues, and a decline in company operations rather than a dicey industry by foreign banks have contributed to commercial credit risks and potential bankruptcy. These undercurrents in the economy have led some viewers to inquire if these incidents are the beginning of full-scale international monetary issues.

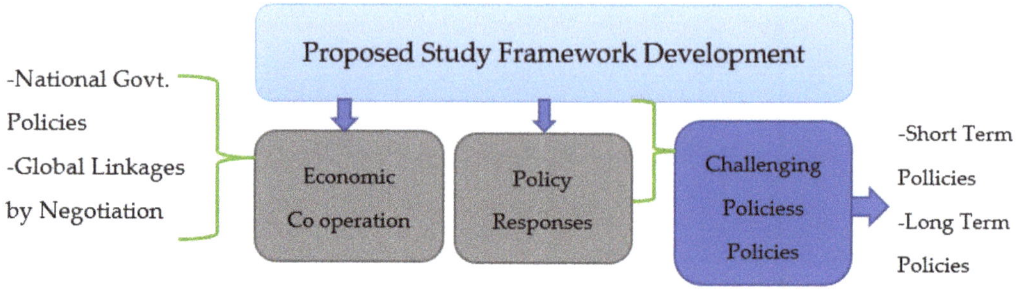

Figure 9. The proposed study framework for major development. Source: author's development (2020).

Early in 2020, economic growth in the world's most-developed and fastest-growing economies was modest but encouraging. Nations exceedingly reliant on trade, such as Germany, Japan, Mexico, Canada, Italy, and South Korea, and exporters, are now reported to be most adversely affected by the sudden decrease in economic activity because of COVID-19. The OECD discovered that China's manufacturing cuts had a global impact on circulation [62]. Because of strong recoveries in a few key economies, the global economy is expected to grow at a rate of 5.6 percent in 2021, the fastest rate of recovery in 80 years. According to the World Bank's June 2021 Global Economic Prospects, many emerging markets and developing nations are still grappling with the COVID-19 pandemic and its consequences. Despite the rebound, worldwide output will be around 2% lower by the end of the year than it was before the pandemic. For nearly two-thirds of emerging markets and developing nations, per capita income losses will not be recovered by 2022. The pandemic effects have reversed poverty-reduction gains and exacerbated insecurity and other long-standing concerns in low-income economies where vaccination has lagged.

According to a new survey, Asia's economy could be in for a slowdown by 2020. In addition, the development ratio in Latin America and the Middle East is expected to be negative in 2020 [62]. These predictions are likely to be revised downward soon because of COVID-19′s influence on international commerce, low energy and product costs, and a rise in the dollar's value.

Another result will be a slow rate of economic growth and a reduction in development. Due to a drop in trade with China and low commodity prices, goods-exporting countries are likely to see a greater decline in development than expected in previous estimates. Simultaneously, many countries are artificially enhancing their expenditure to battle the epidemic and will encounter a sudden decline in their income level, stressing public finances and increasing the chances of sovereign (government) evasion.

When foreign currency is used to indicate debt obligations, deficit crescendos are a particular concern in developing countries (such as U.S. dollars). The value of most developing economies' currencies has plunged since the onset of the epidemic, raising the debt's value in terms of native currency.

Most economies have shown a pledge to or have already instigated strategies to uplift economies affected by the epidemic. Several decisions were made by these countries on the sort of assistance to offer (loans versus direct payments), the amount of assistance required, how to define resources, and the terms to be accorded to funding, if any. To combat the crisis, several unanticipated fiscal and monetary measures have been enacted in countries around the world. These economies are constrained by a lack of financial resources, and the medical care system has become overburdened and unprotected in many cases.

As the COVID-19 issue is global and all countries are affected to a great degree by the downturn or forthcoming recession, global leaders need to overcome the COVID-19 effects swiftly and forcefully on their respective economies and their economic impact. Thus, global action is required that will lay down a proposed framework with international cooperative macroeconomic policies to restore confidence and to address the feasible solution to the existing barriers of the post-COVID crisis to overcome the great recession expected by using collective action with the following research questions:

- How do national governments negotiate and collaborate to overcome sharp drops in income, putting a strain on their public finances, and the looming possibility of sovereign (government) evasions and a 2021–2025 recession?
- How can Pakistan and Canada leverage global cooperation to avoid a sudden drop in income, a strain on their public finances, and an increase in the likelihood of sovereign (government) evasions?

6. The Future Study Recommendation

6.1. Significant Economic Development

The current critical situation demands greater attention paid to the industry, and Pakistan's economic sector consists of small- to medium-sized enterprises (SMEs). The world economy was one in which SMEs received loans to cover their inactivity periods for three months, while special funds were allocated to the larger firms with short-term loans [95]. However, SMEs in Pakistan are facing severe financial difficulties, which are jeopardizing their business prospects [96]. Other governments, such as France and Germany, plan to address the panic-like situation caused by COVID-19 with a liquidity program based on short-term refinancing schemes, accelerated payments, and tax credits [43,44,46]. Similarly, the Government of Pakistan is considering offering tax relief and other packages to compensate for the significant losses inflicted at present [42].

Therefore, Pakistan may follow other countries' strategies to save the economy and avoid the appalling social fallout. Interestingly, China's economic system is to stabilize its economy, and, as one example, Alibaba and J.D. Com's talent-sharing program in China has assisted in reducing the effect of the epidemic. They have recruited short-term workers to temporarily affect businesses. Furthermore, the European Tourism Manifesto Alliance [94,96] advocated for national governments to provide temporary assistance to the tourism and travel industry in the form of short- and medium-term loans. This alliance has proposed lowering or waiving travelers' taxes [97,98]. Several companies in the U.K. now provide online services and virtual sessions to education bodies so they can continue their studies without interruption [99]. The Government of Pakistan and SBP are partnering to provide relief to the industry and the wider community [51].

6.2. Policy Implications for Selected Sectors

At the start of the crisis, many economists felt constrained by the need to respond to the catastrophe. The global synchronized slowdown in economic growth, especially manufacturing and trade, that had preceded the viral breakout, was achieved at this time due to the partial elasticity of the currency and fiscal sustenance under predictable criteria. Liquidity and credit-market problems have also prompted financial analysts to deal with supply-side restraints with a particular range of experience.

Before banking institutions give credit deferment or a financial aid system is in place, job losses may result in missed mortgage and rent payments. Consequently, mortgage evasion may adversely disrupt the mortgage-backed-securities sector and the accessibility of mortgage funds, and it may have a disruptive effect on the country's economic growth rate. Globally, losses in the valuation of many financial markets impact the wealth of households, especially that of fixed-income retirees and those who own securities. Investors who operate on mortgage-backed assets have reduced their holdings. In the current environment, rate-cutting markets have not always routinely handled classic policy instruments like monetary assistance. Such volatility raises the question of what nations should do to fix the international economy's weaknesses.

Although the Pakistani government has done its best, it should still take some measures to tackle this pandemic. Firstly, in this emergency, supportive macroeconomic policies are required to restore trust and economic recovery. Enterprise tax credits should be given so that employers can pay workers their wages and earn an income themselves. Interest-free loans to companies may help address this situation, at least in the short term. We also must provide masks, disinfectants, and medical equipment to manufacturers, even though there are several downsides to doing so. Thirdly, online shopping will grow further and be encouraged to retain social distance and prevent people from risking the spread of COVID-19 when they leave their homes. Finally, a decline in the trade flow of textiles to China, the U.S., and the E.U. could be of great advantage to Pakistan. It can now dictate the price, availability, and routes of low-cost articles according to market conditions. In a nutshell, from the Keynesian theory study, Pakistan already falls into unfavorable circumstances that would lead to deterioration.

Despite the Pakistani government's announcement of compensation packages for catastrophe victims, more substantial measures are still required to get the economy back on track once the crisis has passed. However, the suggested necessary steps are given below:

1. In this emergency, supportive macroeconomic policies are required to restore trust and demand recovery. Hence, the enterprise tax credit may be provided to employers for collecting wages. Interest-free loans to companies may help address the lost income situation.
2. The healthcare industry has made a significant amount of money by mass-producing masks, hand sanitizers, and surgical supplies in large quantities, despite its many shortcomings. The other industry facing a recession could step forward to produce medical equipment. Furthermore, it does not have anywhere near enough hospitals and quarantine facilities, which are urgently needed [79,100,101]. Transmission of the virus can only be controlled by updating and distributing the required medical facilities. For both arrivals and departures, Pakistan needs more screening facilities [77,102].
3. Online shopping can be in vogue and will grow further, keeping in mind the prevailing unpredictable situation. In addition, the tourism industry is now largely redundant, and many people will be unemployed because COVID-19 has natural and long-lasting ramifications for social interaction and entertainment.
4. A decline in the textile sector's trade flow among China, the U.S., and the E.U. can be of great advantage to Pakistan, as Pakistan can manipulate some of the low-cost articles according to her choice.

7. Conclusions

Countries throughout the world are undergoing economic trauma caused by a global pandemic. Millions of people working in various industries are waiting impatiently for COVID-19 to end and never emerge again. Simultaneously, people are trying their level best to cope with this situation, mainly affecting the wage earners and those running small business enterprises. Still, it cannot do so due to curfews and lockdowns. Many industries have fallen prey to this scourge, and it means that GDP, remittances, interest rates, exports, imports, tourism, and health cannot function properly, if at all. The Islamic Republic of Pakistan is no exception, as it is now greatly distressed and experiencing a grave economic crisis due to COVID-19. Pakistan is currently under negative circumstances, according to Keynesian theory, and to avert further deterioration, the government has established special assistance packages for the victims of this disaster. More drastic measures will be required to get the economy back on track once the crisis has passed. Nonetheless, the fiscal policy can be a powerful device to alter the economic condition because the increasing trend in spending or a diminishing tendency in the taxes would be multiplied by encouraging supplementary demand for families' depletion of goods.

This means that reducing COVID-19 transmission while also having a positive impact on the environment, sustainable production, society, the economy, and worldwide prevention measures can be the key objectives of future work. Thus, for the study purpose, the methodology utilized in this exploratory study started with a review of how the COVID-19 pandemics exacerbated global health, identifying the changeable features and structural alterations that made a difference in the built environment, and then analyzed examples of such infections and changes to apply global new rules and regulations, identifying what controllable strategies and anticipations emerge from rethinking sustainable production.

Author Contributions: Conceptualization, K.A. and H.B.; methodology, K.A. and H.B.; software, K.A., H.B. and A.S.A.F.A.; formal analysis, K.A. and H.B.; investigation, K.A., H.B. and A.S.A.F.A.; resources, K.A. and H.B.; writing—original draft preparation, K.A. and H.B.; writing—review and editing, K.A., H.B., A.S.A.F.A. and A.H.A.; visualization, A.S.A.F.A.; supervision, K.A., A.S.A.F.A. and M.K.A.; funding acquisition, H.B., A.H.A., M.K.A. and I.M.M.E. For professional added, R.W. authors. All authors have read and agreed to the published version of the manuscript.

Funding: The APC was funded by Universiti Kebangsaan Malaysia (UKM) (project title: "The effect of Foreign Direct Investment & International Trade on Economics" and code: DCP-2018-002/2, which is headed by Abd Hair Awang.

Institutional Review Board Statement: Not applicable.

Informed Consent Statement: Not applicable.

Data Availability Statement: Not applicable.

Conflicts of Interest: The authors declare no conflict of interest.

References

1. Begum, H.; Alam, A.S.A.F.; Er, A.C.; Ghani, A.B.A. Environmental sustainability practices among palm oil millers. *Clean Technol. Environ. Policy* **2019**, *21*, 1979–1991. [CrossRef]
2. Alam, F.; Begum, H.; Masud, M.M.; Al Amin, A.K.; Filho, W.L. Agriculture Insurance for Disaster Risks Reduction: A case study of Malaysia. *Int. J. Disaster Risk Reduct.* **2020**, *47*, 101626. [CrossRef]
3. Masud, M.M.; Azam, M.N.; Mohiuddin, M.; Banna, H.; Akhtar, R.; Alam, A.S.A.F.; Begum, H. Adaptation barriers and strategies towards climate change: Challenges in the agricultural sector. *J. Clean. Prod.* **2017**, *156*, 698–706. [CrossRef]
4. Masud, S.; Torraca, V.; Meijer, A.H. Modeling infectious diseases in the context of a developing immune system. *Curr. Top. Dev. Biol.* **2017**, *124*, 277–329.
5. Nattrass, B.; Altomare, M. *The Natural Step for Business: Wealth, Ecology & the Evolutionary Corporation*; New Society Publishers: Gabriola Island, BC, Canada, 2009.
6. Miladi, M.; Abdellaoui, K.; Hamouda, A.B.; Boughattas, I.; Mhafdhi, M.; Acheuk, F.; Halima-Kamel, M.B. Physiological, histopathological, and cellular immune effects of Pergularia tomentosa extract on Locusta migratoria nymphs. *J. Integr. Agric.* **2019**, *18*, 2823–2834. [CrossRef]
7. Rees, M.J. *Our Final Hour: A Scientist's Warning: How Terror, Error, and Environmental Disaster Threaten Humankind's Future in This Century—On Earth and Beyond*; Basic Books (AZ): International Society for the Study of Religion, Nature & Culture: Delaware, OH, USA, 2003.
8. Steinfeld, J.L. Changes at HEW. *Science* **1969**, *166*, 819. [CrossRef] [PubMed]
9. The World Bank in China. An Overview; Retrieved 1 June 2020. Available online: https://www.worldbank.org/en/country/china/overview (accessed on 10 December 2020).
10. Maliszewska, M.; Mattoo, A.; Van Der Mensbrugghe, D. *The Potential Impact of COVID-19 on GDP and Trade: A Preliminary Assessment*; The World Bank: Bretton Woods, NH, USA, 2020.
11. Malik, Y.S.; Singh, R.K.; Yadav, M.P. *Recent Advances in Animal Virology*; Springer: Singapore, 2019; Volume XVII, p. 471, ISBN 978-981-13-9072-2. [CrossRef]
12. Malik, Y.S.; Sircar, S.; Bhat, S.; Sharun, K.; Dhama, K.; Dadar, M.; Chaicumpa, W. Emerging novel coronavirus (2019-nCoV)—current scenario, evolutionary perspective based on genome analysis and recent developments. *Vet. Q.* **2020**, *40*, 68–76. [CrossRef]
13. Yadav, M.P.; Singh, R.K.; Malik, Y.S. *Emerging and Transboundary Animal Viral Diseases: Perspectives and Preparedness. Livestock Diseases and Management*; Springer: Singapore, 2020; pp. 1–25, ISBN 978-981-15-0401-3. [CrossRef]
14. Malik, Y.S.; Dhama, K. Zika Virus–An Imminent Risk to the World. *J. Immunol. Immunopathol.* **2015**, *17*, 57–59. [CrossRef]
15. Munjal, A.; Khandia, R.; Dhama, K.; Sachan, S.; Karthik, K.; Tiwari, R.; Malik, Y.S.; Kumar, D.; Singh, R.K.; Iqbal, H.M.N.; et al. Advances in developing therapies to combat Zika virus: Current knowledge and future perspectives. *Front. Microbiol.* **2017**, *8*, 1469. [CrossRef]

16. Singh, R.K.; Dhama, K.; Malik, Y.S.; Ramakrishnan, M.A.; Karthik, K.; Khandia, R.; Tiw Ruchi, A.; Munjal, A.; Saminathan, M.; Sachan, S.; et al. Ebola virus–epidemiology, diagnosis, and control: Threat to humans, lessons learnt, and preparedness plans–an udate on its 40 year's journey. *Vet. Q.* **2017**, *37*, 98–135. [CrossRef]
17. Sylvia Black. *Coronavirus Is Not a Death Sentence—There Is Hope for Those in Grief in Times of This Pandemic*; Sylvia Black: New York, NY, USA, 2020; ISBN 9798639290039.
18. Worldometers, Worldometers, COVID-19 Coronavirus Pandemic. 2020. Available online: https://www.worldometers.info/coronavirus/ (accessed on 5 June 2020).
19. World Health Organization. Novel Coronavirus (2019-nCoV) Advice for the Public. World Health Organization. 2020. Available online: https://www.who.int/emergencies/diseases/novel-coronavirus-2019 (accessed on 19 June 2020).
20. Li, G.; De Clercq, E. Comment: Therapeutic options for the 2019 novel coronavirus (2019-nCoV). *Nat. Rev. Drug Discov.* **2020**, *19*, 149. [CrossRef] [PubMed]
21. Varona, L.; Gonzales, J.R. Dynamics of the impact of COVID-19 on the economic activity of Peru. *PLoS ONE* **2021**, *16*, e0244920. [CrossRef] [PubMed]
22. Lipsitch, M.; Swerdlow, D.L.; Finelli, L. Defining the epidemiology of Covid-19—studies needed. *N. Engl. J. Med.* **2020**, *382*, 1194–1196. [CrossRef]
23. Clarke, L. An introduction to economic studies, health emergencies, and COVID-19. *J. Evid.-Based Med.* **2020**, *13*, 161–167. [CrossRef] [PubMed]
24. Raga, S.; Te Velde, D.W. Economic vulnerabilities to health pandemics: Which countries are most vulnerable to the impact of coronavirus. *Supporting Econ. Transform.* **2020**, *32*. Available online: https://set.odi.org/wp-content/uploads/2020/02/Economic-Vulnerability.pdf (accessed on 11 November 2020).
25. Wang, C.; Cheng, Z.; Yue, X.-G.; McAleer, M. Risk management of COVID-19 by universities in China. *J. Risk Financ. Manag.* **2020**, *13*, 36. [CrossRef]
26. Wang, Q.; Shi, N.; Huang, J.; Cui, T.; Yang, L.; Ai, J.; Jin, H. Effectiveness, and cost-effectiveness of public health measures to control COVID-19: A modeling study. *medRxiv* **2020**. [CrossRef]
27. Glassman, A.; Chalkidou, K.; Sullivan, R. *Does One Size Fit All? Realistic Alternatives for COVID-19 Response in Low-Income Countries*; Blog. Center for Global Development: London, UK, 2020. Available online: www.cgdev.org/blog/does-one-size-fit-all-realistic-alternatives-Covid-19-response-low-income-countries (accessed on 10 December 2020).
28. Peters, D.H.; Hanssen, O.; Gutierrez, J.; Abrahams, J.; Nyenswah, T. Financing common goods for health: Core government functions in health emergency and disaster risk management. *Health Syst. Reform* **2019**, *5*, 307–321. [CrossRef] [PubMed]
29. Callaway, E.J.N. Time to use the p-word? Coronaviruses enter dangerous new phase. *Nature* **2020**, *579*, 1038.
30. Azzopardi, P.J.; Upshur, R.E.; Luca, S.; Venkataramanan, V.; Potter, B.K.; Chakraborty, P.K.; Hayeems, R.Z. Healthcare providers' perspectives on uncertainty generated by variant forms of newborn screening targets. *Genet. Med.* **2020**, *22*, 566–573. [CrossRef] [PubMed]
31. Guerrieri, V.; Lorenzoni, G.; Straub, L.; Werning, I. *Macroeconomic Implications of COVID-19: Can Negative Supply Shocks Cause Demand Shortages?* National Bureau of Economic Research: Cambridge, MA, USA, 2020.
32. Kickbusch, I.; Leung, G.M.; Bhutta, Z.A.; Matsoso, M.P.; Ihekweazu, C.; Abbasi, K. COVID-19: How a virus is turning the world upside down. *BMJ* **2020**, *369*, m1336. [CrossRef] [PubMed]
33. World Health Organization. Novel Coronavirus (2019-nCoV) Situation Reports. 2020. Available online: https://www.who.int/emergencies/diseases/novel-coronavirus2019/situation-reports/ (accessed on 10 December 2020).
34. Government of Pakistan COVID-19 Official Portal. Available online: http://covid.gov.pk/stats/pakistan (accessed on 5 August 2020).
35. Malik, K.; Meki, M.; Morduch, J.; Ogden, T.; Quinn, S.; Said, F. COVID-19 and the Future of Microfinance: Evidence and Insights from Pakistan. *Oxf. Rev. Econ. Policy* **2020**, *36* (Suppl. S1), S138–S168. [CrossRef]
36. Sutradhar, S.R. The impact of remittances on economic growth in Bangladesh, India, Pakistan, and Sri Lanka. *Int. J. Econ. Policy Stud.* **2020**, *14*, 275–295. [CrossRef]
37. Salik, K.M. *Policy Review Remittances and COVID-19: Is Pakistan Ready for a Likely Decline in Flows?* Sustainable Development Policy Institute: Tokyo, Japan, 2020.
38. UNDP. COVID-19: The Looming Crisis in Developing Countries Threatens to Devastate Economies and Ramp Up Inequality. Retrieved 30 April 2020. Available online: https://www.undp.org/content/undp/en/home/news-centre/news/2020/COVID19_Crisis_in_developing_countries_threatens_devastate_economies.html (accessed on 2 May 2020).
39. Robinson, L.A.; Hammitt, J.K.; Jamison, D.T.; Walker, D.G. Conducting benefit-cost analysis in low-and middle-income countries: Introduction to the special issue. *J. Benefit-Cost Anal.* **2019**, *10*, 1–14. [CrossRef]
40. Mamun, M.A.; Ullah, I. COVID-19 suicides in Pakistan, dying off not COVID-19 fear but poverty? The forthcoming economic challenges for a developing country. *Brain Behav. Immun.* **2020**, *87*, 163. [CrossRef]
41. Kabir, M.; Afzal, M.S.; Khan, A.; Ahmed, H. COVID-19 economic cost; impact on forcibly displaced people. *Travel Med. Infect. Dis.* **2020**, *35*, 101661. [CrossRef] [PubMed]
42. Chohan, U.W. Forecasting the Economic Impact of Coronavirus on Developing Countries: Case of Pakistan (28 March 2020). CASS Working Papers on Economics & National Affairs, Working Paper ID: EC016UC (2020). 2020. Available online: https://ssrn.com/abstract=3563616orhttp://dx.doi.org/10.2139/ssrn.3563616 (accessed on 28 March 2020).

43. McKibbin, W.J.; Fernando, R. The global macroeconomic impacts of COVID-19: Seven scenarios. *Asian Econ. Pap.* **2021**, *20*, 1–30. [CrossRef]
44. Elgin, C.; Basbug, G.; Yalaman, A. Economic policy responses to a pandemic: Developing the COVID-19 economic stimulus index. *Covid Econ.* **2020**, *1*, 40–53.
45. Shahbazov, F. China–Pakistan Economic Corridor: An Opportunity for Central Asia? Central Asia-Caucasus Institute. 2017. Available online: http://cacianalyst.org/publications/analytical-articles/item/13449-china-%E2,80 (accessed on 11 November 2020).
46. World Bank Group. Pandemic Preparedness Financing-Status Update. June 2019. Commissioned Paper by the GPMB. 2019. Available online: https://www.who.int/gpmb (accessed on 5 November 2020).
47. MOC. Ministry of Commerce, Pakistan. 2019. Available online: http://www.commerce.gov.pk/ (accessed on 14 December 2020).
48. Nasir, M.; Faraz, N.; Khalid, M. Sectoral analysis of the vulnerably employed: COVID-19 and the Pakistan's labour market. *COVID-19* **2020**, *50*, 90.
49. World Bank National Accounts Data, and OECD National Accounts Data Files. Available online: https://data.worldbank.org/indicator/NY.GDP.MKTP.CD?end=2020&locations=PK&start=2010 (accessed on 14 November 2021).
50. Hassan, S.A.; Zaman, K. RETRACTED: Effect of oil prices on trade balance: New insights into the cointegration relationship from Pakistan. *Econ. Model.* **2012**, *29*, 2125–2143. [CrossRef]
51. Chohan, U.W.; The Trouble with Economics. Daily Times. 25 January 2020. Available online: https://dailytimes.com.pk/545600/the-trouble-with-economics/ (accessed on 5 December 2020).
52. MOPDSI. Ministry of Planning Development & Special Initiatives, Pakistan. 2019. Available online: https://www.pc.gov.pk/ (accessed on 14 December 2020).
53. Zhang, D.; Hu, M.; Ji, Q. Financial markets under the global pandemic of COVID-19. *Financ. Res. Lett.* **2020**, *36*, 101528. [CrossRef]
54. Alam, A.S.A.F.; Siwar, C.; Talib, B.; Islam, R. The analysis of international migration towards economic growth. *Am. J. Appl. Sci.* **2011**, *8*, 782–788. [CrossRef]
55. Alam, A.S.A.F.; Siwar, C.; Talilb, B.; Islam, R. Impacts on international migration and remittances growth. *Am. J. Environ. Sci.* **2011**, *7*, 20–25. [CrossRef]
56. Alam, A.S.A.F.; Siwar, C.; Begum, H.; Er, A.C.; Talib, B.; Elfithri, R. Local Economic Impacts of Human Capital Migration. *Mediterr. J. Soc. Sci.* **2015**, *6*, 563–569. [CrossRef]
57. Amiti, M.; Itskhoki, O.; Konings, J. Importers, exporters, and exchange rate disconnect. *Am. Econ. Rev.* **2014**, *104*, 1942–1978. [CrossRef]
58. SBP. SBP Looks Back on Tough Year for Pakistan. Central Banking Newsdesk. 4 November 2019. Available online: https://www.centralbanking.com/central-banks/economics/macroeconomics/4527276/sbp-looks-back-on-tough-year-for-pakistan (accessed on 4 March 2020).
59. Chohan, U.W. *Public Value and Budgeting: International Perspectives*; Routledge: England, UK, 2019.
60. Chohan, U.W. *The FATF in the Global Financial Architecture: Challenges and Implications*; CASS Working Papers on Economics & National Affairs, EC001UC (2019); SSRN Electronic Journal: Rochester, NY, USA, 2019.
61. The World Bank Group. 2020. Available online: https://www.worldbank.org/en/who-we-are/news/coronavirus-covid19 (accessed on 30 March 2020).
62. OECD. Global Economy Faces the Gravest Threat Since the Crisis as Coronavirus Spreads. 2 March 2020. Available online: http://www.oecd.org/economy/global-economy-faces-gravest-threat-since-the-crisis-as-coronavirus-spreads.htm (accessed on 19 March 2020).
63. Jabbour, C.J.C.; Fiorini, P.D.C.; Wong, C.W.; Jugend, D.; Jabbour, A.B.L.D.S.; Seles, B.M.R.P.; Pinheiro, M.A.; da Silva, H.M.R. First-mover firms in the transition towards the sharing economy in metallic natural resource-intensive industries: Implications for the circular economy and emerging industry 4.0 technologies. *Resour. Policy* **2020**, *66*, 101596. [CrossRef]
64. Abi-Habib, M.; Ur-Rehman, Z.; Mehsud, I.T. God Will Protect Us': Coronavirus Spreads Through an Already Struggling Pakistan. New York Times. 2020. 26 March 2020, Updated 26 June 2020. Available online: https://www.nytimes.com/2020/03/26/world/asia/pakistan-coronavirus-tablighi-jamaat.html (accessed on 26 March 2020).
65. Ahmed, S.H. Pakistan's Economy Battling a Host of Challenges. Express Tribune. 10 February 2020. Available online: https://tribune.com.pk/story/2153367/pakistans-economy-battling-host-challenges (accessed on 5 July 2020).
66. Ahmad, I. COVID-19 and Labour Law: Pakistan. *Ital. Labour Law J.* **2020**, *13*. [CrossRef]
67. Alfaro, L.; Chen, M.X. Surviving the global financial crisis: Foreign ownership and establishment performance. *Am. Econ. J. Econ. Policy* **2012**, *4*, 30–55. [CrossRef]
68. Desai, M.A.; Foley, C.F.; Forbes, K.J. Financial constraints and growth: Multinational and local firm responses to currency depreciations. *Rev. Financ. Stud.* **2008**, *21*, 2857–2888. [CrossRef]
69. Employment Trends Reported by the Pakistan Bureau of Statistics Pakistan Tourism Revenue. 2018. CEIC Data. Available online: https://www.ceicdata.com/en/indicator/pakistan/tourism-revenue (accessed on 5 April 2020).
70. Sareen, S. COVID-19 and Pakistan: The economic fallout. *Obs. Res. Found. Occas. Pap. No* **2020**, *251*, 8.
71. Ministry of Finance, Pakistan. Available online: http://www.finance.gov.pk/ (accessed on 15 March 2020).
72. Begum, H.; Er, A.C.; Alam, A.S.A.F.; Sahazali, N. Tourist's perceptions towards the role of stakeholders in sustainable tourism. *Procedia-Soc. Behav. Sci.* **2014**, *144*, 313–321. [CrossRef]
73. Ozdemir, A.I.; Ar, I.M.; Erol, I. Assessment of blockchain applications in the travel and tourism industry. *Qual. Quant.* **2020**, *54*, 1549–1563. [CrossRef]

74. Israr, M.; Shafi, M.M.; Ahmad, N.; Khan, N.; Baig, S.; Khan, Z.H. Eco tourism in Northern Pakistan and challenges perspective of stakeholders. *Sarhad J. Agric.* **2009**, *25*, 113–120.
75. Asgary, A.; Anjum, M.I.; Azimi, N. Disaster recovery and business continuity after the 2010 flood in Pakistan: Case of small businesses. *Int. J. Disaster Risk Reduct.* **2012**, *2*, 46–56. [CrossRef]
76. Saqlain, M.; Abbas, Q.; Lee, J.Y. A Deep Convolutional Neural Network for Wafer Defect Identification on an Imbalanced Dataset in Semiconductor Manufacturing Processes. *IEEE Trans. Semicond. Manuf.* **2020**, *33*, 436–444. [CrossRef]
77. Waris, A.; Khan, A.U.; Ali, M.; Ali, A.; Baset, A. COVID-19 outbreak current scenario of Pakistan. *New Microbes New Infect.* **2020**, *35*, 100681. [CrossRef]
78. Sethi, B.A.; Sethi, A.; Ali, S.; Aamir, H.S. Impact of Coronavirus disease (COVID-19) pandemic on health professionals. *Pak. J. Med. Sci.* **2020**, *36*, S6. [CrossRef] [PubMed]
79. Raza, S.; Rasheed, M.A.; Rashid, M.K. Transmission potential and severity of COVID-19 in Pakistan. *Preprints* **2020**. [CrossRef]
80. Gali, J. *Introduction to Monetary Policy, Inflation, and the Business Cycle: An Introduction to the New Keynesian Framework [Monetary Policy, Inflation, and the Business Cycle: An Introduction to the New Keynesian Framework]*; Princeton University Press: Princeton, NJ, USA, 2008.
81. Fornaro, L.; Wolf, M. Covid-19 Coronavirus and Macroeconomic Policy, Working Papers 1168, Barcelona Graduate School of Economics. 2020. Available online: https://ideas.repec.org/p/bge/wpaper/1168.html (accessed on 10 June 2020).
82. Lorenzoni, G. A theory of demand shocks. *Am. Econ. Rev.* **2009**, *99*, 2050–2084. [CrossRef]
83. Kilic, K.; Marin, D. How Is COVID-19 Transforming the World Economy? 10 May 2020. Available online: https://voxeu.org/article/how-covid-19-transforming-world-economy (accessed on 5 December 2020).
84. Baker, S.; Bloom, N.; Davis, S.; Terry, S. COVID-Induced Economic Uncertainty and Its Consequences. 13 April 2020. Available online: https://voxeu.org/article/covid-induced-economic-uncertainty-and-its-consequences (accessed on 5 December 2020).
85. The State Bank of Pakistan. 2020. Available online: https://www.sbp.org.pk/m_policy/index.asp (accessed on 25 March 2021).
86. Mahmood, A.; Wang, X.; Shahzad, A.N.; Fiaz, S.; Ali, H.; Naqve, M.; Javaid, M.M.; Mumtaz, S.; Naseer, M.; Dong, R. Perspectives on Bioenergy Feedstock Development in Pakistan: Challenges and Opportunities. *Sustainability* **2021**, *13*, 8438. [CrossRef]
87. Sharif, S. *Impact of COVID-19 Pandemic: Government Relief Package and the Likely Misallocation of Loans in Pakistan*; State Bank of Pakistan: Islamabad, Pakistan, 2020.
88. Shafi, M.; Liu, J.; Ren, W. Impact of COVID-19 pandemic on micro, small, and medium-sized Enterprises operating in Pakistan. *Res. Glob.* **2020**, *2*, 100018. [CrossRef]
89. Sawada, Y. *Sustainable Economic Development in the Post-Covid Era in Asia and the Pacific*; Insitute of Business Administration (IBA): Karachi, Pakistan, 2021.
90. Manzoor, R.; Maken, A.M.; Ahmed, V.; Javed, A. Reforming Trade and Transport Connectivity in Pakistan. *Sukkur IBA J. Manag. Bus.* **2019**, *6*, 45–65. [CrossRef]
91. Sohail, S. Pakistan Employment Trends-2018. Islamabad, Pakistan: Pakistan Bureau of Statistics, Ministry of Statistics, Government of Pakistan. 2019. Available online: https://www.pbs.gov.pk/sites/default/files//Pakistan%20Employment%20Trend%20%20Reprt%202018%20Final.pdf (accessed on 5 December 2020).
92. Augustine, B.D. COVID-19 to Slowdown Pakistan's Economic Growth: Asian Development Bank. 30 April 2020. Available online: https://gulfnews.com/business/covid-19-to-slowdown-pakistans-economic-growth-asian-development-bank-1.70803457 (accessed on 9 March 2020).
93. Anderson, R.M.; Heesterbeek, H.; Klinkenberg, D.; Hollingsworth, T.D. How will country-based mitigation measures influence the course of the COVID-19 epidemic? *Lancet* **2020**, *395*, 931–934. [CrossRef]
94. Dar, M.S.; Ahmed, S.; Raziq, A. Small and medium-sized enterprises in Pakistan: Definition and critical issues. *Pak. Bus. Rev.* **2017**, *19*, 46–70.
95. Bénassy-Quéré, A.; Marimon, R.; Pisani-Ferry, J.; Reichlin, L.; Schoenmaker, D.; Di Mauro, B.W. 16 COVID-19: Europe Needs a Catastrophe Relief Plan. Europe in the Time of COVID-19. 2020, Volume 103. Available online: https://voxeu.org/article/covid-19-europe-needs-catastrophe-relief-plan (accessed on 9 March 2020).
96. Johnson, H.C.; Gossner, C.M.; Colzani, E.; Kinsman, J.; Alexakis, L.; Beauté, J.; Ekdahl, K. Potential scenarios for the progression of a COVID-19 epidemic in the European Union and the European Economic Area, March 2020. *Eurosurveillance* **2020**, *25*, 2000202. [CrossRef] [PubMed]
97. Kickbusch, I.; Leung, G. *Response to the Emerging Novel Coronavirus Outbreak*; BMI Publishing Group: London, UK, 2020.
98. Nicola, M.; Alsafi, Z.; Sohrabi, C.; Kerwan, A.; Al-Jabir, A.; Iosifidis, C.; Agha, M.; Agha, R. The socio-economic implications of the coronavirus pandemic (COVID-19): A review. *Int. J. Surg.* **2020**, *78*, 185–193. [CrossRef]
99. Fernandes, N. Economic Effects of Coronavirus Outbreak (COVID-19) on the World Economy. 22 March 2020. Available online: http://dx.doi.org/10.2139/ssrn.3557504 (accessed on 22 March 2020).
100. Begum, H.; Alam, A.S.A.F.; Filho, W.L.; Awang, A.H.; Ghani, A.B.A. The Covid-19 Pandemic: Are There Any Impacts on Sustainability? *Sustainability* **2021**, *13*, 11956. [CrossRef]
101. Gatto, A.; Sadik-Zada, E.R. Revisiting the East Asian Financial Crises: Lessons from Ethics and Development Patterns. In *Economic Growth and Financial Development*; Springer: Cham, Switzerland, 2021; pp. 23–31.
102. Gatto, A. A pluralistic approach to economic and business sustainability: A critical meta-synthesis of foundations, metrics, and evidence of human and local development. *Corp. Soc. Responsib. Environ. Manag.* **2020**, *27*, 1525–1539. [CrossRef]

sustainability

MDPI

Review

Beyond Diffusion: A Systematic Literature Review of Innovation Scaling

Jessica Breaugh *, Keegan McBride, Moritz Kleinaltenkamp and Gerhard Hammerschmid

Centre for Digital Governance, Hertie School, 10117 Berlin, Germany; Mcbride@hertie-school.org (K.M.); kleinaltenkamp@hertie-school.org (M.K.); hammerschmid@hertie-school.org (G.H.)
* Correspondence: breaugh@hertie-school.org

Abstract: Innovation is essential for our ability to overcome global issues such as climate change, natural resource depletion, and inequality. A central aspect of innovation is the scaling process. While an abundance of studies on innovation scaling exist in many different disciplines, there is a lack of shared understanding of what scaling means and how it can be successfully achieved. This systematic literature review addresses both these issues by reviewing 147 articles on "innovation scaling" making several contributions to research on innovations and innovation scaling. First, in outlining the ontological differences between "diffusion" and "scaling", clear conceptual boundaries are established, which provide clarity and support cross-disciplinary consilience. Second, based on the analysis of articles, eleven common modal contextual factors that influence the outcomes of innovation scaling across contexts and disciplines are presented. Third, an initial theoretical framework of the innovation scaling process is developed, outlining four theoretical propositions. As a fourth contribution, the article establishes a research agenda for the future development of innovation scaling research across many research domains.

Keywords: scaling; upscaling; scaling innovation; innovation diffusion; innovation; systematic literature review

Citation: Breaugh, J.; McBride, K.; Kleinaltenkamp, M.; Hammerschmid, G. Beyond Diffusion: A Systematic Literature Review of Innovation Scaling. *Sustainability* **2021**, *13*, 13528. https://doi.org/10.3390/su132413528

Academic Editors: E. Richard Zada and Andrea Gatto

Received: 9 November 2021
Accepted: 1 December 2021
Published: 7 December 2021

1. Introduction

Innovation constitutes a cross-cutting and multi-disciplinary research field [1–3] spanning sustainability (e.g., sustainable agriculture, green growth, the circular economy, or sustainable urban development), smart cities, management, international development, health, and social welfare literatures. Cross-disciplinary academic interest in the topic stems from the fact that innovation occurs across all aspects of society and when done successfully, may deliver new solutions that address pressing issues and/or increase the efficiency and effectiveness of policies, products, and services [4–6].

The process of innovation is different from the process of invention [6] in that it describes the (collective) development and implementation of new ideas [4,5,7–11]. Innovation is generally understood as a series of steps: Following the initial inception of novelty, ideas are then iteratively tested, refined, and socio-politically legitimated, before they are ultimately implemented [4,5]. Actors' motivations, as well as contextual drivers and hindrances, are central in determining the ultimate outcome of the innovation process [7]. Once an innovation has been implemented, the processes of diffusion and scaling—which both refer to the longitudinal spreading of innovations—take central importance in determining the innovations' wider societal impact [4,7,12–15].

Notably, in the current academic discourse, innovation scaling has received relatively less attention than the other aspects of the innovation process (e.g., inception of novelty or its diffusion). One field where scaling innovation is particularly popular, however, is in sustainability studies, where scaling and innovation are viewed as key components for sustainable management, development, and growth [16–20]. Though popular, a clear and widely agreed upon definition of scaling remains absent, and so too is a comprehensive theoretical framework for understanding the entire scaling process.

For many, the concept of "scaling" revolves around the idea that there are empirical phenomena related to the spreading of innovations that are poorly captured by the population-level perspective of diffusion, which tends to highlight a kind of passive permeation of innovations [15,21–25]. Empirical phenomena considered as instances of "innovation scaling", on the other hand, tend to be characterized by proactive and strategic actors purposefully moving innovations into new contexts and degrees of quality [26–28]. However, when scholars discuss scaling in this way, they tend to do so using relatively shallow or idiosyncratic conceptualizations [29]. Some scholars implicitly or explicitly understand scaling to be a process of vertical movement ("scaling up"), while others understand it as one of horizontal movement ("scaling out/wide") or of qualitative change ("scaling deep") [30]. Furthermore, there lacks a common definition of what it means to "scale" innovations. Rather, the term "scaling" is applied and used differently across different contexts (e.g., private sector compared to public sector) and across areas of research (e.g., social welfare compared to development, management, or public administration research).

This lack of conceptual clarity is problematic. It obstructs the development and exchange of scientific knowledge about the process of innovation scaling. To help effectively address pressing global issues through innovation, scholars must develop a systematic understanding of how high impact innovations may be proactively scaled. To do so, one not only has to integrate the research and knowledge that already exists across research disciplines, one also has to develop a conceptual common ground and shared research agenda that facilitates further advancement and knowledge sharing on the topic of innovation scaling.

In pursuit of these goals, this article presents the results of a systematic review of the innovation scaling literature. The review utilized the PRISMA approach (Preferred Reporting Items for Systematic Reviews and Meta-Analysis) [31,32] and covered 147 research articles. The analysis of these articles was guided by four primary research questions:

1. What does the landscape of research investigating innovation scaling look like?
2. How can innovation scaling be defined?
3. What are the key drivers and barriers of innovation scaling?
4. How can extant research be integrated into an initial overarching framework of innovation scaling?

Fitting the multi-disciplinary relevance of the innovation scaling process, this review covers articles from numerous research fields, including sustainability, development, agriculture, management, and health studies. The innovation scaling research that has been conducted within any one these disciplines has tended to remain largely isolated from the others. While there are fundamental contextual differences between these disciplines, integrating their perspectives into a common framework can help to provide a holistic lens upon the cross-cutting innovation scaling phenomenon. Moreover, where cross-disciplinary commonalities do exist, it is possible to identify essential aspects and building blocks that any conceptualization of innovation scaling must pay respect to and integrate them into an initial theoretical framework.

The results of this systematic literature review facilitate several important theoretical and practical contributions. First, this article presents a cross-disciplinary definition for conceptualizing "innovation scaling" that builds upon extant research while also clearly distinguishing "scaling" from the related concept of "diffusion". In doing so, the article offers new clarity that should aid in facilitating further systematic research and cross disciplinary knowledge exchange relating to innovation scaling. Second, this article reports on the eleven "modal contextual factors" that acted as drivers and/or barriers to the innovation scaling process across the 147 covered research articles. Third, based on our proposed definition and modal factors, the article develops an initial theoretical framework for understanding and studying the innovation scaling process. Fourth, this article presents a comprehensive agenda facilitating further systematic research into this important phenomenon.

2. Innovation Diffusion and Innovation Scaling

"Diffusion" has been identified as one of the core mechanisms in social theories seeking to explain macro-level societal changes [23,24]. Diffusion itself is usually defined as the longitudinal spreading of (new) "ideas, structures, and practices" [24] (p. 335) in social systems [33,34]. Social scientists have researched the diffusion of innovations for more than seven decades (see, e.g., [35,36] for early accounts of diffusion dynamics) using various lenses. Two larger schools of thought may be identified and differentiated: "rational accounts" of diffusion, and "social accounts" of diffusion [25,33].

Rational accounts of diffusion emerged from economic research and make sense of the spreading of innovations by focusing on the benefits that innovations bring vis-à-vis existing ideas, structures, and practices [25]. Some scholars use an evolutionary lens, according to which individual and organizational actors who fail to adopt efficient practices get weeded out [37,38]. Other scholars may adopt a more actor-centric lens, investigating the rational evaluation and decision-making by potential innovation adopters [39–41]. Rational accounts have repeatedly found that the diffusion of innovations can be connected to their (perceived) cost-effectiveness [2,23]. Accordingly, a central aspect of diffusion processes within this school of thought are "information cascades" [42–45] that transmit innovations and their associated utility expectations from one actor to another.

Social accounts of diffusion emerged from sociological research. In opposition to rational accounts, social accounts emphasize the "pressure toward social conformity" [25] (p. 70) that grows as more and more actors in a field adopt an innovation. Drawing on institutional theory [46–49], social accounts of diffusion highlight how group pressures and bandwagon effects may replace rational concerns as an innovation becomes increasingly adopted in a field. These dynamics may even lead actors to adopt new ideas, structures, or practices that are inefficient for them [21,23] simply as a means of maintaining legitimacy [50].

Both rational and social accounts of innovation diffusion tend to assume a population-level perspective that emphasizes interorganizational conditions (information cascades or pressures for social conformity) as the central mechanisms explaining diffusion [25]. In both, innovation diffusion is viewed as a natural phenomenon akin to gravity: It permeates through populations of individual and organizational actors as a passive force that may *prompt* human activity but is hardly *driven by* it. Accordingly, much diffusion research has been focused on measuring the speeds and rates of this force [2,15,35,36,51–53]. This has led to the well-established insight that, "when plotted over time, the cumulative number of adopters of an innovation approximates an S-shaped curve" [54] (p. 544).

One limitation of these perspectives on the spreading of innovations is that they relatively discount the proactive agency and strategic decision-making of individual and organizational actors. Human agency fits into rational and social accounts of innovation diffusion as a *reaction*—i.e., an adoption decision prompted by information cascades and conformity pressures (see also [55])—but seldom as a *proactive* driving force. While [38]'s seminal review of diffusion research noted that "organized dissemination" of innovations falls under the roof of diffusion research, this research has, in practice, paid little attention to such a proactive and purposeful spreading of innovations by actors. Rather, such "evangelism" for specific innovations by individual actors has been characterized as "the reverse of diffusion" [24,56].

This article argues that in order to address pressing global issues through innovation, scholars must develop a systematic understanding not only of how beneficial innovations may passively diffuse through populations, but also of how strategic actors may purposefully and proactively disseminate them. This phenomenon of proactive and strategic dissemination of innovations is increasingly of interest to scholars from across academic disciplines, where it is variously referred to using terms such as "scaling" and "upscaling". This article uses the term "innovation scaling" to refer to this process.

While a cohesive definition and common conceptualization of innovation scaling has not yet manifested, existing definitions gravitate around the understanding that strategic and purposeful actors may indeed be a driving force behind an innovation's movement into new contexts and degrees of quality [26,27,30]. For example, [30] understands scaling to be a deliberate, guided process. According to [26], scaling "is often considered to refer to a series of processes to introduce innovations with demonstrated effectiveness through a program delivery structure with the aim of improving coverage and equitable access to the innovation(s)" [26] (p. 2).

Accordingly, this article views innovation scaling as complementary to diffusion (see Table 1). Research into innovation scaling does benefit from the conceptualizations of field-level rationality and institutional factors developed in innovation diffusion research. At the same time, by investigating not what happens in a field (the focus of diffusion research), but, rather, what actors do, the "agentic account" offered by innovation scaling offers a complementary perspective that may contribute to scholars' overall understanding of how new ideas, structures, and practices spread. This benefit accrues in addition to the distinct practical urgency that exists to develop a systematic understanding of how innovations may be proactively and strategically scaled to address pressing global issues.

Table 1. The relationship between innovation diffusion and innovation scaling research. Source: Authors' own elaboration.

Empirical Phenomenon	**Diffusion**		
Definition	**The longitudinal spreading of (new) ideas, structures, and practices in social systems**		
Research Lenses	Innovation diffusion		Innovation scaling
Focal Entity	Populations/fields		Individual and organizational actors
Conceptualization of Spreading	Passive permeation akin to natural force		Proactively and purposefully driven by strategic actors
Guiding Question	How do innovations diffuse among populations?		How do actors scale innovations?
Schools of Thought	Rational Account	Social Account	Agentic Account
Central Explanatory Mechanism	Interorganizational information cascades	Interorganizational legitimacy pressures	Strategic actors' ability to drive innovation adoption among other actors

Addressing both the theoretical and practical opportunities presented by a systematic review of innovation scaling research, this article now presents the methodological approach and findings of the systematic literature review.

3. Methodological Approach

This article follows the Preferred Reporting Items for Systematic Reviews and Meta-Analyses (PRISMA) methodological approach [31,32]. This approach presents a systematic way of reviewing literature due to its clear step-by-step instructions that can be transparently communicated and replicated. It is widely accepted as a standard for conducting systematic reviews due to its methodological clarity [31]. Figure 1 presents an overview of our application of the PRISMA approach to the innovation scaling literature. The process began with the definition of eligibility criteria that would determine the inclusion of articles in our systematic review.

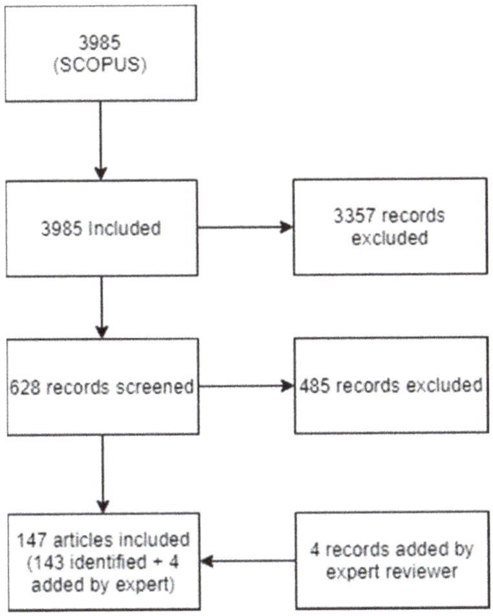

Figure 1. The PRISMA process followed for this literature review. Source: Authors' own elaboration, based on PRISMA guidelines (see [32]).

3.1. Eligibility Criteria

Four criteria were used to determine whether an article was eligible to be included in our systematic review:

Topic: To be eligible for inclusion in the review, an article had to focus on the "scaling" (rather than "diffusion" or "adoption") of a specific innovation or multiple innovations.

Language: To increase the accessibility of the review's findings, only publications written in the English language were eligible.

Publication Year: Only articles published from the year 2000 onwards were included.

Publication Type: To be eligible for inclusion, an article had to be published in a peer-reviewed journal and be a full journal article. This means editors' notes, research commentaries and book reviews were excluded from the review.

Due to the limited quantity of extant research on innovation scaling, and the interdisciplinary nature of this research field, this study included empirical as well as theoretical articles in the review and did not use the field of study as an eligibility criterion.

3.2. Literature Search Strategy

To find the initial sample of articles, a search in the SCOPUS database was conducted, searching articles' titles, abstracts, and keywords using the following query: the word "Scaling", "Scaling Up", "Scaling-Up", "Upscaling", or "Replication" had to be combined with words indicating the theme of "innovation" (see Table 2).

Table 2. Search Query. Source: Authors' own elaboration.

Initial Search Query.
(TITLE-ABS-KEY (("scaling up" OR "scaling-up" OR "upscaling" OR "replication" OR "Scaling") AND innovation) OR TITLE-ABS-KEY ("grow*" innov*"))

Note: the * is a wildcard search function. For example, using the innov* all words starting with innov are captured (i.e., innovation, innovative, innovating).

This initial search yielded 3985 articles. To improve the accuracy of the query, we restricted our review to articles from this sample that had one of the following exact keywords: *"Scaling"*, *"Scaling Up"*, *"Upscaling"*, *"Scaling-up"*, *"Scale-up"*, *or "Innovation"* and applied the inclusion criteria noted above. This resulted in 628 remaining articles.

After this, each article's title, keywords, and abstracts were manually reviewed for relevance. Most articles that were removed at this stage stemmed from the natural sciences. This led to a final sample of 143 eligible articles. An expert reviewed this list of 143 articles and was then asked to suggest any additional articles that may have been falsely excluded. Four additional articles were suggested, thus leading to a final sample of 147 articles. The final list of articles can be found in the Supplementary Materials.

3.3. Coding Process

The 147 articles were coded by multiple coders using a blend of inductive and deductive coding processes. This means that the research began with a predetermined coding scheme informed by the authors' existing understanding of innovation scaling processes (deduction) but used also open codes to capture aspects that were not included in extant typologies and frameworks of innovation scaling (induction). During the process of coding, the coding scheme was regularly and iteratively revised as new open codes emerged. This was done until saturation was reached. In order to ensure intercoder reliability during the initial coding, the same subset of ten articles were coded by all coders at the beginning of the coding process. These results were then compared, and, in case of disagreements, coders discussed and cooperatively refined the deductive codes.

The coding scheme was designed to capture different aspects of the research articles. The codes captured basic information about the articles, such as region of the world, focus, research methodologies. They also captured the exact definition of scaling used (articles that did not provide a clear definition were coded as "no clear definition"), as well as actions and behaviors of strategic actors and contextual factors that were depicted as driving or hindering the scaling process. A factor deemed to be a driver or barrier was coded only once per article.

The coding scheme for drivers and barriers was initially based on the typology of innovation antecedents presented by [57] (p. 155) but further refined and extended through the inductive part of our coding process. Relevant factors not captured by the initial categories were coded with an "open" code, regularly reviewed, discussed, and grouped into new categories. This process ensured that the analysis of the texts was informed by extant knowledge of the innovation process while at the same time being able to surface and systematize relevant aspects typically neglected in innovation scaling typologies and frameworks. Overall, it allowed us to conceptualize eleven modal contextual factors influencing the success of scaling projects as a driver *or* barrier depending on the situation.

For exploring the definitions provided by the articles, the definitions were coded and analyzed based on [30]'s framework differentiating between "scaling out" (the expansion of innovations to a larger group of actors), "scaling up" (the enactment of political and legal change), and "scaling deep" (enacting deep cultural and institutional change). This framework, which the authors became aware in the early phases of the literature review, represents one of the first systematic attempts to present a nuanced and multi-dimensional understanding of scaling. Due to its high relevance and wide breadth, it served as a starting point for the coding of the definitions of scaling.

The coding scheme can be found in the Supplementary Materials.

4. Findings

In this section, we present the results of our systematic review as they correspond to the research questions.

4.1. Landscape of Innovation Scaling Research

The most popular journals included in the analysis were Sustainability, Agricultural Systems, International Journal of Agricultural Sustainability, Journal of Cleaner Production, Environmental Innovation and Societal Transitions, Health Research Policy and Systems, Globalization and Health, and BMC Health Services Research. Twenty-three journals had two articles each included in the review, and eighty journals had only one article each included in the review. The empirical phenomena addressed by the articles, accordingly, were diverse, though they primarily related to health, agriculture, and development. Combined, these areas made-up 78 percent of the included studies. Education, social welfare, business, and public administration were represented as well, but less so.

This result shows that interest in innovation scaling is widespread and multi-disciplinary. However, even though the topic of innovation scaling is cross-cutting, there has been relatively less innovation scaling research in business and public administration journals. This is particularly interesting considering that these journals have a rich history of general innovation scholarship. A possible explanation is that innovation scaling is still a nascent field research. This becomes clear particularly when looking at the yearly number of publications (Figure 2). In this millennium, the first innovation scaling research was published in 2004, but it took until 2010 for a growth trend to emerge. The vast majority of innovation scaling research was published in the last four years.

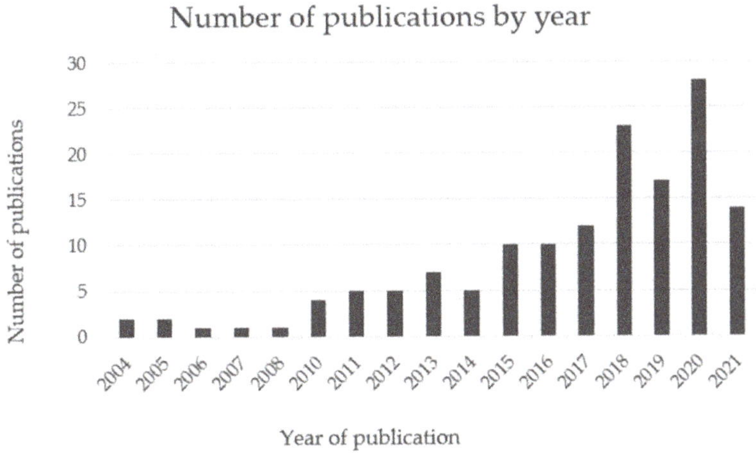

Figure 2. Included publications by year of publishing. Source: Authors' own elaboration.

Methodologically, articles were overwhelmingly qualitative, with a majority of articles utilizing either single or multiple case studies. For the purposes of this review, these case studies provide a large body of empirical evidence that facilitated our identification of the eleven underlying modal contextual factors driving and/or hindering innovation scaling. An overview of both the field of study and the methodology is provided below as Table 3.

Looking to the key sectors involved in innovation scaling, the largest share of articles presented cases in which a mix of different sectors was involved. When only a single sector was involved, the most common one was the public sector, followed by the private sector, and non-profit sector. As there is a large body of literature that focuses on scaling and social enterprises, they were given a separate code. Overall, the majority of articles present research investigating the importance of collaboration and multi-stakeholder involvement for the innovation scaling process.

Geographically, most research focused on Africa, followed by Europe, a Global perspective, Southeast Asia, North America, Asia, Oceania, South America, and then the

Middle East. In some cases, research with multiple case studies had cases from different geographical areas and/or the innovation itself was focused globally, see, e.g., [58].

An overview of the key sectors and the geographical regions is shown below in Table 4.

Table 3. Empirical Areas and Methods in Innovation Scaling Research. Source: Authors' own elaboration.

Empirical Areas	Number
Health	34
Agriculture	32
International Development	49
Environmental Sustainability	19
Education	17
Social Welfare	22
Business Management	14
Public Administration	7
Methodologies	
Qualitative	145
Quantitative	20
Single Case Study	49
Multiple Case Study	46
Survey	7
Systematic Reviews	11
Theory or Frameworks	11

Notes: Total number of articles is 147. Articles were double-coded if they focused on multiple fields of study or combined multiple methods.

Table 4. Key Sectors and Regions Studied Scaling up Research. Source: Authors' own elaboration.

Key Sectors	Number
Cross Sectoral	62
Public Sector	31
Private Sector	13
Social Enterprises	12
Non-profit Sector	10
Regions	**Number**
Africa	46
Europe	33
Global	26
Southeast Asia	23
North America Asia	18 13
Oceania	6
South America	7
Middle East	1

Notes: Number of Articles (N) = 147. For sector analysis, only one code was applied per article. However, 19 articles could not be classified as a distinct sector. These are not included in this table. For the regional analysis, in the event of multiple case studies from different countries, double codes were used. Thus, in the regional analysis 294 regions were coded within the 147 articles.

4.2. *Definitions of Innovation Scaling*

Regarding the conceptualizations and definitions of innovation scaling, somewhat unsurprisingly—and in line with previous research conducted by [57] on innovation—approximately 43 percent of articles did not provide a clear definition of innovation scaling.

Examining the consistency of the definitions, most definitions (61 articles) explicitly or implicitly referred to the concept of scaling out. A much smaller proportion referred to scaling up (19 articles). Scaling deep was referred to the least (10 articles).

Interestingly, 33 articles appeared to use the terminology for their definition incorrectly (e.g., referring to "scaling up", when what they were describing would more accurately have been labelled "scaling out" [30]). This further illustrates the conceptual ambiguity and lack of clarity that has characterized the cross-disciplinary innovation scaling literature. Beyond the themes of expansion (i.e., scaling out), and systems change (i.e., scaling up), we identified two additional themes in the definitions offered. The first is the consistent reference to active and deliberate processes, which supports the above argument that innovation scaling is distinct phenomenon from innovation diffusion. The second is the focus on impact: When characterizing innovation scaling, many articles emphasize that innovation scaling happens for the specific purpose of enhancing the impact of innovations.

4.3. *Drivers and Barriers of Innovation Scaling*

The large number of case studies present in the articles facilitated the empirically grounded identification of eleven modal contextual factors acting either as drivers of barriers to the success of innovation scaling. Four of these are *external factors* (legislative environment; adaptation to local context; collaborations, partnerships, networks; and stakeholder engagement), two are *project factors* (technical factors; and performance management information), and five are *organizational factors* (financial resources; human resources; leadership; risk aversion; time). Tables 5–7 present these factors along with statistics of how common they were across the reviewed articles, and examples for how each modal factor *may either act as a driver or barrier* of innovation scaling.

4.3.1. External Factors

Legislative environment. As a driver, this external factor captures governments who were open to regulatory changes, had clear regulatory frameworks, and/or a coordinated institutional environment. These aspects greatly facilitated the scaling up of innovations. At the same time, this factor could act as a barrier when there was red tape, contradictory policies and regulations, or a lack of a fitting policy arena for the innovation.

Adapting to local context. As a driver, this external factor refers to projects where there was an emphasis on proactively engaging with local conditions, rather than taking a "one size fits all" approach. When projects did not take local conditions into consideration, they ended up meeting resistance in terms of uptake as well as problems in implementing the projects altogether.

Collaboration, Partnerships, Networks. The necessity to collaborate and coordinate was another external factor that seemed to drive scaling up when it was actively done, as it facilitated knowledge sharing, better coordination among actors, and the dissemination of ideas. At the same time, failure to collaborate acted as a barrier to innovation scaling by leading to a lack of coordination.

Stakeholder Engagement. Most articles referring to stakeholder interests as a driver concluded that engaging directly with stakeholders (e.g., ensuring training of users, introducing new products or processes) was a key driver for scaling. On the flipside, failure to do so often led to a lack of buy in, use and/or lack of interest from local communities.

Table 5. Scaling up Drivers and Barriers. Source: Authors' own elaboration, project factors are based on [57].

External Factors		Example	Codes by Theme	Total Codes
Legislative Environment	Drivers	Open to regulatory changes, clear regulatory frameworks, coordinated institutional environments	31	48
	Barriers	Red tape, contradictory policies, innovation lacked regulatory understanding, "limited institutional anchoring"	17	
Adaptation to local context	Drivers	Projects willing and able to adapt to local circumstances both regulatory and cultural/social context (e.g., widening the scope of eligible participants, engaging with local knowledge), used to identify potential bottlenecks/pressure points.	38	51
	Barriers	Not taking the local context (local population constraints, legislation, culture or social norms) into consideration led to problems and resistance, did not account for how to flexibility needed to adapt the programs.	13	
Collaboration, Partnerships, Networks	Drivers	Collaboration across stakeholders lead to clearer visions, continuous dialogues, learning, better coordination, and knowledge sharing, identifying key partners who could help push various project component, formal and informal networks helped to disseminate ideas, and/or organize many stakeholders.	63	80
	Barriers	Lack of coordination body, limited capacity to coordinate, lack of access to networks and partnerships, lack of coordinated expansion	17	
Stakeholder Engagement	Drivers	Training and capacity building of users, using large scale stakeholder engagement throughout the process, and specifically community engagement for introducing new products, services or technologies.	50	64
	Barriers	Lack of training, lack of buy in, no interest, stakeholders saw no benefit, complete lack of local involvement.	14	

4.3.2. Project Factors

Technical factors. Technical factors were a driver when innovations were complementary to existing systems, or adapted to existing, accepted, standards. They were a barrier when the innovation was too advanced for the users or when there was a general lack of technical capacity.

Performance management information. Pilot projects, performance evaluations and the use of performance management were seen as key drivers in scaling. They were used to identify critical weakness of projects, and to develop more reflective learning as projects begin to scale. As barriers, a lack of information and performance information stalled or derailed some projects as the implementers ran the risk duplicating errors or approaches that may have already failed in the past.

Table 6. Scaling up Drivers and Barriers. Source: Authors' own elaboration, project factors are based on DeVries et al. [57].

Project Factors		Example	Codes by Theme	Total Codes
Technical Factors	Drivers	Widely accepted standards, complementary systems, using technology to bring the prices down.	13	30
	Barriers	Too advanced technologies to be used in the local context, lack of technical capacity (i.e., data storage, interoperability), user (un) friendliness.	17	
Performance Management Information	Drivers	Evaluation reports were used as part of a proof of concept (to gain support), identifying critical points, performance information use for feedback, reflective learning, adjusting processes.	29	40
	Barriers	Lacking information about good practice, performance information needed to understand how the scaling up process was happening, how they could be improved, and built upon.	11	

4.3.3. Organizational Factors

Financial Resources. One of the most commonly identified factors influencing the success of innovation scaling were financial resources, both in terms of projects costs of implementation as well as the cost of usage for the end user. Having stable financial resources for the entire duration of the scaling phase, including accurate budgeting forecasting, was a critical driver. As a barrier, a lack of financial recourses became a strain to many innovation scaling projects. These included fragmented budgets or a financial unviability of products.

Human Resources. As a driver, the capacity to hire staff and recruit the "right" people for a particular team was identified to contribute to the success of innovation scaling projects. In some cases, this was very much intertwined with the availability of financial resources. As a barrier, factors such as high turnover and a lack of skills and training resulted in projects failing to launch, or projects failing the scale effectively because the skills were either spread too thin or were not adequate at all.

Leadership. Having political commitment for the projects scaling phase, either at the local or national level, acted as a driver of innovation scaling. Not only did this provide legitimacy for the projects, it also often facilitated other factors such as funding and converging regulatory frameworks. By the same token, a lack of political commitment, especially in more hostile environments, created serious problems related to accessibility and acceptability in the local communities. Internal to the projects themselves, clear leadership skills, including adaptive leadership skills and the ability to negotiate, appeared as critical drivers. As a barrier, a lack of leadership and management skills created leadership vacuums, leading to projects either not realizing their intended outcomes, or not actually being completed in the first place.

Table 7. Scaling up Drivers and Barriers. Source: Authors' own elaboration, project factors are based on [57].

Organizational Factors		Example	Codes by Theme	Total Codes
Financial Resources	Drivers	Secure funding, diverse funding, subsidized means for end users to access the innovation	49	84
	Barriers	Lack of stable funding, fragmented budgets, no long-term funding solutions (e.g., only "startup" funds), financial cost for end user not viable.	35	
Human Resources	Drivers	Capacity building opportunities and training of staff, having the right team that understand and can work within project constraints.	38	69
	Barriers	High turnover, lack of staff, skills, and training.	31	
Leadership	Drivers	Senior leadership/political commitment, general leadership, and management skills, engaging local leaders, adaptive leadership	38	54
	Barriers	Lack of leadership and management skills, management instability	16	
Risk Disposition	Drivers	Reframing risks	2	10
	Barriers	Professional resistance of leaders who did not endorse the projects, risks associated with the uncertainty of outcomes. These uncertainties stemmed from contextual factors (e.g., weather, technological reliability), but also due to lack of consistent performance information.	8	
Time	Drivers	Optimal timing, accurate time forecasting and planning necessary time to see results.	6	9
	Barriers	Not enough time to commit to the project, or the projects themselves did not have enough time to develop and adapt before their funding ended.	3	

Risk Disposition. The concept of risk disposition was primarily identified as a barrier to scaling processes: Professional resistance of leaders, or uncertainly of outcomes, led to hamstringing a project by leading to a lack of action. On the flipside, being able to address these risks directly through increased stakeholder engagement and reframing was also noted as a driver of scaling processes.

Time. The concept of time as a driver refers primarily to the ability to accurately forecast not only each project stage, but also the ability for projects to be implemented in a "timely manner" or at the "optimal time". In the cases where time was perceived as a barrier, this was mostly related to ill- planned projects (e.g., too short project cycles), or projects that were tied to hard-to-achieve external funding deadlines. Some articles also noted as a barrier the underestimation of the time for implementation that may occur within project teams.

To examine the consistency of the modal contextual factors and their manifestation as either driver of barrier, the coded segments were compared across different sectors and regions of the world. Based on this analysis, the top three to four drivers and barriers are presented in Tables 8 and 9. The drivers and barriers are ordered by based on the number of codes assigned to all articles for each respective sector/region. It is important to keep in mind that as this is initial exploratory research, this article cannot make a statement on the statistical significance of these observations, but future research may wish to explore this in greater detail.

The results show that many drivers manifest in a similar fashion across sectors and regions. For example, regardless of sector, collaboration, networking building, and adaptation to local contexts were key drivers. At the same time, some clear differences do emerge in terms of sector. For example, financial resources did not appear to be a salient driver of scaling (3.6%) in the private sector, while it was one of the most common drivers for non-profit (15.6%) and social enterprises (13.8%). Additionally, stakeholder engagement and user engagement appear to be strong drivers of scaling in the private (10.7%) and public (12.5%) sectors, but they did not appear to be as salient in the non-profit (6.3%) and social enterprise literature (5.2%).

In terms of barriers, financial resources, human resources, and a lack of coordination appear to be clear barriers across sectors. However, lack of stakeholder engagement appeared as a more common barrier for the non-profit sector (15%) and social enterprises (10.3%) compared to the public sector (5.4%) or the private sector (0%), while technical barriers appeared to be more problematic in the public sector (10.8%) and projects that are cross sectoral (11%) compared to the non-profit sector (5%) or social enterprises (3.4%).

Table 8. Modal Contextual Factors as Drivers and Barriers by Sector. Source: Authors' own elaboration.

Sector	Most common Drivers	Most Common Barriers
Non-Profit N = 10	Financial Resources Human Resources * Government Regulation * Collaborations and Partnerships *	Financial Resources * Lack of Stakeholder Engagement * Leadership ** Human Resources ** Lack of Collaboration **
Private N = 13	Collaborations and Partnerships Stakeholder Engagement * Leadership *	Lack of Collaboration * Government Regulations * Financial Resources *
Public N = 31	Stakeholder Engagement Financial Resources * Collaborations and Partnerships *	Financial Resources * Leadership * Human Resources ** Technical Resources **
Social Enterprises N = 22	Collaborations and Partnerships * Financial Resources * Human Resources	Financial Resources Lack of Stakeholder Engagement * Lack of Collaboration * Leadership * Human Resources * Government Regulation *
Mixed Collaboration N = 62	Collaborations and Partnerships Financial Resources * Stakeholder Engagement *	Human Resources * Financial Resources * Technical Barriers

Notes: The modal contextual factors are presented based on frequency of codes, where the top 3 most coded modal contextual factors are presented. Some sectors may have more modal contextual factors presented because these factors had an equal number of coded segments. * Refers to modal contextual factors with the same number of codes. ** Refers to modal contextual factors with the same number of codes.

Examining the modal contextual factors as drivers and barriers by the most commonly studied regions, collaboration, partnerships, networks were identified as particularly salient (ranging from 11 to 17%). Stakeholder engagement appeared to be more salient in cases from Asia, Southeast Asia, Africa, and Europe (ranging between 8.5% and 13.5%), compared to North America (4.3%). Adapting to local context was more often referenced as a driver of scaling in North America (15.2%), Europe and Asia (both 8.5 and 8.1%), compared to Southeast Asia (2.6%). A lack of collaboration and coordination appear to be larger barriers in Southeast Asia (18.9%) compared to other regions (ranging from 0 to 10%). Government regulations were particularly salient in North America (16.7%) compared to the other regions (which ranged from 5% to 10%), financial resources were most commonly cited as barrier in Southeast Asia (21.6%), Europe (17.3%) and Africa (12.5%) and human resources were relatively more salient in North America (25%) and Southeast Asia (18.9%), and Africa (15%).

Table 9. Modal Contextual Factors as Drivers and Barriers by Region. Source: Authors' own elaboration.

Region	Most Coded Drivers	Most Coded Barriers
Asia N = 13 *	Collaborations and Partnerships Stakeholder Engagement Leadership * Human Resources *	Lack of Stakeholder Engagement Lack of Performance Information * Failure to Adapt * Time * Lack of Collaboration * Leadership * Human Resources * Technical Barriers * Government Regulations *
Southeast Asia N = 22	Collaborations and Partnerships Financial Resources Stakeholder Engagement *	Financial Resources Lack of Collaboration * Human Resources *
North America N = 18	Adopting to Local Context * Collaborations and Partnerships * Financial Resources	Human Resources Failure to Adapt * Government Regulations * Financial Resources *
Europe N = 33	Collaborations and Partnerships Financial Resources Stakeholder Engagement * Adopting to Local Context * Leadership *	Financial Resources Technical Barriers Government Regulations
Africa N = 46	Collaborations and Partnerships * Financial Resources * Stakeholder Engagement *	Human Resources Financial Resources Leadership * Lack of Collaboration *

Notes: The themes presented based on frequency of codes, where the top 3 most coded theme is presented, some sectors may have more themes presented as these themes had the same number of coded segments. The category 'other' is not included in this table. * Refers to themes with the same number of codes. ** Refers to themes with the same number of codes. *** N, refers to the number of papers coded in this region.

5. Discussion

Drawing on the results of this systematic literature review, it is possible to start to develop a strong foundation for the discussion and theorization of innovation scaling. This foundation will be elaborated in this section and become the basis of four concrete contributions of this research. Three are discussed in this section, and the fourth is presented in the conclusion.

The first contribution is related to the conceptual differentiation between the topics of innovation diffusion and innovation scaling. While scaling and diffusion have often been used as synonyms, in this article it is argued that there are clear theoretical differences between the two terms that must be considered. While the general social science phenomenon of "diffusion" can be understood as the longitudinal spreading of (new) ideas, structures, and practices in social systems, based on extant theory and the systematic literature review, it is possible to differentiate between innovation diffusion and innovation scaling.

Innovation diffusion views spreading as a passive process where innovations permeate naturally throughout a population or field. Innovation scaling views spreading as a proactive and deliberate process driven by individual and organizational actors. As the results presented above show, research into the latter cuts across disciplines, is nascent but rapidly growing, emphasizes the increasing of an innovations social impact, yet currently lacks theoretical clarity. Drawing on extant theory and the systematic review presented here, it is possible for this article to offer a clear baseline conceptualization of the innovation scaling phenomenon:

Innovation scaling refers to the process of proactively, strategically, and/or deliberately spreading an innovation. Individual and organizational actors engage in innovation scaling in order to enhance the beneficial impact of a given innovation by replicating it.

Working from this understanding, it is possible to explore the drivers and barriers of innovation scaling that have been identified captured by extant research. This is the second contribution of this article. Analyzing the reported drivers and barriers, it was possible to identify 11 modal contextual factors influencing the success of innovation scaling projects as *either drivers or barriers*, and which could be further categorized into three primary groups (external factors, project factors, organizational factors). The fact that two of the three most commonly identified drivers appear to be inter-personal in nature highlights that innovation scaling is characterized by a stark inter-dependency of actors. These actors are not only the recipients of the innovations, but also employees, citizens, organizations, regulators, and governments alike. Therefore, innovation scaling must progress utilizing a systems perspective that considers not only the scalability of the innovation itself, but also its impact and relevance to the population in which it is to be scaled.

Based on these insights and extant research, it is possible to derive an initial overarching framework of the innovation scaling process. This is the third contribution of this article. As described above, one of the clearest and most popular existing frameworks for understanding innovation scaling is a typology developed by Moore et al. [30], who distinguish three types of scaling: Scaling up, scaling deep, and scaling wide. Moore et al. [30] argued that specific strategies and processes could be identified for each type of scaling depending on one's goals, but also acknowledged the cross-cutting nature of these strategies. The approach of distinguishing between strategies for different types of scaling can also be seen in previous models and frameworks [19], or in papers that are attempting to develop "success strategies" for different types of scaling [17].

Based on the above analysis of how innovation scaling may be conceptualized and how relates to the modal contextual factors, the authors of this paper would go a step further and argue that scaling deep, scaling wide, and scaling up are indeed important, but are best conceptualized not as distinct *types* of innovation scaling but rather as different *facets* of the same fundamental scaling process.

To successfully scale an innovation may at one stage require the legislative environment to change (typically related to "scaling up"), and at another stage the engagement with a wider range of stakeholders may become critical (typically related to "scaling wide"). These appear to be less distinct types of scaling processes and rather different aspects of the same fundamental scaling process that become salient at different points in time or in different situations. Furthermore, it is also possible that when one engages in scaling wide, it leads to a change in the legislative environment (scaling up) (this holds true for other combinations as well), thus demonstrating the close linked nature of these processes.

It is also important to recognize that the relationship between these facets of innovation scaling and the modal contextual factors is *bilateral*. The presence of modal contextual factors affects the outcomes (ultimate success or failure) of an innovation scaling project and leads to the enactment of specific facets of the innovation scaling process (i.e., wide, up, out). At the same time, the enactment of these facets may in turn change the modal contextual factors, such that they may flip from being a barrier (e.g., unsupportive legislative environment) to a driver (e.g., supportive legislative environment). Within innovation scaling processes, modal contextual factors and facets of scaling therefore are likely to exhibit a co-dependent dynamic in which both change across time.

These insights facilitated the development of an initial overarching framework of innovation scaling processes, which is presented in Figure 3. The framework proposes three essential components of a scaling process: modal contextual factors, facets of innovation scaling, and eventual outcomes. Each component is co-dependent on the others, both *influencing* them and being *influenced by* them.

Explicating Figure 3, four initial propositions delineating the innovation scaling process can be proposed:

Proposition 1—Innovation scaling is a multi-faceted process. Depending on the specific situation, the three facets of scaling out, scaling up and scaling deep may become enacted to differing degrees. It is possible that they become enacted to equal degrees, or that one or two facets overshadow the others.

Proposition 2—The multi-faceted innovation scaling process both affects, and is affected by, eleven modal contextual factors: financial resources, human resources, leadership, risk disposition, time, technical factors, performance management information, legislative environment, adaptation to local context, collaboration, and stakeholder engagement. Depending on the specific situation, these modal contextual factors may either act as drivers or barriers of the scaling process.

Proposition 3—The modal contextual factors and innovation scaling facets exist in a dynamic interplay. A specific arrangement of modal contextual factors may prompt the enactment of a certain arrangement of innovation scaling facets, which in turn may shift the modal contextual factors (e.g., changing one from acting as a barrier to acting as a driver).

Proposition 4—The dynamic interplay between modal contextual factors and innovation scaling facets determines the ultimate outcome of the innovation scaling process, i.e., whether it is a failure or success.

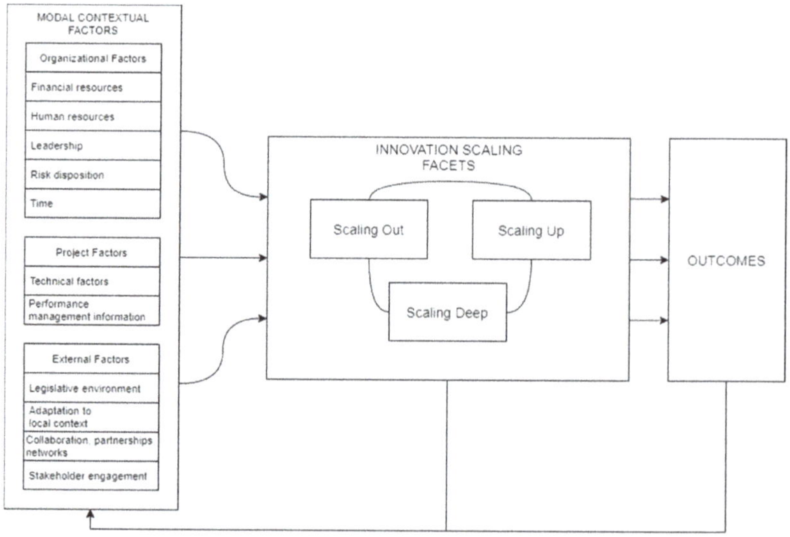

Figure 3. Scaling Innovations Theoretical Framework. Source: own elaboration based on own analysis.

As this framework is derived from a cross-disciplinary literature review, the propositions are likely to be applicable across many domains. At the same time, further research is needed to explore the framework in more depth, after which our initial propositions may need to be revisited. The next section presents promising avenues for such research and the limitations of the study.

6. Conclusions

The concept of innovation scaling is of high interest to a wide number of disciplines. This article sought to address the conceptual ambiguity that exists in the field using a cross disciplinary systematic literature review. As a result of this research, four critical contributions are made to the scaling innovations literature. The first contribution outlined presents the conceptual differences between the concepts of innovation diffusion and innovation scaling. The second contribution identified eleven modal contextual factors that influence the outcomes of innovation scaling. These factors were consistent across many innovation projects, suggesting an element of universality across research domains and contexts. The third contribution presented a framework of innovation scaling accompanied by four theoretical propositions that emerged therefrom. The framework developed consists of three main components: the modal contextual factors, the facets of innovation scaling, and the outcomes. The co-dependency of these three components highlights the complexity of the scaling process. This is by no means a finished framework—but the beginnings of a larger research agenda in understand how innovations scale of which we invite scholars to participate in.

In making these contributions, this systematic literature review facilitates a burgeoning agenda for future research—this is the fourth core contribution of this article.

The theoretically and empirically grounded framework of innovation scaling points to the multifaceted relationship between action and modal contextual factors in the scaling process. Where much extant theory of innovation scaling had treated contextual factors primarily as *antecedents* of the scaling process influencing outcomes (see, e.g., de Vries [57]), the analysis presented above suggests that contextual factors exist in a dynamic relationship with innovation scaling efforts. Overall, the findings and framework call to attention Gidden's [59]'s seminal structuration theory explicating the dynamic relationship between agency (the actors doing the innovation scaling as a strategic project) and structure (contextual factors). As such, a first line of future research is related to the theoretical understanding that contextual factors such as the legislative environment are not *antecedents* but *changeable* modal factors. As such, new questions come into view. For example, how do innovations that are scaled in different contexts, change modal contextual factors from hindering their deliberate spreading to supporting it? How can strategic actors engaged in innovation scaling projects make such changes to the modal contextual factors? Which facets of innovation scaling are enacted as part of these processes? Answering research questions like these will require in-depth qualitative studies utilizing process analysis methods [60] to explicate how the scaling unfolded.

A second promising avenue for research regards the variability of innovations during the scaling process. Extant theoretical frameworks of innovation scaling treat the innovations themselves as unchanging. The assumption is that, as strategic actors spread innovations into new contexts, the innovations maintain their original form and function. However, looking to the results of this literature review—specifically the prevalence of the external factor "adaptation to local context"—this assumption appears flawed. Indeed, within the literature on innovation diffusion (which, as discussed above, also investigates the spreading of innovations) it has become well established that the *adoption* of innovations by new actors frequently also requires the *adaptation* of those innovations [25,61–64]. Innovation diffusion research, due to its population-level perspective, has focused on how *adopters* may adapt innovations when information cascades and/or legitimacy pressures bring those innovations to their attention. Mirroring these investigations, further empirical and theoretical research is needed to systematically investigate and explicate how the actors attempting to scale innovations may deliberately adapt them to make them more palatable for the targeted audience—or in contrast, study the resistance to adaption of an innovation. As this article demonstrates, some research is already paying explicit attention to this factor. Accordingly, this article has sought to acknowledge the potentially central role of adaption in the proposed framework of innovation scaling. But further research is needed to develop a more fined grained understanding of adaption within innovation scaling by systematically importing more theoretical insights from innovation diffusion research on this topic and engaging in empirical research that probes the process of deliberate adaptation in-depth.

Despite the benefits and rigor of the PRISMA approach, several limitations should be acknowledged. First, this research focused primarily on articles that referred directly to the concept of "scaling". This means that it is possible that articles outlining innovations that scaled but were not described as such were not captured by this review. Related to this, in the process of deciding upon the filters for this systemic review, and to reduce the sample to a manageable number, keywords were used. As keywords are not standardized, some relevant articles may not have been captured. While a thorough review of keywords from well-known studies examining innovation scaling was conducted, it is still possible that some articles were missed. This limitation was somewhat minimized by the manual review of the final list of included articles by an expert in the field. Third, there is always a risk of coding error. This was minimized though various coding strategies developed between the coders. In addition, due to the transparency of the PRISMA process as well as making the coding scheme available in the appendix, any reader can assess the internal validity of the results presented. Fourth, as with all systematic reviews there is a threat of survivorship bias. Articles included in this study almost all focus on successful innovations that have been scaled up or are in the process of scaling up. Finally, by taking a multi-disciplinary approach, we run the risk of oversimplifying the scaling processes across different domains. While the review did show many similarities between different disciplines and areas of innovation, any form of theoretical framework must always be critically reviewed and applied in a case-by-case fashion.

Despite these limitations, several implications emerge from this research. For anyone who is interested in the topic of scaling, this paper provides the most comprehensive overview currently available in the academic literature. Thus, for future research focusing on scaling or innovation scaling, this paper will be a key starting point in guiding this research. Second, this paper develops a strong foundation for future cross disciplinary research on the topic of innovation scaling. This is accomplished by setting a common vocabulary and establishing clear definitions of scaling and diffusion. In doing so, this paper offers a strong initial attempt at both enabling future researchers to establish consilience between one another and encouraging cross-disciplinary learning. Third, by identifying commonly occurring contextual modal factors that affect the scaling process, it is possible to begin to develop an understanding of how each of these is likely to *affect* the scaling process itself. Even more interesting, however, is that the identification of these factors should enable the development of new strategies and approaches to innovation scaling. Finally, the implications of our theoretical model would suggest that scaling needs to be understood as a continually changing process, both in terms of the interdependency regarding the type of innovation scaling (up, out, or deep), but also the contextual factors that influence this possibility in the first place.

Beyond the theoretical and academic contributions, practitioners, too may find this article relevant through its clarification of scaling and the identification of eleven driver and barriers of scaling (identified as modal contextual factors) that influence the success of scaling projects. The conceptualization of scaling as a multi-faceted process also offers practitioners and scholars alike a new way to look at and understand how to scale a given innovation. Situating this within the developed framework offers a new tool to understand either why a given innovation did scale, or why the outcome of scaling did not go as planned. Overall, this article provides a strong foundation for future theoretical exploration and development of the concept of innovation scaling across disciplines.

Supplementary Materials: The following are available online at https://www.mdpi.com/article/10.3390/su132413528/s1, File S1: Coding Schema, File S2: Articles included within the review.

Author Contributions: Conceptualization, J.B., K.M. and G.H.; writing—original draft preparation, J.B., K.M., M.K., G.H.; writing—review and editing, J.B., K.M., M.K., G.H. All authors have read and agreed to the published version of the manuscript.

Funding: This research received no external funding.

Institutional Review Board Statement: Not applicable.

Informed Consent Statement: Not applicable.

Data Availability Statement: Data has been made available as Supplementary Material.

Acknowledgments: The authors would like to think Sona Jose and Ann-Marie Rittler for their valuable assistance with the research in this article. Thanks are also extended to those who provided feedback on an earlier draft of this article presented at the EGPA 2021 conference and at the Hertie School Centre for Digital Governance Research Colloquium as well as McKinsey Germany for their engagement in a joint knowledge project on scaling digital innovation in the public sector.

Conflicts of Interest: The authors declare no conflict of interest.

References

1. Downs, G.W.; Mohr, L.B. Conceptual Issues in the Study of Innovation. *Adm. Sci. Q.* **1976**, *21*, 700. [CrossRef]
2. Rogers, E.M. *Diffusion of Innovations*, 4th ed.; The Free Press: New York, NY, USA, 1995; ISBN 978-0029266717.
3. Kelly, P.; Kranzberg, M. *Technological Innovation: A Critical Review of Current Knowledge*, 1st ed.; San Francisco Press: San Francisco, CA, USA, 1978; ISBN 0911302344.
4. Van de Ven, A.H. Central Problems in the Management of Innovation. *Manag. Sci.* **1986**, *32*, 590–607. [CrossRef]
5. Thompson, V.A. Bureaucracy and Innovation. *Adm. Sci. Q.* **1965**, *10*, 1–20. [CrossRef]
6. Schumpeter, J.A. *Capitalism, Socialism, and Democracy*, 2nd ed.; Harper and Brothers: New York, NY, USA, 1942.
7. Mohr, L.B. Determinants of Innovation in Organizations. *Am. Polit. Sci. Rev.* **1969**, *63*, 111–126. [CrossRef]
8. Blau, J.R.; McKinley, W. Ideas, Complexity, and Innovation. *Adm. Sci. Q.* **1979**, *24*, 200. [CrossRef]
9. Zaltman, G.; Duncan, R.; Hobek, J. *Innovations and Organizations*; John Wiley & Sons: New York, NY, USA, 1973.
10. Hage, J.T. Organizational Innovation and Organizational Change. *Annu. Rev. Sociol.* **1999**, *25*, 597–622. [CrossRef]
11. Gupta, A.K.; Tesluk, P.E.; Taylor, M.S. Innovation At and Across Multiple Levels of Analysis. *Organ. Sci.* **2007**, *18*, 885–897. [CrossRef]
12. Fliegel, F.C.; Kivlin, J.E. Attributes of Innovations as Factors in Diffusion. *Am. J. Sociol.* **1966**, *72*. [CrossRef] [PubMed]
13. Meyer, A.D.; Goes, J.B. Organizational Assimilation of Innovations: A Multilevel Contextual Analysis. *Acad. Manag. J.* **1988**, *31*, 897–923. [CrossRef]
14. Damanpour, F. Organizational Innovation: A Meta-Analysis of Effects of Determinants and Moderators. *Acad. Manag. J.* **1991**, *34*, 555–590. [CrossRef]
15. Rogers, E.M. *Diffusion of Innovations*, 3rd ed.; Free Press: New York, NY, USA, 1983; ISBN 9780029266502.
16. Kaan Kuguoglu, B.; van der Voort, H.; Janssen, M.; Gasco-Hernandez, M.; Chen, T.; Comi, A. The Giant Leap for Smart Cities: Scaling Up Smart City Artificial Intelligence of Things (AIoT) Initiatives. *Sustainability* **2021**, *13*, 12295. [CrossRef]
17. Jha, S.K.; Bhawe, N.; Satish, P.; Schouten, J.; Press, M.; Dufault, B.; Rosen, M.A. Scaling Social Enterprises through Product Diversification. *Sustainability* **2021**, *13*, 11660. [CrossRef]
18. Pitt, H.; Jones, M. Scaling up and out as a Pathway for Food System Transitions. *Sustainability* **2016**, *8*, 1025. [CrossRef]
19. Frake, A.N.; Messina, J.P. Toward a Common Ontology of Scaling Up in Development. *Sustainability* **2018**, *10*, 835. [CrossRef]
20. Ilieva, R.T.; Hernandez, A. Scaling-Up Sustainable Development Initiatives: A Comparative Case Study of Agri-Food System Innovations in Brazil, New York, and Senegal. *Sustainability* **2018**, *10*, 4057. [CrossRef]
21. Abrahamson, E. Managerial Fads and Fashions: The Diffusion and Rejection of Innovations. *Acad. Manag. Rev.* **1991**, *16*, 586–612. [CrossRef]
22. Wolfe, R.A. Organizational Innovation: Review, Critique, and Suggested Research Directions. *J. Manag. Stud.* **1994**, *31*, 405–431. [CrossRef]
23. Strang, D.; Macy, M.W. In Search of Excellence: Fads, Success Stories, and Adaptive Emulation. *Am. J. Sociol.* **2001**, *107*, 147–182. [CrossRef]
24. Davis, G.F.; Marquis, C. Prospects for Organization Theory in the Early Twenty-First Century: Institutional Fields and Mechanisms. *Organ. Sci.* **2005**, *16*, 332–343. [CrossRef]
25. Ansari, S.M.; Fiss, P.C.; Zajac, E.J. Made to Fit: How Practices Vary As They Diffuse. *Acad. Manag. Rev.* **2010**, *35*, 67–92. [CrossRef]
26. Willis, C.D.; Riley, B.L.; Stockton, L.; Abramowicz, A.; Zummach, D.; Wong, G.; Robinson, K.L.; Best, A. Scaling up Complex Interventions: Insights from a Realist Synthesis. *Heal. Res. Policy Syst.* **2016**, *14*, 88. [CrossRef]
27. Brown, L. A Lasting Legacy? Sustaining Innovation in a Social Work Context. *Br. J. Soc. Work* **2015**, *45*, 138–152. [CrossRef]
28. Horn, P. Enabling Participatory Planning to Be Scaled in Exclusionary Urban Political Environments: Lessons from the Mukuru Special Planning Area in Nairobi. *Environ. Urban.* **2021**, *33*, 519–538. [CrossRef]

29. Nguyen, D.T.K.; McLaren, L.; Oelke, N.D.; McIntyre, L. Developing a Framework to Inform Scale-up Success for Population Health Interventions: A Critical Interpretive Synthesis of the Literature. *Glob. Health Res. Policy* **2020**, *5*, 1–11. [CrossRef]
30. Moore, M.-L.; Riddell, D.; Vocisano, D. Scaling out, Scaling up, Scaling Deep: Strategies of Non-Profits in Advancing Systemic Social Innovation. *J. Corp. Citizsh.* **2015**, *58*, 67–85. [CrossRef]
31. Liberati, A.; Altman, D.G.; Tetzlaff, J.; Mulrow, C.; Gøtzsche, P.C.; Ioannidis, J.P.A.; Clarke, M.; Devereaux, P.J.; Kleijnen, J.; Moher, D. The PRISMA Statement for Reporting Systematic Reviews and Meta-Analyses of Studies That Evaluate Health Care Interventions: Explanation and Elaboration. *J. Clin. Epidemiol.* **2009**, *62*, e1–e34. [CrossRef] [PubMed]
32. Moher, D.; Liberati, A.; Tetzlaff, J.; Altman, D.G. The PRISMA Group Preferred Reporting Items for Systematic Reviews and Meta-Analyses: The PRISMA Statement. *PLoS Med.* **2009**, *6*, e1000097. [CrossRef] [PubMed]
33. Strang, D.; Soule, S.A. Diffusion in Organizations and Social Movements: From Hybrid Corn to Poison Pills. *Annu. Rev. Sociol.* **1998**, *24*, 265–290. [CrossRef]
34. Katz, E. Theorizing Diffusion: Tarde and Sorokin Revisited. *Ann. Am. Acad. Pol. Soc. Sci.* **1999**, *566*, 144–155. [CrossRef]
35. Ryan, B.; Gross, N. The Diffusion of Hybrid Seed Corn in Two Iowa Communities. *Rural Sociol.* **1943**, *8*, 15–24.
36. Hagerstrand, T. Aspects of the Spatial Structure of Social Communication and the Diffusion of Information. *Pap. Reg. Sci.* **1966**, *16*, 27–42. [CrossRef]
37. Katz, M.L.; Shapiro, C. R and D Rivalry with Licensing or Imitation. *Am. Econ. Rev.* **1987**, *77*, 402–420.
38. Mansfield, E. Technical Change and the Rate of Imitation. *Econometrica* **1961**, *29*, 741. [CrossRef]
39. Chandler, A.D. *Strategy and Structure Chapters in the History of the American Industrial Enterprise*, 1st ed.; MIT Press: Cambridge, MA, USA, 1962; ISBN 9780262030045.
40. Teece, D.J. The Diffusion of an Administrative Innovation. *Manag. Sci.* **1980**, *26*, 464–470. [CrossRef]
41. Williamson, O.E. Transaction-Cost Economics: The Governance of Contractual Relations. *J. Law Econ.* **1979**, *22*, 233–261. [CrossRef]
42. Banerjee, A.V. A Simple Model of Herd Behavior. *Q. J. Econ.* **1992**, *107*, 797–817. [CrossRef]
43. Bikhchandani, S.; Hirshleifer, D.; Welch, I. A Theory of Fads, Fashion, Custom, and Cultural Change as Informational Cascades. *J. Polit. Econ.* **1992**, *100*, 992–1026. [CrossRef]
44. Bikhchandani, S.; Hirshleifer, D.; Welch, I. Learning from the Behavior of Others: Conformity, Fads, and Informational Cascades. *J. Econ. Perspect.* **1998**, *12*, 151–170. [CrossRef]
45. Rao, H.; Greve, H.R.; Davis, G.F. Fool's Gold: Social Proof in the Initiation and Abandonment of Coverage by Wall Street Analysts. *Adm. Sci. Q.* **2001**, *46*, 502–526. [CrossRef]
46. DiMaggio, P.J.; Powell, W.W. The Iron Cage Revisited: Institutional Isomorphism and Collective Rationality in Organizational Fields. *Am. Sociol. Rev.* **1983**, *48*, 147. [CrossRef]
47. Scott, W.R. *Institutions and Organizations Ideas, Interests, and Identities*, 4th ed.; SAGE: Thousand Oaks, CA, USA, 2013; ISBN 9781483321912.
48. Sturdy, A. The Adoption of Management Ideas and Practices: Theoretical Perspectives and Possibilities. *Manag. Learn.* **2004**, *35*, 155–179. [CrossRef]
49. Tolbert, P.S.; Zucker, L.G. The Institutionalization of Institutional Theory. In *The Handbook of Organization Studies*; Clegg, S.R., Hardy, C., Nord, W.R., Eds.; SAGE: Thousand Oaks, CA, USA, 1996; pp. 170–185. ISBN 1446237192.
50. Tolbert, P.S.; Zucker, L.G. Institutional Sources of Change in the Formal Structure of Organizations: The Diffusion of Civil Service Reform, 1880–1935. *Adm. Sci. Q.* **1983**, *28*, 22. [CrossRef]
51. Brown, L. *Innovation Diffusion: A New Perspective*; Methuen: London, UK, 1981; ISBN 9780416742701.
52. Mahajan, V.; Peterson, R.A. *Models for Innovation Diffusion*; SAGE: Thousand Oaks, CA, USA, 1985; ISBN 0803921365.
53. Abrahamson, E. Management Fashion. *Acad. Manag. Rev.* **1996**, *21*, 254–285. [CrossRef]
54. Cool, K.O.; Dierickx, I.; Szulanski, G. Diffusion of Innovations within Organizations: Electronic Switching in the Bell System, 1971–1982. *Organ. Sci.* **1997**, *8*, 543–559. [CrossRef]
55. O'Mahoney, J. The Diffusion of Management Innovations: The Possibilities and Limitations of Memetics. *J. Manag. Stud.* **2007**, *44*, 1324–1348. [CrossRef]
56. Stinchcombe, A.L. 14. New Sociological Microfoundations for Organizational Theory: A Postscript. In *Social Structure and Organizations Revisited*; Lounsbury, M., Ventresca, M.J., Eds.; Research in the Sociology of Organizations; Emerald Group Publishing Limited: Bingley, UK, 2002; Volume 19, pp. 415–433. ISBN 978-0-76230-872-9/978-1-84950-149-1.
57. De Vries, H.; Bekkers, V.; Tummers, L. Innovation in the Public Sector: A Systematic Review and Future Research Agenda. *Public Adm.* **2016**, *94*, 146–166. [CrossRef]
58. Braund, P.; Schwittay, A. Scaling Inclusive Digital Innovation Successfully: The Case of Crowdfunding Social Enterprises. *Innov. Dev.* **2015**, *6*, 15–29. [CrossRef]
59. Giddens, A. *The Constitution of Society: Outline of the Theory of Structuration*; University of California Press: Berkley, CA, USA, 1984; ISBN 9780520052925.
60. Langley, A. Strategies for Theorizing from Process Data. *Acad. Manag. Rev.* **1999**, *24*, 691–710. [CrossRef]
61. Strang, D.; Kim, Y.-M. The Diffusion and Domestication of Managerial Innovations: The Spread of Scientific Management, Quality Circles, and TQM between the United States and Japan. In *The Oxford Handbook of Work and Organization*; Ackroyd, S., Batt, R., Thompson, P., Tolbert, P., Eds.; Oxford University Press: Oxford, UK, 2004.

62. Robertson, M.; Swan, J.; Newell, S. The Role of Networks in the Diffusion of Technological Innovation. *J. Manag. Stud.* **1996**, *33*, 333–359. [CrossRef]
63. Zbaracki, M.J. The Rhetoric and Reality of Total Quality Management. *Adm. Sci. Q.* **1998**, *43*, 602–636. [CrossRef]
64. Frenkel, M. The Politics of Translation: How State-Level Political Relations Affect the Cross-National Travel of Management Ideas. *Organization* **2005**, *12*, 275–301. [CrossRef]

sustainability

MDPI

Review

A Review on Energy and Renewable Energy Policies in Iran

Saeed Solaymani

Department of Economics, Faculty of Administration and Economics, Arak University, Arak 38156879, Iran; s-solaymani@araku.ac.ir or saeedsolaymani@gmail.com; Tel.: +98-8632621000

Abstract: Iran, endowed with abundant renewable and non-renewable energy resources, particularly non-renewable resources, faces challenges such as air pollution, climate change and energy security. As a leading exporter and consumer of fossil fuels, it is also attempting to use renewable energy as part of its energy mix toward energy security and sustainability. Due to its favorable geographic characteristics, Iran has diverse and accessible renewable sources, which provide appropriate substitutes to reduce dependence on fossil fuels. Therefore, this study aims to examine trends in energy demand, policies and development of renewable energies and the causal relationship between renewable and non-renewable energies and economic growth using two methodologies. This study first reviews the current state of energy and energy policies and then employs Granger causality analysis to test the relationships between the variables considered. Results showed that renewable energy technologies currently do not have a significant and adequate role in the energy supply of Iran. To encourage the use of renewable energy, especially in electricity production, fuel diversification policies and development program goals were introduced in the late 2000s and early 2010s. Diversifying energy resources is a key pillar of Iran's new plan. In addition to solar and hydropower, biomass from the municipal waste from large cities and other agricultural products, including fruits, can be used to generate energy and renewable sources. While present policies indicate the incorporation of sustainable energy sources, further efforts are needed to offset the use of fossil fuels. Moreover, the study predicts that with the production capacity of agricultural products in 2018, approximately 4.8 billion liters of bioethanol can be obtained from crop residues and about 526 thousand tons of biodiesel from oilseeds annually. Granger's causality analysis also shows that there is a unidirectional causal relationship between economic growth to renewable and non-renewable energy use. Labor force and gross fixed capital formation cause renewable energy consumption, and nonrenewable energy consumption causes renewable energy consumption.

Keywords: energy demand; renewable energy; biomass; energy policies

Citation: Solaymani, S. A Review on Energy and Renewable Energy Policies in Iran. *Sustainability* **2021**, *13*, 7328. https://doi.org/10.3390/su13137328

Academic Editors: E. Richard Zada and Andrea Gatto

Received: 18 May 2021
Accepted: 24 June 2021
Published: 30 June 2021

Publisher's Note: MDPI stays neutral with regard to jurisdictional claims in published maps and institutional affiliations.

1. Introduction

The close and high relationship between production growth and energy consumption growth in the economy refers to the dependence of the economy on energy [1]. Thus, the economy is not only sensitive to energy supply and price shocks, but any initiative to conserve energy can have an impact on the performance of the economy.

Iran was recognized as a country with significant oil reservoirs in the world when it discovered the first oil well in the Middle East in 1908 at Masjid Solaiman. With the subsequent exploration and extraction of oil reserves, the Iranian economy, like the world economy, has increasingly become dependent on crude oil consumption and export revenue for industrial growth. However, the dependency on oil has declined with time in Iran as well as other countries due to the use of other energy sources, such as natural gas and renewable energies. For example, the share of oil consumption in total final energy demand has declined significantly from 91% in 1980 to 43% in 2018 (Figure 1). This led to increased use of natural gas in the country, from 7% in 1980 to 56% in 2018. While oil consumption has declined and been substituted by natural gas, the share of oil in total consumption is still high and the use of renewable energy resources is low due to low production. The share of renewable energy consumption in total final energy demand declined from 1.5% in 1980 to 0.58% in 2018 (Figure 1). This shows that the country has not paid more attention to the generation of renewable energy sources through investment and providing some stimulations for private investors. One reason is the abundance of oil and gas resources and their low prices compared to their international prices. However, the country needs to achieve the diversity of energy resources and energy security, which are essential steps toward sustainable economic growth and social development. While Iran is endowed with abundant natural resources, it faces insecurity about the challenges of reducing energy resources, protecting the environment and sustainable development. Considering the diversity of climates in different regions and the cultivation of a variety of crops and food products, the country has a substantial potential for diversification of the energy mix, particularly the renewable energy base. In addition to the forests of the north and west of the country, which represent about 7% of the country's area, Iran also has many other products and agricultural wastes that can use to produce biofuel. Within the existing framework of government organizations and institutions, which supervises planting, farming, harvesting and, in some cases, distribution and research and development, efforts to strengthen cooperation and coordination among them will facilitate the transmission and implementation of sustainable development plans.

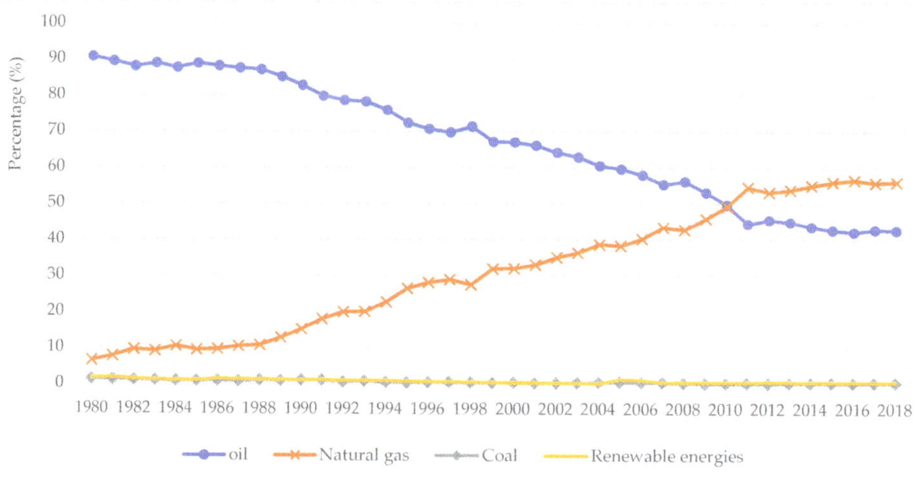

Figure 1. The share of consumption of the various energy sources in total final energy demand in Iran. Source: [2].

To predict the energy mix, which includes all kinds of fossil fuel resources and available renewable options, it is necessary to take into account current and future energy demand and key figures relating to current energy resources. This study aims to review the status and current trends in potential resources and to investigate main policies in Iran to suggest some solutions to help the government to achieve its sustainable energy security goals. These suggestions can simplify the application of sustainable inventions, and formally examines how biomass can contribute to the structure of the energy mix in Iran.

The contribution of this study is that it investigates all kinds of energy sources in Iran which compare to other studies that focus on renewable energy sources or specific energy. Along with reviewing current trends and policies, it estimates the relationship between energy demand and economic growth using an econometric model. Therefore, it makes a significant contribution to the literature as an in-depth study of all kinds of renewable energy sources and their policies since the late 1940s.

The rest of the study is organized as the next section introduces the research methodology. Section 3 provides an overview of Iran's current energy demand and structure. Section 4 discusses Iran's current energy supply situation. Section 5 looks at the current and future state of renewable energy sources and their policies, and Section 6 summarizes the analyses.

2. Research Method

This study uses two methodologies, a review of the current situation of energy and energy policies in Iran and an econometric method. The review method describes the facts and the general energy trends, particularly renewable energy, using the library research method based on Iran's energy statistics. The econometric method uses a Granger causality technique to analyze the relationship between pairs of variables in the model. The data used in this study were collected from a variety of sources, such as the energy balances of the Ministry of Energy, documents and programs of the Plan and Budget Organization and the Research Center of Iran's Parliament.

Econometric Model

In the literature, many studies have investigated the relationship between energy and economic growth. Some studies have focused on the relationship between energy demand and industry growth in Iran, such as Mozayani et al. [3] and Charatin and Goltbar [4]. They found a positive relationship between energy use and the growth in value-added of the industrial and transport sectors.

According to Abdoli and Hammami [5], a 1% increase in energy consumption will increase Iran's economic growth by 0.548%. Furthermore, the simulated projections made by Mirzaei and Bakri [6] show that Iran's total energy consumption will reach 2150 million barrels by 2025; with an annual growth rate of 4.3%, that quantity was equal to 1910 million barrels in 2010. Ghaseminejad et al. [7] also predicted that a 1% increase in energy consumption would lead to a 1.29% increase in long-run agricultural growth. GDP growth also means more purchasing power and higher living standards, and ultimately poverty reduction [8]. This condition leads to increased car ownership and demand for gasoline, natural gas and electricity.

Furthermore, many studies applied the Granger causality technique to investigate the relationship between energy and other variables. For example, Tugcu and Topcu [9] introduced GDP as a function of labor, fiscal capital, energy and other variables and estimated this function by applying some techniques such as Granger causality. Rahman and Velayutham [10] also used this function in causality analysis but separated energy into renewable and nonrenewable sources. Sunde [11] also performed a Granger causality analysis between energy consumption and economic growth for the Southern Africa Development Community. Other studies that have used economic growth as a function of primary inputs and energy in the causality context are Aydin [12], Rahman and Abul Kashem [13] and Akadiri et al. [14].

Therefore, based on the above literature, a Cobb–Douglass production function is considered as a function of primary inputs, labor and capital, and energy, which can be separated into nonrenewable and renewable energy sources.

$$\text{GDP} = f(\text{LF,K,NREN, REN}) \tag{1}$$

where GDP is gross domestic product (IRR billion based on 2010 prices), LF denotes labor force (1000 people), K shows capital which in this study is gross fixed capital formation (GFKF) (USD million based on 2010 prices), NREN is the non-renewable energy (ktoe), and REN denotes renewable energy (ktoe). The natural logarithmic form of Equation (1) is as follows:

$$LGDP_t = (LLF_t,\ LGFKF_t, LNREN_t, LREN_t) \tag{2}$$

All possible pairs of variables in Equation (2) will be estimated based on the following equations, which have the Vector Autoregression (VAR) form.

$$GDP_t = \sum_{i=1}^{p} \propto_i LLF_{t-i} + \sum_{j=1}^{q} \beta_j GDP_{t-j} + \varepsilon_{GDP_t} \quad (3)$$

$$LLF_t = \sum_{i=1}^{p} \gamma_i LLF_{t-i} + \sum_{j=1}^{q} \delta_j GDP_{t-j} + \varepsilon_{LF_t} \quad (4)$$

where p and q are the maximum number of lags that each variable has taken. The reported F statistics are the Wald statistics for the joint hypothesis, i.e., $H_0 : \sum_{i=1}^{p} \propto_i = 0$. Causality methods are very important in energy economics in defining whether energy conservation policies have a negative impact on economic growth or not [15]. Engel and Granger [16] argued that if two variables are cointegrated, there will always be a vector error correction pattern between them. As a result, this type of model can be used to investigate the Granger causality relationship between variables.

3. Current Energy Supply

3.1. General Overview

Many studies have investigated different aspects of the energy system in Iran [17–19]. In terms of energy supply, Figure 2 represents the main component of the primary energy supply of various types of energy used in Iran. Natural gas has remained the largest contributor to Iran's total final energy consumption since 2003, followed by crude oil. Over the past decade, the share of natural gas has increased by 61.9% from 110.8 Mtoe (5 × 109 GJ) in 2008 to 179.3 Mtoe (3.5 × 109 GJ) in 2017. The share of oil and petroleum products decreased by 12.3% from 87.8 Mtoe (3.9 × 109 GJ) in 2008 to 77 Mtoe (3.5 × 109 GJ) in 2017. Interestingly, as a result of the abundance of fossil fuels in the country, the share of renewable energy sources in the same decade increased by 131.9% from 1.17 Mtoe (0.01 × 109 GJ) in 2008 to 2.7 Mtoe (0.12 × 109 GJ) in 2017. This may occur because of rising world crude oil prices, high concerns about the adverse environmental effects of fossil fuels or more use of these resources in remote areas of Iran.

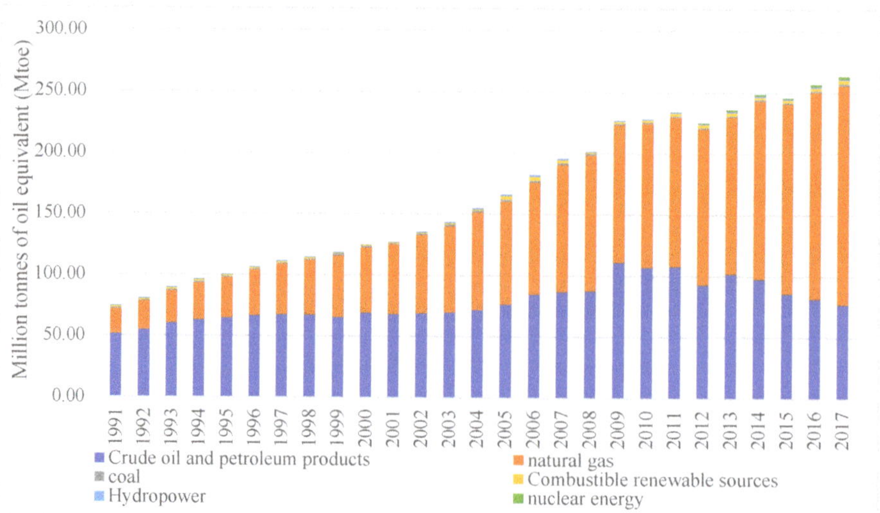

Figure 2. Primary energy supply based on different energy sources in Iran. Source: Ministry of Energy [2].

In spite of various oil and energy laws and regulations, there is still no integrated, comprehensive, purposeful, long-term and coordinated study for the entire country. In this regard, Ghasemi et al. [20] stress the design of a new energy development mechanism, reducing political and economic risks to oil companies, establishing a system for monitoring and evaluating national energy information and the legal mechanism for negotiations, and finally, a specific fiduciary with the ability to pool resources to implement the strategies.

Despite several changes and revisions to energy policies in various documents, there has been a steady trend in economic growth, albeit with some challenges. Real gross domestic product (GDP) (at 2004 constant prices) grew on average by 4.6% from 1991 to 2001 and grew by 6.9% during the 2000s, even though remarkable negative growth rates can be observed in some years of these two decades [21]. In subsequent years, with the exception of negative growth rates in 2012, 2013 and 2015, real GDP growth (based on 2011 constant prices) fluctuated between 4% (in 2011) and 6.6% (in 2016) and in 2017 reached 3.5% [21]. As Figure 3 shows, an increasing trend can be observed in the energy demand of four major energy consumers (residential, public and commercial, industry, transport and agriculture). Total energy consumption in 2017 was about 185.98 million tonnes of oil equivalent (Mtoe; one million tonnes of oil equivalent (Mtoe) = 44.76 GJ (GJ) [2].) (8.3×10^9 GJ), which is the most consumed in industry and transportation [2]. While over the past few decades the transport sector has contributed more to total energy consumption than that of other sectors, the overall increase in the energy demand in recent years occurred with the increase in the share of the industrial sector relative to the other sectors.

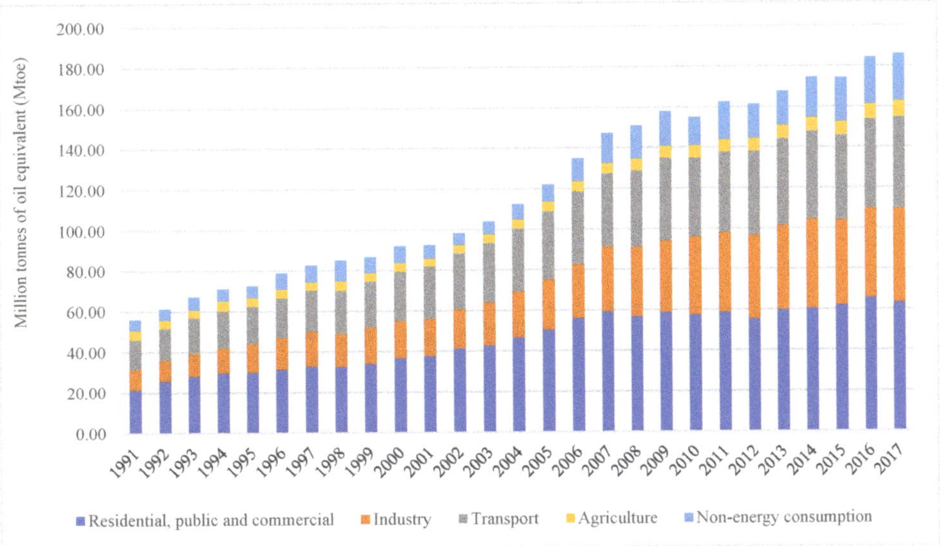

Figure 3. Final energy demand in various economic sectors in Iran. Source: Ministry of Energy [2].

3.2. Crude Oil

As of March 2017, Iran's total reserves of extractable liquid hydrocarbons, including crude oil and gas condensate, were estimated to be 160.1 billion barrels, an increase of 2.9% from the previous year, and the country's natural gas reserves in 2017 were estimated at 33.3 trillion cubic meters, a decrease of 1.2% from the previous year. Besides this, in December 2017, total international oil reserves, including crude oil and gas condensate, were estimated at 1696.6 billion barrels [2].

The Ministry of Petroleum's strategies in the production and supply of crude oil in the horizon of 2041 can be mentioned as follows: (1) increasing the recovery factor for the country's oil fields using over-harvesting and conservation production methods, (2) maximum development and exploitation of all oil and gas fields, (3) the expansion of oil and gas exploration to support the country's oil and gas production, (4) increasing capacity and maintaining the share of oil production in OPEC and the global market and (5) collecting and converting the associated crude oil gases at maximum capacity, taking into account economic and environmental considerations [22]. These strategies will be implemented through national and international investments and partnerships. The goal of the Ministry of Petroleum and the entire energy mix of the country is to continue efforts to increase the country's oil reserves and energy security. Moreover, the efforts of the National Oil Company and the Ministry of Petroleum should be to strengthen its global position in international markets so that it can play an important role in the international arena under appropriate circumstances. With regards to unconventional energy sources, scientific studies show that success in this area requires very close cooperation between national and international companies and the training of the labor force.

Iran's most active oil and gas exploration and production areas are located in the south and west of the country. Most of the discovered fields are in the Zagros and the Persian Gulf regions (southwest of the country). Fields have also been discovered or are being explored in the north of central Iran, the south of the Caspian Sea and Kopedagh. Iran has 20 joint oil fields with neighboring countries, including five with Iraq, four with Saudi Arabia, one with Qatar, one with Oman, seven with the UAE, one with Turkmenistan and one with Kuwait [23]. The country needs the local and international capital and technology to use these enormous oil and gas resources.

The actions taken by the National Iranian Oil Company to supply crude oil to the energy stock market are a big step towards the maximum participation of private investors in the upstream oil and gas industries. The first shipment of Iranian light crude oil was offered in the international energy stock exchange ring in November 2017, and according to the plans, the gradual and regular supply of this valuable shipment of hydrocarbon was provided in the stock exchange [24].

Iran—taking advantage of its geographic location, being among the top oil and gas producers in the region, with the existence of a vast market for energy consumption and long and important transport pipelines that connect the southwest of the county and Persian Gulf (in the south) to the Caspian Sea and neighboring countries in the north, such as Russia and Turkey, and southwest to the east—is becoming an important regional storage and trade center, and can also use existing local resources to reduce its dependency on foreign distillates. The capacity of crude oil, petroleum products and gas condensate storage tanks at refineries across the country at the end of 2017 equaled 18,951, 45,058 and 5147 thousand barrels, respectively.

The country's practical refining capacity is about 1.8 million barrels per day (bbl/d) in the 10 largest refineries. This capacity will increase with the completion of two projects of 0.4 million barrels per day at the Persian Gulf Star in the south and 0.2 million barrels per day in the Kermanshah province. It is projected to increase the capacity of current refineries by 0.5 million barrels per day or to improve the quality of their products in the future [2].

We can conclude that Iran has a significant potential capacity for crude oil and natural gas reserves, its transport and storage. It can increase the weak flexibility of the energy system by constructing more transition lines and braking swap with its neighbors [25]. This makes a significant contribution to the export and transport of crude oil regionally and internationally.

3.3. Natural Gas

As of December 2017, 22 onshore and offshore gas fields were active, including 18 onshore and four offshore gas fields [26]. The natural gas transmission system plays a decisive role in the supply of gas to consumers. The main components of this system are gas transmission lines, gas pressure-raising and reducing facilities and distribution networks [27]. At the end of 2017, more than 38.4 thousand kilometers of gas transmission lines have been constructed and 288 turbochargers are operating in 81 gas pressure boosting stations in the country [28].

Natural gas accounts for the largest share (88.4%) of total fuel consumption in the country's power plants [2]. According to the country's energy policy, efforts have been made to use natural gas in power plants because of ease of operation, lower maintenance costs and less adverse environmental effects [27]. The largest consumer of natural gas is the residential sector (39.1%), followed by industry (24.9%) and petrochemical (21.8%) [26]. The study conducted by Rezaei et al. [29] showed that Iran can become a gas hub in the region if the internal and external facilities and barriers to the advancement of this industry are removed. The South Pars gas field is the largest in the world. It is located at the depth of 3000 m below the bottom of the Persian Gulf is jointly in the territorial waters of Iran and Qatar. The area of this field is 9700 square kilometers, of which 3700 square kilometers are in the territorial waters of Iran and the rest are in the territorial waters of Qatar [28]. The field is mainly developed by local companies. The Iranian share of the field is 14 trillion cubic meters of in situ gas and 18 billion barrels of gas condensate (9 billion barrels of recoverable gas), which represents 50% of Iran's proven natural gas reserves and 8% of the world's gas reserves [26]. One of the goals of the National Iranian Gas Company is to increase the export of natural gas to regional and international markets. In this regard, Iran has been exporting gas to neighboring countries for many years, having enough gas for domestic consumption and export. Turkey, Armenia, Azerbaijan, Nakhchivan and Iraq are the current markets of Iranian gas exports [27]. Iran also imports natural gas from Turkmenistan and Azerbaijan, as well as exports to Turkey, Nakhchivan, Azerbaijan, Armenia and Iraq [2]. The country's natural gas imports in 2017 reached about 3.9 billion cubic meters and its exports reached 13.2 billion cubic meters [28].

This leads us to the conclusion that natural gas is a major contributor to final energy consumption. Natural gas is important as a stabilizer for the energy mixes, which are dominated by renewable energies. Due to the lack of storage possibilities for renewable energy, natural gas, as the least polluting fossil fuel, fills the gaps in the renewable energy supply [30]. The country with high reserves from this source can be one of the world's leading exporters of natural gas. If the country wants to be effective in the international natural gas market, despite the use of advanced technologies and more investment, it must provide clear policies in this industry and try to reduce its barriers at the regional and international levels. Hafezi et al. [31] believe that Iran needs to plan for raising production capacity and considering return project in the short-run.

3.4. Coal

Iran's total definite coal reserves measured by 203 mines in 2017 were 1143 million tons with a majority of coking coal [2]. They are located in Mazandaran, Gilan, Golestan, Kerman, Semnan, Northern Khorasan, Southern Khorasan, Razavi Khorasan, Tehran, Alborz, West Azerbaijan and East Azerbaijan [26]. Isfahan Steel is the largest coal user in Iran, with an annual consumption of one million tons [27]. Other coking and tar refining units also use coal. Despite the relatively large coal resources in Iran, coal mining is unfortunately weak and coal imports are therefore high. This is due to the lack of an appropriate mining strategy, old mining technology, the lack of public sector support to these mines for the equipment and modernization of machines and the use of modern technologies [28]. These challenges have reduced the attractiveness of investment in the mining sector, particularly in coal mines. At the same time, developed countries have defined a clear picture for their mines and, through the adoption of strategic policies, determine the direction of investment and development of mines and mining industries.

Due to the abundance of oil and gas resources in Iran and its general policies, there is no desire to establish and invest in the construction of coal-fired power plants. The only coal-fired power plant project is underway in Tabas and its implementation and operation have begun [2]. For this reason, from an energy perspective, coal has not been more considered in Iran, and the main use of coking coal is attributed only to the Isfahan steel plant, where steel and steel products are produced using the blast furnace method. According to the National Energy Document, diversifying the country's energy portfolio is necessary to achieve the sustainability and development of the country's energy structure. Therefore, coal as an abundant and relatively reliable source with a stable price in the country can play the most important role in achieving this goal.

We can deduce that while the country's oil and gas resources are high, it makes no sense to import coal from other countries with a high inventory of coal resources. Therefore, there is a need to amend the upstream energy documents to include coal sources in these documents in order to develop sustainable and effective policies for this industry.

3.5. Wind Enrgy

Wind energy is produced by generating electrical energy from wind or airflow, which occurs naturally in the earth's atmosphere, with windmills or wind turbines. Wind energy usage in Iran goes back to windmills in 200 BC. Regarding wind energy potential, 1.3% of Iran's land (2.1 million hectares) has an average annual wind speed of 8 m per second and above, which makes these areas capable of exploiting this energy source [32]. According to the Atlas of the Renewable Energy and Energy Efficiency Organization (SATBA), the amount of wind energy that can be extracted in the country is estimated at 18,000 MW in all surveyed areas [33]. Meanwhile, a total of 7801.7 MW represents the nominal capacity of all power plants in the country in 2017. The provinces with strong wind energy potential in the country are Gilan, Southern Khorasan (especially Doruh and Khaf regions), Sistan and Balochistan (especially the northern region of Zabol), Semnan and Qazvin, where most of the country's wind farms have been built so far [26]. Zabol is the most appropriate area for large scale wind turbine establishment [34,35]. The first large-scale wind turbine was installed in 1994 in Manjil, Gilan Province, in two phases with generating capacities of 30 and 60 MW, which were completed in subsequent years [2]. In 2020, the farm capacity reached 92.5 MW [33]. Another windfarm is Binalood in Khorasan Razavi Province, which has a capacity of 28.2 MW [2].

Iran plans to increase the share of non-hydro renewable resources in its total electricity generation capacity to 5% (about 4 GW) by 2021 [33]. Aryanpur et al. [36] argued that Iran should increase the share of non-hydro renewables to 32% by 2050. This will not only make it possible to use alternative gas (or liquefied petroleum gas) in other applications that have higher economic benefits, but also contribute to improved air quality in large cities. Wind energy not only can help Iran's energy security, independence and climate goals in the future, but it can also turn a serious energy supply problem into an opportunity in the form of trade interests, technology research, exports and employment. The future of Iran's economy can be planned based on recognized and predictable electricity costs because that electricity comes from indigenous energy and is free from all the security, political, economic and environmental problems associated with oil and gas. The economic growth and population rise show that energy demand is increasing rapidly in Iran. It is estimated that approximately 10 million watts of energy are continually available in the world's wind [37] while the total installed global capacity was 60.4 GW in 2019, an increase of 19% from 2018 [38]. Razavieh et al. [39] recommended the annual average wind energy density of 388 W m^{-2} is suitable for large-scale wind.

Currently, the Ministry of Energy is offering a 20-year Feed-in-Tariff (FIT) contract for renewable energy at rates higher than the selling price of electricity to end consumers. In other words, for power plants with a capacity of less than 10 MW, the rate for wind energy is IRR 7644 per kilowatt-hour, IRR 8918 per kilowatt-hour for solar and geothermal energy and an average of IRR 6930 per kilowatt-hour for various biomass power plants [32]. In recent years, local and foreign investors have installed approximately 350 MW of renewable energy in Iran through the electricity purchase agreement mechanism, while several other energy farms with a total capacity of approximately 700 MW are in various stages of development [32].

Figure 4 represents the trends of wind energy generation and the capacity of wind power plants in Iran. In recent years, there has been significant growth in wind energy production: 186 GWh in were produced in 2014, 223 GWh in 2015, 250 GWh in 2016, 308 GWh in 2017 and 320 GWh in 2018. However, that is far from what the Ministry of Energy wants to achieve. Achieving an acceptable share of the expected target (4 GW by 2021) requires extensive research and significant operational action, and extensive research has been conducted on various dimensions of wind energy in Iran.

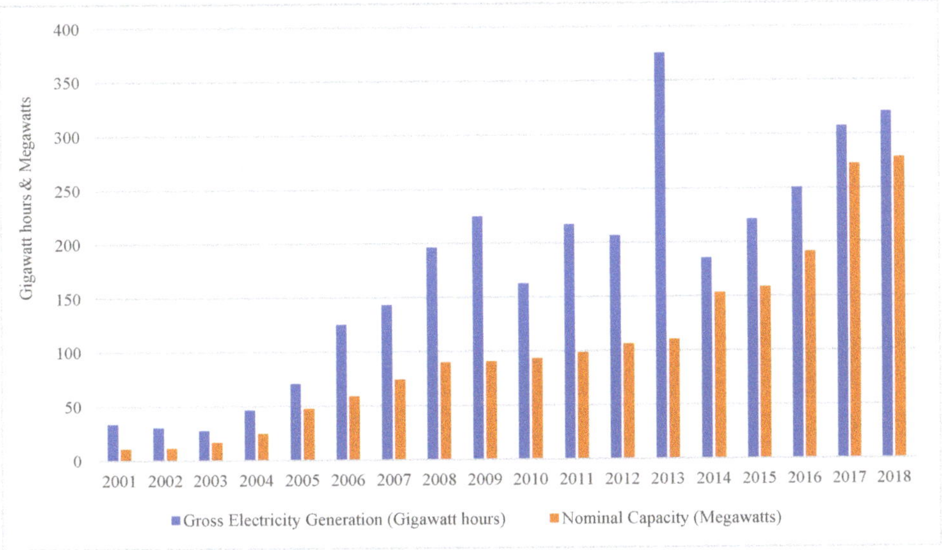

Figure 4. Electricity generation and capacity of wind power plants from 2009 to 2010 in Iran. Source: [26].

We can conclude that Iran's electricity capacity is high and this can help to increase the share of wind energy in the total primary supply of energy. To achieve long-term electricity targets, it is necessary to provide incentives to private investors and to put in place clear and stable policies. The country's electricity demand is growing rapidly due to economic and population growth. Therefore, Iran must make an important decision for the future of the country's environment and electricity generation, particularly through the use of renewable energy resources.

3.6. Hydroelectric Energy

With average rainfall one-third of the global average, Iran is one of the top arid and semi-arid countries in the world [2]. Accordingly, the inadequate temporal and spatial distribution of rainfall is perhaps the most important reason to build dams in Iran. For this reason, hydropower cannot be considered a reliable and sustainable source of electricity generation in the country—but hydropower plants play an effective role in controlling the frequency of the grid.

Hydroelectric facilities in Iran are more than centers for producing renewable energy. Hydroelectric dams also play an important role in socio-economic development, such as irrigating agricultural fields, providing adequate water supply, especially during drought seasons, controlling the situation during the monsoon period and the improvement of major waterways. Currently, electricity is the only renewable energy that is commercially available on a large scale in Iran. The hydropower resources and potential are estimated at 26,000 MW, most of which is provided by the Karun, Karkheh and Dez rivers [40].

Hydroelectricity accounts for about 98.8% of total renewable energy in the country [27]. As of 2017, a total of 15,329.1 GWh of hydropower was generated, which has a share of 5% of total electricity generated and 0.30% of the total energy supply in the country, while on the global scale, hydropower contributes 16.6% of the total energy produced [28]. Table 1 shows total electricity generation and its share of total energy supply in Iran and selected countries and regions in 2016. As can be seen, Iran derives a small portion of its electricity from hydropower compared to other countries and regions.

Table 1. Total electricity and hydropower generation and share of hydropower in different countries and regions in 2016 (TWh).

Country/Region	Total Generation	Hydropower	Share (%)
Iran	308.3	15.4	5
Turkey	274.4	67.2	24.5
Pakistan	114	36.6	32.1
Europe	4941.5	823.1	16.7
Asia	9796.9	1587.9	16.2
Middle East	1147.2	21.2	1.9
United States of America	4322	292.1	6.8
World	25,082	4170	16.6

Source: [26].

The amount of electricity production by hydropower is not stable and depends on the amount of annual rainfall. Therefore, it is used only to help the system in peak times. Since Iran is a country with an abundance of fossil fuels, the choice of the type of power plant seems to be based only on the primary investment and the availability of its primary inputs, which is pointed out in some studies. For example, based on various indicators, Manzoor and Rahimi [41] showed that Iran's priorities for construction and investment in electricity generation and power plants in the future include, in order, wind energy, hydropower, photovoltaic energy, combined-cycle power plants, nuclear power plants and thermal power plants.

4. Renewable Energy Sources

The Ministry of Energy, as the governing body, is responsible for all issues related to electricity and renewable energies, such as developing, planning, organizing, legislating and implementing policies and initiatives related to renewable energy. Although the Office of New Energy was established in previous years before 1995 in the Ministry of Energy, Iran's New Energy Organization was established in 1995, and in 2000 it was officially supervised by the Ministry of Energy to reduce the harmful effects of fossil fuel used on the environment and the diversity of the country's energy sources for sustainable development. However, in 2016, the Renewable Energy and Energy Efficiency Organization was formed by integrating the New Energies Organization and the Energy Efficiency Organization. As a result of the process of structural change that took place in 2000, green technology was added to the ministry's energy portfolio to reflect the growing emphasis on sustainability and renewable energy options. As discussed earlier and summarized in Table 2, a shift towards renewable energy is essential to meet increased demand while maintaining environmental and energy security. Moreover, rural areas have used low levels of renewable energy [42]. International sanctions, emphasis on the development of conventional energies, lack of adequate government policies on the development of renewable energy and the lack of sustainable energy security and environmental policies are impediments to the development of renewable energy [43].

Table 2. Potential of various types of renewable energy sources in Iran.

Renewable Energy	Potential (MW)
Hydropower	26,000
Solar energy	86,198.2
Wind energy	18,000
Biomass and biogas	19.04
Geothermal energy	187

Source: [2].

4.1. Energy Diversity Policy and Other Energy Policies

Because of its diverse environment and climate, Iran is can obtain energy from a variety of renewable sources. Unfortunately, policies and measures associated with renewable energy sources have been neglected due to more government attention on fossil energy fuels because of their abundance as sources of revenue and given high past investments. As well, there was no specific program for renewable energy resources in the country's policy documents until the late 1990s. According to Pirasteh et al. [44], due to the low price of energy from fossil sources in Iran, managers, civil servants, craftspeople and consumers have little interest in the use of renewable energy sources.

Iran's general policy, which was introduced in 2000, is the first document concerning these types of energies. In this document, two of the twelve articles address the issue of renewable energy sources [45]. These two articles emphasize the diversity of the country's energy resources with regards to environmental issues, trying to increase the share of renewable energy with the priority of hydroelectricity and attempting to acquire technology and technical knowledge of new energies and power plants such as wind and solar, fuel cells and geothermal energy in the country [45]. In the law of the Fifth Five-Year Economic Development Plan (2011–2015), only the replacement of fossil fuels and renewable energies with firewood is sufficient and no achievable goal has been considered for it. The first document that highlighted the achievement of a specific target for renewable energy is the law of the Sixth Five-Year Economic Development Plan (2017–2021). It obliges the government to increase the share of renewable and clean power plants with the priority of nongovernmental investment (domestic and foreign investors) with the maximum use of domestic capacity to reach at least 5% of the country's electricity capacity by the end of the implementation of the program law.

In 2011, the Energy Model Reform Act was endorsed. Chapter 10 and Articles 61 and 62 of this act outline the use and investment in renewable energy and nuclear energy. Under this law, the Ministry of Energy is required to support the expansion of the use of renewable energy sources, including wind, solar, geothermal, small-scale hydropower plants (up to ten megawatts), marine and biomass (including agricultural and forest wastes, municipal waste and wastewater, industrial waste, livestock waste, biogas and biomass). To facilitate and consolidate these issues, it can do so through the relevant organization to make a long-run contract for guaranteeing the purchase of non-governmental renewable electricity producers. The Ministries of Energy and Petroleum are also obliged to publicly announce the necessary supports for promoting the economical use of renewable energy sources in separated systems from the grid, such as solar water heaters, solar baths, air pumps, wind turbines, photovoltaic systems, gas extraction from biomass sources and cost efficiency in supply and distribution of fossil fuels publicly announces the necessary support. Accordingly, the Ministry of Energy has a plan to increase the production capacity of hydropower and renewable power plants by 5% by 2021. Under this plan, a small-scale renewable energy program has been launched to encourage the private sector to invest in small-scale power generation projects using biomass, biogas, mini-hydroelectric, solar and wind energy.

The impact of renewable energy policies and electricity efficiency in recent years has resulted in the construction of 134 power plants and 4038 generators for consumers [26]. This has resulted in the generation of 5035 million kWh electricity, the reduction of 468 million kWh in losses in the electricity network and 3417 thousand tons of CO_2 emissions (which itself reduces the external effects of reducing CO_2 emissions), and the non-emission of 21.3 thousand tons of other local pollutants [27]. Improving energy structure and energy efficiency in Iran can significantly reduce the level of CO_2 emissions [46]. It also saves 1108 million liters of water consumption, creates 31,271 jobs and attracts IRR 143.5 trillion of nongovernmental investments [33]. Meanwhile, SATBA has 78 power plants and 1000 generators for subscribers under construction. Figure 5 shows the share of renewable energy capacity and electrical energy efficiency up to July 2020. The construction of these power plants will not only create jobs and generate revenue for renewable energy businesses but will also increase tax revenues for the government [33].

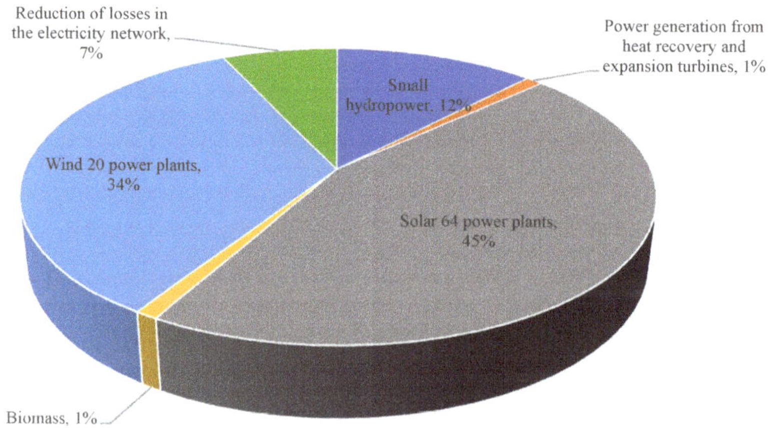

Figure 5. Share of renewable energy capacity and electrical energy efficiency in Iran. Source: [32].

Potential sources of renewable energy in Iran are summarized in Table 2. In high-rainfall and mountainous regions of Iran, large rivers' adequate water levels promote the development of hydropower plants. Moreover, the high share of desert and arid areas, which provides more than 300 sunny days per year, makes solar energy a suitable option as an important source of renewable energy. Furthermore, the existence of large biomass resources in Iran can be used to produce electricity and fuel. Although Iran ratified the Kyoto Protocol and the Paris Agreement, the government should seek to create clean and sustainable fuels to not only guarantee energy security independence but also to maximize available natural resources. Evidence has predicted energy independence in Iran by using non-fossil fuels. For example, Aghahosseini et al. [47] showed that an energy system based on 100% renewable resources is not a dream but is a safe and low-cost option up to 2030.

High energy losses and efforts to address this gap (for example, the use of sunlight in buildings) have overshadowed the use of renewable energy. Waste reduction and the use of renewable energy sources can lead to significant increases in energy efficiency across the country's sectors. Relatively cheap fuel prices and low private and public investment in renewable energy production are the main factors behind the low share of renewable energy in the total energy mix of the country. The National Renewable Energy Policy can be introduced as an important step in increasing the investment and extraction of renewable energy in the total energy mix. Furthermore, to increase the use of local renewable energy sources, increasing the share of renewable energy in the composition of electricity generation can be achieved through facilitating the growth of the renewable industry, ensuring the reasonable cost of renewable energy production and creating public awareness of the importance of sustainable energy and clean technology.

4.2. Small-Scale Hydropower

Small-scale hydropower plants have received considerable attention in recent years, particularly in Iran's water-rich regions, because of relatively low capital and maintenance costs. These types of dams also reduce the environmental effects of clearing land and forests to build a reservoir of large dams. The construction of small hydropower plants is suggested in mountain areas, because of the high costs of medium-sized electricity transmission lines, long distances between villages and high energy losses. These kinds of dams have been considered because of providing cost-effectiveness benefits, no environmental pollution, long life, fast and easy installation and non-use of fossil fuels [48].

A small-scale hydroelectric dam can be defined as a facility that generates up to 10 megawatts of electricity. In 2020, about 14 small-scale hydropower plants in 10 provinces of the country with a total generating capacity of 261 MW began to operate. Most of these projects are located in the northern regions and the water-rich slopes of the Zagros, including the provinces of Gilan, Mazandaran, Lorestan, Kohkiluyeh and Boyer-Ahmad. Haji Ghafouri Bukani [49] believes that including hydropower programs in the country's energy development programs will lead to economic prosperity in villages, increase industrial and commercial activities in villages and prevent villagers from migrating to cities. This is consistent with the results of the study conducted by Sovacool and Walter [50]. The reason for this is the highly suitable zones for the construction of such power plants and their economic and social characteristics. Therefore, to reduce the use of fossil energy sources, it is suggested to use of this type of power plant in water-rich areas.

4.3. Solar Energy

Iran is a rich country in solar energy. The country's priority for renewable energy sources is solar energy, averaging 300 sunny days per year [51]. The average daily sunlight in Iran is about 5.5 to 8.5 kWh per square meter, particularly in the central regions [19,52]. The technical potential for solar thermal energy is estimated to be 91,000 TWh [53]. July, August and September are the months with the highest solar radiation of 1050 h, and January, February and March have the lowest solar radiation of 500 h [52]. The installed capacity of cumulative renewable energy sources is projected to be 2.8 GW by 2030. This represents an investment of more than USD 2.8 trillion from 2010 to 2030 [52].

The central, southern, eastern and southeastern parts of Iran, such as the provinces of Yazd, Kerman, South Khorasan, Zahedan and Kerman, have high solar radiation throughout the year. Cities in the north of the country present the greatest potential for investment in photovoltaic energy [54].

Photovoltaic (PV) systems similar to small hydroelectric power stations can meet needs with off-grid power supply systems (electricity supply for residential houses, nomadic tents, rural cottages) and generally meet the electrical needs of areas without a national electricity grid. As a result of the very high cost of mass production of electricity from photovoltaic systems, commercial supply is not yet possible. Currently, the majority of photovoltaic users are high-income groups, representing a very small percentage of the total population. Furthermore, to encourage the private sector collaboration to invest in photovoltaic systems, the government announced incentives and preferential prices in 2015 and guaranteed the purchase of renewable electricity produced. The adoption of these policies brought the private sector's contribution to electricity production from 8.71 MW in 2015 to 300.86 MW in 2018 [33]. Solar energy has numerous applications such as cooking and drying food, air conditioning and refrigeration, space heating and water heating, and industrial applications, i.e., solar thermal energy generation [55,56].

The localizing of the production of photovoltaic systems, which in turn lowers end-user prices, raising public awareness of the use of renewable energies (for traditional and industrial applications) and government incentives can all help reduce the use of fossil fuels. Gorjian et al. [57] and Edalati et al. [58] state that technology, weather conditions for photovoltaic installation, government policies and the lack of a specific roadmap are the most significant barriers in the development of PV systems in Iran.

4.4. Biomass

Biomass is defined as all plant resources and plant residues and human biological activities capable of generating energy. Wood, human waste, sawdust, animal waste, food waste and wastewater are utilized for biomass. The process of converting biomass and other renewable resources into energy is called biotechnology [59]. This energy is mostly used for cooking, heating, lighting and electricity production. The focus of the National Renewable Energy Policy is to emphasize the importance of transitioning to renewable energy sources that have the least impact on the environment. In these conditions, biomass has become a very attractive and practical option. Municipal waste that has remained from previous years is the largest source of biomass in Iran (Table 3).

Table 3. Biomass power plants in operation in 2018.

Type of Power Plant	Province	Year of Operation	Capacity (MW)
Extraction of methane gas and construction of biogas power plant to generate electricity from municipal waste	Razavi Khorasan	2009	0.6
	Fars	2009	1.06
	Tehran	2014	1.9
	Tehran	2010	4
	Tehran	2014	3
Total			10.56

Source: [32].

The German DLR (Deutsches Zentrum für Luft- und Raumfahrt) center in 2002 estimated Iran's biomass energy potential at about 3500 MW by 2050 [60]. Moreover, the potential to generate electricity from municipal waste in Iran's provinces is predicted in Table 4. Since electricity generation from waste is a costly method, the government guaranteed to buy the electricity generated by these power plants at a reasonable price to encourage investment in this field.

Table 4. Biomass energy potential in the selected provinces in 2018.

City	Ardabil	East Azerbaijan	West Azerbaijan	Fars	North Khorasan	Razavi Khorasan	South Khorasan	Gilan	Sistan and Baluchestan	Kerman	Bushehr	Hormozgan	Total Potential (Theory)	Total Economic Potential with 10% Capacity
No. of sites	4	8	5	6	2	8	4	5	4	4	1	5		
Production (MW)	10	25	15	30	1	65	2	15	10	1	3	25	208	20

Source: [29].

Due to climatic diversity and water availability in most agricultural regions, over 80% of agricultural products are agronomic products and the rest are horticultural products. Major agricultural products in Iran include wheat (13.26%), corn (10.83%), sugarcane (6.66%), beets (6.37%) and Lucerne (6.18%) from 16.5 million hectares of cultivated land (14.7 million hectares of arable land and 1.8 million hectares of garden lands, i.e., 8.78% of the total land area of Iran) [61]. The amount of waste produced from the agricultural products will be around 24 million tons per year, taking into account the average of 30% of the waste for these crops. Given that an average of 450 cubic meters of biogas waste is produced per ton [62]. About 10,800 million cubic meters of biogas from these wastes will be produced, which is a huge source for use in simultaneous heat and power (CHP) plants. The findings from Samadi et al. [63] showed that the total energy obtained from agricultural waste using degassing technology was 341,290 Terajoules (TJ) and the amounts of electricity and heat generated by this energy were 66,075 and 399,112 Terajoules (TJ), respectively.

Finally, one of the proposed projects for the effective use of renewable energies is wind–solar power plants, which provide higher wind energy during the wind season and higher solar energy during the solar season [64]. The production of valuable gas for clean thermal and electrical energy, along with many other benefits, encourages energy policymakers to consider the construction of biomass power plants as a necessity in energy supply policies. The Renewable Energy and Energy Efficiency Organization of Iran (SATBA) plan is focusing on the participation of small and medium-sized companies to develop innovation, knowledge and business activities to create a sustainable production system in the high-value industry. In this regard, the government has a significant role to promote accessibility, affordability and sustainability of energy security by preparing appropriate renewable energy development [65]. Therefore, using a waste stream should also include food processing waste or post-harvest agricultural residues such as fruits.

4.5. Biofuels

Biofuels are types of renewable energies derived from organic matter (living matter or biomaterials) and, in fact, their energy content is from biomass, which can be found in the form of liquids (bioethanol and biodiesel), solids (dry plant material) and gases (biogas). Biofuels, such as biodiesel and bioethanol, can be used in motor vehicles due to their high energy content and compatibility with fuel infrastructure, and therefore, are more important [66]. The diversity of Iran's land and climate causes the production of various energy sources suitable for the production of liquid biofuels. The current bioethanol sources in Iran are sugarcane molasses and sugar beet which are easily available [67]. The use of biofuels in Iran is relatively low. South Khorasan is the best place for biogas establishment [68]. Although efforts have been made to include renewable energy sources in electricity generation, renewable fuel transportation has not been given any priority because of higher transportation fuel demand than electricity generation and high fuel subsidies to the market. It is worthy to note that the removal of energy subsidies can decline energy consumption, which in turn decreases CO_2 emissions in the country [69].

Having the potential to use different types of biomass as raw materials can be very attractive for investment and the government can use various conversion methods to fit the needs and resources that existed in each region. Despite the progress in advanced biofuel conversion techniques and the high potential of using different biomass due to their abundance, currently, there is no large-scale or even small-scale production of any type of biofuel in Iran. Therefore, we reviewed studies that have examined the potential of producing various biofuels in the country.

Abyaz et al. [66] analyzed the use of lignocellulosic materials and suggested that the amount of bagasse produced in seven units of the sugarcane and ancillary industries development plan of Khuzestan during a year is equal to 231 million tons; this amount of bagasse can produce 57,750 million liters of bioethanol annually. Hajinejad and Katooli [65] also showed that the use of Nowruzak in the edible oil extraction industry is possible and its residues can be used directly in the biofuels industry with an average oil content of 56%. Nowruzak has the potential to produce 3348.8 L per hectare of biodiesel. Ghobadian [70] also believes that there are about 17.86 million tons of agricultural residues in Iran, of which about 5 billion liters of bioethanol can be produced annually. According to the estimated amount of various biomass sources reported in Figure 6, such as vegetable oils (i.e., palm, jatropha, castor, cellulose, algae), the production of bioethanol in Iran, which can be an optimal alternative fuel for gasoline engines using alcoholic gasoline, in various ratios can be predicted until 2026.

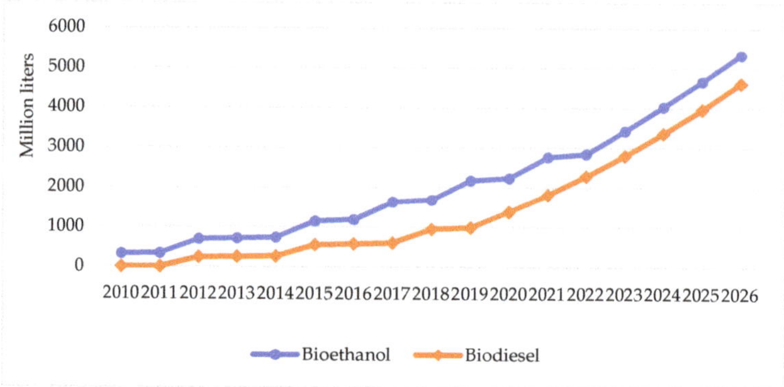

Figure 6. Prospects for liquid biofuels share in Iranian transport. Source: [70].

According to Table 5, it is predicted that Iran's potential in 2018 to produce bioethanol from grain residues, sugar beet molasses and sugarcane as well as some fruits such as apples, grapes and dates is about 4.827 billion liters per year.

Table 5. Biomass energy potential (ethanol) of agricultural residues in Iran in 2018.

	Production (Million Tons)	Waste (%) *	Waste (Million Tons)	Conversion Factor (Liters per Ton) *	Total Bioethanol (Billion Liters)
Wheat	13	50	6.5	400	2.6
Sugarcane molasses	0.3	100	0.3	300	0.09
Rice	3	25	0.75	400	0.3
Barley	3	20	0.6	350	0.21
Corn	10	30	3	360	1.08
Potatoes	5	30	1.5	110	0.165
Dates	1.3	40	0.52	360	0.1872
Beet molasses	0.25	100	0.25	300	0.075
Grapes	3	30	0.9	70	0.063
Apple	2.9	30	0.87	65	0.0566
Total	42.75		15.49		4.827

* The waste percentage and conversion factor is taken from [70].

Moreover, based on the calculated made in Table 6, it is predicted that Iran's potential in 2018 to produce biodiesel from oilseeds residues, such as cotton, soybean, rapeseed, sesame and corn, is about 526 thousand tons per year.

Table 6. Production potential of biodiesel from the major edible oilseeds in Iran in 2018.

Oilseeds	Area under Cultivation (Hectares)	Production (Tons)	Oil Content (%) *	Returns (Tons of Biodiesel per Year)
Cotton	70,800	165,295	20	33,059
Soy	40,327	83,303	20	16,660.6
Rapeseed	191,251	329,843	40	131,937.2
Sesame	30,017	30,649	50	15,324.5
Corn	126,971	946,021	10	94,602.1
Olive	90,000	102,648	20	20,529.6
Sunflower	10,756	11,960	50	5980
Safflower	5239	4470	32	1430.4
Almond	200,000	142,000	55	78,100
Walnut	166,000	200,000	60	120,000
Hazelnut	27,000	15,000	55	8250
Total	958,361	2,031,189	—	525,873.4

* The oil content is obtained from [71].

Sepaskhah [72] believes that the use of fodder and sugar and starch products to produce biogas due to the lack of organic matter in Iranian soils and the lack of water and fodder is unreasonable. It is worthy to note that the country must not use agricultural products for the production of bio-energies while it needs food and organic matter. If there is agricultural waste, the first priority is its use to strengthen the organic matter of Iranian soils. These residues can be collected and their contained energy can be extracted with conventional and advanced technologies. Moreover, fossil fuels are finite and their consumption harms the environment and countries must concentrate their long-run policies and efforts on using more renewable, clean and sustainable energy sources. In addition to these cases, as Rashidikia and Moradi [73] stated that these resources provide many national advantages, such as (1) increasing job opportunities in rural areas, which reduces the migration of indigenous people to cities and other regions, (2) increasing the public health of rural communities and (3) providing new investment opportunities for indigenous people.

Finally, it can be concluded that although world oil prices are uncertain and fluctuate widely over time, the fact that Iran is heavily subsidizing all types of energies for public consumption leads to the introduction of biofuels with a slightly higher cost in the household energy basket mix, which would be a viable option to reduce government subsidies. In the case of sufficient production and availability of biofuels in the country, the use of biofuels in cities requires the obligation and acceptance by the consumer. Thus, it is necessary to increases public awareness to use renewable energies.

4.6. Geothermal

The Meshkinshahr area was introduced as the first priority for geothermal exploratory studies and the first geothermal wells were drilled in this area in 2004 [74]. The capacity of the project is 25 MW [2]. Since the only investor is the government in this field, it is necessary to provide some incentives for private investors. The hot underground water temperature of 85 °C is accessible in geothermal areas [75]. The geothermal heating system is used in food production, at the tourist center, for district heating, at greenhouses and at fish farms [75,76]. The main geothermal development barriers are lack of legislation, poor human resources management and improper technology transfer [77].

5. Econometric Results

This section presents the estimated results of the econometric model. To reduce the number of tables in the main text, we provided a descriptive statistics table and a stationary table in Appendix A. The unit root results show that we cannot reject the null hypothesis of unit roots for all variables in the model; the variables are stationary after difference 1 and they are integrated in order 1, I(1).

Another test is co-integration analysis; in this study, we used the Johansen test. The cointegration results show that the model has one cointegrating vector (Table 7). Therefore, we can conclude that a cointegration relationship exists among variables.

Table 8 reports the results for the Granger causality analysis. The results fail to reject the null hypothesis that economic growth does not Granger-cause labor force and gross fixed capital formation, but rejects it that economic growth Granger-causes non-renewable energy and renewable energy. This means that there is unidirectional causality from economic growth to renewable and non-renewable energy and its inverse is not correct. These findings are consistent with the findings of the study conducted by Khoshnevis and Shakouri [78]. Moreover, Li and Solaymani [79] argued that economic growth is the main contributor to energy consumption in Malaysia in the short- and long-run. Razmi et al. [80] also showed that renewable energy consumption cannot Granger-cause economic growth in Iran. The results also cannot reject the null hypothesis that labor force does not Granger-cause gross fixed capital formation and non-renewable energy, but rejects that labor force Granger-causes economic growth and renewable energy. This means that there exists a one-way causality from labor force to real GDP and from labor force to renewable energy, which their inverses are not correct. This finding supports the results of the study conducted by Salmanzadeh and Fatemi [81]. Regarding the variable gross fixed capital formation, we cannot reject the null hypothesis that gross fixed capital formation does not Granger-cause real GDP, labor force and non-renewable energy, but it can be rejected that gross fixed capital formation Granger-causes renewable energy. The null hypothesis fails to reject when testing Granger causality from non-renewable energy to real GDP, labor force and gross fixed capital formation, but does not reject runs from non-renewable energy to renewable energy. This means that a unidirectional causality exists between non-renewable energy and renewable energy. Finally, the null hypothesis of no renewable energy Granger-causes real GDP, labor force, gross fixed capital formation and non-renewable energy cannot be rejected.

Table 7. Johansen co-integration test.

Unrestricted Cointegration Rank Test (Trace)				
Hypothesized No. of CE(s)	**Eigenvalue**	**Trace Statistic**	**0.05 Critical Value**	**Prob. ****
None *	0.501697	78.37974	69.81889	0.0088
At most 1	0.300879	43.5524	47.85613	0.1197
At most 2	0.257474	25.65581	29.79707	0.1393
Unrestricted Cointegration Rank Test (Maximum Eigenvalue)				
Hypothesized No. of CE(s)	**Eigenvalue**	**Max-Eigen Statistic**	**0.05 Critical Value**	**Prob. ****
None *	0.501697	34.82734	33.87687	0.0384
At most 1	0.300879	17.8966	27.58434	0.5038
At most 2	0.257474	14.88488	21.13162	0.297

Source: estimated results. * shows rejection of null hypothesis at 5% level. ** Mackinnon-Haung-Michelis p-value.

Table 8. Results of the Granger causality test.

Null Hypothesis	F-Stat. [Prob.]	Null Hypothesis	F-Stat. [Prob.]
LGDP → LLF	0.563 [0.573]	LNREN → LLF	1.517 [0.230]
LLF → LGDP	2.448 [0.097] ***	LLF → LNREN	1.313 [0.279]
LGFKF → LGDP	0.286 [0.753]	LREN → LLF	0.909 [0.410]
LGDP → LGFKF	1.303 [0.282]	LLF → LREN	5.649 [0.007] *
LNREN → LGDP	1.893 [0.162]	LNREN → LGFKF	0.773 [0.468]
LGDP → LNREN	2.664 [0.081] ***	LGFKF → LNREN	0.588 [0.560]
LREN → LGDP	0.183 [0.834]	LREN → LGFKF	1.213 [0.307]
LGDP → LREN	3.692 [0.033] **	LGFKF → LREN	3.829 [0.029] **
LGFKF → LLF	0.416 [0.662]	LREN → LNREN	0.597 [0.555]
LLF → LGFKF	0.816 [0.448]	LNREN → LREN	3.664 [0.034] **

*, **, *** denotes significant level at 1%, 5% and 10%, respectively. Source: estimated results.

6. Conclusions

This study, using a review methodology, investigated current and future energy demands and existing renewable energy resource policies in Iran by employing the latest available data from the Ministry of Energy, Ministry of Petroleum and national laws and documents. Then, using Granger causality analysis, it estimated the direction of relationships among considered variables.

The econometric results showed that there exists a unidirectional causality relationship between economic growth and renewable and renewable energy demand. This means that economic growth helps the development of renewable and non-renewable energy consumptions, but other sides are not correct. Labor force and gross fixed capital formation do not Granger-cause non-renewable energy demand, but they cause renewable energy demand. Finally, renewable energy demand does not Granger-cause non-renewable energy demand, while non-renewable energy demand Granger-causes renewable energy demand.

The study's analyses showed that Iran can access sustainable, clean and renewable energy sources. It is logical that fossil fuels have better and more valuable applications than heating and lighting. Therefore, Iran has acceptable and compelling incentives to pursue renewable and sustainable options, although it remains a leading exporter of crude oil and is exploring and developing new oil fields. The annual cost of subsidies for energy commodities for public consumption in 2009 was about IRR 442 thousand billion. High fuel prices harm food prices and other macroeconomic indices, and the government should look for better options to reduce public discontent. Although international agreements and environmental policies, such as the Kyoto and Paris agreements, emphasize reducing CO_2 emissions and Iran is a party to some of these agreements, there is still a commitment to reducing CO_2 emissions and preparing Iran's infrastructure for clean energy use. The overall outlook for the future of renewable energies seems unclear, at least regarding the effectiveness of government policies. Although some policies seem to be of particular benefit to some people (for example, the use of solar panels in remote rural areas), the overall prospects of these policies have been minimal. At present, renewable energy technologies do not play an important and appropriate role in Iran's energy supply. This also has been experienced in another high energy resources country, i.e., Saudi Arabia [82]. The main barriers to renewable energy development in Iran include the absence of suitable and effective government policies, international sanctions and low public awareness.

The current Iran development plan (Sixth Economic, Social and Cultural Development Plan, 2017–2021) aims to increase the share of renewable energy sources in the electricity generation mix to 5%. However, the evidence and data from the Renewable Energy and Energy Efficiency Organization indicate a severe lag in the goal of this program. With a closer look at the energy mix, the decision to make greater use of fossil fuels, especially natural gas, when it is less environmentally friendly seems to contradict the goal of reducing greenhouse gas emissions, as well as weakening energy security by relying on fossil resources. The development of renewable energy sources under the guidelines of the current Iranian program can be considered a step in the right direction, albeit a more active position needs to be taken to develop potential renewable energies, such as biomass. Implementation of biomass exploitation techniques under the current framework can be facilitated in relevant government organizations such as SATBA, the Niroo Research Institute, large city municipalities and the Agricultural Research, Education and Extension Organization. A country with many fossil fuel resources needs appropriate policies and guidelines for the future of renewable energy development and reduced international barriers to the use of advanced technologies and the experiences of other countries in the use of renewable energies. At the local level, it needs to use crude oil export revenues to make a significant investment in various and high potential renewable energies for the future without fossil energy resources. It is also necessary to reduce energy subsidies, raise public awareness and offer substantial and suitable incentives to motivate private investors. Salam and Khan [83] also showed that increasing public awareness through media is crucial for renewable energy development in Saudi Arabia. The use of the wind–solar hybrid plant is also suggested to generate both energies more effectively than single power plants. The hybrid renewable energy system is also suggested by Mohammed [84] for Iraq and by Marchenko and Solomin [85] for Russia.

Funding: This research received no external funding.

Institutional Review Board Statement: Not applicable.

Informed Consent Statement: Not applicable.

Data Availability Statement: Most of the data used in this study come from the energy balance. Please refer to the following link: https://moe.gov.ir/ (accessed on 25 June 2021).

Conflicts of Interest: The author declares no conflict of interest.

Appendix A

Table A1. Descriptive statistics.

	LGDP (IRR Billion)	LLF (1000 People)	LGFKF (USD Million)	LNREN (ktoe)	LREN (ktoe)
Mean	15.09776	9.492841	10.74469	10.78031	7.042203
Median	15.02835	9.493683	10.88931	10.95113	6.900351
Maximum	15.75293	10.0981	11.84446	12.04142	8.356657
Minimum	14.19673	8.824737	9.019639	8.690626	6.365443
Std. Dev.	0.392414	0.41071	0.739127	0.97287	0.525495

Source: estimated results.

Table A2. Results for the Unit root test.

Variable	Dickey Fuller GLS		Phillips–Perron		Augmented Dickey Fuller	
	Level	First Difference	Level	First Difference	Level	First Difference
LGDP	0.104	−5.098 *	−0.558	−5.098 *	−0.743	−5.071 *
LLF	0.430	−2.433 **	0.759	−4.497 *	0.170	−2.601 ***
LGFKF	0.987	−6.103 *	−0.044	−6.012 *	0.110	−6.054 *
LNREN	0.849	−2.451 *	2.456	−5.496 *	3.006	−2.686 ***
LREN	-0.467	−6.194 *	0.020	−5.151 *	−0.473	−6.174 *

*, **, *** denotes the level of significant at 1%, 5% and 10%, respectively. Source: estimated results.

References

1. Mohammadi, V.; Mozaffari Shamsi, H.; Asadi, F. Investigating the relationship between economic growth, energy consumption and human development in selected MENA countries (MENA). *Iran. J. Energy Econ.* **2019**, *8*, 811–850.
2. Ministry of Energy. *Energy Balance 2017*; Ministry of Energy: Tehran, Iran, 2020.
3. Mozayani, A.H.; Essari Arani, A.; Afsharian, B.; Rasouli, A. Redefinition of the Relation between Energy Consumption and Economic Growth in Iran: Markov Switching Approach. *Econ. Modeling* **2015**, *9*, 67–89.
4. Cheratian, I.; Goltabar, S. *Energy Consumption and Regional Economic Growth: The Case of Iranian Manufacturing Sector*; MPRA Paper 78315; University Library of Munich: Munich, Germany, 2017.
5. Abdouli, M.; Hammami, S. Exploring links between FDI inflows, energy consumption, and economic growth: Further evidence from MENA countries. *J. Econ. Dev.* **2017**, *42*, 95–117. [CrossRef]
6. Mirzaei, M.; Bekri, M. Energy consumption and CO_2 emissions in Iran, 2025. *Environ. Res.* **2017**, *154*, 345–351. [CrossRef] [PubMed]
7. Ghaseminejad, A.A.; Hosseini, S.; Moghaddasi, R. How energy consumption is related to agricultural growth and export: An econometric analysis on Iranian data. *Energy Rep.* **2019**, *5*, 50–53.
8. Aghaei, M.; Lawell, C.Y.C.L. Energy, economic growth, inequality, and poverty in Iran. *Singap. Econ. Rev.* **2020**. [CrossRef]
9. Tugcu, C.T.; Topcu, M. Total, renewable and non-renewable energy consumption and economic growth: Revisiting the issue with an asymmetric point of view. *Energy* **2018**, *152*, 64–74. [CrossRef]
10. Rahman, M.M.; Velayutham, E. Renewable and non-renewable energy consumption-economic growth nexus: New evidence from South Asia. *Renew. Energy* **2020**, *147*, 399–408. [CrossRef]
11. Sunde, T. Energy consumption and economic growth modelling in SADC countries: An application of the VAR Granger causality analysis. *Int. J. Energy Technol. Policy* **2020**, *16*, 41–56. [CrossRef]
12. Aydin, M. Natural gas consumption and economic growth nexus for top 10 natural Gas–Consuming countries: A granger causality analysis in the frequency domain. *Energy* **2018**, *165*, 179–186. [CrossRef]
13. Rahman, M.M.; Abul Kashem, M. Carbon emissions, energy consumption and industrial growth in Bangladesh: Empirical evidence from ARDL cointegration and Granger causality analysis. *Energy Policy* **2017**, *110*, 600–608. [CrossRef]
14. Akadiri, S.S.; Lasisi, T.T.; Uzuner, G.; Akadiri, A.C. Examining the causal impacts of tourism, globalization, economic growth and carbon emissions in tourism island territories: Bootstrap panel Granger causality analysis. *Curr. Issues Tour.* **2020**, *23*, 470–484. [CrossRef]
15. Al-Mulali, U.; Solarin, S.A.; Ozturk, I. Biofuel energy consumption-economic growth relationship: An empirical investigation of Brazil. *Biofuels Bio-Prod. Biorefining* **2016**, *10*, 753–775. [CrossRef]
16. Engle, R.F.; Granger, C.W.J. Co-integration and error correction: Representation, estimation, and testing. *J. Econom.* **1987**, *55*, 251–276. [CrossRef]

17. Mollahosseinia, A.; Hosseinib, S.A.; Jabbaric, M.; Figolid, A.; Rahimpour, A. Renewable energy management and market in Iran: A holistic review on current state and future demands. *Renew. Sustain. Energy Rev.* **2017**, *88*, 774–788. [CrossRef]
18. Alamdari, P.; Mirhosseini, O.N.M. Assessment of wind energy in Iran: A review. *Renew. Sustain. Energy Rev.* **2012**, *16*, 836–860. [CrossRef]
19. Najafi, G.; Ghobadian, B.; Mamat, B.; Yusaf, T.; Azmi, W.H. Solar energy in Iran: Current state and outlook. *Renew. Sustain. Energy Rev.* **2015**, *49*, 931–942. [CrossRef]
20. Ghasemi, A.; Taklif, A.; Mohammadi, T.; Mohammadian, F. Analysis of Iran's energy strategies on the horizon of 2041 in the framework of global energy scenarios. *Iran. J. Energy Econ.* **2017**, *6*, 121–189.
21. Central Bank of Iran. *National Accounts*; Central Bank of Iran: Tehran, Iran, 2020.
22. Vice President of Legal Affairs. *Resolution of the High Energy Council Regarding the National Document of the Country's Energy Strategy*; Vice President of Legal Affairs: Tehran, Iran, 2016.
23. Ghaedizadeh, N.; Khalili Dizaji, H. Status of Iran's Common Oil and Gas Reservoirs from the Perspective of International Law. *Mon. Oil Gas Explor. Prod.* **2014**, *110*, 3–8.
24. National Iranian Oil Company. *Supply of 6 Million Barrels of Crude Oil and Gas Condensate in the Energy Exchange*; National Iranian Oil Company: Tehran, Iran, 2019; Available online: https://www.nioc.ir/portal/home/?news/100193/359783/360628/ (accessed on 25 June 2021).
25. Dudlák, T. After the sanctions: Policy challenges in transition to a new political economy of the Iranian oil and gas sectors. *Energy Policy* **2018**, *121*, 464–475. [CrossRef]
26. Ministry of Energy. *Energy Statistics and Charts of Iran and the World 2017*; Ministry of Energy: Tehran, Iran, 2020.
27. Ministry of Energy. *An Overview of 30 Years of the Country's Energy Statistics*; Ministry of Energy: Tehran, Iran, 2020.
28. Ministry of Energy. *Time Series for Iran's Energy Balances*; Ministry of Energy: Tehran, Iran, 2020.
29. Rezaei, T.; Takif, A.; Ghasemi, A. Feasibility study of gas hub formation using SWOT approach. *Iran. J. Energy Econ.* **2018**, *8*, 69–102.
30. Sadik-Zada, E.R.; Gatto, A. Energy Security Pathways in South East Europe: Diversification of the Natural Gas Supplies, Energy Transition, and Energy Futures. In *From Economic to Energy Transition. Energy, Climate and the Environment*; Mišík, M., Oravcová, V., Eds.; Palgrave Macmillan: Cham, Switzerland, 2021. [CrossRef]
31. Hafezi, R.; Akhavan, A.N.; Pakseresht, S. Projecting plausible futures for Iranian oil and gas industries: Analyzing of historical strategies. *J. Nat. Gas Sci. Eng.* **2017**, *39*, 15–27. [CrossRef]
32. Azadi, P.; Nezam Sarmadi, A.; Shirvani, T. *The Outlook for Natural Gas, Electricity, and Renewable Energy in Iran*; The Stanford Iran 2040 Project, Working Paper No. 3; Stanford University: Stanford, CA, USA, 2017.
33. Renewable Energy and Energy Efficiency Organization (SATBA). *Annual Report on Renewable Energy and Employment-2018*; SATBA: Tehran, Iran, 2020.
34. Pishgar-Komleh, S.H.; Akram, A. Evaluation of wind energy potential for different turbine models based on the wind speed data of Zabol region, Iran. *Sustain. Energy Technol. Assess.* **2017**, *22*, 34–40. [CrossRef]
35. Fazelpour, F.; Markarian, E.; Soltani, N. Wind energy potential and economic assessment of four locations in Sistan and Balouchestan province in Iran. *Renew. Energy* **2017**, *109*, 646–667. [CrossRef]
36. Aryanpur, V.; Atabaki, M.S.; Marzband, M.; Siano, M.P.; Ghayoumi, K. An overview of energy planning in Iran and transition pathways towards sustainable electricity supply sector. *Renew. Sustain. Energy Rev.* **2019**, *112*, 58–74. [CrossRef]
37. Najafi, G.; Ghobadian, B. LLK1694-wind energy resources and development in Iran. *Renew. Sustain. Energy Rev.* **2011**, *15*, 2719–2728. [CrossRef]
38. Global Wind Energy Council. *Annual Global Wind Report (2019)*; Global Wind Energy Council: Brussels, Belgium, 2020.
39. Razavieh, A.; Sedaghat, A.; Ayodele, R.; Mostafaeipour, A. Worldwide wind energy status and the characteristics of wind energy in Iran, case study: The province of Sistan and Baluchestan. *Int. J. Sustain. Energy* **2017**, *36*, 103–123. [CrossRef]
40. Skandari Shabani, S.; Maki, S.V. Hydropower plants. In Proceedings of the 3rd International Conference on Science and Engineering, Istanbul, Turkey, 2 June 2016; Available online: https://civilica.com/doc/491476 (accessed on 25 June 2021).
41. Manzoor, D.; Rahimi, A. Prioritization of power plants in Iran using multidisciplinary decision models. *Iran. J. Energy Econ.* **2015**, *4*, 191–215.
42. Rezaei, R.; Ghofranfarid, M. Rural households' renewable energy usage intention in Iran: Extending the unified theory of acceptance and use of technology. *Renew. Energy* **2018**, *122*, 382–391. [CrossRef]
43. Jahangoshai Rezaee, M.; Yousefi, S.; Hayati, J. Root barriers management in development of renewable energy resources in Iran: An interpretative structural modeling approach. *Energy Policy* **2019**, *129*, 292–306. [CrossRef]
44. Pirasteh, G.; Ashrafizadeh, A.; Salehi Dezfuli, M.M. Energy policies and their impact on the development of renewable energy. In Proceedings of the Fifth Conference on Renewable, Clean and Efficient Energy, Tehran, Iran, 6 March 2013.
45. Mistry of Energy. *General Policies, Country's General Policies*; Mistry of Energy: Tehran, Iran, 2000.
46. Solaymani, S. A CO_2 Emissions Assessment of the Green Economy in Iran. *Greenh. Gases Sci. Technol.* **2020**, *10*, 390–407. [CrossRef]
47. Aghahosseini, A.; Bogdanov, D.; Ghorbani, N.; Breyer, C. Analysis of 100% renewable energy for Iran in 2030: Integrating solar PV, wind energy and storage. *Int. J. Environ. Sci. Technol.* **2018**, *15*, 17–36. [CrossRef]
48. Torabi, G.; Khoshnavaz, M.; Safari Karchani, A.; Ghadimi, A.A. *Potential Assessment and Identification of Potential Points for Installation of Very Small Hydropower Plants in Lorestan Province*; Niroo Research Institute: Tehran, Iran, 2009.

49. Hajighafouri Bukani, N. Investigation of hydropower plants in Iran and some countries of the world. *Mech. Eng.* **2015**, *103*, 36–46.
50. Sovacool, B.K.; Walter, G. Major hydropower states, sustainable development, and energy security: Insights from a preliminary cross-comparative assessment. *Energy* **2018**, *142*, 1074–1082. [CrossRef]
51. Shahsavari, A.; Yazdi, F.T.; Yazdi, H.T. Potential of solar energy in Iran for carbon dioxide mitigation. *Int. J. Environ. Sci. Technol.* **2019**, *16*, 507–524. [CrossRef]
52. Dehghan, A.A. Status and potentials of renewable energies in Yazd Province-Iran. *Renew. Sustain. Energy Rev.* **2011**, *15*, 1491–1496. [CrossRef]
53. Sharbafian, N. Estimation of Technical and Economic Potential of Thermal Solar Energy in Iran: A Strategy for Sustainable Development of Solar Energy. *Energy Econ. Stud.* **2007**, *4*, 35–53.
54. Gorgani Firouzjah, K. Assessment of small-scale solar PV systems in Iran: Regions priority, potentials and financial feasibility. *Renew. Sustain. Energy Rev.* **2018**, *94*, 267–274. [CrossRef]
55. Sansaniwal, S.K.; Sharma, V.; Mathur, J. Energy and exergy analyses of various typical solar energy applications: A comprehensive review. *Renew. Sustain. Energy Rev.* **2018**, *82*, 1576–1601. [CrossRef]
56. Badran, O.O. Study in industrial applications of solar energy and the range of its utilization in Jordan. *Renew. Energy* **2001**, *24*, 485–490. [CrossRef]
57. Gorjian, S.; Zadeh, B.N.; Eltrop, L.; Shamshiri, R.R.; Amanlou, Y. Solar photovoltaic power generation in Iran: Development, policies, and barriers. *Renew. Sustain. Energy Rev.* **2019**, *106*, 110–123. [CrossRef]
58. Edalati, E.; Ameri, M.; Iranmanesh, M.; Sadeghi, Z. Solar photovoltaic power plants in five top oil-producing countries in Middle East: A case study in Iran. *Renew. Sustain. Energy Rev.* **2017**, *69*, 1271–1280. [CrossRef]
59. Wyman, C.E.; Goodman, B.J. Biotechnology for production of fuels, chemicals, and materials from biomass. *Appl. Biochem. Biotechnol.* **1993**, *39*, 41. [CrossRef]
60. SATBA. *Report 4: Biomass Energy 1*; SATBA: Tehran, Iran, 2005. Available online: http://www.dres-isf.ir/home/phocadownload/%D8%B2%DB%8C%D8%B3%D8%AA%20%D8%AA%D9%88%D8%AF%D9%87.pdf (accessed on 25 June 2021).
61. Ministry of Agriculture Jihad. *Annual Report 2019*; Ministry of Agriculture Jihad: Tehran, Iran, 2019.
62. Mohammadi Moghanaki, M.; Ghobadian, B.; Janzadeh Gulogah, R.; Najafi, G.H. Feasibility study of the use of agricultural waste in the simultaneous production of electricity and heat in Iran. In Proceedings of the Second International Conference on Recycled Waste and Biomass Management, Tehran, Iran, 14–17 November 2012.
63. Samadi, S.H.; Ghobadian, B.; Nosrati, M. Prediction and estimation of biomass energy from agricultural residues using air gasification technology in Iran. *Renew. Energy* **2020**, *149*, 1077–1091. [CrossRef]
64. Rezaei, M.; Mostafaeipour, A.; Qolipour, M.; Tavakkoli Moghaddam, R. Investigation of the optimal location design of a hybrid wind-solar plant: A case study. *Int. J. Hydrog. Energy* **2018**, *43*, 100–114. [CrossRef]
65. Solaymani, S. A demand-side assessment of sustainable energy security in Iran. *Int. J. Energy Water Resour.* **2020**, *4*, 307–320. [CrossRef]
66. Abyaz, A.; Najafi, Z.; Afra, E. Evaluation of biofuel production infrastructures from sugarcane bagasse in Iran. In Proceedings of the National Conference on Wood and Lignocellulose Products, Gonbad-e Kavus, Iran, 8 March 2016.
67. Hajinezhad, A.; Katooli, M.H. Study on nutritional values of a novel plant in Iran-Nowruzak- and its application in producing biofuels. *Sustain. Chem. Pharm.* **2018**, *10*, 112–117. [CrossRef]
68. Kheybari, S.; Rezaie, F.M. Selection of biogas, solar, and wind power plants' locations: An MCDA approach. *J. Supply Chain. Manag.* **2020**, *1*, 45–71.
69. Solaymani, S. Energy subsidy reform evaluation research—Reviews in Iran. *Greenh. Gases Sci. Technol.* **2021**. [CrossRef]
70. Ghobadian, B. Liquid biofuels potential and outlook in Iran. *Renew. Sustain. Energy Rev.* **2012**, *16*, 4379–4384. [CrossRef]
71. Madadi, M.; Aqleem Abbas, Z. Green biodiesel production potential from oil seeds in Iran. *Int. J. Life Sci. Sci. Res.* **2017**, *3*, 895–904. [CrossRef]
72. Sepaskhah, A. Sustainable production of bioethanol in relation to soil and water resources in Iran. *Strateg. Res. Agric. Sci. Nat. Resour.* **2017**, *2*, 24–29.
73. Rashidikia, M.; Moradi, M. Biofuel for the optimal use of agricultural waste and its role in sustainable development. *Stud. Nat. Resour. Environ. Agric.* **2015**, *2*, 17–18.
74. Yousefi, H.; Roumi, S.; Ármannsson, H.; Noorollahi, Y. Cascading uses of geothermal energy for a sustainable energy supply for Meshkinshahr City, Northwest, Iran. *Geothermics* **2019**, *79*, 152–163. [CrossRef]
75. Roumi, S.; Yousefi, H.; Noorollahi, Y.; Bekhrad, K.; Zarrouk, S.J. Direct Applications of Geothermal Energy; Economic and Environmental Assessment: Meshkin-Shahr, Iran. In Proceedings of the 42nd Workshop on Geothermal Reservoir Engineering, Stanford, CA, USA, 13–15 February 2017.
76. Pourkiaei, S.M.; Pourfayaz, F.; Shirmohammadi, R.; Moosavi, S.; Khalilpoor, N. Potential, current status, and applications of renewable energy in energy sector of Iran: A review. *Renew. Energy Res. Appl.* **2020**. [CrossRef]
77. Noorollahi, Y.; Shabbir, M.S.; Siddiqi, A.F.; Ilyashenko, L.K.; Ahmadi, E. Review of two decade geothermal energy development in Iran, benefits, challenges, and future policy. *Geothermics* **2019**, *77*, 257–266. [CrossRef]
78. Khoshnevis, S.; Shakouri, B. Renewable energy, nonrenewable energy consumption, and economic growth. *Energy Sources Part B Econ. Plan. Policy* **2017**, *12*, 1038–1045. [CrossRef]

79. Li, Y.; Solaymani, S. Energy consumption, technology innovation and economic growth nexuses in Malaysian. *Energy* **2021**, *232*, 121040. [CrossRef]
80. Razmi, S.F.; Bajgiran, B.R.; Behname, M.; Salari, T.E.; Razmi, S.M.J. The relationship of renewable energy consumption to stock market development and economic growth in Iran. *Renew. Energy* **2020**, *145*, 2019–2024. [CrossRef]
81. Salmanzadeh, M.N.; Fatemi, G.S.M.T. The causal relationship among electricity consumption, economic growth and capital stock in Iran. *J. Policy Modeling* **2019**, *41*, 1230–1256. [CrossRef]
82. Ahssein Amran, Y.H.; Mugahed Amran, Y.H.; Alyousef, R.; Alabduljabbar, H. Renewable and sustainable energy production in Saudi Arabia according to Saudi Vision 2030; Current status and future prospects. *J. Clean. Prod.* **2020**, *247*, 119602. [CrossRef]
83. Salam, M.A.; Khan, S.A. Transition towards sustainable energy production—A review of the progress for solar energy in Saudi Arabia. *Energy Explor. Exploit.* **2018**, *36*, 3–27. [CrossRef]
84. Mohammed, O.H. Optimal Hybridisation of a Renewable System to Fulfill Residential Electrical Load: In Mosul, Iraq. In Proceedings of the 2018 Third Scientific Conference of Electrical Engineering (SCEE), Baghdad, Iraq, 19–20 December 2018. [CrossRef]
85. Marchenko, O.V.; Solomin, S.V. Efficiency of Hybrid Renewable Energy Systems in Russia. *Int. J. Renew. Energy Res.* **2017**, *7*, 1561–1569.

MDPI
St. Alban-Anlage 66
4052 Basel
Switzerland
Tel. +41 61 683 77 34
Fax +41 61 302 89 18
www.mdpi.com

Sustainability Editorial Office
E-mail: sustainability@mdpi.com
www.mdpi.com/journal/sustainability

www.ingramcontent.com/pod-product-compliance
Lightning Source LLC
Chambersburg PA
CBHW061545220926
51850CB00005B/257